SOLID WASTES

ENGINEERING PRINCIPLES AND MANAGEMENT ISSUES

**McGRAW-HILL SERIES IN WATER RESOURCES
AND ENVIRONMENTAL ENGINEERING**
Ven Te Chow, Rolf Eliassen, and Ray K. Linsley
Consulting Editors

Bailey and Ollis: Biochemical Engineering Fundamentals

Bockrath: Environmental Law for Engineers, Scientists, and Managers

Canter: Environmental Impact Assessment

Chanlett: Environmental Protection

Graf: Hydraulics of Sediment Transport

Hall and Dracup: Water Resources Systems Engineering

James and Lee: Economics of Water Resources Planning

Linsley and Franzini: Water Resources Engineering

Linsley, Kohler, and Paulhus: Hydrology for Engineers

Metcalf and Eddy, Inc.: Wastewater Engineering:
Collection, Treatment, Disposal

Nemerow: Scientific Stream Pollution Analysis

Rich: Environmental Systems Engineering

Schroeder: Water and Wastewater Treatment

Tchobanoglous, Theisen, and Eliassen: Solid Wastes:
Engineering Principles and Management Issues

Walton: Groundwater Resources Evaluation

Wiener: The Role of Water in Development:
An Analysis of Principles of Comprehensive Planning

SOLID WASTES

ENGINEERING PRINCIPLES AND MANAGEMENT ISSUES

GEORGE TCHOBANOGLOUS
Professor of Environmental Engineering
University of California, Davis

HILARY THEISEN
Chief, Division of Solid Waste Management
Department of Public Works
Sacramento County, California

ROLF ELIASSEN
Chairman of the Board
Metcalf & Eddy, Inc.
and
Professor Emeritus
Stanford University

McGRAW-HILL BOOK COMPANY

New York   St. Louis   San Francisco   Auckland
Bogotá   Düsseldorf   Johannesburg   London   Madrid
Mexico   Montreal   New Delhi   Panama   Paris
São Paulo   Singapore   Sydney   Tokyo   Toronto

SOLID WASTES: ENGINEERING PRINCIPLES AND MANAGEMENT ISSUES

3 4 5 6 7 8 9 0 FGRFGR 7 8 3 2 1 0

This book was set in Times Roman by Maryland Composition Incorporated.
The editors were B. J. Clark and J. W. Maisel;
the designer was Nicholas Krenitsky;
the production supervisor was Leroy A. Young.
The drawings were done by Danmark & Michaels, Inc.
Fairfield Graphics was printer and binder.

Unless otherwise noted, photographs were taken by George Tchobanoglous.

Library of Congress Cataloging in Publication Data

Tchobanoglous, George.
    Solid wastes.

    (McGraw-Hill series in water resources and environmental engineering)
    Includes bibliographical references and index.
    1. Refuse and refuse disposal. I. Theisen, Hilary, joint author. II. Eliassen, Rolf, joint
author. III. Title.
TD791.T38   628'.44   76-41825
ISBN 0-07-063235-9

TO
ROSEMARY
ANNETTE
MARY

Contents

Preface   **xiii**

PART I PERSPECTIVES

Chapter 1 Solid Wastes—A Consequence of Life   **3**

1-1 The Impacts of Solid Waste Generation   3
1-2 Waste Generation in a Technological Society   4
1-3 Quantities of Wastes   7
1-4 Projections for the Future   10
1-5 Future Challenges and Opportunities   12
1-6 Discussion Topics   14
1-7 References   14

Chapter 2 The Evolution of Solid Waste Management   **15**

2-1 Historical Development   16
2-2 Functional Elements   21

2-3 Solid Waste Management Systems   27
2-4 Solid Waste Management Planning   34
2-5 Discussion Topics   37
2-6 References   37

Chapter 3   Legislation and Governmental Agencies   **39**

3-1 Legislation   39
3-2 Governmental Agencies   43
3-3 Discussion Topics   46
3-4 References   47

PART II   ENGINEERING PRINCIPLES

Chapter 4   Generation of Solid Wastes   **51**

4-1 Sources and Types of Solid Wastes   51
4-2 Composition of Municipal Solid Wastes   54
4-3 Generation Rates   64
4-4 Discussion Topics and Problems   73
4-5 References   76

Chapter 5   Onsite Handling, Storage, and Processing   **77**

5-1 Public Health and Aesthetics   78
5-2 Onsite Handling   78
5-3 Onsite Storage   83
5-4 Onsite Processing of Solid Wastes   90
5-5 Discussion Topics and Problems   101
5-6 References   102

Chapter 6   Collection of Solid Wastes   **103**

6-1 Collection Services   104
6-2 Collection Systems, Equipment, and Labor Requirements 108
6-3 Analysis of Collection Systems   121
6-4 Collection Routes   140
6-5 Advanced Techniques of Analysis   151
6-6 Discussion Topics and Problems   155
6-7 References   158

Chapter 7   Transfer and Transport   **160**

7-1 The Need for Transfer Operations   160
7-2 Transfer Stations   163
7-3 Transport Means and Methods   177
7-4 Location of Transfer Stations   185

7-5 Discussion Topics and Problems   197
7-6 References   199

Chapter 8   Processing Techniques and Equipment   201

8-1 Purposes of Processing   202
8-2 Mechanical Volume Reduction   203
8-3 Chemical Volume Reduction   211
8-4 Mechanical Size Reduction   221
8-5 Component Separation   228
8-6 Drying and Dewatering   248
8-7 Discussion Topics and Problems   252
8-8 References   253

Chapter 9   Recovery of Resources, Conversion Products, and Energy   255

9-1 Materials Processing and Recovery Systems   256
9-2 Recovery of Chemical Conversion Products   265
9-3 Recovery of Biological Conversion Products   282
9-4 Recovery of Energy from Conversion Products   295
9-5 Materials and Energy Recovery Systems Flowsheets   301
9-6 Discussion Topics and Problems   310
9-7 References   313

Chapter 10   Disposal of Solid Wastes and Residual Matter   315

10-1 Site Selection   317
10-2 Landfilling Methods and Operations   321
10-3 Reactions Occurring in Completed Landfills   326
10-4 Gas and Leachate Movement and Control   333
10-5 Design of Landfills   344
10-6 Ocean Disposal of Solid Wastes   369
10-7 Discussion Topics and Problems   370
10-8 References   373

Chapter 11   Hazardous Wastes   375

11-1 Identification of Hazardous Wastes   375
11-2 Classification of Hazardous Wastes   377
11-3 Regulations   381
11-4 Generation   383
11-5 Onsite Storage   385
11-6 Collection   387
11-7 Transfer and Transport   389
11-8 Processing   389
11-9 Disposal   393

11-10 Planning   394
11-11 Discussion Topics and Problems   395
11-12 References   395

PART III   MANAGEMENT ISSUES

Chapter 12   Planning in Solid Waste Management   **399**

12-1 Important Considerations in the Planning Process   400
12-2 Programs and Plans   404
12-3 Planning Study Methodology   407
12-4 The Decision Process   410
12-5 Discussion Topics   413
12-6 References   414

Chapter 13   Choices in Onsite Handling, Storage, and Processing   **415**

13-1 Management Issues and Concerns   416
13-2 Inventory and Data Accumulation   422
13-3 Case Studies in Onsite Handling, Storage, and Processing   426
13-4 Discussion Topics   431
13-5 References   432

Chapter 14   Collection Alternatives   **433**

14-1 Management Issues and Concerns   433
14-2 Inventory and Data Accumulation   441
14-3 Case Studies in Collection   447
14-4 Discussion Topics   456
14-5 References   457

Chapter 15   Transfer and Transport Options   **459**

15-1 Management Issues and Concerns   459
15-2 Inventory and Data Accumulation   466
15-3 Case Studies in Transfer and Transport   470
15-4 Discussion Topics   481
15-5 References   482

**Chapter 16   Choices in Processing and in Materials and
                      Energy Recovery**   **483**

16-1 Management Issues and Concerns   484
16-2 Inventory and Data Accumulation   493
16-3 Case Studies in Processing and Recovery   495
16-4 Discussion Topics   506
16-5 References   507

Chapter 17   Disposal—The "No Alternative" Option   508

17-1 Management Issues and Concerns   508
17-2 Inventory and Data Accumulation   519
17-3 Case Studies in Disposal   522
17-4 Discussion Topics   526
17-5 References   527

Chapter 18   Plan Development, Selection, and Implementation   528

18-1 Evaluation and Development of Alternatives   528
18-2 Program and Plan Selection   531
18-3 Development of Implementation Schedules   532
18-4 Case Studies   533
18-5 Discussion Topics   558
18-6 References   559

Appendixes   560

A. Glossary   560
B. Public Information Programs   569
C. Statistical Analysis of Solid Waste Generation Rates   580
D. Typical Cost Data and Cost-estimating Procedures for Equip-
      ment Used in Solid Waste Management Systems   587
E. Metric Conversion Factors   594

Indexes   597

   Name Index
   Subject Index

Preface

Solid waste management is concerned with the generation, onsite storage, collection, transfer and transport, processing and recovery, and disposal of the solid wastes from a technological society. Today solid waste management is a multidisciplinary activity that is based on engineering principles, but also involves economics, urban and regional planning, and the social sciences, among other fields. Traditional engineering approaches, which in the past often neglected public attitudes and concerns, are no longer acceptable. Technological advances have resulted in a change in emphasis in many important areas. While public health and economics remain primary considerations, great emphasis is now placed on environmental (particularly aesthetic) constraints. The relationship between resource depletion and the disposal of solid wastes is also coming under scrutiny.

It is the objective of this book, therefore, to bring together a wide body of knowledge concerning the rapidly changing and expanding field of solid waste management and to present it in a format that will be useful as a text for students and as a reference work for practicing professionals in a variety of fields. The major focus is specifically on residential and commercial solid wastes, although industrial and other types of wastes are also discussed.

The book is divided into three parts that deal with perspectives, engineering principles, and management issues. To understand the overall field of solid waste management, it is important to know about the nature and generation of solid wastes, how the field has evolved, and the legislative framework in which solid waste management activities are now conducted. These are the main topics of Part I.

Basic engineering principles are presented in Part II. Because solid waste management is a dynamic field, there is no "best method" for the solution of all the problems that arise. In each situation engineering principles must be applied to evaluate equipment and facility options, to make operational choices, and to develop management systems. Only in this way can the impact of alternative courses of action be determined. Then, to provide the decision-maker with a basis for selection, proposed engineering alternatives can be assessed and ranked in terms of cost-effectiveness.

Important management issues that must be considered in the development and operation of solid waste management systems are discussed in Part III. For example, resource recovery is now a vital concern of the public as well as of industry and government agencies, and it must be considered in any solid waste management plan for a community, county, or region. Yet the relationships between solid waste management and resource recovery and reuse are not well defined. There are many management and engineering determinants that the student and professional must consider. The fluctuation of prices that industry will pay for recovered materials can have a significant impact on the viability of an engineering decision to build or not to build a solid waste processing or recovery plant.

The authors hope that from this book the student, as well as the professional (engineer, planner, economist, attorney, or other specialist), will obtain a perspective of the immense problems involved in solid wastes, will learn that there are engineering and economic solutions to these problems, and will recognize that the people of the community must be included in the decision-making process if rational, economic, and socially acceptable answers to the problems are to be found.

For easy reference, design data and other useful information, selected from a variety of sources or developed specially for this text, are presented in more than 80 tables. To increase the usefulness of this text for both teachers and students, more than 45 example problems and case studies are included. The example problems are worked out in step-by-step format, and all the necessary units are presented in the computational steps. To test and further extend the student's knowledge, numerous discussion topics and problems are included at the end of each chapter.

Although the United States is moving toward the adoption and use of a single set of international metric units, known as the SI system (Système

International d'Unités), the basic units in this text are the United States customary units. The reason is that customary units will probably continue to be used for some time, especially with respect to many of the basic quantities involved in the analysis of solid waste management systems. To assist in the transition, however, metric conversions are included at the bottom of each data table, metric equivalents are provided for the final answers to the example problems, and some of the homework problems are given in metric units.

Many people have contributed to the development of this textbook. Their help is acknowledged gratefully. Members of the 1975 fall semester solid waste management class of the University of California at Davis reviewed Part II and provided valuable editorial comments on the presentation of the information. Justus P. Allen, Harvey F. Collins, Larry J. Karns, Patrick L. Maxfield, Frank X. Reardon, and Edward D. Schroeder reviewed various portions of the manuscript and made helpful contributions. Richard A. Mills, Jack C. Scroggs, and Sam A. Vigil critiqued portions of the manuscript and assisted with many of the example and homework problems. Jan C. Cudrnak helped prepare the figures. The Secretariat in Davis typed the final manuscript. Finally, to Marcella S. Tennant, who served as technical editor, we owe a debt of gratitude beyond measure. Her concern for logic and clarity is reflected throughout the text.

George Tchobanoglous
Hilary Theisen
Rolf Eliassen

SOLID WASTES

ENGINEERING PRINCIPLES AND MANAGEMENT ISSUES

PART I

PERSPECTIVES

What are solid wastes? What are the impacts of solid waste generation? What is the magnitude of the problem? What does the future hold with respect to solid waste generation? What are the future challenges and opportunities for change? How did the field of solid waste management evolve? Why are the various activities associated with waste generation, onsite storage, collection, transfer and transport, processing and recovery, and disposal identified as functional elements? What are the day-to-day responsibilities of an operating agency? What is meant by the term *comprehensive planning* as applied to solid waste management? Which legislation at the federal level has affected the field of solid waste management? Which governmental agencies are responsible for administering the applicable legislation? And what are the impacts at the local level?

The answers to these questions are discussed in Part I. They are also the essence of an introductory understanding of the field of solid waste management. The story of Part I is the story of progress in this field from the use of horse-drawn carts to specially designed motor vehicles. It is also the story of progress from open dumps (which became hazards to public health and sites on which burning and horrendous air pollution took place) to the development of mechanized landfill methods for the control of the vectors of disease and land reclamation, followed by the more sophisticated practices of solid waste management and resource recovery that are used today.

1

Solid Wastes- A Consequence of Life

Solid wastes are all the wastes arising from human and animal activities that are normally solid and that are discarded as useless or unwanted. The term as used in this text is all-inclusive, and it encompasses the heterogeneous mass of throwaways from the urban community as well as the more homogeneous accumulations of agricultural, industrial, and mineral wastes. In an urban setting, the accumulation of solid wastes is a direct consequence of life. From this consequence has evolved what is today (1976) in the United States a $3 billion to $4 billion per year activity associated with the management of these wastes.

To introduce the reader to the field of solid waste management, in this chapter an overview of the following topics is presented: (1) public health and ecological impacts of solid wastes, (2) the generation of solid wastes in a technological society, (3) the magnitude of the solid waste problem in terms of the quantities generated, (4) projections for the future, and (5) future challenges and opportunities with respect to solid waste management.

1-1 THE IMPACTS OF SOLID WASTE GENERATION

From the days of primitive society, humans and animals have used the resources of the earth to support life and to dispose of wastes. In early times, the disposal of human and other wastes did not pose a significant problem, for the population was small and the amount of land available for the assimilation of wastes was large. Nowadays we speak of recycling the energy and fertilizer values of solid wastes, but the farmer in ancient times probably made a bolder attempt at this. Indications of recycling may still be seen in the primitive, yet sensible, agricultural practices in many of the developing nations where farmers recycle solid wastes for fuel or fertilizer values.

Problems with the disposal of wastes can be traced from the time when humans first began to congregate in tribes, villages, and communities and the accumulation of wastes became a consequence of life (see Fig. 1-1). Littering of food and other solid wastes in medieval towns—the practice of throwing wastes into the unpaved streets, roadways, and vacant land— led to the breeding of rats, with their attendant fleas carrying the germs of disease, and the outbreak of plague. The lack of any plan for the management of solid wastes led to the epidemic of plague, the Black Death, that killed half of the Europeans in the fourteenth century and caused many subsequent epidemics and high death tolls. It was not until the nineteenth century that public health control measures became a vital consideration to public officials, who began to realize that food wastes had to be collected and disposed of in a sanitary manner to control the vectors of disease.

The relationship between public health and the improper storage, collection, and disposal of solid wastes is quite clear. Public health authorities have shown that rats, flies, and other disease vectors breed in open dumps, as well as in poorly constructed or poorly maintained housing, in food storage facilities, and in many other places where food and harborage are available for rats and the insects associated with them. The U.S. Public Health Service (USPHS) has published the results of a study [3] tracing the relationship of 22 human diseases to improper solid waste management. Data are also available to show that the illness-accident rate for sanitation workers engaged in the collection and disposal of solid wastes is several times higher than that for industrial employees [3].

Ecological impacts, such as water and air pollution, also have been attributed to improper management of solid wastes. For instance, liquid from dumps and poorly engineered landfills has contaminated surface waters and groundwaters. In mining areas the liquid leached from waste dumps may contain toxic elements, such as copper, arsenic, and uranium, or may contaminate water supplies with unwanted salts of calcium and magnesium. While the capacity of nature to dilute, disperse, degrade, absorb, or otherwise dispose of its unwanted residues in the atmosphere, in the waterways, and on the land is well known, humans cannot stress these natural capacities with their unwanted residues too much or an ecological imbalance will be imposed on the biosphere.

1-2   WASTE GENERATION IN A TECHNOLOGICAL SOCIETY

The development of a technological society in the United States can be traced to the beginnings of the Industrial Revolution in Europe; unfortu-

FIG. 1-1 Solid waste problems are not new. (B.C. by permission of Johnny Hart and Field Enterprises, Inc.)

nately, so can a major increase in solid waste disposal problems. In fact, in the latter part of the nineteenth century, conditions were so bad in England that an urban sanitary act was passed in 1888 prohibiting the throwing of solid wastes into ditches, rivers, and waters. This preceded by about 11 years the enactment of the Rivers and Harbors Act of 1899 in the United States, which was intended to regulate the dumping of debris in navigable waters and adjacent lands.

Thus, along with the benefits of technology have also come the problems associated with the disposal of the resultant wastes. To understand the nature of these problems, it will be helpful to examine the flow of materials and the associated generation of wastes in a technological society and to consider the direct impact of technological advances on the design of solid waste facilities.

Materials Flow and Waste Generation

An indication of how and where solid wastes are generated in our technological society is shown in the simplified materials flow diagram presented in Fig. 1-2. Solid wastes (debris) are generated at the start of the process, beginning with the mining of raw material [6]. The debris left from strip-mining operations, for example, is well known to everyone. There-

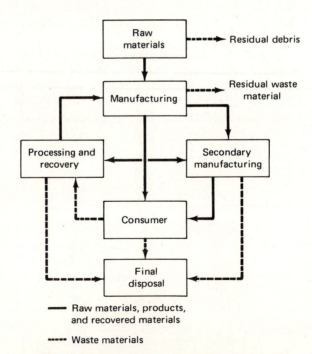

FIG. 1-2   Materials flow and the generation of solid wastes in a technological society.

after, solid wastes are generated at every step in the process as raw materials are converted to goods for consumption.

It is apparent from Fig. 1-2 that one of the best ways to reduce the amount of solid wastes that must be disposed of is to limit the consumption of raw materials and to increase the rate of recovery and reuse of waste materials. Although the concept is simple, effecting this change in a modern technological society has proved extremely difficult. This subject is considered further in the last section of this chapter.

The Effect of Technological Advances

Modern technological advances in the packaging of goods create a constantly changing set of parameters for the designer of solid waste facilities. Of particular significance are the increasing use of plastics and the use of frozen foods, which reduce the quantities of food wastes in the home but increase the quantities at agricultural processing plants. The acceptance of so-called TV dinners, for example, results in almost no wastes in the homes except for the packaging materials [1]. These continuing changes present problems to the facilities designer because engineering structures for the processing of solid wastes involve such large capital expenditures that they must be designed to be functional for approximately 25 years. Thus, the engineers responsible for the design of solid waste facilities must be aware of trends, even though they cannot be clairvoyant in the prediction of changes in technology that will affect the characteristics of solid wastes 25 years hence.

On the other hand, every possible prediction technique must be used in this ever-changing technological society so that flexibility and utility can be designed into the facilities. Ideally, a facility should be functional and efficient over its useful life, which should coincide with the maturity of the bonds that were floated to pay for it. But important questions arise: Which elements of society generate the greatest quantities of solid waste and what is the nature of these wastes? Also, how can the quantities be minimized? What is the role of resource recovery?

1-3 QUANTITIES OF WASTES

Everyone is familiar with solid wastes, especially those generated in municipalities, such as food wastes and rubbish, abandoned vehicles, demolition and construction wastes, street sweepings, and garden wastes. Far greater amounts, however, result from agricultural, industrial, and mineral sources.

Estimated Total and Per Capita Quantities

Although the data are varied, recent estimates indicate that an average of 4.4 billion tons of solid wastes is generated each year in the United States alone. Of this total, municipal wastes represent approximately 230 million tons; industrial wastes, 140 million tons; and agricultural wastes, 640 million tons. By far the greatest amount of solid wastes comes from mines and minerals and from animal wastes, each with an average of 1.7 billion tons/yr. The total amount generated from all sources by the year 2000 may approach 12 billion tons/yr.

Looking at just the urban and industrial wastes, the generation rate in the United States is approximately 3,600 lb/capita/yr. In comparison, other industrialized countries have lower rates, but they have similar problems. Based on rough estimates, Japan is the closest to the United States with an average of 800 lb. The rate in the Netherlands is over 600 lb; in West Germany, about 500 lb. From these figures it can be concluded that in these countries either the rate of consumption of goods is lower or a more serious effort is made to recover and reuse the wastes.

Data from Recent Surveys

Many surveys of solid waste generation have been made by consulting engineers and planners in their studies for municipalities and regional authorities. State and federal agencies, particularly the USPHS and the U.S. Environmental Protection Agency (EPA), have also been active in this area. There are so many reported values that any one for a particular category may be disputed because of the aforementioned impact of technological developments, the marketing of consumer products and their packaging, and commercial and industrial practices. For instance, institutions having large computers, with their long printouts, have had an appreciable impact on the generation of waste paper in certain communities. For these reasons it is imperative that surveys be made for any specific municipality or region to determine the ranges of values of solid wastes generated from municipal and industrial sources.

U.S. Public Health Service Data   In 1968 the USPHS published data obtained in its National Survey of Community Solid Waste Practices [4]. The average generation rates for urban sources in the United States are shown in Table 1-1.

It should be emphasized that these are yearly averages and that the actual generation rates for a given city vary with the seasons (garden clippings, leaves, Christmas paper and gift containers, etc., as discussed later in this chapter). The exact amounts for a given city may also be far above average, such as the vast amount of paper discarded in Washington, D.C., state capitals, and large commercial and industrial centers. Again, beware of averages in the design of facilities, but be guided by them in analyzing the results for a specific location! Detailed values of solid waste generation are presented in Chap. 4.

Environmental Protection Agency   The EPA has continued the studies of

TABLE 1-1 AVERAGE PER CAPITA QUANTITIES OF SOLID WASTES COLLECTED FROM URBAN SOURCES IN THE UNITED STATES, 1968*

Source	lb/capita/day
Combined residential and commercial	4.29
Industrial	1.90
Institutional	0.16
Demolition and construction	0.72
Street and alley cleanings	0.25
Tree and landscaping	0.18
Park and beach	0.15
Catch basin	0.04
Sewage treatment plant solids	0.50
Total	8.19†

? DIFF. THAN DATA IN TABLE 4-12.

* Adapted from Ref. 4.
† As reported in Table 4-12, the corresponding total per capita quantities for all areas (7.92 lb/capita/day) are somewhat lower than those from urban areas.
Note: lb/capita/day × 0.4536 = kg/capita·day

the USPHS and in 1971 published a report [5] to the Congress containing estimates of present and future solid waste generation. The estimate of waste generation in the United States by component is shown in Table 1-2. Note that the quantities shown as disposed are greater than those shown as generated. The difference is attributed to the increase of moisture content in the disposed wastes and to the measure of the wastes generated on a dry basis. Excluded from this table are treatment plant sludges, demolition and construction wastes, and such special wastes as abandoned automobiles.

Monthly and Seasonal Variations

Solid wastes from residential sources—potentially one of the principal public health problems in any community—vary considerably in composition and quantity. The authors have found significant variations, depending on the economic status, ethnic composition, and social habits of neighborhoods (e.g., backyard burning of paper and leaves). Quantities also vary with seasons, the horticultural choices of neighborhoods, the geographical characteristics of the land, rainfall, climate, and the habits of the people—what they eat, drink, and the packaged materials they buy. The list is virtually endless.

These data are important in the design and operation of any solid waste facility. The data shown in Fig. 1-3, obtained in 1940 in a comprehensive survey [2] in New York City, are useful in illustrating the variation in the composition of wastes generated on a monthly and seasonal basis. As shown, these data were influenced greatly by the use of coal for

TABLE 1-2   COMPONENTS OF MUNICIPAL SOLID WASTES GENERATED IN THE UNITED STATES, 1971*

	Total generated		Total disposed	
Component	Tons, millions	Percent	Tons, millions	Percent
Paper	39.1	31.3	47.3	37.8
Glass	12.1	9.7	12.5	10.0
Metal	11.9	9.5	12.6	10.1
Ferrous	10.6	8.5	—	—
Aluminum	0.8	0.6	—	—
Other nonferrous	0.5	0.4	—	—
Plastic	4.2	3.4	4.7	3.8
Rubber and leather	3.3	2.6	3.4	2.7
Textiles	1.8	1.4	2.0	1.6
Wood	4.6	3.7	4.6	3.7
Food	22.0	17.6	17.7	14.2
Subtotal	99.0	79.2	104.8	83.9
Yard wastes	24.1	19.3	18.2	14.6
Miscellaneous inorganics	1.9	1.5	2.0	1.5
Total	125.0	100.0	125.0	100.0

* Adapted from Ref. 5.
Note: tons × 907.2 = kg

residential heating in 1940, a condition that may recur as the importation of foreign oil and liquid propane gas becomes more restrictive and the plentiful coal resources of the United States are utilized.

Similar data to those shown in Fig. 1-3 must be obtained for the design of any solid waste operation so that all its components may be flexible enough to contend with the ever-changing waste loads discarded by the residents, as well as by commercial and industrial generators. The components of solid wastes generated in New York City in 1940 are not indicative of the components to be expected in 1976, any more than those of the year 2000 will be similar to those of today; however, from the past, the future must be predicted.

1-4   PROJECTIONS FOR THE FUTURE

Whether the USPHS values in Table 1-1 or the EPA values in Table 1-2 are used to make projections for the future, the figures are necessarily based on assumed annual growth rates of population and on human consumption and disposal of rejected products. The EPA has assumed that there will be an increase in industrially prepared foods (with solid wastes

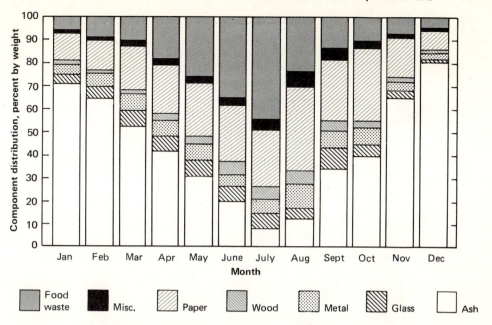

FIG. 1-3   Monthly distribution of solid waste components arriving at a sanitary landfill in New York, 1940 [2].

produced in the agricultural or industrial category at remote locations) and an increase in packaging wastes. On the other hand, yard wastes, which amount to 14.6 percent of the wastes disposed (Table 1-2), will not increase as rapidly in the future because housing types are expected to change from single homes to apartment dwellings.

In view of the many factors that influence predictions for the future, ranges must be used. Because future estimates are necessary in the design of solid waste processing and disposal facilities, the design engineer must attempt to predict facility use for at least 25 years to justify the capital expenditures involved. To assist the engineer in design projections, the EPA has developed Table 1-3 by assuming three different growth rates of solid waste generation.

From experience during the 1960s, solid waste generation might be expected to increase at a rate of 4.5 percent/yr. A more reasonable figure, based on the greater awareness of the public with respect to resource recovery and recycling, might be 3.5 percent. If major efforts at resource recovery and recycling were to be effective, the rate of increase might be as low as 2.5 percent.

The economic impact of solid waste management costs can be projected by using the values in Table 1-3 and by applying some estimated national average unit collection and disposal costs developed by the EPA. Data from the EPA studies, along with estimates for 1980 and 1985, are presented in Table 1-4. Thus, solid waste management is expected to

TABLE 1-3   PROJECTED TOTAL SOLID WASTE
QUANTITIES, 1980 to 1990*,†

Assumed annual compound growth, percent	Tons, millions		
	1980	1985	1990
2.5 (low)	155	175	200
3.5 (medium)	170	200	230
4.5 (high)	185	230	290

* 1971 base = 125 million tons (see Table 1-2)
† From Ref. 5.
Note: tons × 907.2 = kg

become a $4 billion to $5 billion industry by 1985 (based on 1971 dollars).
Depending on the rate of inflation, these values could change significantly.

1-5 FUTURE CHALLENGES AND OPPORTUNITIES

The multibillion-dollar industry of solid waste management can be supported only by the public which is responsible for the generation of the vast tonnage of wastes—about 200 million tons/yr, as shown in Table 1-4. Public attitudes must be aroused in an attempt to reduce the economic burden being placed on society. National concern must transcend both the concept of what the public can afford to pay and the question of why the public cannot insist on reducing the economic burden by whatever individual or societal action is necessary.

Unfortunately, the standard of living in the United States is inevitably tied to the generation of solid wastes—the squandering of natural resources from this country and abroad, the one-time use of materials of so many types, and the philosophy of wastefulness and rapid obsolescence of products. It is reasonable to presume that a departure from this philosophy of wastefulness might reduce the tonnage of wastes to be managed. This concept inevitably leads to the need for resource recovery and the recycling of recovered materials to the mainstream of industry. Furthermore, the habits of people must change through their own volition, guided by conservation groups and information made available through industrial and governmental agencies. Efforts must be made to reduce the quantity of both materials used in packaging and obsolescent goods, and to begin the process of recycling at the source—the home, office, or factory—so that so many materials will not become a part of the disposable solid wastes of a city. This is an alternative that will conserve resources and that also has economic viability.

Another alternative is to continue the wasteful practices of modern industrial society and pay the penalty. As indicated in Table 1-4, the difference between the high and low figures for 1985 could be $1 billion. It

TABLE 1-4 COSTS OF SOLID WASTE COLLECTION AND DISPOSAL IN U.S., 1971 TO 1985[*]

Item	1971 (estimated)	1980 (projected)			1985 (projected)		
		Low	Medium	High	Low	Medium	High
Collected wastes (tons, millions)[†]	120	150	160	175	165	190	220
Unit costs ($/ton)							
Collection	18	18	18	18	18	18	18
Disposal	4	5	5	5	5	5	5
Total	22	23	23	23	23	23	23
Total national costs, millions of dollars (1971)							
Collection	2,160	2,700	2,880	3,150	3,150	3,420	3,960
Disposal	480	750	800	875	875	950	1,100
Total	2,640	3,450	3,680	4,025	4,025	4,370	5,060

[*] Adapted from Ref. 5.
[†] It is assumed that 95 percent of the projected waste generation (Table 1-3) will be collected.
Note: tons × 907.2 = kg
          $/ton × 0.0011 = $/kg

could be even greater if conservation were to become a way of life for citizens, commerce, and industry.

Still another alternative is to continue the poor solid waste management practices that prevail in many parts of the country. This alternative carries with it a severe economic penalty in the abuse of land and the profligate use of increasingly scarce resources of materials, energy, work force, and money. It is not acceptable to society. Its counterpart lies at the heart of progressive solid waste management that is the subject of all the chapters in this book.

1-6 DISCUSSION TOPICS

1-1. Discuss the various factors that may account for the large differences in the generation of solid wastes in municipalities of several industrial nations around the world.

1-2. As a community leader, what steps could you take or what advice could you give to reduce the economic impact of solid waste generation in your city?

1-3. Gas and oil are becoming expensive and scarce commodities, yet they are clean fuels that produce no solid wastes. What other sources of energy for home heating could you suggest for two different regions (such as the New England states and Texas or Arizona) and still minimize the generation of solid wastes?

−1-4. What would be the impact of the monthly variation in the composition of solid wastes, shown in Fig. 1-2, on the operation of the solid waste management system?

1-5. Discuss the major factors that have influenced the changes that have taken place in the composition of solid wastes since the turn of the century. Do you feel that the changes in the composition of solid wastes will be significant in the next 10, 25, or 50 years? Explain.

1-6. What is your present concept of resource recovery? How can it affect the costs of solid waste disposal?

− 1-7. What is being done in your community for the recycling of bottles, cans, and paper? In your opinion, is the program successful? How can it be improved and what agency should take the lead in this improvement?

1-8. What is your opinion of the Oregon plan (all soft drink and beer containers must be returnable or reusable)? Do you think such a plan would work in your state? (Reference: *The Wall Street Journal,* Friday, January 9, 1976.)

1-7 REFERENCES

1. Darnay, A. and W. E. Franklin: *The Role of Packaging in Solid Waste Management 1966 to 1976,* U.S. Department of Health, Education, and Welfare, Public Health Service, Publication SW-5c, Rockville, Md., 1969.

2. Eliassen, R.: Decomposition of Landfills, *American Journal of Public Health,* vol. 32, no. 3, 1942.

3. Hanks, T. G.: *Solid Waste/Disease Relationships,* U.S. Department of Health, Education, and Welfare, Solid Wastes Program, Publication SW-1c, Cincinnati, Ohio, 1967.

4. Muhich, A. J., A. J. Klee, and P. W. Britton: *Preliminary Data Analysis, 1968 National Survey of Community Solid Waste Practices,* U.S. Department of Health, Education, and Welfare, Public Health Service, Publication 1867, Washington, D.C., 1968.

5. *Resource Recovery and Source Reduction,* Second Report to Congress, U.S. Environmental Protection Agency, Publication SW-122, Washington, D.C., 1974.

6. *Surface Mining and Our Environment,* A Special Report to the Nation, U.S. Department of the Interior, Washington, D.C., 1967.

2

The Evolution of Solid Waste Management

Solid waste management may be defined as that discipline associated with the control of generation, storage, collection, transfer and transport, processing, and disposal of solid wastes in a manner that is in accord with the best principles of public health, economics, engineering, conservation, aesthetics, and other environmental considerations, and that also is responsive to public attitudes. In its scope, solid waste management includes all administrative, financial, legal, planning, and engineering functions involved in the whole spectrum of solutions to problems of solid wastes thrust upon the community by its inhabitants. The solutions may involve complex interdisciplinary relationships among such fields as political science, city and regional planning, geography, economics, public health, sociology, demography, communications, and conservation, as well as engineering and materials science.

The purpose of this chapter is twofold: to provide a general introduction to the field of solid waste management and to serve as a basis for understanding the interrelationships of the engineering fundamentals presented in Part II and the management aspects presented in Part III. The information presented in four sections covers: (1) a brief history of the evolution and development of this discipline, (2) a description of the functional elements of solid waste management sytems, (3) a brief discussion of financial and other aspects of solid waste management systems which are not covered in detail in this book, and (4) a brief discussion of solid waste management planning.

2-1   HISTORICAL DEVELOPMENT

> To describe the characteristics of the different classes of refuse, and to draw attention to the fact that, if a uniform method of nomenclature and record of quantities handled could be kept by the various cities, then the data obtained and the information so gained would be a material advance toward the sanitary disposal of refuse. Such uniformity would not put any expense upon cities, and direct comparisons and correct conclusions could be made for the benefit of others. [9]

This statement of objectives in itself does not seem too unusual, until it is realized that it was written in 1906 by H. de B. Parsons in his book entitled "The Disposal of Municipal Refuse" [9]. In reviewing this book, which may have been the first to deal solely with the subject of solid wastes from a rigorous engineering standpoint, we note that many of the basic principles and methods underlying what is known today as the field of solid waste management, are not new and were well known even then. For example, although the motor truck has replaced the horse-drawn cart (see Fig. 2-1), the basic methods of solid waste collection remain the same; they continue to be essentially labor intensive. The development of uniform data for purposes of comparison is still an important need.

(a)

FIG. 2-1 Evolution of vehicles used for the collection of solid wastes. (a) Horse-drawn cart, about 1900 [9]. (b) Solid-tire motor truck, about 1925 [7]. (c) Modern collection vehicle equipped with compaction mechanism, 1976.

Early Disposal Practices

The most commonly recognized methods for the final disposal of solid wastes at the turn of the century were: (1) dumping on land, (2) dumping in water, (3) plowing into the soil, (4) feeding to hogs, (5) reduction, and (6) incineration [9, 10]. Not all these methods were applicable to all types of wastes. Plowing into the soil was used for food wastes and street sweepings. Feeding to hogs and reduction were used specifically for food wastes [9].

(b)

(c)

Dumping on Land   Because it was a simple task to haul solid wastes to the edge of town and dump them there, open dumps became a common method of disposal for urban communities, and burning on these dumps was a common practice (see Fig. 2-2). Open dumps also attracted flies and rats that spread diseases. This haphazard disposal became a matter of great concern to public health authorities who were given responsibility for the control of solid wastes.

Over the years, the vector control divisions of state health departments and the USPHS achieved outstanding success in the control of vector-borne diseases through the development, promotion, and enforcement of sanitary landfill practices, as well as the storage, collection, and transportation of solid wastes.

Dumping in Water   Although this method was used by some coastal cities, it was not favored because the pollution consequences were well recognized. The disfigurement of Coney Island Beach in New York City became a case in point [9]. Nevertheless, the practice continued until 1933 when it was finally prohibited by the United States Supreme Court.

Plowing Into the Soil   As mentioned, this plowing into the soil was a method of disposal used for food wastes and street sweepings. Because of the large land requirements and the fact that the food wastes had to be separated from other wastes, this method was not used extensively, but interest in it has been rekindled in the 1970s.

Feeding to Hogs   Food wastes frequently were fed to hogs on farms close to urban areas, such as those in Los Angeles County and in the mud flats of New Jersey. Food wastes from New York City were fed to hogs at the smelly farms across the river in Secaucus, New Jersey. Unfortunately, because of this practice trichinosis became widespread when contaminated pork scraps were fed to hogs in recycled food wastes, which reinfected other hogs and people who consumed their meat. As much as 16 percent of the United States population was infected by eating uncooked pork from hogs fed on food wastes in the first third of this century. Nonetheless, this practice continued well into the middle half of this century. It is still used in some isolated areas in the United States, but under controlled conditions of cooking and feeding.

Reduction   Food waste reduction, a method no longer used, was a rendering process by which the raw wastes were treated to separate them into solid and liquid portions and to recover the grease contained in one or both portions [10]. The solid portion was known as "tankage." Several processes were developed and used [9]. The recovered grease was used to

(a)

(b)

FIG. 2-2 Burning at open dumps. (a) Open dump located in ravine. (b) Open dump located in flat area. (California Department of Public Health, Bureau of Vector Control.)

make pomades and the cheaper grades of perfumery as well as wagon grease.

Incineration   Although incineration was considered to be a final disposal method at the turn of the century, it is now considered to be either a volume reduction or an energy conversion process. Because little has changed in the application of this process, the reader is referred to Chaps. 8 and 9 for a more detailed discussion.

The Beginnings of Solid Waste Management

The beginning of solid waste management can be traced to antiquity and the practice of recycling human wastes. One of the early attempts to manage solid wastes in the United States took place at the turn of the century when New York City built a wooden-crib bulkhead around Rikers Island in the East River and filled in the area behind the marshes with ashes, rubbish, and street sweepings [9]. The wastes were transferred from wagons to scows (barges) in New York City, towed to the island, unloaded by clamshell dredges onto a conveyor or into rail cars, and distributed. Clearly some sort of management plan was required to accomplish this large operation. Here again, it is evident that solid waste management planning is not new.

Enlightened solid waste management, with emphasis on controlled tipping (now known as "sanitary landfilling"), can be traced to the early 1940s in the United States and to a decade earlier in the United Kingdom [8]. New York City, under the leadership of Mayor La Guardia, and Fresno, California, with its health-minded Director of Public Works, Jean Vincenz, were the pioneers in the sanitary landfill method for large cities. During World War II, the U.S. Army Corps of Engineers, under the direction of Jean Vincenz, who then headed its Repairs and Utilities Division in Washington, modernized its solid waste disposal programs to serve as model landfills for communities of all sizes. The medical department of the Army, through Col. W. A. Hardenbergh of the Sanitary Corps' engineering group, took an active part in vector control and the prevention of the spread of disease to the troops by helping to sponsor the sanitary landfill program.

But municipalities did not follow these programs with consistency. The California Department of Public Health, along with several other progressive state health departments, established standards for municipal sanitary landfills and carried out aggressive campaigns for the elimination of conventional dumps. Still, in 1965, after a thorough review of solid waste management practices in the United States, Congress concluded:

> . . .that inefficient and improper methods of disposal of solid waste result in scenic blights, create serious hazards to public health, including pollution of air and water resources, accident hazards, and increase in rodent and insect vectors of disease, have an adverse

effect on land values, create public nuisances, otherwise interfere with community life and development; . . . that the failure or inability to salvage and reuse such materials economically results in the unnecessary waste and depletion of natural resources; . . . [6]

Congress also found that the continuation of the trend of population concentration in metropolitan and urban areas had presented these communities with serious financial and administrative problems in the collection, transportation, and disposal of solid wastes.

2-2 FUNCTIONAL ELEMENTS

The problems associated with the management of solid wastes in today's society are complex because of the quantity and diverse nature of the wastes, the development of sprawling urban areas, the funding limitations for public services in many large cities, the impacts of technology, and the emerging limitations in both energy and raw materials. As a consequence, if solid waste management is to be accomplished in an efficient and orderly manner, the fundamental aspects and relationships involved must be identified and understood clearly.

In this text, the activities associated with the management of solid wastes from the point of generation to final disposal have been grouped into the six functional elements identified in Fig. 2-3 and illustrated in photographs in Fig. 2-4. By considering each functional element separately, it is possible, (1) to identify the fundamental aspects and relation-

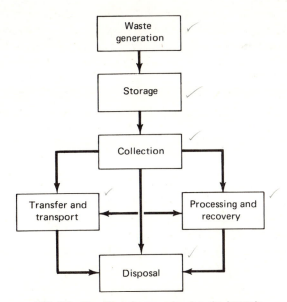

FIG. 2-3 Simplified diagram showing the interrelationships of the functional elements in a solid waste management system.

(a)

(b)

(c)

FIG. 2-4 Pictorial representation of the functional elements in a solid waste management system. (a) Generation. (b) Onsite storage. (c) Collection. (d) Transfer and transport. (e) Processing and recovery. (f) Disposal.

(d)

(e)

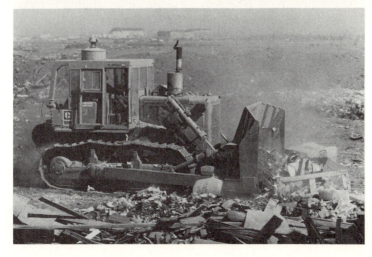

(f)

ships involved in each element and (2) to develop, where possible, quantifiable relationships for the purposes of making engineering comparisons, analyses, and evaluations. This separation of functional elements is important because it allows the development of a framework within which to evaluate the impact of proposed changes and future technological advancements. As an example, while the means of transport in the collection of solid wastes has changed from the horse-drawn cart to the motor vehicle (see Fig. 2-1), the fundamental method of collection—that is, the manual physical handling required—remains the same (see Chap. 6).

To solve specific solid waste problems, the various functional elements are combined in what usually is known as a solid waste management system. In most cities, a solid waste management system comprises four functional elements: waste generation, onsite storage, collection, and disposal. Therefore, one of the goals of solid waste management is the optimization of these systems to provide the most efficient and economic solution, commensurate with all constraints imposed by the users of the system and those affected by it or controlling its use.

The individual functional elements are described in the following discussion. Each one is considered in detail in Part II. The purpose of the following discussion is to introduce the reader to the physical aspects of solid waste management and to establish a useful framework within which to view the activities associated with management of solid wastes.

Waste Generation

Waste generation encompasses those activities in which materials are identified as no longer being of value and are either thrown away or gathered together for disposal. For example, the wrapping of a candy bar usually is considered to be of little further value to the owner once the candy is consumed, and more often than not it is just thrown away, especially outdoors. What is important in waste generation is to note that there is an identification step and that this step varies with each individual.

Because waste generation is at present an activity that is not very controllable, it is often not considered to be a functional element. In the future, however, more control will probably be exercised over the generation of wastes. For example, from the standpoint of economics, the best place to sort waste materials for recovery purposes is at the source of generation. Homeowners are becoming more aware of the importance of separating newspaper and cardboard, tinned steel cans, and aluminum cans and bottles.

Onsite Storage

Although solid wastes from urban sources may constitute only 5 percent of the nation's solid wastes, their management requires a large and continuous effort. The reason is that they are visible heterogeneous wastes that are generated, for the most part, where people live and in areas with limited storage space. These wastes cannot be tolerated for long on individual premises because of their biodegradability, and they must be moved within a reasonable time—usually less than 8 days.

The cost of providing storage for solid wastes at the source normally is borne by the householder or apartment owner in the case of individuals, or by the management of commercial and industrial properties. Onsite storage is of primary importance because of the aesthetic consideration, public health, and economics involved. Unsightly make-shift containers and even open ground storage, both of which are undesirable, are often seen at many residential and commercial sites.

Collection

The functional element of collection, as used in this book, includes not only the gathering of solid wastes, but also the hauling of wastes after collection to the location where the collection vehicle is emptied. As shown in Fig. 2-3, this location may be a transfer station, a processing station, or a landfill disposal site. In small cities where final disposal sites are nearby, the hauling of wastes is not a serious problem. In large cities, however, where the haul to the point of disposal often may be greater than 10 mi, the haul may have serious economic implications [7].

The solution to the problem of long-distance hauling is complicated by the fact that the motor vehicles that are well adapted to long-distance hauling are not well suited or particularly economical for house-to-house collection. Consequently, in most cases, supplemental transfer and transport facilities and equipment are needed.

As noted in Chap. 1 (Table 1-4), collection accounts for close to 80 percent of the total annual cost ($2.64 billion in 1971) of urban solid waste management. This service may cost the individual homeowner $30/yr or more, depending on the number of containers and frequency of collection. Typically collection is provided under various management arrangements, ranging from municipal services to franchised services conducted under various forms of contracts. In several parts of the country, large solid waste disposal companies, with contracts in many cities, own and operate collection vehicles and landfill disposal sites.

Collection services for industries vary widely. Some industrial wastes are handled like residential wastes; some companies have disposal sites on their own properties that use conveyor belts or water slurry transport. The latter is used for mineral wastes and agricultural wastes in many cases. Each industry requires an individual solution to its waste problems.

Transfer and Transport

The functional element of transfer and transport involves two steps: (1) the transfer of wastes from the smaller collection vehicle to the larger transport equipment and (2) the subsequent transport of the wastes, usually over long distances, to the disposal site. The transfer usually takes place at a transfer station. Although motor vehicle transport is most common, rail cars or barges are also used to transport wastes.

For example, in San Francisco the collection vehicles, which are relatively small because of the need to maneuver in city streets, haul their loads to a transfer station at the southern boundary of the city. At the transfer station, the wastes unloaded from the collection vehicles are reloaded into large tractor-trailer trucks. The loaded trucks are then driven to a disposal site located about 40 mi away in another county.

Processing and Recovery

The functional element of processing and recovery includes all the techniques, equipment, and facilities used both to improve the efficiency of the other functional elements and to recover usable materials, conversion products, or energy from solid wastes.

In the recovery of materials, as an example, separation operations have been devised to recover valuable resources from the mixed solid wastes delivered to transfer stations or solid waste processing plants. These operations include size reduction and density separation by air classifiers. Further separation may include magnetic devices to pull out iron, eddy-current separators for aluminum, and screens for glass. Flotation, inertial separation, and other metallurgical industry unit operations may also be used. The selection of any recovery process is a function of economics—cost of separation versus value of the recovered-materials products. Because prices fluctuate so sharply, high and low price estimates must be considered in any economic analysis.

Currently, many of the unit operations and processes for solid wastes are undergoing extensive development by equipment manufacturers and by the EPA through its research, development, and demonstration grant programs. Many of the older methods have been found unsatisfactory from one or more standpoints—public health, economics, environmental problems, as well as the exhaustion of available land and consequent restrictions placed on its use by planning authorities.

Disposal

The final functional element in the solid waste management system depicted in Fig. 2-3 is disposal. Disposal is the ultimate fate of all solid wastes, whether they are residential wastes collected and transported directly to a landfill site, semisolid wastes (sludge) from municipal and industrial treatment plants, incinerator residue, compost, or other substances from the various solid waste processing plants that are of no further use to society.

Thus, land-use planning becomes a primary determinant in the selection, design, and operation of landfill operations. In many cities this involves the planning commissions of the city, county, or other regional planning authority. Environmental impact statements (see Chap. 3) are required for all new landfill sites to ensure compliance with public health, aesthetics, and future use of land. A modern sanitary landfill is not a dump. It is a method of disposing of solid wastes on land without creating nuisances or hazards to public health, such as the breeding of rats and insects and the contamination of groundwater, or public safety.

Engineering principles must be followed to confine the wastes to the smallest possible area, to reduce them to the lowest practical volume by compaction at the site, and to cover them after each day's operation to reduce exposure to vermin. After the entire area is filled, an earth cover at least 2 ft thick must be applied to the surface, and reapplied if uneven settlement takes place during decomposition of the underlying organic matter. This decomposition is anaerobic and consequently has a very slow reaction rate. One of the dangers of the biodegradation is the generation of methane gas. Even though it is formed at a slow rate, it can accumulate under buildings [5] and thus must be vented to the atmosphere. Attempts are being made in some cities with large sanitary landfills to capture the methane for energy.

One of the most important concepts is to plan for the final use of the reclaimed land. Many golf courses have been established on old landfills. Parks, open storage areas, and athletic fields occupy the sites of many former landfills. These must be planned properly so that no buildings are located over the decomposing solid wastes. Planning must take place ahead of filling so that any building areas will be filled with earth only.

To remove the stigma of filling lowlands with solid wastes, the City of New York has placed large signs adjacent to highways and landfill sites. These signs indicate that land is being filled for a future golf course, under the direction of the Department of Parks (the planners), the Department of Sanitation (the solid waste disposers), and the Department of Public Works (which furnishes the sludge from its sewage treatment plants). Thus, three agencies of the city are "killing three birds with one stone" and benefiting the public by careful planning and operation.

2-3 SOLID WASTE MANAGEMENT SYSTEMS

The six functional elements that constitute solid waste management systems have been identified and discussed in the previous section. These elements, as noted previously, are the main subjects of this book. The objective in this section is to describe briefly certain other practical aspects associated with solid waste management systems that are not covered in detail in this text. These include financing; operations; equipment management; personnel; reporting, cost accounting, and budgeting; contract administration; ordinances and guidelines; and public communications. For additional information on these subjects, Refs. 2 and 3 are recommended.

Typically, in most cities and counties of the United States, the responsibility for the operation or contract supervision of solid waste management systems is under the jurisdiction of the department of public works. Other departments that have been given this responsibility include engineering, sanitation, streets and sanitation, public utilities, and health. To understand the interrelationship of the day-to-day concerns of these departments, it will be helpful first to consider the organizational structure of some typical solid waste management agencies.

Organizational Structure

The principal reason for developing an organizational structure is to identify the relationships and responsibilities of the individuals charged with accomplishing the stated objectives or goals of the organization. This is especially important for a solid waste management agency where the tasks to be accomplished are so varied.

The complexity of the structure varies according to the size of the organization, as illustrated by the organization charts for the Department of Sanitation of the City of New York and the Division of Solid Waste Management of Sacramento County, California, shown in Figs. 2-5 and 2-6, respectively.

Financing

The principal methods of financing solid waste management systems include (1) general property taxes, (2) separate property taxes, (3) service charges or fees, (4) the utility concept, (5) container charges, (6) special assessments, and (7) miscellaneous revenues [3]. Data on the methods used to finance solid waste collection in 396 cities in the United States are reported in Table 2-1. A discussion of the advantages and disadvantages of these methods may be found in Ref. 3.

Operations

The number of operating divisions depends on the size of the solid waste management agency. For example, in New York City, as shown in Fig.

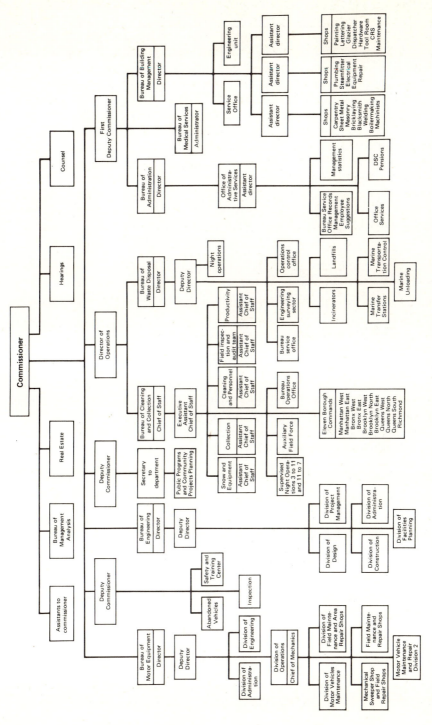

FIG. 2-5 Organization chart for Department of Sanitation, New York. (Reprinted with permission from American Public Works Association, Institute for Solid Wastes [3].)

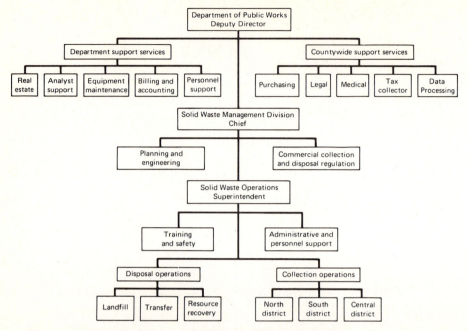

FIG. 2-6 Organization chart for the Division of Solid Waste Management, Sacramento County, California. (Division of Solid Waste Management, Sacramento County.)

2-5, separate bureaus are responsible for motor equipment, engineering, cleaning and collection, waste disposal, administration, medical services, and building management. By contrast, in a small community, one or two individuals may be responsible for all the operations.

Equipment Management

Proper equipment management involves more than maintenance. It also includes equipment replacement analysis, development of equipment specifications, and, in larger organizations, the design of special equipment features. In view of soaring equipment costs, equipment management has become a critical and specialized activity in many solid waste management agencies and requires the employment of qualified people.

Personnel

The successful operation of any solid waste management agency will depend, to a large extent, on the selection, training, and supervision of its

TABLE 2-1 METHODS USED TO FINANCE SOLID WASTE COLLECTION IN 396 UNITED STATES CITIES IN 1973*

Method	Total	Percent of cities, population (000)								
		0–5	5–10	10–25	25–50	50–100	100–250	250–500	500–1,000	Over 1,000
General tax	54.3	66.7	41.7	50.0	49.1	60.0	56.7	50.0	66.7	80.0
Service charge	30.8	16.7	50.0	34.4	34.9	28.0	30.0	25.0	13.3	20.0
Tax plus charge	12.6	16.7	8.3	14.1	13.2	11.0	13.3	14.3	13.3	0.0
Other	2.3	0.0	0.0	1.6	2.8	1.0	0.0	10.7	6.7	0.0

* Reprinted with permission from American Public Works Association, Institute for Solid Wastes [3].

personnel. In most agencies, personnel management is concerned with (1) employee morale and (2) wages, hours, and working conditions. Details concerning personnel management, policies, and practices are considered in Ref. 3.

Reporting, Cost Accounting, and Budgeting

It has been said that "there is probably no other field of endeavor of its magnitude where reporting has been so meager in the past as in solid waste collection" [3]. In general, this statement applies to the entire field of solid waste management. With rising costs, however, more emphasis is being placed on reporting and cost accounting in an effort to assess the true costs of the various activities.

To aid local solid waste management agencies, the EPA has developed a collection management information system called COLMIS. This cost-oriented system was developed for use with computers and was designed to accomplish two objectives: (1) to provide managers at all levels with tools to facilitate the decision-making process and (2) to provide the necessary information for detailed evaluation of the needs for, and effects of, change in operational procedures [1]. The use of COLMIS and of a variety of other computer programs that have been developed has not been extensive. In the future, however, as data reporting procedures become more standardized, it is anticipated that wider use will be made of these computer programs.

Contract Administration

The collection of solid wastes in the United States is accomplished by municipal public agencies, by private organizations under contract to governmental agencies, by private collection organizations, or by various combinations of these. The extent to which each type of arrangement is now used is summarized in Table 2-2. The selection of municipal, contract, or private collection must be based on a valid comparison of costs. Unfortunately, because of poor past cost accounting procedures, the necessary information on which to base such a comparison is seldom available. In many cities contract collection has almost become institutionalized.

Ordinances and Guidelines

The development of effective ordinances and guidelines for all the functional elements identified in Fig. 2-3 is today an important responsibility of any

TABLE 2-2 TYPE OF RESIDENTIAL SOLID WASTE COLLECTION IN 661 NORTH AMERICAN CITIES IN 1973*

Method	Population (000)								
	5–10	10–25	25–50	50–100	100–500	Over 500	Not stated	Total	Percent
Municipal	11	51	73	64	43	14	1	257	39
Contract	11	19	41	22	9	1	3	106	16
Private	2	12	34	25	5	2	3	83	12
Municipal and private	2	14	30	23	32	3	—	104	16
Municipal and contract	1	4	7	14	7	5	—	38	6
Municipal, contract and private	1	3	5	9	4	6	—	28	4
Contract and private	1	8	14	9	7	4	2	45	7
Total	29	111	204	166	107	35	9	661	100

* Reprinted with permission from American Public Works Association, Institute for Solid Wastes [3].

solid waste management agency. Although ordinances vary from state to state and from location to location, most of them include provisions dealing with the following topics: (1) definitions, (2) responsibility for administration, (3) onsite storage, (4) collection, (5) transportation, (6) processing, (7) disposal, (8) special problems, (9) financing, and (10) violations and penalties. Representative ordinances, model contracts, and guidelines may be found in Refs. 2 and 3.

Public Communications

Because employees of solid waste management agencies often have direct contact with the public, it is especially important that they be aware of the need to establish proper relationships with the people they serve. This can be achieved through appropriate training programs. Further, it is equally important that the public be aware of the activities of the solid waste management agency. The means, methods, and techniques for accomplishing this objective are many and varied. The approach selected will depend on the nature of the information to be disseminated. An example of an effective information pamphlet may be found in Appendix B.

2-4   SOLID WASTE MANAGEMENT PLANNING

Solid waste management planning may be defined as the process by which workable alternative programs and plans are developed to solve solid waste problems. In most situations the alternative programs and plans must be presented to the public and to decision-makers for consideration, selection, and adoption. At present the interrelationships among the many technical, economic, environmental, social, and political factors involved in these problems are not well defined. Because the details of solid waste management planning are covered in Part III, the following discussion is intended to serve only as a brief introduction to this subject.

Definition of Terms

In discussing planning in the field of solid waste management (as well as other subjects), the terms functional element, system(s), programs, alternatives, and plans are often used. It is therefore important to have a clear understanding of these terms as they are used in this text.

Functional Element   As noted previously in this chapter, the term functional element is used to describe the various activities associated with the management of solid wastes from the point of generation to final disposal. In general, a functional element represents a physical activity. The six functional elements used throughout this book are waste generation, onsite storage, collection, transfer and transport, processing and recovery, and disposal.

System   The assemblage of one or more of the functional elements to achieve a given objective or goal is known as a solid waste management system.

Program   In the field of solid waste management planning, the term program encompasses all the activities associated with the development of a solution to a problem or problems within a functional element of a solid waste management system. Programs dealing with specific problems related to a functional element may or may not involve policy issues and objectives. If they do, they must be presented to the appropriate decision-makers, such as the members of a city council or a county board.

Alternatives   The term alternative is used to describe various groupings of programs as presented in plans for the purpose of making comparisons. By comparing alternatives composed of individual programs, it is possible for decision-makers to assess the impacts and risks involved in selecting a given alternative. Often the alternative selected will be made up of programs taken from one or more of the proposed alternatives.

Plans   Solid waste management plans are developed to define and establish objectives and policies. Plans may be developed to deal with problems at any level—city or county, subregional or regional, state, or federal. Typically, a local plan will encompass one or more functional elements and one or more program areas. For example, a local plan for collection may involve only the program area associated with the setting of rates.

Preparation of Comprehensive Plans

In the mid-1970s, planners and consulting engineers in many parts of the United States were engaged in regional solid waste management studies as part of the charge of Public Law 92-208, which was designed to achieve comprehensive regional solutions to environmental problems. The federal government and state governments stressed the need for regional planning of solid waste management because the expenses of many smaller communities were becoming too great and there was a tendency toward duplication of facilities.

Objectives and Goals of Comprehensive Plans   The objectives of regional plans vary depending on the size of area involved, the nature of the problems, and the number of overlapping jurisdictions. Often, as discussed in Chap. 3, the overall guiding objectives or goals are specified by legislation. Typically, the objectives of regional comprehensive plans will be stated in broadly defined language. For example, consider the following statement of objectives taken from the report dealing with the preparation of a countywide comprehensive plan for Los Angeles County [4].

> The major objectives of the plan are to preserve and enhance our land, air and water environment; to conserve our resources; and to provide for the protection of the public health, safety and welfare. To accomplish these objectives, it is intended that the plan shall bring about improvements where needed in the management of solid wastes, with the least possible disruption of proven methods of operation.

In most cases the goals of comprehensive planning will vary with local or regional conditions. Because today's solutions will seldom meet tomorrow's needs, the most important goal should be the development of plans in which an unknown future is assumed and a means is provided whereby the changing aspirations and capabilities of society can be accommodated.

Regional Management Problems   Most solid waste management problems know no political boundaries. Many of our larger cities are running out of land for disposal of raw or processed wastes. For example, San Francisco ran out of land many years ago and used land in San Mateo County, about 10 mi south of the city, for filling and disposal of its solid wastes. When that land was filled, as mentioned previously, the city had to resort to contracting with the second county to the south at a site in Mountain View, Santa Clara County, about 40 mi south of San Francisco. The city also has been negotiating with the Western Pacific Railroad to haul solid wastes to deserts about 350 mi to the east where cut-and-cover landfill would be used. The District of Columbia has considered hauling wastes by rail to Virginia. The city of Philadelphia has studied the filling of strip mines in northeastern Pennsylvania.

Other industrialized nations have similar problems. Even such seemingly impenetrable barriers as the Berlin Wall are open to vehicles carrying solid wastes from West Berlin to East Berlin to fill an abandoned quarry on East German territory and to generate revenue for that government. However, a fence surrounds the road leading from the new hole in the Berlin Wall all the way to the quarry, as well as the quarry itself. Such are the political ramifications that face the solid waste management team called upon to develop comprehensive plans.

Overlapping Responsibilities and Jurisdictions   The number of agencies with responsibilities and overlapping jurisdictions that must be considered in the development of a comprehensive plan makes the task a staggering undertaking. To cite one example, consider Los Angeles County, which is the most populous and complex county in the United States. This county has 78 incorporated cities with a total population of about 6 million and another million living in unincorporated areas. To quote the County

Engineer, Harvey T. Brandt [4]:

> Because of the existing multijurisdictional complexity and fragmented authority in solid waste management, it can readily be seen that preparation of a countywide, comprehensive plan faces many problems. Preservation of home rule to the maximum possible degree is a prime consideration. Presently, each city independently regulates waste collection services within its boundaries, in some cases providing collection with their own forces while others license, contract, or franchise collection to private enterprises.
>
> Similarly, in unincorporated territory some areas have formed County-controlled garbage disposal districts where collection is performed by contractors. Other areas are served by licensed haulers on a free lance basis. Likewise a few cities operate their own landfill disposal sites, while other public landfills are operated by County Sanitation Districts or by private enterprise.

2-5 DISCUSSION TOPICS

2-1. From historical records, develop a brief chronology of the disposal methods used in your community during the past 50 years. Identify, where possible, the major events that led to the abandonment of one given method in favor of another.

2-2. Obtain the organization chart for the agency responsible for solid waste management in your community. How does it compare to those shown in Figs. 2-5 and 2-6? Does it seem adequate?

2-3. Contact your local solid waste management agency and determine the types of programs now used to train new employees. In your view as a resident of the community, are they adequate to cover most of the situations that are likely to be encountered on a routine basis? If not, what other items or topics should be included in the training program?

2-4. Explain why collection costs amount to 60 to 80 percent of the total costs of solid waste management.

2-5. In your opinion, what effect do the ownership and operation of landfill disposal sites by private contractors (as compared to public agencies) have on the economics, efficiency, and environmental aspects of operation?

2-6. Why have solid waste management practices been so slow in developing? What changes can you see taking place in the future?

2-7. Identify and discuss briefly the issues that you feel will be important in the field of solid waste management in the late 1970s and early 1980s.

2-8. Discuss the possibilities and advantages of a regional solid waste management plan for the cities and towns in your area, or in a metropolitan area near your home or school.

2-6 REFERENCES

1. *A Collection Management Information System for Solid Waste Management (COLMIS)*, U.S. Environmental Protection Agency, Publication SW-57c, Washington, D.C., 1974.
2. American Public Works Association: "Municipal Refuse Disposal," 3d ed., Public Administration Service, Chicago, 1970.
3. American Public Works Association, Institute for Solid Wastes: "Solid Waste Collection Practice," 4th ed., American Public Works Association, Chicago, 1975.
4. Brandt, H. T.: Preparation of a Comprehensive Solid Waste Management Plan, *Public Works,* vol. 106, no. 5, 1975.

5. Eliassen, R.: Housing Construction on Refuse Landfills, *Engineering News-Record,* vol. 138, no. 18, 1947.
6. Eliassen, R.: *Solid Waste Management: A Comprehensive Assessment of Solid Waste Problems, Practices, and Needs,* Office of Science and Technology, Executive Office of the President, Washington, D.C., 1969.
7. Hardenbergh, W. A.: "Municipal Sanitation," International Textbook, Scranton Pa., 1928.
8. Jones, B. B. and F. Owen: "Some Notes on the Scientific Aspects of Controlled Tipping," Henry Blacklock and Co., Ltd., Printers, Manchester, England, 1934.
9. Parsons, H. de B.: "The Disposal of Municipal Refuse," 1st ed., Wiley, New York, 1906.
10. *Report of a Study of the Collection and Disposal of City Wastes in Ohio,* Supplement to the Twenty-Fifth Annual Report of State Board of Health of the State of Ohio, Heer Printing, Columbus, Ohio, 1911.

3
Legislation and Governmental Agencies

Because so much of the current activity in the field of solid waste management, especially with respect to resources recovery, is a direct consequence of recent legislation, the purpose of this chapter is (1) to review the principal legislation that has affected the entire field of solid waste management and (2) to examine the role of the several governmental agencies responsible for administering the applicable legislation. This information provides some perspective on the political environment in which solid waste management planning is now conducted and introduces the reader to some of the more important requirements associated with the preparation of planning reports and to the several governmental agencies that may have an impact on any plan related to solid waste management. No attempt is made to identify or discuss governmental agencies at the state level because they are so varied and are still undergoing considerable reorganization.

3-1 LEGISLATION

Environmental legislation has become increasingly restrictive as public health agencies, conservationists, and concerned citizens have pressured Congress and the state legislatures to take action. Federal agencies have taken the lead. As mentioned in Chap. 1, in 1899 the Rivers and Harbors Act directed the U.S. Army Corps of Engineers to regulate the dumping of debris in navigable waters and adjacent lands. Many USPHS (U.S. Public Health Service) regulations were placed in force to permit the federal government to regulate the interstate transport of solid wastes, particularly food waste that was fed to hogs, in an attempt to control trichinosis.

Solid Waste Disposal Act, 1965

Modern solid waste legislation dates from 1965 when the Solid Waste Disposal Act, Title II of Public Law 89-272, was enacted by Congress. The intent of this act was to

1. Promote the demonstration, construction, and application of solid waste management and resource recovery systems which preserve and enhance the quality of air, water, and land resources

2. Provide technical and financial assistance to states and local governments and interstate agencies in the planning and development of resource recovery and solid waste disposal programs

3. Promote a national research and development program for improved management techniques, more effective organizational arrangements, and new and improved methods of collection, separation, recovery, and recycling of solid wastes, and the environmentally safe disposal of nonrecoverable residues

4. Provide for the promulgation of guidelines for solid waste collection, transport, separation, recovery, and disposal systems

5. Provide for training grants in occupations involving the design, operation, and maintenance of solid waste disposal systems

Enforcement of this act became the responsibility of the USPHS, an agency of the Department of Health, Education, and Welfare, and the Bureau of Mines, an agency of the Department of the Interior. The USPHS had responsibility for most of the municipal wastes generated in the United States. The Bureau of Mines was charged with supervision of solid wastes from mining activities and the fossil-fuel solid wastes from power plants and industrial steam plants.

President Johnson and his Presidential Scientific Advisory Committee were not satisfied that legislation alone would accomplish the goal of regulation of urban, commercial, industrial, agricultural, and mineral wastes, as specified in the 1965 act. Therefore, in 1968, the President directed that a special study be made of the national problem of solid waste management by the White House staff with the assistance of representatives of the USPHS, The Department of Agriculture, the Department of Defense, and the Department of the Interior. The report that resulted was submitted by the Executive Office of the President to Congress with the demand—which was subsequently met—for adequate staffing, funding, and action by the responsible federal agencies and Congress.

Resources Recovery Act, 1970

The Solid Waste Disposal Act of 1965 was amended by Public Law 95-512, the Resources Recovery Act of 1970. This act directed that the emphasis of the national solid waste management program should be shifted from disposal as its primary objective to that of recycling and reuse of recoverable materials in solid wastes, or to the conversion of wastes to energy. The USPHS, through its National Office of Solid Waste Management, was directed to prepare a report on the *Recovery and Utilization of Municipal Solid Waste* [2], which was completed in 1971. By that time the EPA (Environmental Protection Agency) had been formed by Presidential order under Reorganizational Plan No. 3 of 1970, and all solid waste management activities were transferred from the USPHS to the EPA. Many other reports on various phases of solid waste management have been published since then, including the yearly reports to Congress on resources recovery [7, 8, 9] and the basic reference report, *Decision-Makers Guide in Solid Waste Management* [1].

Another feature of the 1970 act was the mandate of Congress to the Secretary of Health, Education, and Welfare to prepare a report on the treatment and disposal of hazardous wastes, including radioactive, toxic chemical, biological, and other wastes of significance to the public health and welfare. Previously, the Atomic Energy Act of 1954 had authorized the U.S. Atomic Energy Commission to manage all radioactive wastes generated by the Commission and the nuclear power industry.

The report to Congress in response to the 1970 act was prepared by the Office of Solid Waste Management Programs of the EPA, and it was submitted on June 30, 1973. This report, entitled *Disposal of Hazardous Wastes* [6], is a complete treatise on all aspects of the disposal of hazardous wastes and should be referred to by all industries and agencies active in this field. Several states have used it as a basis for their regulations. For example, the California Department of Health published in February 1975 a comprehensive document entitled *Hazardous Waste Management—Laws, Regulations, and Guidelines for the Handling of Hazardous Wastes* [4].

National Environmental Policy Act, 1969

The National Environmental Policy Act (NEPA) of 1969 is an all-encompassing congressional law. It affects all projects that have some federal funding or that come under the regulation of federal agencies. Although the act has some shortcomings and has caused delay in the completion of some projects, it has served a useful purpose in giving the public an opportunity to participate in the decision-making process.

The act specified the creation of the Council on Environmental Quality in the Executive Office of the President. This body has the authority to force every federal agency to submit to the council an Environmental Impact Statement on every activity or project which it may sponsor or over which it has jurisdiction. The project cannot proceed until the Environmental Impact Statement has been approved by the council. For example, if a sanitary landfill is planned on marshland adjacent to an estuary (such as San Francisco Bay or the Delaware River Estuary), the EPA, the Army Corps of Engineers, the Fish and Wildlife Service, and other affected agencies will have to participate in the review and approval of an Environmental Impact Report, which must be prepared by the agency planning the landfill site. The federal lead agency on any particular project (in this case, probably the EPA) would be responsible for preparing an Environmental Impact Statement from the Environmental Impact Report of the municipality. The Environmental Impact Statement, in turn, is sent to the Council on Environmental Quality.

The preparation of environmental impact reports practically has become a new interdisciplinary profession. Every conceivable effect on the environment must be taken into account because the Environmental Impact Statement becomes a legal document that may have to be defended in court. Consulting engineers and planners have large and diversified staffs to serve municipal, county, regional, and state agencies that must prepare environmental impact reports for solid waste management facilities. Many environmental consulting firms have been created to serve planning agencies with experts in fields such as ecology, land-use planning, aquatic and terrestrial biology, soil science, economics and sociology, to name just a few specialties needed for these reports. It is important that each new solid waste management project have a well-conceived, competently prepared, and thoroughly justified environmental impact report to inform the public, to invite its participation, and to seek its support. This will ensure that the project will be approved under the terms of the National Environmental Policy Act and any state and regional laws.

Miscellaneous Laws and Executive Orders

Many other laws apply to the environmental control of solid waste management problems. These include the Noise Pollution and Abatement Act of 1970, a federal law that limits environmental noise among workers employed in all industries, including many of the facilities of solid waste systems, as well as the public, because many noisy operations may be involved—from collection to disposal. The Clean Air Act of 1970 (Public Law 91-604) is very important where dust, smoke, and gases discharged from solid waste operations are involved. Many old incinerators have been shut down because stack emissions were exceeding newly established limits. Composting plants have been shut down because of odor emissions beyond the control of the operators. Air pollution control is discussed in subsequent chapters of this book.

Many executive orders from the President also control the activities of the Council on Environmental Quality and the administration of the National Environmental Policy Act. Also, many states have adopted their own laws and restrictive covenants and have established new agencies for the control of solid waste management systems. Thus, in the planning and design of these facilities, the consulting engineer and/or staff planner for the solid waste management agency must seek legal advice from attorneys who are qualified in the specialties and vagaries of environmental law. This field is in a state of continuous flux.

Similarly, in the financing of new solid waste management systems, the advice of financial consultants should be sought. Many types of governmental grants, from both the federal government and state agencies, might apply to any particular project. These grants and financial aids are also subject to change and are followed closely by financial consultants.

3-2 GOVERNMENTAL AGENCIES

As indicated above, the various laws, regulations, and executive orders have created a divided responsibility among many federal departments and agencies for the regulation and financing of solid waste management. Some of the more significant agencies and their separate impacts are described in the following discussion.

Environmental Protection Agency

The EPA was created by presidential order in 1970 to become the central or lead agency for the control of pollution of the nation's air, water, and land resources. It took over the responsibilities of the USPHS for air pollution control, quality of water supply, and solid waste management. The former Federal Water Pollution Control Administration was abolished and incorporated into the EPA.

The organization chart is very complex, as with most governmental agencies and large corporations. Of particular interest to solid waste management are certain staff offices, including the Office of Solid Waste Management Programs, the Office of Air Programs, the Office of Water Programs, the Office of Noise Control, and the Office of Resources Management. Many documents have been published by these offices, including guidelines for many phases of solid waste management.

An important element of the EPA is its organization of 10 regional offices across the country. Each office has a representative of the Office of Solid Waste Management Programs who can be most helpful in supplying the latest information on technology, laws, funding of projects, and cooperation with state agencies. These regional offices are located in Boston, New York, Philadelphia, Atlanta, Chicago, Dallas, Kansas City, Denver, San Francisco, and Seattle.

Energy Research and Development Administration

The Energy Research and Development Administration (ERDA) was created by an act of Congress and an executive order of the President and was activated on January 19, 1975. It took over most of the activities of the Atomic Energy Commission, which was abolished when the ERDA was created. As President Ford said at its inception,

> What is envisioned is nothing less than a complete energy research and development organization. It will be one which will fill in the gaps in our present research efforts and provide a balanced national research program. It will give proper emphasis to each energy source according to its potential and its readiness for practical use.

Primary research and development facilities include the national laboratories formerly operated by the Atomic Energy Commission, as well as energy research centers formerly under the Bureau of Mines of the Department of the Interior. From the National Science Foundation, the ERDA took over responsibility for solar and geothermal power development programs. Automotive power systems research programs were transferred from the EPA into the ERDA. In addition to having its own powerful laboratories, the ERDA carries out programs by grants and contracts with universities, industrial research groups, and other governmental agencies, such as the EPA.

Again, there is a complicated organizational structure, with several activities associated with solid wastes. Under the assistant administrator of environment and safety falls the responsibility for waste management, one of the most important segments of which is the final disposal of high-level solid wastes containing billions of curies of radioactivity of long half-lives. These wastes originate from the nuclear fuel reprocessing used to recover the unused uranium and the newly generated plutonium. The final product is a solid waste that is highly toxic and must be stored for thousands of years before it loses its radioactivity, although it will never be lost entirely—*never* being a term loosely used to indicate a million years.

However, the mass and volume of the solid wastes produced will be relatively small when compared with urban and industrial wastes. It has been estimated that the volume of concentrated solid wastes produced from all nuclear power reactors for the next 25 years will be less than 500,000 ft^3, or the equivalent of a building 200 ft square by 12.5 ft high. Of course, the solid wastes will be sealed in stainless steel containers, perhaps 10 ft long by 1 ft in diameter. Storage will not be in a building, but in an underground concrete structure that can be monitored and guarded for a thousand years (longer than any government now in existence). The ERDA is also investigating the possibility of storage in deep caverns excavated out of solid granite or in deep salt domes that are impervious to water [5, 10].

The ERDA also has a large staff under the assistant administrator for conservation. Among other responsibilities, the Large Scale Conservation Demonstrations branch is interested in the conversion of solid wastes to energy. The EPA is involved in this activity as well. It has been estimated that from 10 to 14 percent of the electrical energy of a city could be generated by burning its solid wastes. Other estimates indicate that if all the solid wastes from the 11 largest cities in the United States were subjected to biochemical processing to produce methane, about 700 billion ft³/yr would result, or about 3 percent of the nation's needs.

The assistant administrator for fossil energy of the ERDA has responsibility for the new technologies of coal liquification and coal gasification, which may produce large quantities of solid wastes in the form of ash. Oil shale development will also lead to almost unbelievable quantities of solid wastes. There is more oil entrapped in the shales underlying Colorado than in all the oil fields of the Middle East. The temptation to extract these oils from deep rocks is great, particularly to free the country of its dependence on Middle East oil. But the penalties are also great when it is realized that even "good" shale yields only 25 gal/ton of rock. Thus, a recovery plant capable of producing 200,000 bbl/day of oil—the size of a large oil refinery—would also produce 336,000 tons/day of solid wastes. These wastes would be of a powdery nature and almost impossible to dispose of without damage to the environment. This becomes a tremendous research problem for the ERDA and the country.

U.S. Army Corps of Engineers

The responsibility of the U.S. Army Corps of Engineers in the field of solid waste management has been mentioned previously. As cities near navigable waters exhaust their available land for the final disposal of solid wastes, they may try to encroach on tidal estuaries or flood plains of rivers, but they will have to face up to severe criticisms of their environmental impact reports by government and conservation groups. The Coastal Zone Management Act of 1972 declared it a national policy to preserve and protect the resources of the nation's coastal zone. This act recognizes waste disposal as a "competing demand" on coastal zone lands that has caused serious environmental losses.

Department of Labor

The Occupational Safety and Health Act of 1970 directed the Secretary of Labor to set mandatory standards to protect the occupational health and safety of all employers and employees of businesses engaged in interstate commerce. This applies particularly to resource recovery plants that will be involved in materials shipments across state lines. Many states have adopted strict standards of their own, following the lead of the Occupational Safety and Health Act. Thus, occupational safety becomes a significant problem in the design of solid waste management facilities, which in the past have been subject to high accident rates.

Department of Transportation

Stringent regulations have been placed on the transportation of hazardous wastes. Containers must be labeled, and the wastes must be placed in specially designed and approved containers. The Coast Guard controls marine shipment of solid wastes of a hazardous nature.

Interstate Commerce Commission

Freight rates are controlled by the Interstate Commerce Commission. Many differences exist in the rates for the shipment of virgin materials, such as pulpwood and paper pulp, as opposed to higher rates for reclaimed materials. The secondary-materials industry has been struggling to obtain a favorable rate structure so that resource recovery and recycling can have economic significance.

Department of Health, Education, and Welfare

Public health effects of solid waste disposal facilities—particularly land disposal sites where vector control is necessary for the prevention of disease transmission—come within the province of the Department of Health, Education, and Welfare. Complete cooperation with state departments of public health is maintained through regional offices.

3-3   DISCUSSION TOPICS

3-1.   In view of the multiplicity of federal agencies having "a piece of the action" in solid waste management, would you favor a study by the various congressional committees having jurisdiction over these agencies to evaluate means of consolidating the individual agency activities and responsibilities? Why? If not, why not?

3-2.   Knowing that congressional bills are written by the professional staffs of Senate and House committees, assume that you are a member of the staff of the Senate committee on public works. What would be the principal features of a new bill you would write for the Senators to submit to Congress to consolidate responsibilities within specific agencies for a rational solid waste management program?

3-3.   If you were to choose a career in solid waste management in a governmental agency, what level of government—city, county, regional, state, federal—would you select to give you breadth of experience and a fair degree of responsibility for taking constructive action? Why?

3-4.   Discuss your opinion of the value of an environmental impact report for every major decision of a solid waste management agency—city, county, or regional—affecting its plans for transportation, processing, or disposal of solid wastes. Is the cost justified in your estimation?

3-5.   Identify the principal state and local laws and agencies important in solid waste management in your area.

3-6.   Which agency of your state government has jurisdiction over the promulgation of codes and guidelines for solid waste management? Does this agency have an adequate staff to assist municipalities and enforce regulations? What is your opinion of the literature they distribute to cities?

3-7. Where is the regional office of the EPA in your area? What is it doing to assist your state government in its attempts to improve solid waste management in the cities of your state?

3-8. Has your departmental library received an adequate number and variety of EPA publications on solid waste management? If not, can a team project be organized to obtain these publications from the EPA regional office? What is your opinion of these publications?

3-4 REFERENCES

1. Colonna, R. A. and C. McLaren: *Decision-Makers Guide in Solid Waste Management,* U.S. Environmental Protection Agency, Publication SW-127, Washington, D.C., 1974.

2. Drobny, N. L., H. E. Hull, and R. F. Testin: *Recovery and Utilization of Municipal Solid Waste,* U.S. Environmental Protection Agency, Publication SW-10c, Washington, D.C., 1971.

3. Eliassen, R.: *Solid Waste Management: A Comprehensive Assessment of Solid Waste Problems, Practices, and Needs,* Office of Science and Technology, Executive Office of the President, Washington, D.C., 1969.

4. *Hazardous Waste Management: Laws, Regulations, and Guidelines for the Handling of Hazardous Wastes,* California Department of Public Health, Sacramento, 1975.

5. Pittman, F. K.: Management of Radioactive Wastes, *Water, Air, and Soil Pollution,* vol. 4, no. 3, 1975.

6. *Report to Congress: Disposal of Hazardous Wastes,* U.S. Environmental Protection Agency, Publication SW-115, Washington, D.C., 1974.

7. *Resource Recovery and Source Reduction,* First Report to Congress, U.S. Environmental Protection, Agency, Publication SW-118, Washington, D.C., 1973.

8. *Resource Recovery and Source Reduction,* Second Report to Congress, U.S. Environmental Protection Agency, Publication SW-122, Washington, D.C., 1974.

9. *Resource Recovery and Source Reduction,* Third Report to Congress, U.S. Environmental Protection Agency, Publication SW-161, Washington, D.C., 1975.

10. Schneider, K. J.: High Level Wastes, in L. A. Sagen (ed), "Human and Ecological Effects of Nuclear Power Plants," Charles C Thomas, Springfield, Ill., 1974.

PART II

ENGINEERING PRINCIPLES

Perspectives of the field of solid waste management were discussed in Part I. The purpose of Part II is to present, discuss, and illustrate engineering principles of solid waste management. The chapters in Part II are organized in a logical sequence starting with solid waste generation and continuing through onsite storage, collection, transfer and transport, process and recovery, and disposal. The last chapter is reserved for the important subject of hazardous wastes.

Although much is known about the engineering aspects of solid waste management, the field is very dynamic and much remains to be learned, especially in developing areas such as materials and energy recovery. New technologies and equipment are also being developed constantly in other areas of the field. By devoting a separate chapter to each of the functional elements that make up solid waste management systems, it is possible to identify the fundamental aspects of each one and to delineate the interrelationships involved, to the extent that they are known.

Mastery of the engineering principles presented in Part II is fundamental to the understanding and assessment of existing operations and systems, to the evaluating of impacts of new and proposed technologies, and to the proper selection and analysis of alternatives in the development of new systems. The ability to measure the impact of alternative courses of action is vital in the management of these systems and in the decision-making process considered in Part III.

4

Generation of Solid Wastes

Solid wastes, as noted previously, include all solid or semisolid materials that the possessor no longer considers of sufficient value to retain. Collectively, they form the fundamental concern in all activities encompassed in solid waste management—whether the planning level involved is local, subregional or regional, or state and federal. For this reason, it is important to know as much about these materials as possible.

The purpose of this chapter is threefold: to identify the sources and types of solid wastes, to examine the physical and chemical composition of solid wastes, and to discuss solid waste generation rates and the influencing factors involved. Information presented in this chapter will also have application throughout the remainder of this text.

4-1 SOURCES AND TYPES OF SOLID WASTES

Knowledge of the sources and types of solid wastes, along with data on the composition and rates of generation, is basic to the design and operation of the functional elements associated with the management of solid wastes.

Sources of Solid Wastes

Sources of solid wastes are, in general, related to land use and zoning. Although any number of source classifications can be developed, the following categories have been found useful: (1) residential, (2) commercial, (3) municipal, (4) industrial, (5) open areas, (6) treatment plants, and (7) agricultural. Typical waste generation facilities, activities, or locations

TABLE 4-1 TYPICAL SOLID WASTE GENERATING FACILITIES, ACTIVITIES, AND LOCATIONS ASSOCIATED WITH VARIOUS SOURCE CLASSIFICATIONS

Source	Typical facilities, activities, or locations where wastes are generated	Types of solid wastes
Residential	Single-family and multifamily dwellings, low-, medium-, and high-rise apartments, etc.	Food wastes, rubbish, ashes, special wastes
Commercial	Stores, restaurants, markets, office buildings, hotels, motels, print shops, auto repair shops, medical facilities and institutions, etc.	Food wastes, rubbish, ashes, demolition and construction wastes, special wastes, occasionally hazardous wastes
Municipal*	As above*	As above*
Industrial	Construction, fabrication, light and heavy manufacturing, refineries, chemical plants, lumbering, mining, power plants, demolition, etc.	Food wastes, rubbish, ashes, demolition and construction wastes, special wastes, hazardous wastes
Open areas	Streets, alleys, parks, vacant lots, playgrounds, beaches, highways, recreational areas, etc.	Special wastes, rubbish
Treatment plant sites	Water, waste water, and industrial treatment processes, etc.	Treatment plant wastes, principally composed of residual sludges
Agricultural	Field and row crops, orchards, vineyards, dairies, feedlots, farms, etc.	Spoiled food wastes, agricultural wastes, rubbish, hazardous wastes

* The term *municipal* normally is assumed to include both the residential and commercial solid wastes generated in the community.

associated with each of these sources are presented in Table 4-1. The types of wastes generated, which are discussed next, are also identified.

Types of Solid Wastes

The term solid wastes is all-inclusive and encompasses all sources, types of classifications, composition, and properties. Wastes that are discharged may be of significant value in another setting, but they are of little or no value to the possessor who wants to dispose of them. To avoid confusion, the term refuse, often used interchangeably with the term solid wastes, is not used in this text.

As a basis for subsequent discussions, it will be helpful to define the various types of solid wastes that are generated (see Table 4-1). It is important to be aware that the definitions of solid waste terms and the classifications vary greatly in the literature. Consequently, the use of published data requires considerable care, judgment, and common sense. The following definitions are intended to serve as a guide and are not meant to be arbitrary or precise in a scientific sense.

Food Wastes Food wastes are the animal, fruit, or vegetable residues resulting from the handling, preparation, cooking, and eating of foods (also called *garbage*). The most important characteristic of these wastes is that they are highly putrescible and will decompose rapidly, especially in warm weather. Often, decomposition will lead to the development of offensive odors. In many locations, the putrescible nature of these wastes will significantly influence the design and operation of the solid waste collection system. In addition to the amounts of food wastes generated at residences, considerable amounts are generated at cafeterias and restaurants, large institutional facilities such as hospitals and prisons, and facilities associated with the marketing of foods, including wholesale and retail stores and markets.

Rubbish Rubbish consists of combustible and noncombustible solid wastes of households, institutions, commercial activities, etc., excluding food wastes or other highly putrescible materials. Typically, combustible rubbish consists of materials such as paper, cardboard, plastics, textiles, rubber, leather, wood, furniture, and garden trimmings. Noncombustible rubbish consists of items such as glass, crockery, tin cans, aluminum cans, ferrous and other nonferrous metals, and dirt.

Ashes and Residues Materials remaining from the burning of wood, coal, coke, and other combustible wastes in homes, stores, institutions, and industrial and municipal facilities for purposes of heating, cooking, and disposing of combustible wastes are categorized as ashes and residues. Residues from power plants normally are not included in this category. Ashes and residues are normally composed of fine, powdery materials, cinders, clinkers, and small amounts of burned and partially burned materials [1]. Glass, crockery, and various metals are also found in the residues from municipal incinerators.

Demolition and Construction Wastes Wastes from razed buildings and other structures are classified as demolition wastes. Wastes from the construction, remodeling, and repairing of individual residences, commercial buildings, and other structures are classified as construction wastes. These wastes are often classified as rubbish. The quantities produced are difficult to estimate and variable in composition, but may include dirt, stones, concrete, bricks, plaster, lumber, shingles, and plumbing, heating, and electrical parts.

Special Wastes Wastes such as street sweepings, roadside litter, litter from municipal litter containers, catch-basin debris, dead animals, and abandoned vehicles are classified as special wastes. Because it is impossible to predict where dead animals and abandoned automobiles will be found, these wastes are often identified as originating from nonspecific diffuse sources. This is in contrast to residential sources, which are also diffuse but specific in that the generation of the wastes is a recurring event.

Treatment Plant Wastes   The solid and semisolid wastes from water, waste water, and industrial waste treatment facilities are included in this classification. The specific characteristics of these materials vary, depending on the nature of the treatment process. At present, their collection is not the charge of most municipal agencies responsible for solid waste management. In the future, however, it is anticipated that their disposal will become a major factor in any solid waste management plan (see Chap. 12).

Agricultural Wastes   Wastes and residues resulting from diverse agricultural activities—such as the planting and harvesting of row, field, and tree and vine crops, the production of milk, the production of animals for slaughter, and the operation of feedlots—are collectively called *agricultural wastes*. At present, the disposal of these wastes is not the responsibility of most municipal and county solid waste management agencies. However, in many areas the disposal of animal manure has become a critical problem, especially from feedlots and dairies.

Hazardous Wastes   Chemical, biological, flammable, explosive, or radioactive wastes that pose a substantial danger, immediately or over time, to human, plant, or animal life are classified as hazardous. Typically, these wastes occur as liquids, but they are often found in the form of gases, solids, or sludges.

In all cases, these wastes must be handled and disposed of with great care and caution. Because of the specialized nature of these wastes, their management is considered in Chap. 11, which deals specifically with hazardous wastes.

4-2   COMPOSITION OF MUNICIPAL SOLID WASTES

Information on the composition of solid wastes is important in evaluating alternative equipment needs, systems, and management programs and plans. For example, if the solid wates generated at a commercial facility consist of only paper products, the use of special processing equipment, such as shredders and balers, may be appropriate. Separate collection may also be considered if the city or collection agency is involved in a paper-products recycling program. Evaluation of the feasibility of incineration depends on the chemical composition of the solid wastes.

The physical and chemical composition of municipal solid wastes is discussed in this section. Possible future changes in composition are also described. The discussion is limited to an analysis of municipal solid wastes because consideration of the composition of all types of wastes would add little useful information and is beyond the scope of this text, which deals principally with the management of municipal solid wastes. It is important to note, however, that the fundamentals of analysis presented are applicable to all types of solid wastes. Additional details on the various physical, chemical, and microbiological methods of testing for solid wastes are presented in Ref. 2.

Physical Composition

Information and data on the physical composition of solid wastes are important in the selection and operation of equipment and facilities (see Chaps. 5 through 8), in assessing the feasibility of resource and energy recovery (see Chap. 9), and in the analysis and design of disposal facilities (see Chap. 10). The individual components that make up municipal solid wastes, and the moisture content and density of solid wastes, are described in the following discussion.

Individual Components   Components that typically make up most municipal solid wastes and their relative distribution are reported in Table 4-2. Although any number of components could be selected, those listed in Table 4-2 have been selected because they are readily identifiable and consistent with component categories reported in the literature and because they have proved adequate for the characterization of solid wastes for most applications. The data in Table 4-2 are derived from both the literature and the authors' experience.

TABLE 4-2   TYPICAL PHYSICAL COMPOSITION OF MUNICIPAL SOLID WASTES

Component	Percent by weight			
	Range	Typical	Packaging materials*	Davis, California†
Food wastes	6–26	15	—	9.5
Paper	25–45	40 }	55.8	43.1
Cardboard	3–15	4 }		6.5
Plastics	2–8	3	3.6	1.8
Textiles	0–4	2	0.4	0.2
Rubber	0–2	0.5	—	0.8
Leather	0–2	0.5	—	0.7
Garden trimmings	0–20	12	—	14.3
Wood	1–4	2	7.8	3.5
Glass	4–16	8	18.1	7.5
Tin cans	2–8	6	14.3	5.2
Nonferrous metals	0–1	1	—	1.5
Ferrous metals	1–4	2	—	4.3
Dirt, ashes, brick, etc.	0–10	4	—	1.1
		100		

* From Ref. 4.
† Based on measurements made over a 5-yr period (1971 to 1975).

The percentages of municipal solid waste components vary with location, the season, economic conditions, and many other factors. For this reason, if the distribution of components is a critical factor in a particular management decision process, a special study should be undertaken, if possible, to assess the actual distribution. Even then it may still be impossible to obtain an accurate assessment unless a prohibitively large number of samples are analyzed.

A common failing in many engineering studies is to spend far too much money collecting data that may never be used. This is especially true with regard to distribution data on solid waste components. For example, if glass is not to be recovered, it is not especially important to know if the amount is 7 percent as opposed to 8 percent (see Table 4-2). Unless there is some specific reason for which a more detailed distribution of components must be known, the data presented in Table 4-2 may be used for most management studies.

Determination of Components in the Field   Because of the heterogeneous nature of solid wastes, determination of the composition is not an easy task. Strict statistical procedures are difficult, if not impossible, to implement. For this reason, a more generalized field procedure, based on common sense and random-sampling techniques, has been developed for determining composition.

The procedure involves unloading a quantity of wastes in a controlled area of a disposal site that is isolated from winds and separate from other operations. A representative residential sample might be a truckload resulting from a typical daily collection in a residential area. A mixed sample from an incinerator storage pit or the discharge pit of a shredder would also be representative. Common sense is important in selecting the load to be sampled. For example, a load containing the weekly accumulation of yard wastes (leaves) during autumn would not be typical.

To ensure that the results obtained are sound statistically, a large enough sample must be obtained. It has been found that measurements made on a sample size of about 200 lb vary insignificantly from measurements made on samples of up to 1,700 lb taken from the same waste load [6]. The authors have obtained similar results in field studies performed in Hawaii and at Davis, California.

To obtain a sample for analysis, the load is first quartered. One part is then selected for additional quartering until a sample size of about 200 lb is obtained. It is important to maintain the integrity of each selected quarter, regardless of the odor or physical decay, and to make sure that all the components are measured. Only in this way can some degree of a randomness and unbiased selection be maintained. Additional information has been published by the American Public Works Association.

Moisture Content   The moisture content of solid wastes usually is expressed as the weight of moisture per unit weight of wet or dry material. In

TABLE 4-3 TYPICAL DATA ON MOISTURE CONTENT OF MUNICIPAL SOLID WASTE COMPONENTS

	Moisture, percent	
Component	Range	Typical
Food wastes	50–80	70
Paper	4–10	6
Cardboard	4–8	5
Plastics	1–4	2
Textiles	6–15	10
Rubber	1–4	2
Leather	8–12	10
Garden trimmings	30–80	60
Wood	15–40	20
Glass	1–4	2
Tin cans	2–4	3
Nonferrous metals	2–4	2
Ferrous metals	2–6	3
Dirt, ashes, brick, etc.	6–12	8
Municipal solid wastes	15–40	20

the wet-weight method of measurement, the moisture in a sample is expressed as a percentage of the wet weight of the material; in the dry-weight method, it is expressed as a percentage of the dry weight of the material. In equation form, the wet-weight moisture content is expressed as follows:

$$\text{Moisture content } (\%) = \left(\frac{a - b}{a} \right) 100 \qquad (4\text{-}1)$$

where a = initial weight of sample as delivered
b = weight of sample after drying

Typical data on the moisture content for the solid waste components in Table 4-2 are given in Table 4-3. For most municipal solid wastes, the moisture content will vary from 15 to 40 percent, depending on the composition of the wastes, the season of the year, and the humidity and weather conditions, particularly rain. The use of the data in Table 4-3 to estimate the overall moisture content of solid wastes is illustrated in Example 4-1.

EXAMPLE 4-1 *Estimation of Moisture Content of Typical Municipal Solid Wastes*

Estimate the overall moisture content of a sample of solid wastes with the typical composition given in Table 4-2.

Solution
1. Set up the computation table (see Table 4-4).
2. Determine the dry weight of solid waste components using the following relationship:

$$\text{Dry weight, lb} = (100 - \text{moisture content, \%})$$
$$\text{(as-delivered weight)}$$

3. Determine the moisture content of the solid waste sample using Eq. 4-1.

$$\text{Moisture content (\%)} = \left(\frac{100 - 78.1}{100}\right)100 = 21.9$$

Density Density data are often needed to assess the total mass and volume of water that must be managed. Unfortunately, there is little or no uniformity in the way solid waste densities have been reported in the literature. Often, no distinction has been made between uncompacted or compacted densities. Typical densities for various wastes as found in

TABLE 4-4 DETERMINATION OF MOISTURE CONTENT FOR SOLID WASTES SAMPLE IN EXAMPLE 4-1

Component	Percent by weight	Moisture content, percent	Dry weight*, lb
Food wastes	15	70	4.5
Paper	40	6	37.6
Cardboard	4	5	3.8
Plastics	3	2	2.9
Textiles	2	10	1.8
Rubber	0.5	2	0.5
Leather	0.5	10	0.4
Garden trimmings	12	60	4.8
Wood	2	20	1.6
Glass	8	2	7.8
Tin cans	6	3	5.8
Nonferrous metals	1	2	1.0
Ferrous metals	2	3	1.9
Dirt, ashes, brick, etc.	4	8	3.7
Total	100		78.1

$$\text{Moisture content (\%)} = \left(\frac{100 - 78.1}{100}\right)100 = 21.9$$

* Based on an as-delivered sample weight of 100 lb.

TABLE 4-5 TYPICAL DENSITIES OF MUNICIPAL SOLID WASTES BY SOURCE*

	Density, lb/yd³	
Source	Range	Typical
Residential (uncompacted)		
Rubbish†	150–300	220
Garden trimmings	100–250	175
Ashes	1,100–1,400	1,250
Residential (compacted)		
In compactor truck	300–750	500
In landfill (normally compact)	600–850	750
In landfill (well compacted)	1,000–1,250	1,000
Residential (after processing)		
Baled	1,000–1,800	1,200‡
Shredded, uncompacted	200–450	360
Shredded, compacted	1,100–1,800	1,300‡
Commercial-industrial (uncompacted)		
Food waste (wet)	800–1,600	900
Combustible rubbish	80–300	200
Noncombustible rubbish	300–600	500

* Adapted in part from Ref. 10.
† Does not include ashes.
‡ Low pressure compaction, less than 100 lb/in².
Note: lb/yd³ × 0.5933 = kg/m³
       lb/in² × 6.895 = kN/m²

containers are reported by source in Table 4-5. Corresponding data for the solid waste components in Table 4-2 are given in Table 4-6.

Because the densities of solid wastes vary markedly with geographic location, season of the year, and length of time in storage, great care should be used in selecting typical values. Municipal solid wastes as delivered in compaction vehicles have been found to vary from 300 to 700 lb/yd³; a typical value is about 500 lb/yd³.

Chemical Composition

Information on the chemical composition of solid wastes is important in evaluating alternative processing and recovery options. For example, consider the incineration process. Typically, wastes can be thought of as a combination of semimoist combustible and noncombustible materials. If solid wastes are to be used as fuel, the four most important properties to be known are:

1. Proximate analysis
   a. Moisture (loss at 105°C for 1 h)

TABLE 4-6   TYPICAL DENSITIES OF MUNICIPAL SOLID WASTE COMPONENTS AS DISCARDED*

	Density, lb/ft³	
Components	Range	Typical†
Food wastes	8–30	18.0
Paper	2–8	5.1
Cardboard	2–5	3.1
Plastics	2–8	4
Textiles	2–6	4
Rubber	6–12	8
Leather	6–16	10
Garden trimmings	4–14	6.5
Wood	8–20	15.0
Glass	10–30	12.1
Tin cans	3–10	5.5
Nonferrous metals	4–15	10.0
Ferrous metals	8–70	20
Dirt, ashes, brick, etc.	20–60	30

* Uncompacted.
† Based on measurements made over a 5-yr period (1971 to 1975) at Davis, California.
Note: lb/ft³ × 16.019 = kg/m³

  b. Volatile matter (additional loss on ignition at 950°C)
  c. Ash (residue after burning)
  d. Fixed carbon (remainder)
 2. Fusing point of ash
 3. Ultimate analysis, percent of C (carbon), H (hydrogen), O (oxygen), N (nitrogen), S (sulfur), and ash
 4. Heating value

A proximate analysis for the combustible components of municipal solid wastes as discarded is presented in Table 4-7.

TABLE 4-7   TYPICAL PROXIMATE ANALYSIS FOR MUNICIPAL SOLID WASTES

	Value, percent	
Component	Range	Typical
Moisture	15–40	20
Volatile matter	40–60	53
Fixed carbon	5–12	7
Glass, metal, ash	15–30	20

Representative data on the ultimate analysis of typical municipal waste components in Table 4-2 are presented in Table 4-8. If Btu values are not available, the approximate Btu value can be determined by using Eq. 4-2, known as the modified Dulong formula [4], and the data in Table 4-8.

$$\text{Btu/lb} = 145.4C + 620(H - \tfrac{1}{8}O) + 41S \qquad (4\text{-}2)$$

where C = carbon, percent
     H = hydrogen, percent
     O = oxygen, percent
     S = sulfur, percent

Typical data on the inert residue and calorific values for municipal wastes are reported in Table 4-9. As shown, the calorific values are on an as-discarded basis. The Btu values in Table 4-9 may be converted to a dry basis by using Eq. 4-3.

$$\text{Btu/lb (dry basis)} = \text{Btu/lb (as discarded)}\left(\frac{100}{100 - \%\ \text{moisture}}\right) \qquad (4\text{-}3)$$

The corresponding equation for the Btu per pound on an ash-free dry basis is

Btu/lb (ash-free dry basis)

$$= \text{Btu/lb (as discarded)}\left(\frac{100}{100 - \%\ \text{ash} - \%\ \text{moisture}}\right) \qquad (4\text{-}4)$$

The use of the data in Table 4-9 in computing the energy content of a municipal solid waste is illustrated in Example 4-2.

TABLE 4-8 TYPICAL DATA ON ULTIMATE ANALYSIS OF THE COMBUSTIBLE COMPONENTS IN MUNICIPAL SOLID WASTES

Component	Percent by weight (dry basis)					
	Carbon	Hydrogen	Oxygen	Nitrogen	Sulfur	Ash
Food wastes	48.0	6.4	37.6	2.6	0.4	5.0
Paper	43.5	6.0	44.0	0.3	0.2	6.0
Cardboard	44.0	5.9	44.6	0.3	0.2	5.0
Plastic	60.0	7.2	22.8	—	—	10.0
Textiles	55.0	6.6	31.2	4.6	0.15	2.5
Rubber	78.0	10.0	—	2.0	—	10.0
Leather	60.0	8.0	11.6	10.0	0.4	10.0
Garden trimmings	47.8	6.0	38.0	3.4	0.3	4.5
Wood	49.5	6.0	42.7	0.2	0.1	1.5
Dirt, ashes, brick, etc.	26.3	3.0	2.0	0.5	0.2	68.0

TABLE 4-9   TYPICAL DATA ON INERT RESIDUE AND ENERGY CONTENT OF MUNICIPAL SOLID WASTES

Component	Inert residue, percent*		Energy, Btu/lb†	
	Range	Typical	Range	Typical
Food wastes	2–8	5	1,500– 3,000	2,000
Paper	4–8	6	5,000– 8,000	7,200
Cardboard	3–6	5	6,000– 7,500	7,000
Plastics	6–20	10	12,000–16,000	14,000
Textiles	2–4	2.5	6,500– 8,000	7,500
Rubber	8–20	10	9,000–12,000	10,000
Leather	8–20	10	6,500– 8,500	7,500
Garden trimmings	2–6	4.5	1,000– 8,000	2,800
Wood	0.6–2	1.5	7,500– 8,500	8,000
Glass	96–99+	98	50– 100	60
Tin cans	96–99+	98	100– 500	300
Nonferrous metals	90–99+	96	—	—
Ferrous metals	94–99+	98	100– 500	300
Dirt, ashes, brick, etc.	60–80	70	1,000– 5,000	3,000
Municipal solid wastes			4,000– 5,500	4,500

* After complete combustion.
† As-discarded basis.
Note: Btu/lb × 2.326 = kJ/kg

EXAMPLE 4-2   *Estimation of Energy Content of Typical Municipal Solid Wastes*

Determine the energy value of typical municipal solid wastes with the average composition shown in Table 4-2.

Solution

1. Assume the heating value will be computed on an as-discarded basis.

2. Determine the energy value using a computation table (see Table 4-10).

3. As shown in Table 4-10, the as-discarded energy content of the wastes would be 4,762 Btu/lb (11,053 kJ/kg). This value compares well with the typical value given in Table 4-9.

Future Changes in Composition

In terms of solid waste management planning, knowledge of future trends in the composition of solid wastes is of great importance. For example, if a paper recycling program were instituted on the basis of current distribution data and if paper production were to be eliminated in the future, such a

TABLE 4-10 COMPUTATION OF ENERGY CONTENT FOR MUNICIPAL SOLID WASTES IN EXAMPLE 4-2

Component	Solid wastes*, lb	Energy*, Btu/lb	Total energy, Btu
Food wastes	15	2,000	30,000
Paper	40	7,200	288,000
Cardboard	4	7,000	28,000
Plastics	3	14,000	42,000
Textiles	2	7,500	15,000
Rubber	0.5	10,000	5,000
Leather	0.5	7,500	3,750
Garden trimmings	12	2,800	33,600
Wood	2	8,000	16,000
Glass	8	60	480
Tin cans	6	300	1,800
Nonferrous metals	1	—	—
Ferrous metals	2	300	600
Dirt, ashes, brick, etc.	4	3,000	12,000
Total	100		476,230

$$\text{Energy content} = \frac{476,230 \text{ Btu}}{100 \text{ lb}} = \frac{4,762 \text{ Btu}}{lb} \left(\frac{11,053 \text{ kJ}}{kg}\right)$$

* Data were derived from Tables 4-2 and 4-9.
Note: Btu/lb × 2.326 = kJ/kg

program would more than likely become a costly "white elephant." Although this case is extreme, it nevertheless illustrates the point that future trends must be assessed carefully in long-term planning. Another important question is whether the quantities are actually changing or only the reporting system has improved.

Food Wastes The quantity of residential food wastes collected has changed significantly over the years as a result of technical advances and changes in public attitude. Two technological advances that have had a significant effect are the use of the home grinder and the development of the food processing and packaging industry.

Recently, because the public has become more environmentally aware and concerned and because the effects of inflation have become more widespread, a trend has developed toward the use of more raw, rather than processed, vegetables. While it would appear that such a trend would increase the quantity of food wastes collected, this is not necessarily the case. Alternative uses for food wastes, such as composting, have served to offset any increases in quantities produced.

Paper   The percentage of paper found in solid wastes has increased greatly in recent years. If we consider that the percentage increase in the quantity of paper produced in the United States amounts to more than twice the percentage increase in population for the years 1950 to 1962 [1], the reason for the increase in waste paper is clear. Without intervening legislation, it is expected that this trend will continue for some time in the future.

Plastics   The percentage of plastics in solid wastes has also increased significantly during the past 20 years. Future conditions (economic and political) surrounding the oil-producing industry will affect the production of plastics. The extent of any impacts is at present unknown.

4-3 GENERATION RATES

The subject of solid waste generation rates has caused considerable confusion because of the different methods of measurement and the different waste classifications adopted for reporting data. The reason for measuring generation rates is to obtain data that can be used to determine the total amount of wastes to be managed. Therefore, in any solid waste management study, extreme care must be exercised in allocating funds and deciding what actually needs to be known.

Because of the importance of being able to assess the quantity of solid wastes generated, separate discussions in this section are devoted to:

1. Measures of quantities
2. Statistical analysis of generation rates
3. Expressions for unit generation rates
4. Methods used to determine generation rates
5. Typical generation rates
6. Factors that affect generation rates

Measures of Quantities

Both volume and weight are used for the measurement of solid waste quantities. Unfortunately, the use of volume as a measure of quantity can be extremely misleading. For example, a cubic yard of loose wastes represents a different quantity than a cubic yard of wastes that have been compacted in a packer truck, and each of these is different from a cubic yard of wastes that have been compacted further in a landfill. Accordingly, if volume measurements are to be used, the measured volumes must be related to the degree of compaction of the wastes.

To avoid confusion, solid waste quantities should be expressed in terms of weight. Weight is the only accurate basis for records because tonnages can be measured directly, regardless of the degree of compaction.

The use of weight records is also important in the transport of solid wastes because the quantity that can be hauled usually is restricted by highway weight limits rather than by volume.

Statistical Analysis of Generation Rates

In developing solid waste management systems, it is often necessary to determine the statistical characteristics of solid waste generation. For example, for many large industrial activities it would be impractical to provide container capacity to handle the largest conceivable quantity of solid wastes to be generated in a given day. The container capacity to be provided must be based on a statistical analysis of the generation rates and the characteristics of the collection system. The statistical measures that must be considered include the mean, mode, median, standard deviation, and coefficient of variation. The definition of these measures and their application are delineated in Appendix C.

Expressions for Unit Generation Rates

In addition to knowing the sources and composition of the solid wastes that must be managed, it is equally important to be able to develop meaningful units of expression for the quantities generated. Because different units of expression must be used for different generation sources, each source is discussed separately. It is noted, however, that unit generation data available for commercial and industrial activities are presently sparse. Consequently, it has been found expedient in many cases to use the same units for these activities as those used for residential wastes, as opposed to the more rational units in the following discussion. The most comprehensive waste records now available are those kept at local landfills, transfer stations, or processing stations, and usually it is impossible to separate the sources from which the wastes were derived.

Residential   Because of the relatively stable nature of residential wastes in a given location, the most common unit expression used for their generation rates is pounds per capita per day. However, should the waste composition vary significantly from typical municipal wastes (Table 4-2), the use of pounds per capita per day may be misleading, especially when quantities are being compared.

Commercial   In the past, commercial waste generation rates have also been expressed in pounds per capita per day. Although this practice has been continued as an expedient, it adds little useful information about the nature of solid waste generation at commercial sources. A more meaningful approach would be to relate the quantities generated to the number of customers, the dollar value of sales, or some similar unit. Use of such factors would allow comparisons to be made throughout the country.

Industrial   Ideally, wastes generated from industrial activities should be expressed on the basis of some repeatable measure of production, such as pounds per automobile for an automobile assembly plant or pounds per case for a packaging plant. When and if such data are developed, it will be possible to make meaningful comparisons between similar industrial activities throughout the country.

Agricultural   Where adequate records have been kept, solid wastes from agricultural activities are now most often expressed in terms of some repeatable measure of production, such as pounds of manure per 1,400-pound cow per day and pounds of waste per ton of raw product. At present, quantification of the solid wastes generated from agricultural activities associated with field and row crops is difficult because little useful information has been gathered in the past.

Methods Used to Determine Generation Rates

Methods commonly used to assess the per capita generation of solid wastes are (1) a load-count analysis, (2) weight-volume analysis, and (3) materials-balance analysis. In reviewing the information presented in this discussion, it will be helpful to remember that most measurements of generation rates do not represent what they are reported or assumed to represent. In the case of predicting residential waste generation rates, the measured rates seldom reflect the true rate because there are so many confounding factors, such as onsite storage and the use of alternative disposal locations, that make the true rate difficult to assess. Most solid waste generation rates reported in the literature are actually collection rates and not generation rates.

Load-Count Analysis   In this method, the number of individual loads and the corresponding vehicle characteristics are noted over a specified time period. If scales are available, weight data are also recorded. Unit generation rates are determined by using the field data and, where necessary, published data. This method is illustrated in Example 4-3.

EXAMPLE 4-3   *Estimation of Unit Solid Waste Generation Rates for a Residential Area*

From the following data estimate the unit waste generation rate for a residential area consisting of approximately 1,000 homes. The observation location is a local transfer station, and the observation period is 1 wk.

1.   Number of compactor truck loads = 10
2.   Average size of compactor truck = 20 yd³

3. Number of flatbed loads = 10

4. Average flatbed volume = 1.5 yd³

5. Number of loads from individual residents' private cars and trucks = 20

6. Estimated volume per domestic vehicle = 8 ft³

Solution

1. Set up the computation table (see Table 4-11).

2. Determine the unit waste generation rate based on the assumption that each household is comprised of 3.5 people.

$$\text{Unit rate} = \frac{72,850 \text{ lb/wk}}{(1,000 \times 3.5)(7 \text{ days/wk})}$$

$$= 3.0 \text{ lb/capita/day } (1.36 \text{ kg/capita·day})$$

Comment The difficulty in using such data is knowing whether they are truly representative of what needs to be measured. For example, how many loads were hauled elsewhere? How much material was stored on the homeowner's premises? All such questions tend to confound the observed data in the statistical sense.

Weight-Volume Analysis Although the use of detailed weight-volume data obtained by weighing and measuring each load will certainly provide better information on the density of the various forms of solid wastes at a given location, the question remains: Is this what is needed in terms of survey results?

Materials-Balance Analysis The only way to determine the generation and movement of solid wastes with any degree of reliability is to perform a detailed materials balance analysis for each generation source, such as an individual home or a commercial or industrial activity. Because of the high

TABLE 4-11 ESTIMATION OF UNIT SOLID WASTE GENERATION RATES IN EXAMPLE 4-3

Item	Number of loads	Average volume, yd³	Unit weight*, lb/yd³	Total weight, lb
Compactor truck	10	20	350	70,000
Flatbed truck	10	1.5	150	2,250
Individual private vehicle	20	0.30	100	600
Total, lb/wk				72,850

* Estimated by using data given in Table 4-4 and some limited onsite weight measurements.
Note: yd³ × 0.7646 = m³
lb/yd³ × 0.5933 = kg/m³
lb × 0.4536 = kg

expense and the large amount of work involved, however, this method of analysis should be used only in special situations.

The approach to be followed in the preparation of a materials balance analysis is as follows. First, draw a system boundary around the unit to be studied (see Fig. 4-1). Second, identify all the activities that cross or occur within the boundary and affect the generation of wastes. Third, if possible, identify the rate of generation associated with these activities. Fourth, using a materials balance, determine the quantity of wastes generated, collected, and stored. A simplified materials-balance analysis is illustrated in Example 4-4.

EXAMPLE 4-4 *Materials-Balance Analysis*

A cannery receives on a given day 12 tons of raw produce, 5 tons of cans, 0.5 ton of cartons, and 0.3 ton of miscellaneous materials. Its output includes 10 tons of processed produce, the remaining becoming part of the waste water. Four tons of the cans are stored for future use, and the remainder are used to package the produce. About 3 percent of the cans used are damaged and recycled. The cartons are also used for packaging, except for 3 percent which become damaged and are incinerated with other paper wastes. Of the miscellaneous materials, 75 percent become paper wastes that are incinerated, and the remainder are disposed of by the municipal collection agency. Draw a materials flow diagram for this activity.

Solution

1. Each day the cannery receives
    12 tons of raw produce
    5 tons of cans

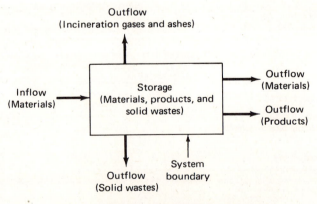

FIG. 4-1 Materials balance definition sketch.

0.5 ton of cartons

0.3 ton of miscellaneous materials

2. As a result of internal activity,

10 tons of product are produced, and the remainder of the produce is discharged to the sewer

4 tons of cans are stored, and the remainder is used

3 percent of the cans used are damaged and recycled

3 percent of the cartons are damaged and incinerated, and the remainder is used

75 percent of the miscellaneous materials become paper wastes which are incinerated, and the remainder is disposed of

3. Determine the required quantities:

Wastes generated = (12 − 10) tons = 2 tons (1,814 kg)

Cans damaged and recycled = (0.03)(5 − 4) tons = 0.03 ton (27 kg)

Cans used in product = (1 − 0.03) ton = 0.97 ton (880 kg)

Cartons incinerated = (0.03)(0.5 ton) = 0.015 ton (14 kg)

Cartons used in product = (0.5 − 0.015) ton = 0.485 ton (440 kg)

Miscellaneous incinerated = (0.75)(0.3 ton) = 0.225 ton (204 kg)

Miscellaneous disposed of = (0.3 − 0.225) ton = 0.075 ton (68 kg)

Total incinerated = (0.015 + 0.225) ton = 0.240 ton (218 kg)

Total produce = (10 + 0.97 + 0.485) tons = 11.455 tons (10,392 kg)

4. Neglecting the amount of materials discharged in the incinerator stack gases, draw a materials flow diagram (see Fig. 4-2).

Comment This simple example was presented to illustrate some of the computations involved in the preparation of a materials-balance analysis. If the internal processing activities are more complex, the amount of work involved in arriving at a materials balance obviously could become prohibitively expensive.

Typical Generation Rates

Perhaps the most comprehensive information on unit quantities of solid wastes generated in the United States was obtained by the U.S. Depart-

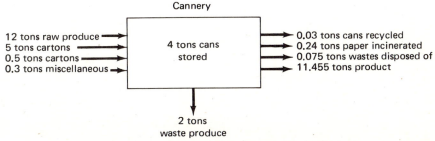

FIG. 4-2 Materials balance definition sketch for Example 4-4.

TABLE 4-12   PER CAPITA SOLID WASTES COLLECTED IN THE UNITED
STATES, 1968*   ➤ AVG. OF [ALL] AREAS

? DIFF. THAN
DATA IN
TABLE 1-1

Source or composition	Population reporting (000)	lb/capita/day
Combined household and commercial†	46,970	4.05
Industrial	29,330	1.86
Institutional	20,533	0.24
Demolition and construction	23,697	0.66
Street and alley cleanings	35,340	0.25
Tree and landscaping	25,890	0.18
Park and beach	17,230	0.16
Catch basin	22,010	0.04
Sewage treatment plant solids	20,504	0.47
Total		7.92

* Adapted from the 1968 National Survey of Community Solid Waste Practices [7].
† The data for combined household and commercial solid wastes subsequently have been
lowered on the basis of estimates derived from production statistics. In a population of 207
million, the as-disposed amount is estimated to be about 3.316 lb/capita/day.
Note: lb/capita/day × 0.4536 = kg/capita·day

ment of Health, Education, and Welfare in the 1968 National Survey of
Community Solid Waste Practices, which covered a population of about
92.5 million [7]. Even though local quantities may vary significantly, the
data gathered in this survey, listed in Table 4-12, may be used as a guide.

Data collected in a California survey conducted in 1968 are reported in
Table 4-13 [9]. Although at first glance the California quantities appear to
be at variance with the United States data, they are, in fact, quite similar if

TABLE 4-13   SOLID WASTES GENERATED IN
CALIFORNIA, 1968*

Source	Generation rate	
	tons/yr	lb/capita/day
Municipal	22.9 × 10⁶	6.5
Industrial	13.7 × 10⁶	3.9
Agricultural	34.9 × 10⁶	9.8
Total	71.5 × 10⁶	20.2

* Adapted from California study [9].
Note: tons/yr × 907.2 = kg/yr
          lb/capita/day × 0.4536 = kg/capita·day

the individual factors leading to the differences are analyzed. If the California agricultural component (9.8 lb/capita/day) is added to the United States value, the total value would be about 18 lb/capita/day. If it is also noted that the industrial component for California is approximately twice the value given in the United States data, then the overall quantities become quite comparable.

When information on the unit waste generation quantities is not available, the generalized data in Table 4-14 may be used. As shown, data are not presented for agricultural and special wastes because they vary significantly with geographic location. Waste generation rates for selected industrial and agricultural sources in California are reported in Table 4-15.

Factors that Affect Generation Rates

Factors that influence the quantity of wastes generated include geographic location, season of the year, frequency of collection, the use of home grinders, the habits and economic status of the people, the extent of salvage and recycle operations, legislation, and public attitudes. All these factors are important in planning for solid waste management. Broad generalizations are of little or no value, however; the impact of several factors must be evaluated separately in each situation. Specific situations where such evaluations may be required are discussed in Part III.

Geographic Location   The influence of geographic location is related primarily to the different climates that can influence both the amount of

TABLE 4-14   TYPICAL PER CAPITA SOLID
WASTE GENERATION RATES

	Unit rate, lb/capita/day	
Source	Range	Typical
Municipal*	2.0–5.0	3.5
Industrial	1.0–3.5	1.9
Demolition	0.1–0.8	0.6
Other municipal†	0.1–0.6	0.4
Subtotal		6.4
Agricultural		—‡
Special wastes		—‡

* Includes residential and commercial.
† Excludes water, waste water, and industrial treatment plant wastes which must be estimated separately for each location.
‡ Must be estimated separately for each location.
Note: lb/capita/day × 0.4536 = kg/capita·day

TABLE 4-15   UNIT SOLID WASTE GENERATION RATES FOR SELECTED INDUSTRIAL AND AGRICULTURAL SOURCES

Source	Unit	Range
Industrial		
Canned and frozen foods	tons/ton of raw product	0.04–0.06
Printing and publishing	tons/ton of raw paper	0.08–0.10
Automotive	tons/vehicle produced	0.7–0.9
Petroleum refining	tons/employee/day	0.04–0.05
Rubber	tons/ton of raw rubber	0.01–0.3
Agricultural*		
Manures		
Chickens (fryers)	tons/1,000 birds/yr	4–5
Hens (layers)	tons/1,000 birds/yr	45–50
Cattle	lb/head/day	85–120
Fruit and nut crops	tons/acre/yr	1.3–2.5
Field and row crops	tons/acre/yr	1.5–4.5

* Adapted from state of California data [3].
Note: tons × 907.2 = kg
     lb × 0.4536 = kg

certain types of solid wastes generated and the collection operation. Substantial variations in the amount of yard and garden wastes generated in various parts of the country are also related to climates. For example, in the warmer southern areas where the growing season is considerably longer than in the northern areas, yard wastes are collected not only in considerably greater amounts but also over a longer period of time.

Because of the variations in the quantities of certain types of solid wastes generated under varying climates, special studies should be conducted when such information will have a significant impact on the system. Often, the necessary information can be obtained from a load-count analysis.

Season of the Year   The quantities of certain types of solid wastes are also affected by the season of the year. For example, the quantities of food wastes are affected by the growing season for vegetables and fruits. (See also the previous paragraph, Geographic Location.)

Frequency of Collection   In general, it has been observed that where unlimited collection service is provided, more wastes are collected. This observation should not be used to infer that more wastes are generated. For example, if a homeowner is limited to one or two containers per week, he or she may, because of limited container capacity, store newspapers or other materials in the garage or storage area; with unlimited service, the homeowner would tend to throw them away. In this situation the quantity of wastes generated may actually be the same, but the quantity collected is considerably different. Thus, the fundamental question of the effect of collection frequency on waste generation remains unanswered.

Use of Home Grinders   While the use of home grinders definitely reduces the quantity of food wastes collected, it is not clear whether they affect quantities of wastes generated. Because the use of home grinders varies widely throughout the country, the effects of their use must be evaluated separately in each situation if such information is warranted.

Characteristics of Population   It has been observed that the characteristics of the population influence the quantity of solid wastes generated. For example, the quantities of yard wastes generated on a per capita basis are considerably greater in many of the wealthier neighborhoods than in other parts of town.

Extent of Salvage and Recycling   The existence of salvage and recycling operations within a community definitely affects the quantities of wastes collected. Whether such operations affect the quantities generated is another question. Until more information is available, no definite statement can be made on this issue.

Legislation   Perhaps the most important factor affecting the generation of certain types of wastes is the existence of local, state, and federal regulations concerning the use and disposal of specific materials. Legislation dealing with packaging and beverage container materials is an example.

Public Attitudes   Ultimately, as noted in Part I, significant reductions in the quantities of solid wastes that are generated will occur when and if people are willing to change—on their own volition—their habits and lifestyles to conserve national resources and to reduce the economic burdens associated with the management of solid wastes.

4-4   DISCUSSION TOPICS AND PROBLEMS

4-1.   In your first position as a junior city engineer, you are assigned by your supervisor to report on the generation rates and composition of solid wastes for various sources of your community. How would you go about it? If these data were needed in 30 days and thus you had no time to assess seasonal effect, how would you estimate this factor?

4-2.   Obtain data on the percentage distribution of components for the solid wastes in your community or a nearby community. How do they compare with the typical values given in Table 4-2? Explain any major differences. If the individual component values are not within the ranges given in Table 4-2, explain why.

4-3.   Using the data reported in Table 4-6, determine the as-discarded density of the solid wastes from the city of Davis, California, as reported in Table 4-2.

4-4.   Derive an approximate chemical formula for a waste comprised of the following components, using the data given in Tables 4-3 and 4-8.

Component	Percent by weight
Food waste	15
Paper	35
Cardboard	7
Plastic	5
Textiles	3
Rubber	3
Leather	2
Garden trimmings	20
Wood	10

4-5.   Estimate the as-discarded energy content for a waste with the composition given in Problem 4-4. Use the typical data given in Table 4-9. What is the energy content on a moisture-free basis?

4-6.   Consider a household that generates a certain amount of wastes per day. Of this amount, bottles and cans represent 20 percent (by weight) and are recycled by the family. The paper wastes (32 percent) are burned in a backyard incinerator. The rest of the wastes are put into containers for collection. On a given day, 20 lb of consumer goods (food, newspapers, magazines, etc.) is brought into the house. The family consumes 7 lb of food that day, and 5 lb of food is stored. The magazines received represent 5 percent of the paper wastes of the day, and they are not thrown away. Draw a materials flow diagram of this problem and calculate the amount of solid wastes disposed of during this day.

4-7.   The residential and commercial solid wastes of a city of 25,000 people are collected on Tuesday and Saturday mornings. The volume of wastes collected has been recorded for 1 yr, and the data are given below. Prepare a frequency histogram for each collection day. Find the mean, median, mode, standard deviation, and coefficient of variation for each distribution (see Appendix C). Discuss briefly the nature of the distribution and its significance.

Generation rate, yd³/collection day	Frequency Tuesday	Saturday
800–900	0	0
900–1,000	0	0
1,000–1,100	4	1
1,100–1,200	9	3
1,200–1,300	14	4
1,300–1,400	11	9
1,400–1,500	7	11
1,500–1,600	4	10
1,600–1,700	2	7
1,700–1,800	0	4
1,800–1,900	1	2
1,900–2,000	0	1

4-8. What conclusions can be drawn from frequency plots (histograms) of solid waste generation?

4-9. Given the following daily solid waste generation data for a period of 10 days, determine the type of distribution, mean, standard deviation, and coefficient of variation.

Generation rate, yd³/day	
34	170
48	120
290	75
61	110
205	90

4-10. The shape of a solid waste generation frequency curve reflects the nature of the generating facility. From the frequency curves shown in Fig. 4-3 what can be deduced about the facilities' activity and operation?

4-11. One of the first steps in conducting a solid waste management study is the identification of factors contributing to the generation of solid wastes now and in the future. In outline form, list the factors that affect the generation of municipal,

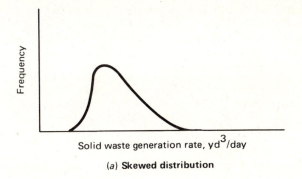

Solid waste generation rate, yd³/day

(a) **Skewed distribution**

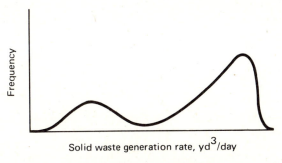

Solid waste generation rate, yd³/day

(b) **Bimodal distribution**

FIG. 4-3 Frequency distribution curves for solid waste generation rates for Prob. 4-10.

industrial, and agricultural solid wastes in your community, and list those that may affect generation in the future.

4-12. Describe the general trends you would expect in the future in the generation of the following types of wastes in your community: food wastes, paper, plastics, rags and leather, and garden trimmings. What effect will improved reporting techniques have on your answer?

4-5 REFERENCES

1. American Public Works Association: "Municipal Refuse Disposal," 3d ed., Public Administration Service, Chicago, 1970.

2. Bender, D. F., M. L. Peterson, and H. Stierli (eds.): *Physical, Chemical, and Microbiological Methods of Solid Waste Testing,* U.S. Environmental Protection Agency, National Environmental Research Center, Cincinnati, Ohio, 1973.

3. California Solid Waste Management Board: *Solid Waste Generation Factors in California,* Technical Information Series, Bulletin 2, State of California, Sacramento, 1974.

4. Darnay, A. and W. E. Franklin: *The Role of Packaging in Solid Waste Management 1966 to 1976,* U.S. Department of Health, Education, and Welfare, Public Health Service, Publication SW-5c, Rockville, Md., 1969.

5. Kaiser, E. R.: Chemical Analyses of Refuse Compounds, in *Proceedings of National Incinerator Conference,* ASME, New York, 1966.

6. Klee, A. J. and D. Carruth: Sample Weights in Solid Waste Composition Studies, ASCE, *Journal of the Sanitary Engineering Division,* vol. 96, no. 5A, 1970.

7. Muhich, A. J., A. J. Klee, and P. W. Britton: *Preliminary Data Analysis, 1968 National Survey of Community Solid Waste Practices,* U.S. Department of Health, Education, and Welfare, Public Health Service, Publication 1867, Washington, D.C., 1968.

8. *Solid Waste Reclamation Study: Phase I and II,* Department of Public Works, County of Sacramento, Sacramento, Calif., 1973.

9. *Status of Solid Waste Management,* vol. 1, California Solid Waste Planning Study, California Department of Public Health, Berkeley, 1968.

10. Tchobanoglous, G. and G. Klein: *An Engineering Evaluation of Refuse Collection Systems Applicable to the Shore Establishment of the U.S. Navy,* Sanitary Engineering Research Laboratory, University of California, Berkeley, 1962.

5

Onsite Handling, Storage, and Processing

The handling, storage, and processing of solid wastes at the source before they are collected is the second of the six functional elements in the solid waste management system. Because this element can have a significant effect on public health, on subsequent functional elements, and on public attitudes concerning the operation of the system, it is important to understand what it involves.

This chapter includes (1) a brief description of the public health and aesthetic aspects and (2) detailed discussions of onsite handling, storage, and processing methods and equipment, with particular emphasis on residential sources of waste generation.

While residential dwellings and building types can be classified in various ways, a classification based on the number of stories is adequate for the purpose. The three classifications most often used and adopted for this text are: low-rise, under four stories; medium rise, from four to seven stories; and high-rise, over seven stories [4]. In the discussion of onsite storage, low-rise residential dwellings are further subdivided into the following categories: single-family detached; single-family attached, such as row or townhouses; and multifamily, of which garden apartments are typical.

It is noted that onsite processing may take place at any time before collection (before, during, or after storage) and is therefore discussed as appropriate throughout the chapter. The main discussion of processing is presented in the last section of the chapter.

Hazardous wastes are discussed separately in Chap. 11 because of their critical importance and specialized requirements.

5-1 PUBLIC HEALTH AND AESTHETICS

Although residential solid wastes account for a very small proportion of the total wastes generated in the United States (5 percent or less), they are perhaps the most important because they are generated in areas with limited storage space. As a result, they can have significant public health and aesthetic impacts.

Public health concerns are related primarily to the infestation of areas used for the storage of solid wastes with vermin and insects that often serve as potential reservoirs of disease. By far the most effective control measure for both rats and flies is proper sanitation. Typically, this involves the use of containers with tight lids, the periodic washing of the containers as well as of the storage areas, and the periodic removal of biodegradable materials (usually within less than 8 days), which is especially important in areas with warm climates. An excellent description of solid waste–disease relationships may be found in Ref. 6.

Aesthetic considerations are related to the production of odors and the unsightly conditions that can develop when adequate attention is not given to the maintenance of sanitary conditions. Most odors can be controlled through the use of containers with tight lids and with the maintenance of a reasonable collection frequency. If odors persist, the contents of the container can be sprayed as a temporary expedient. To maintain aesthetic conditions, the container should be scrubbed and washed periodically.

5-2 ONSITE HANDLING

Onsite handling refers to the activities associated with the handling of solid wastes until they are placed in the containers used for their storage before collection. Depending on the type of collection service, handling may also be required to move the loaded containers to the collection point and to return the empty containers to the point where they are stored between collections. The persons responsible for onsite handling and the auxiliary equipment and facilities that normally are used at the sources where wastes are generated are listed in Table 5-1. The onsite handling methods used at residential and commercial sources are described in the following discussion. Handling methods at the other sources are variants of these same methods, but they will not be discussed because they tend to be source-specific.

TABLE 5-1 TYPICAL EQUIPMENT USED FOR THE ONSITE HANDLING OF SOLID WASTES BY WASTE SOURCES

Source	Persons responsible	Auxiliary equipment and facilities
Residential		
Low-rise	Residents, tenants	Household compactors, small-wheeled handcarts
Medium-rise	Tenants, building maintenance crews, janitorial services, unit managers	Gravity chutes, service elevators, collection carts, pneumatic conveyors
High-rise	Tenants, building maintenance crews, janitorial services	Gravity chutes, service elevators, collection carts, pneumatic conveyors
Commercial	Employees, janitorial services	Wheeled or castored collection carts, container trains, burlap drop cloths, service elevators, conveyors, pneumatic conveyors
Industrial	Employees, janitorial services	Wheeled or castored collection carts, container trains, service elevators, conveyors
Open areas	Owners, park officers	Vandalproof containers
Treatment plant sites	Plant operators	Various conveyors and other manually operated equipment and facilities
Agricultural	Owners, workers	Varies with the individual commodity

Residential

Because of the significant differences in the solid waste handling operations for low-rise dwellings and medium- and high-rise apartments, each is considered separately in the following discussion.

Low-Rise Dwellings As shown in Table 5-1, the residents or tenants of low-rise dwellings are responsible for placing solid wastes that are generated and accumulated at various locations in and around their dwellings in the storage containers. The types of containers used and their locations are described in the following section on storage. Household compactors are now available to reduce the volume of wastes to be collected. In many locations handcarts are used to transport loaded containers to the pickup point.

Medium- and High-Rise Apartments Handling methods in most medium-

(a)

FIG. 5-1 Chute used for solid wastes in older medium-rise apartment in San Francisco. (Solid waste chutes usually are installed internally in modern apartment buildings.) (a) Chute attached to exterior of apartment building. (b) Waste hopper at intermediate floor level. (c) Chute discharge into storage container in apartment basement.

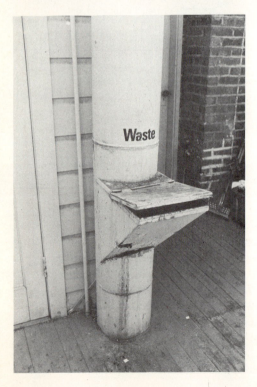

(b)

(c)

rise apartment buildings are similar to those used for low-rise dwellings and high-rise apartments, or various combinations of the two. The methods vary depending on the age and location of the buildings. In some of the older medium-rise apartment buildings, it is still common practice for the tenants to place containers outside their doors and to have the waste collector pick up wastes from each floor.

In high-rise apartment buildings (higher than seven stories) the most common methods of handling solid wastes involve one or more of the following: (1) wastes are picked up by building maintenance personnel or porters from the various floors and taken to the basement or service area; (2) wastes are taken to the basement or service area by tenants; or (3) wastes, usually bagged, are placed by the tenants in specially designed vertical chutes (usually circular) with openings located on each floor (see Fig. 5-1).

Where kitchen grinders are used, food wastes and other grindable materials are ground and discharged to the waste-water collection system. Newspapers may be bundled and put out for pickup by salvage handlers or city crews, or they may be taken by the tenants, bundled or loose, to the service area for pickup or disposal. Bulky items usually are taken to the service area by the tenants. Building maintenance crews are responsible for the handling or processing of the wastes accumulated in the service area. Where processing equipment and facilities, such as incinerators and compactors, are used in conjunction with solid waste chutes, the building maintenance personnel are responsible for handling the incinerator residues and/or the compressed wastes, as well as the other materials brought to the service areas by the tenants.

Chutes for use in apartment buildings are available in diameters from 12 to 36 in; the most common size is 24 in in diameter. All the available chutes can be furnished with suitable intake doors, either side- or bottom-hinged, for installation on various floor levels [4]. Draft baffles at the intake stations, door locks, sprinklers, disinfection systems, sound insulation, and roof vents are among the many accessories that are available. The use of a disinfecting and sanitizing unit is recommended because, in general, it has been found that the cleanliness of the chute and the absence of odors depend to a large extent on their use [4]. In designing chutes for high-rise buildings, variations in the rate at which solid wastes are discharged must be considered. Typical discharge rates in apartments with chutes are shown in Fig. 5-2. In sizing chutes it is common to assume (1) that the bulk density of the solid wastes is equal to 150 lb/yd³, (2) that all

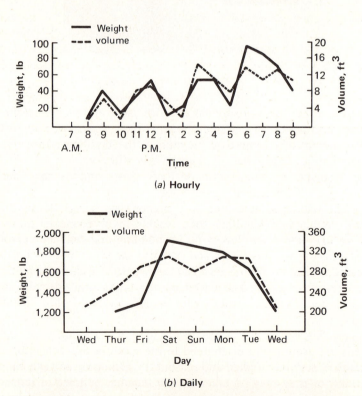

FIG. 5-2 Typical variation in solid waste discharge rates for high-rise apartments with chutes [8].

the daily wastes will be discharged within a 4-h period, and (3) that between 1 and 2 lb of wastes will be generated by each tenant each day.

In some of the more recent apartment building developments, underground pneumatic transport systems have been used in conjunction with the individual apartment chutes (see Fig. 5-3). The underground pneumatic systems are used to transport the wastes from the chute discharge points in each building to a central location for storage in large containers or for onsite processing. Both air pressure and vacuum transport systems have been used in this application.

Commercial

In most office and commercial buildings, solid wastes that accumulate in the individual offices or work locations usually are collected in relatively large containers mounted on rollers. Once filled, these containers are removed by means of the service elevator, if there is one, and emptied into (1) large storage containers or (2) compactors used in conjunction with the storage containers or (3) stationary compactors that can compress the material into bales or into specially designed containers, or (4) other processing equipment.

Because many large offices and commercial buildings were designed without adequate provision for the onsite storage of solid wastes, the storage and processing equipment now used is often inadequate and tends to create handling problems. A common handling method in such situations is to empty the contents of the containers used to collect the wastes from the individual offices onto burlap drop cloths. Once the drop cloth is loaded, the corners are knotted together and taken to the basement where they are piled until collection. Although not especially attractive from a materials handling point of view, this method works and is used widely.

5-3   ONSITE STORAGE

Factors that must be considered in the onsite storage of solid wastes include (1) the type of container to be used, (2) the container location, (3) public health and aesthetics, and (4) the collection methods to be used. The first two factors are described in the following discussion. Public health and aesthetic aspects were discussed earlier in this chapter. The various collection methods are discussed in Chap. 6. Additional details may be found in Refs. 1 to 4.

Containers

To a large extent, the types and capacities of the containers used depend on the characteristics of the solid wastes to be collected, the collection frequency, and the space available for the placement of containers. The types and capacities of containers now commonly used for onsite storage

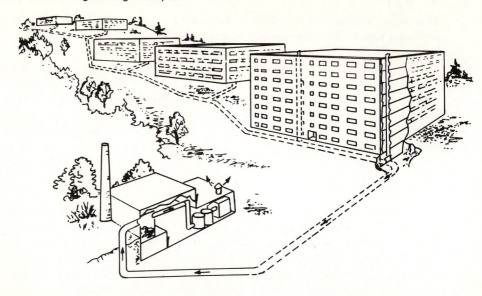

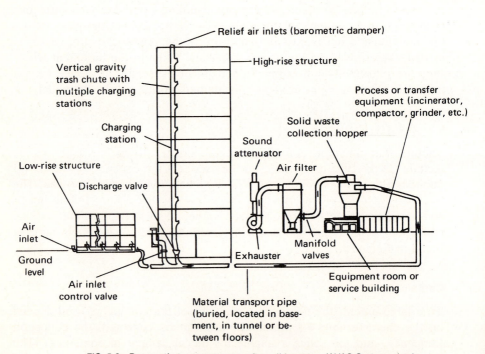

FIG. 5-3 Pneumatic transport system for solid wastes. (AVAC Systems, Inc.)

of solid wastes are summarized in Table 5-2. Typical applications and limitations of the containers are reported in Table 5-3. Some of the more common types of containers are shown in Figs. 5-4 and 5-5.

Low-Rise Dwellings Because solid wastes are collected manually from most residential low-rise detached dwellings, the containers should be light enough to be handled easily by one collector when full. Injuries to collectors have resulted from handling containers that were loaded too heavily. In general, the upper weight limit should be between 40 and 65 lb. The 30-gal galvanized metal or plastic container has proved to be the least expensive means of storage for low-rise dwellings.

FIG. 5-4 Containers used for the onsite storage of solid wastes at low-rise dwellings.

TABLE 5-2 DATA ON THE TYPES AND SIZES OF CONTAINERS USED FOR THE ONSITE STORAGE OF SOLID WASTES

Type	Capacity			Dimensions*	
	Unit	Range	Typical	Unit	Typical
Small					
Container, plastic or galvanized metal	gal	20–40	30	in	20D × 26 H (30 gal)
Barrel, plastic, aluminum, or fiber	gal	20–65	30	in	20D × 26H (30 gal)
Disposable paper bags					
Standard	gal	20–55	30	in	15W × 12d × 43H (30 gal)
Leak-resistant	gal	20–55	30	in	as above
Leakproof	gal	20–55	30	in	as above
Disposable plastic bag				in	18W × 15d × 40H (30 gal)
Medium					
Container	yd³	1–10	4	in	72W × 42d × 65H (4 yd³)
Large					
Container					
Open top, roll off (also called debris boxes)	yd³	12–50	—†	ft	8W × 6H × 20L (35 yd³)
Used with stationary compactor	yd³	20–40	—†	ft	8W × 6H × 18L (30 yd³)
Equipped with self-contained compaction mechanism	yd³	20–40	—†	ft	8W × 8H × 22L (30 yd³)
Container, trailer-mounted					
Open top	yd³	20–50	—†	ft	8W × 12H × 20L (35 yd³)
Enclosed, equipped with self-contained compaction mechanism	yd³	20–40	—†	ft	8W × 12H × 24L (35 yd³)

* D = diameter, H = height, L = length, W = width, d = depth
† Size varies with waste characteristics and local site conditions.
Note: gal × 0.003785 = m^3
     in × 2.54 = cm
     yd³ × 0.7646 = m^3
     ft × 0.3048 = m

TABLE 5-3 TYPICAL APPLICATIONS AND LIMITATIONS OF CONTAINERS USED FOR THE ONSITE STORAGE OF SOLID WASTES

Container type	Typical applications	Limitations
Small		
Container, plastic or galvanized metal	Very low-volume waste sources, such as individual homes, walkways in parks, and small isolated commercial establishments; low-rise residential areas with setout collection service	Containers are damaged over time and degraded in appearance and capacity; containers add extra weight that must be lifted during collection operations; containers are not large enough to hold bulky wastes.
Disposable paper bags	Individual homes with packout collection service; can be used alone or as a liner inside a household container; low- and medium-rise residential areas	Bag storage is more costly; if bags are set out on streets or curbside, dogs or other animals tear them and spread their contents; paper bags themselves add to the waste load.
Disposable plastic bags	Individual homes with setout collection service; can be used alone or as a liner inside a household container; for cold climates, bags are useful in holding wet garbage inside household containers as well as in commercial containers; low-, medium-, and high-rise residential areas; commercial areas; and industrial areas	Bag storage is more costly; bags tear easily, causing litter and unsightly conditions; bags become brittle in very cold weather, causing breakage; plastic lightness and durability causes later disposal problems.
Medium		
Container	Medium-volume waste sources that might also have bulky wastes; location should be selected for direct-collection truck access; high-density residential areas; commercial areas; industrial areas	Snow inside the containers forms ice and lowers capacity while increasing weight; containers are difficult to get to after heavy snows.
Large		
Container, open top	High-volume commercial areas; bulky wastes in industrial areas; low-density rural residential areas; location should be within a covered area but with direct-collection truck access	Initial cost is high; snow inside containers lowers capacity.
Container, used with stationary compactor	Very high-volume commercial areas; location should be outside buildings with direct-collection truck access	Initial cost is high; if container is compacted too much, it is difficult to unload it at the disposal site.

(a)

(b)

FIG. 5-5 Containers used for the onsite storage of solid wastes at apartment and commercial facilities. (a) Low-rise apartments. (b) Commercial establishment.

The choice of container materials depends on the preferences of the individual homeowner. Galvanized metal containers tend to be noisy when being emptied and, in time, can be damaged so that a proper lid seal cannot be achieved. Although less noisy in handling, some containers constructed of plastic materials tend to crack under exposure to the ultraviolet rays of the sun and to freezing temperatures, but the more expensive plastic containers apparently do not present these problems.

Temporary and disposable containers are commonly used when curb service is provided and the homeowner is responsible for placing accumulated wastes on the curb for collection (see Chap. 6). Paper bags, cardboard boxes, plastic containers and bags, and wooden boxes are routinely used as temporary and disposable containers. Under normal circumstances, temporary containers are removed along with the wastes. The principal problem in the use of temporary containers is the difficulty involved in loading them. Paper and cardboard containers tend to disintegrate because of the leakage of liquids. In extremely warm areas where disposable plastic bags are used for lawn trimmings, plastic containers frequently stretch or break at the seams when the collector lifts the loaded bag. Such breakage is potentially hazardous and may lead to injuries to the collector because of the presence of glass and sharp or otherwise dangerous items in the wastes.

With the widespread availability of paper and plastic products, the use of container liners is now common. All types of thicknesses and grades of material are available. Here again, in most areas, it has been left to the homeowner to decide what type of liner to use, if any. A disadvantage in the use of liners is that if the wastes are to be salvaged for metals or glass, or if they are to be incinerated, it is necessary to break up the liner bags in a preprocessing step. Thus, although their use may be a convenience for the homeowner, liners may not be ideal from the standpoint of materials recovery and recycling.

In low-rise apartment complexes, a number of different storage containers have been used. The two most common types are (1) individual plastic or galvanized metal containers and (2) large portable or fixed containers. Where apartments are grouped in close proximity, containers assigned to the individual apartments are often located in a common area.

Although individual containers are used in some low-rise apartment buildings, the most common practice is to use one or more large containers for a group of apartments. Typically, containers are kept in enclosed areas with easy access to a nearby street. Often, the container enclosures are covered. In most locations the containers are equipped with castors or rollers so that they can be moved easily for emptying into collection vehicles or onsite processing equipment.

Medium- and High-Rise Apartments Where solid waste chutes are available, separate storage containers are not used. In some older medium- and high-rise apartments without chutes, wastes are stored in containers on the premises between collections. The most common means of storage used for wastes accumulated from the individual apartments include (1) large open-top containers, (2) enclosed storage containers or disposable bags used in conjunction with stationary compactors, (3) containers equipped with a self-contained compaction mechanism, and (4) special containers used in conjunction with processing equipment.

Commercial   Storage techniques used in commercial facilities depend, to a large extent, on the internal methods used for collecting the wastes produced at various locations within the facility and the available space (see Fig. 5-5). Typically, large open-top containers are used. The use of containers equipped with compaction mechanisms or in conjunction with stationary compactors is increasing. Where a considerable amount of recoverable material is generated, special onsite processing equipment may also be used.

Container Locations

Between collections, containers used in low-rise detached dwellings usually are placed (1) at the sides or rear of the house, (2) in alleys where alley collection is used, (3) in the garage or, where available, some common location specifically designated for that purpose. Although ground-level bins have been used, they are not recommended. When two or more dwellings are located in close proximity, a concrete pad may be constructed at some convenient location between them. The pad may be either open or surrounded by a wooden enclosure. Unless enclosed pads are supervised closely, however, unsightly and unsanitary conditions may develop.

The location of containers at existing commercial and industrial facilities depends on both the location of available space and service-access conditions. In many of the newer designs, specific service areas have been included for this purpose. Often, because the containers are not owned by the commercial or industrial activity, the locations and types of containers to be used for onsite storage must be worked out jointly between the building owners and the public or private collection agency.

5-4   ONSITE PROCESSING OF SOLID WASTES

Grinding, sorting, compaction, shredding, composting, and hydropulpery are all onsite processing methods used to (1) reduce the volume, (2) alter the physical form, or (3) recover usable materials from solid wastes. Typical onsite processing operations and facilities are listed by source in Table 5-4. Onsite processing as applied to residential sources and large commercial and industrial sources is discussed in this section. Additional equipment and process details may be found in Chap. 8. Because the proper selection of onsite processing equipment is of great importance,

TABLE 5-4 TYPICAL OPERATIONS AND FACILITIES USED FOR ONSITE PROCESSING OF SOLID WASTES BY WASTE SOURCES

Source	Persons responsible	Operations and facilities
Residential		
Low-rise	Residents, tenants	Grinding, sorting, compaction, composting, incineration
Medium- and high-rise	Tenants	Grinding, sorting (paper), compaction
	Building maintenance crews	Compaction, incineration, shredding, hydropulping
Commercial	Janitorial services	Sorting, compaction, shredding, incineration, hydropulping
Industrial	Janitorial services	Sorting, compaction, shredding, incineration, hydropulping
Open areas	Owners, park operators	Compaction, incineration
Treatment plant sites	Plant operators	Dewatering facilities
Agriculture	Owner, workers	Varies with individual commodity

however, the key factors that should be considered in its selection are summarized in Table 5-5.

Low-Rise Dwellings

The most common onsite processing operations used at low-rise detached residential dwellings include grinding, sorting, compaction, composting, and incineration.

Grinding   In the past 20 years the use of home grinders has gained such wide acceptance that, in some areas, nearly all the new homes are equipped with them. Home grinders are used primarily for wastes from the preparation, cooking, and serving of foods, and they cannot be used for large bones or other bulky items.

Functionally, grinders render the material that passes through them suitable for transport through the sewer system. Because the organic material added to sewage has resulted in overloading many treatment facilities, it has been necessary, in some communities, to forbid the installation of grinders in new developments until additional treatment capacity becomes available.

In terms of the collection operation, the use of home grinders does not have a significant impact on the volume of solid wastes collected. Even the weight difference is not major. In some cases where grinders are used, it has been possible to increase the time period between collection pickups because wastes that might readily decay are not stored.

TABLE 5-5   FACTORS THAT SHOULD BE CONSIDERED IN EVALUATING ONSITE PROCESSING EQUIPMENT*

Factor	Evaluation
Capabilities	What will the device or mechanism do? Will its use be an improvement over conventional practices?
Reliability	Will the equipment perform its designated functions with little attention beyond preventive maintenance? Has the effectiveness of the equipment been demonstrated in use over a reasonable period of time or merely predicted?
Service	Will servicing capabilities beyond those of the local building maintenance staff be required occasionally? Are properly trained service personnel available through the equipment manufacturer or the local distributor?
Safety of operation	Is the proposed equipment reasonably foolproof so that it may be operated by tenants or building personnel with limited mechanical knowledge or abilities? Does it have adequate safeguards to discourage careless use?
Ease of operation	Is the equipment easy to operate by a tenant or by building personnel? Unless functions and actual operations of equipment can be carried out easily, they may be ignored or "short-circuited" by paid personnel, and most often by "paying" tenants.
Efficiency	Does the equipment perform efficiently and with a minimum of attention? Under most conditions, equipment that completes an operational cycle each time it is used should be selected.
Environmental effects	Does the equipment pollute or contaminate the environment? Where possible, equipment should reduce environmental pollution presently associated with conventional functions.
Health hazards	Does the device, mechanism, or equipment create or amplify health hazards?
Aesthetics	Does the equipment and its arrangement offend the senses? Every effort should be made to reduce or eliminate offending sights, odors, and noises.
Economics	What are the economics involved? Both first and annual costs must be considered. Future operation and maintenance costs must be assessed carefully. All factors being equal, equipment produced by well-established companies, having a proved history of satisfactory operation, should be given appropriate consideration.

* Adapted from Ref. 4.

Sorting   The sorting or separation of wastepaper, aluminum cans, and glass by hand at the household is one of the most positive ways to achieve the recovery and reuse of materials. Once the waste is sorted into separate containers, the biggest problem facing the homeowner is what to do with the wastes until they are collected or taken to a local recycling center. The effect of home recovery operations is considered in Example 5-1.

EXAMPLE 5-1   *Effect of Home Recovery on Energy Content of Collected Solid Wastes*

Using the typical percentage distribution data given in Table 4-2 (Chap. 4), estimate the number of British thermal units per pound (Btu/lb) of the remaining solid wastes if 90 percent of the cardboard and 60 percent of the paper were recovered by the homeowner.

Solution

1. From Table 4-10 the total energy content of 100 lb of solid wastes, with the composition given in Table 4-2, is equal to 476,230 Btu.
2. Determine the energy content and weight of 90 percent of the cardboard in the original sample.

    Energy content, 90% cardboard
    $\quad$ = 0.90(28,000 Btu), see Table 4-10
    $\quad$ = 25,200 Btu

    Weight, 90% cardboard
    $\quad$ = 0.90(4 lb), see Table 4-10
    $\quad$ = 3.6 lb

3. Determine the energy content and weight of 60 percent of the paper in the original sample.

    Energy content, 60% paper
    $\quad$ = 0.60(288,000 Btu), see Table 4-10
    $\quad$ = 172,800 Btu

    Weight, 60% paper
    $\quad$ = 0.60(40 lb), see Table 4-10
    $\quad$ = 24 lb

4. Determine the total energy content, weight, and energy content per pound of the original sample after cardboard and paper have been recovered.

    Total energy after recovery
    $\quad$ = (476,230 − 25,200 − 172,800) Btu
    $\quad$ = 278,230 Btu

    Total weight after salvage
    $\quad$ = (100 − 3.6 − 24) lb
    $\quad$ = 72.4 lb

    Energy content per lb after recovery
    $$= \frac{278,230 \text{ Btu}}{72.4 \text{ lb}}$$
    $\quad$ = 3,843 Btu/lb (8,939 kJ/kg) vs. 4,762 Btu/lb (11,076 kJ/kg) in original sample

Comment   In this example, the removal by weight of approximately 28 percent of the wastes reduced the per-pound energy content of the original sample by approximately 20 percent. If the solid wastes were to be incinerated for conversion to power, the economic feasibility of the recovery operation would have to be evaluated to determine whether it is cost-effective.

Compaction   Within the past few years, a number of small compactors designed for home use have appeared on the market. Manufacturers' claims for these units in terms of the compaction ratio usually are based on the compaction of loose paper and cardboard. Although it is possible to reduce the original volume of wastes placed in them by up to 70 percent, they can be used only for a small proportion of the wastes actually generated. The effect of the use of home compactors on the volume of wastes collected is illustrated in Example 5-2.

EXAMPLE 5-2   *Effect of Home Compactors on Volume of Collected Solid Wastes*

Assume that home compaction units are to be installed in Davis, California. Estimate the volume reduction that could be achieved in the solid wastes collected if the compacted density is equal to 20 lb/ft³ and the data given in Tables 4-2 and 4-6 are applicable.

Solution

1. Set up a computation table (see Table 5-6) to determine the volume of wastes as discarded in containers, using the data given in Tables 4-2 and 4-6.

2. Determine the volume of compacted wastes, excluding garden trimmings, wood, ferrous metals, and dirt, ashes, brick, etc.

$$\text{Compacted volume} = \frac{(100 - 14.3 - 3.5 - 4.3 - 1.1)\ \text{lb}}{20\ \text{lb/ft}^3}$$

$$= \frac{76.8\ \text{lb}}{20\ \text{lb/ft}^3} = 3.8\ \text{ft}^3$$

3. Determine the volume reduction for the compressible material.

$$\text{Volume reduction} = \frac{(12.4 - 3.8)\ \text{ft}^3}{12.4\ \text{ft}^3}\ 100$$

$$= 69\ \text{percent}$$

4. Determine the overall volume reduction achieved with a home compactor, taking into account garden trimming, wood, ferrous metals, dirt, ashes, and brick, etc.

Overall volume reduction

$$= \frac{(15.09)\ \text{ft}^3 - (2.2 + 0.23 + 0.22 + 0.04 + 3.8)\ \text{ft}^3}{15.09\ \text{ft}^3}\ 100$$

$$= \frac{15.09 - 6.49}{15.09}\ 100$$

$$= 57\ \text{percent}$$

TABLE 5-6 VOLUME DETERMINATION FOR SOLID WASTES AS DISCARDED IN CONTAINERS FOR EXAMPLE 5-2

Component	Weight, lb	Density, lb/ft³	Volume, ft³
Food wastes	9.5	18	0.53
Paper	43.1	5.1	8.45
Cardboard	6.5	6.2*	1.05
Plastics	1.8	4	0.45
Textiles	0.2	4	0.05
Rubber	—	8	—
Leather	1.5	10	0.15
Garden trimmings	14.3†	6.5	(2.20)
Wood	3.5†	15	(0.23)
Glass	7.5	12.1	0.62
Tin cans	5.2	5.5	0.95
Nonferrous metals	1.5	10	0.15
Ferrous metals	4.3†	20	(0.22)
Dirt, ashes, brick, etc.	1.1†	30	(0.04)
Total	100		15.09
			12.40‡

* Cardboard partially compressed by hand before being placed in container.
† Components usually not placed in home compactors.
‡ Total not including items in parentheses.
Note: lb × 0.4536 = kg
lb/ft³ × 16.019 = kg/m³
ft³ × 0.02833 = m³

Comment When the overall volume reduction is assessed, it becomes apparent that the effectiveness of a home compactor is reduced. This is especially true as the percentages of the components not compacted, such as garden trimmings, increase.

The use of home compactors may also be counterproductive from the standpoint of subsequent processing operations. For example, if the wastes are to be separated mechanically into components (see Chap. 8), the compacted wastes will have to be broken up again before sorting. Also, by compacting, the wastes may become so saturated with the liquids present in the food wastes that the recovery of paper or other components may not be feasible.

Composting In the 1970s home composting as a means of recycling organic materials increased in popularity. It is an effective way of reducing the volume and altering the physical composition of solid wastes while at the same time producing a useful by-product. A variety of methods are used, depending on the amount of space available and the wastes to be composted. Additional process details may be found in Chap. 9. In terms of the overall waste management problems facing most cities, the impact of home composting on the volume of solid wastes to be handled is negligible.

Incineration   Until recently, home incineration—burning combustible materials in fireplaces and burning rubbish in crude backyard incinerators—was a common practice. Backyard incineration is now banned in many parts of the country. The effect of the elimination of backyard burning on the quantity of wastes collected is shown in Fig. 5-6. The magnitude of the effect will, however, vary significantly with location.

The design of small outdoor and indoor incinerators has improved, however. The simplest outdoor incinerator consists of a metal drum with holes punched near the bottom. The more elaborate units are lined with refractory brick and are equipped with cast-iron grates and small chimneys. Some indoor incinerators are provided with an auxiliary fuel source. The appearance of the better indoor incinerator is similar to that of other household appliances, such as water heaters. Additional details on home incinerators may be found in Refs. 1 and 2.

The impact of the use of home incinerators on disposal varies, depending on the methods used for the disposal of municipal solid wastes. In the case of municipal incineration, the widespread use of home incinerators may necessitate the use of auxiliary fuels, special designs, or

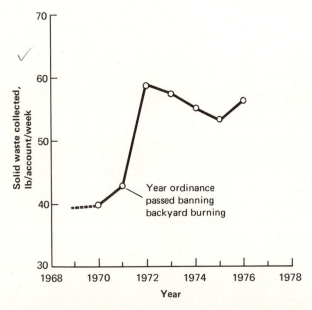

FIG. 5-6   Effect of the elimination of backyard burning on the quantity of solid wastes collected. (Division of Solid Waste Management, Sacramento County, Sacramento, California. Data based on 85,000 accounts.)

operating procedures [1]. In the case of land disposal, the use of home incinerators may reduce both the putrescible components and the total volume of solid wastes to be disposed.

Medium- and High-Rise Apartments

The processing operations used at medium- and high-rise apartment buildings for wastes accumulated from individual apartments include compaction, incineration, shredding and pulping, as well as grinding and sorting as previously described for low-rise dwellings. Still other handling facilities are reviewed in Ref. 3. The mechanical details of these operations may be found in Chap. 8.

Compaction   To reduce the volume of solid wastes that must be handled, compaction units commonly are installed in large apartment buildings. Typically, a compactor is installed at the bottom of a solid waste chute (see Fig. 5-7). Wastes falling through the chute activate the compactor by means of photoelectric cells or limit switches. Once these switches are activated, the wastes are compressed. Depending on the design of the compactor, the compressed wastes may be formed into bales or extruded and loaded automatically into metal containers or paper bags.

When a bale has been formed or a container or bag is filled, the compactor shuts down automatically and a warning light turns on. The operator must then tie and remove the bale from the compactor, or remove the full bag and replace it with an empty one. In some applications the use of completely automatic equipment may be warranted. In sizing compaction equipment for use in conjunction with solid waste chutes in apartments, it is common to use the same assumptions used in sizing the chutes, as discussed previously.

While the use of compactors reduces the bulk volume of the wastes to be handled, it must be remembered that the weight remains the same. Typically, the compacted volume will vary from 20 to 60 percent or less of the original volume. Compacted solid wastes are acceptable where disposal is by landfilling. Where incinerators are used, the compacted wastes must be broken up to avoid delayed combustion in the furnace and high losses of unburned combustible materials. Unless the wastes are broken up, it is impossible to recover individual components from compacted wastes. All these factors must be considered when the use of onsite compactors is being evaluated.

Incineration   By reducing the volume to 10 percent or less and the weight to 25 percent or less of the wastes charged, incineration can achieve a significant reduction in the handling and storage requirements at the source as well as in the pickup and disposal facilities.

Two types of incinerators are used, depending on the method of charging: flue-fed and chute-fed. In the flue-fed type, wastes are charged

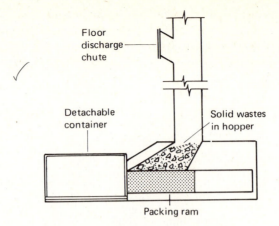

(*a*) **Start of compaction cycle**

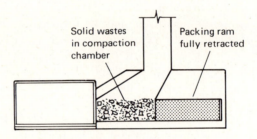

(*b*) **Loading of compaction chamber**

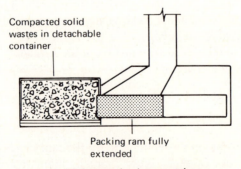

(*c*) **Compaction into container**

FIG. 5-7 Chute-fed compactor installation used for apartments.

through doors on each floor directly into the refractory flue, the bottom of which opens directly into the top of the furnace combustion chamber (see Fig. 5-8). In the chute-fed type, wastes are charged through hopper doors on each floor into a metal chute, and they collect in a basement hopper. The wastes are then either manually or mechanically transferred into the furnace.

Shredding and Pulping   Shredding and pulping are alternative processing operations that have been used, both in conjunction with the previous methods and by themselves, for reducing the volume of wastes that must be handled. Where shredding is used alone without the addition of water, the volume of wastes has often been observed to increase.

A typical pulverizer system is shown in Fig. 5-9. Wastes from the various floors are discharged into a chute that discharges into a pulper tank. In the pulper tank, water is added and maintained at the proper level. The wastes are shredded into a pulp by the teeth in the impeller plate located at the bottom of the pulper tank. Nonpulpable components, such as metals and glass, are discharged into a collection chamber after being reduced in size. The pulp slurry passes through a sizing ring where oversized material is not allowed to pass out of the pulper. The slurry is then discharged to the dewatering press where it is picked up by a helical screw contained in a perforated housing. As the material is moved up by the screw, water is extracted and the semidry pulp is discharged into a container. The extracted water is returned to the pulper tank [8]. Excess water containing some residual pulp usually is discharged to the waste-water collection system.

Although the system works well and the volume of solid wastes is reduced, it is expensive. Special equipment may be required to remove and empty the full pulp containers. An alternative is to discharge the pulped material to the local sewer. This is often done in small operations where a pulper is used to destroy outdated confidential documents. Because the discharge of pulped material increases the organic loading on local treatment facilities, the use of pulverizers may be restricted if treatment capacity is limited.

Commercial-Industrial Facilities

For the most part, the onsite processing operations carried out at commercial-industrial facilities are similar to those described for residential sources. Differences occur primarily in industrial facilities. Because most of the processes tend to be industry-specific, however, no attempt has been made to document the various processes that have been used. Additional processing methods applicable to large industries are discussed in Chap. 8.

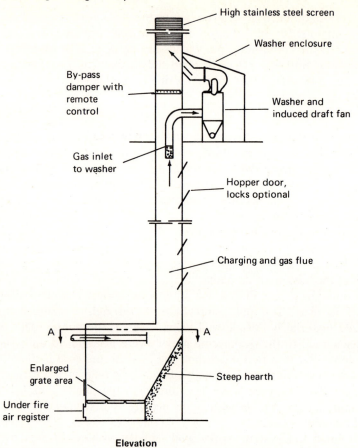

High stainless steel screen

Washer enclosure

By-pass
damper with
remote
control

Washer and
induced draft fan

Gas inlet
to washer

Hopper door,
locks optional

Charging and gas flue

A   A

Enlarged
grate area

Steep hearth

Under fire
air register

Elevation

Outside over fire air
maniford and fan

Inside
over fire air
manifold and fan

Section A–A

FIG. 5-8   Single-flue incinerator with washer or precipitator on roof. (Adapted
from Ref. 7.)

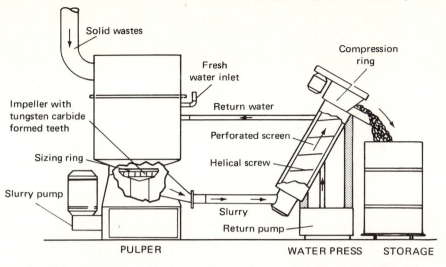

FIG. 5-9   Solid waste pulverizer for use in apartment complexes. (AMSCO-WASCON Systems, Inc.)

5-5   DISCUSSION TOPICS AND PROBLEMS

5-1. Drive around your community and make a brief survey of the different types of solid waste containers now used.

5-2. Obtain a component distribution of the solid wastes generated in your community, and determine the percentage reduction that could be achieved if home compactors were installed. Assume the density of the compacted wastes to be 20 lb/ft^3. Compare your answer with that derived in Example 5-2.

5-3. The kitchen grinder in a detached single-family dwelling has broken. Assuming that it will take 1 wk to repair the unit, estimate the increase in both the volume and the weight of solid wastes to be collected. Assume that the family has four members and that the collection frequency is once per week.

5-4. Assuming that the daily quantities of solid wastes generated at a commercial facility are distributed normally (see Appendix C), with a mean value of 10 yd^3 and a standard deviation of 7 yd^3, what would be the size of the container you would recommend for this facility? What are the important tradeoffs in the selection of container size?

5-5. Using the data presented in Fig. 5-2, estimate the size of a container to be used with a gravity chute for a 24-story apartment building with 192 individual living units if the container will be emptied daily. Assume that the average occupancy rate for each living unit is 3.2 persons.

5-6. What compactor volume displacement (e.g., capacity), expressed in terms of cubic yards per hour, would you recommend for use in the 24-story apartment building of Problem 5-5?

5-7. As a consulting engineer, you have been commissioned to develop a comprehensive solid waste system for a community interested in achieving a greater recovery and reuse of its solid wastes. Two of the possible alternatives are separation at the home or separation at the disposal site. What are the important factors that must be considered in evaluating these two alternatives?

5-8. List the advantages and disadvantages associated with the home separation of solid wastes, and devise a workable separation scheme for the home separation of paper, aluminum cans, and colored glass. Suggest and discuss any possible implementation problems with your plan.

5-6   REFERENCES

1. American Public Works Association: "Municipal Refuse Disposal," 3d ed., Public Administration Service, Chicago, 1970.
2. American Public Works Association, Institute for Solid Wastes: "Solid Waste Collection Practice," 4th ed., American Public Works Association, Chicago, 1975.
3. Connelly, J. A. (ed.): *Abstracts: Selected Patents on Refuse Handling Facilities for Buildings,* U.S. Department of Health, Education, and Welfare, Public Health Service, Publication 1793, Cincinnati, Ohio, 1968.
4. Greenleaf/Telesca, Planners, Engineers, and Architects: *Solid Waste Management in Residential Complexes,* U.S. Environmental Protection Agency, Publication SW-35c, Washington, D.C., 1971.
5. *Guidelines for Local Governments on Solid Waste Management,* U.S. Environmental Protection Agency, Publication SW-17c, Washington, D.C., 1971.
6. Hanks, T. G.: *Solid Waste/Disease Relationships,* U.S. Department of Health, Education, and Welfare, Solid Wastes Program, Publication SW-1c, Cincinnati, Ohio, 1967.
7. Meissner, H. G.: B. Multiple Dwellings, in R. C. Corey (ed.), "Principles and Practices of Incineration," Wiley-Interscience, New York, 1969.
8. *Solid Waste Management in High-Rise Dwellings, A Consideration,* U.S. Environmental Protection Agency, Publication SW-27c1, Washington, D.C., 1972.

6
Collection of Solid Wastes

Collection of solid wastes in urban areas is difficult and complex because the generation of residential and commercial-industrial solid wastes is a diffuse process that takes place in every home, every apartment building, and every commercial and industrial facility as well as in the streets, parks, and even the vacant areas of every community. The mushroomlike development of suburbs all over the country has further complicated the collection task.

As the generation patterns become more diffuse and the total quantity of wastes increases, the logistic problems associated with collection become more complex. Although these problems have always existed to some degree, they have now become more critical because of the high cost of fuel and labor. Of the total amount of money spent for the collection, transportation, and disposal of solid wastes in 1975, approximately 60 to 80 percent was spent on the collection phase. This fact is important because a small percentage improvement in the collection operation can effect a significant saving in the overall cost.

In view of the importance of the collection operation, it is discussed in detail in this chapter. The information is presented in five parts: (1) the types of collection services that are provided, (2) the types of collection systems and some of the equipment now used as well as the associated labor requirements, (3) an analysis of collection systems, including the component relationships that can be used to quantify collection operations, (4) the general methodology involved in setting up collection routes, and (5) some of the more advanced techniques of analysis that can be used to evaluate collection operations.

6-1   COLLECTION SERVICES

The term collection, as noted in Chap. 2, includes not only the gathering or picking up of solid wastes from the various sources, but also the hauling of these wastes to the location where the contents of the collection vehicles are emptied. Unloading of the collection vehicle is also considered part of the collection operation. While the activities associated with hauling and unloading are similar for most collection systems, the gathering or picking up of wastes will vary with the characteristics of the facilities, activities, or locations where wastes are generated (see Table 4-1) and the ways and means used for the onsite storage of accumulated wastes between collections.

The various types of collection services now used for residential and commercial-industrial sources are described in this section. The other sources considered in Table 4-1 are not discussed separately because collection services for them are site-specific and, for the most part, variants of those used for the residential and commercial-industrial sources.

Residential Collection Service

Residential collection service varies depending on the type of dwelling unit. Collection for low-rise detached dwellings and collection for medium- and high-rise apartments are considered separately.

Low-Rise Detached Dwellings   The most common types of residential services used in various parts of the country for low-rise detached dwellings include (1) curb, (2) alley, (3) setout-setback, (4) setout, and (5) backyard carry. The characteristics of these services are compared in Table 6-1, and additional details may be found in Refs. 1 and 2.

Where curb service is used, the homeowner is responsible for placing the containers to be emptied at the curb on collection day and for returning the empty containers to their storage location until the next collection. Where alleys are part of the basic layout of a city or a given residential area, alley storage of containers used for solid wastes is common. In setout-setback service, containers are set out from the homeowner's property and set back after being emptied by additional crews that work in conjunction with the collection crew responsible for loading the collection vehicle. Setout service is essentially the same as setout-setback service, except that the homeowner is responsible for returning the containers to their storage location. In backyard carry service, the collection crew is responsible for entering the homeowner's property and removing the wastes from their storage location.

Methods of loading the collection vehicle may be classified as either manual or mechanical. Methods commonly used for residential wastes include (1) the direct lifting and carrying of containers, (2) the rolling of

TABLE 6-1 COMPARISON OF RESIDENTIAL COLLECTION SERVICES*

Considerations	Type of service				
	Curb	Alley	Setout-Setback	Setout	Backyard carry
Requires homeowner cooperation:					
To carry full containers	Yes	Optional	No	No	No
To carry empty containers	Yes	Optional	No	Yes	No
Requires scheduled service for homeowner cooperation	Yes	No	No	Yes	No
Poor aesthetically:					
Spillage and litter problem	High	High	Low	High	Low
Containers visible	Yes	No	No	Yes	No
Attractive to scavengers	Yes	Highest	No	No	No
Prone to upsets	Yes	Yes	No	Yes	No
Average numbers of persons in crew required for efficiency	1 to 3 persons	1 to 3 persons	3 to 7 persons	1 to 5 persons	3 to 5 persons
Crew time	Low	Low	Great	Medium	Medium
Collector injury rate due to lifting and carrying	Low	Low	High	Medium	High
Trespassing complaints	Low	Low	High	High	High
Special considerations		Requires alleys and vehicles that can maneuver in them, less prone to block traffic; high vehicle and container depreciation rate			Requires wheeled caddy to roll filled barrels or the use of burlap carry cloth or hand-carry bin, works best with driveway
Cost due to crew size and time requirements	Low cost	Low cost	High cost	Medium cost	Medium cost

* Adapted from Ref. 6.

FIG. 6-1   Tote container being carried to collection vehicle.

loaded containers on their rim, (3) the use of small lifts for rolling the containers to the collection vehicle, and (4) the use of large containers (referred to as "tote" containers) or drop cloths (often called tarps) into which wastes from small containers are emptied before being carried (see Fig. 6-1) and/or rolled to the collection vehicle.

Where collection vehicles with low loading heights are used, wastes are transferred directly from the containers in which they are stored or carried to the collection vehicle by the collection crew (see Fig. 6-2). In some cases, where open-body trucks are used, crew members stationed on the truck will lift the loaded container onto the truck with the help of the

FIG. 6-2 Contents of tote container being emptied into collection vehicle. (Note method used to lift and unload tote container equipped with wheels.)

collectors on the ground, empty the container, and return it to the collectors on the ground. In other cases, collection vehicles are equipped with auxiliary containers into which the wastes are emptied. The auxiliary containers are emptied into the collection vehicle by mechanical means.

A recent innovation that has become quite popular involves the use of small satellite vehicles. Wastes are emptied into the satellite vehicles which are equipped with large containers. When loaded, these vehicles are driven to the collection truck where the containers are emptied into the truck by mechanical means (see Fig. 6-3). Additional details may be found in Ref. 4.

Medium- and High-Rise Apartments   Most of the methods of collecting solid wastes from medium- and high-rise apartment buildings are essentially the same as those used for commercial-industrial sources (see following discussion). In some older sections of some cities, it is still common practice to have wastes from the individual floors picked up by collectors using tote containers (see Fig. 6-4) and/or drop cloths.

Commercial-Industrial Service

The collection service provided to large apartment buildings and commercial activities typically is centered around the use of large moveable and stationary containers and large stationary compactors (see Chap. 5). Compactors are of the type that can be used to compress material directly into large containers or to form bales that are then placed in large containers. Because the collection of industrial solid wastes is so location-dependent, it is difficult to define any representative type of service. In general, the service is tailored to each individual activity and is based on the use of large containers and/or stationary compactors.

In commercial-industrial service where the use of containers provided with rollers or castors is common, loaded containers are rolled manually to the collection vehicle and emptied mechanically (see Fig. 6-5). Otherwise, because of the weight involved, direct pickup methods are used. Where large drop-box containers are used, the entire loading operation is by mechanical means (see following section).

6-2   COLLECTION SYSTEMS, EQUIPMENT, AND LABOR REQUIREMENTS

Over the past 10 years a wide variety of collection systems and equipment have been used for the collection of solid wastes. Some of the more common types now used and the corresponding labor requirements for these systems are described in this section.

Types of Collection Systems

Solid waste collection systems may be classified from several points of view, such as the mode of operation, the equipment used, and the types of

(a)

(b)

FIG. 6-3 Satellite vehicle collection system. (a) Loading of satellite vehicle equipped with 2-yd^3 container. (Note that in high-wind conditions, blowing of wastes is a problem.) (b) Mechanical unloading of contents of satellite vehicle container.

FIG. 6-4   Collector climbing staircase to pick up wastes from individual
apartments, San Francisco.

wastes collected. In this text, collection systems have been classified
according to their mode of operation into two categories: (1) hauled
container systems and (2) stationary container systems [14]. The individual
systems included in each category lend themselves to the same method of
engineering and economic analysis.

Hauled Container Systems (HCS)   These are collection systems in which the
containers used for the storage of wastes are hauled to the disposal site,
emptied, and returned to either their original location or some other
location.

(a)

(b)

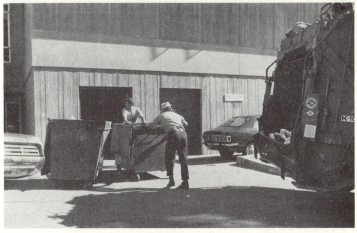

(c)

Fig. 6-5 Emptying sequence for containers used at commercial complex. (*a*)
Loaded container is attached to collection vehicle. (*b*) Contents of container are
emptied mechanically. (*c*) Empty container is returned to onsite location.

Stationary Container Systems (SCS)   These are collection systems in which the containers used for the storage of wastes remain at the point of generation, except for occasional short trips to the collection vehicle.

Hauled Container Systems and Equipment

Hauled container systems are ideally suited for the removal of wastes from sources where the rate of generation is high because relatively large containers are used (see Table 6-2). The use of large containers eliminates handling time as well as the unsightly accumulations and unsanitary conditions associated with the use of numerous smaller containers. Another advantage of hauled container systems is their flexibility: containers of many different sizes and shapes are available for the collection of all types of wastes.

Because containers used in this system usually must be filled manually, the use of very large containers often leads to low-volume utilization unless loading aids, such as platforms and ramps, are provided. In this context, container utilization may be defined as the fraction of the total container volume actually filled with wastes.

While hauled container systems have the advantage of requiring only one truck and driver to accomplish the collection cycle, each container

TABLE 6-2 TYPICAL DATA ON CONTAINER CAPACITIES AVAILABLE FOR USE WITH VARIOUS COLLECTION SYSTEMS

	Collection	Typical range of container capacities*, yd^3
Vehicle	Container type	
Hauled container systems		
Hoist truck	Used with stationary compactor	6–12
Tilt-frame	Open top, also called debris boxes	12–50
	Used with stationary compactor	15–40
	Equipped with self-contained compaction mechanism	20–40
Truck-tractor	Open-top trash-trailers	15–40
	Enclosed trailer-mounted containers equipped with self-contained compaction mechanism	20–40
Stationary container systems		
Compactor, mechanically loaded	Open top and enclosed top and side-loading	1–8
Compactor, manually loaded	Small plastic or galvanized metal containers, disposable paper and plastic bags	20–55 (gal)

* See Table 5-2 for typical container dimensions.
Note: yd^3 × 0.7646 = m^3
        gal × 0.003785 = m^3

picked up requires a round trip to the disposal site (or other destination point). Therefore, container size and utilization are of great economic importance. Further, when highly compressible wastes are to be collected and hauled over considerable distances, the economic advantages of compaction are obvious.

There are three main types of hauled container systems: (1) hoist truck, (2) tilt-frame container, and (3) trash-trailer. Typical data on the collection vehicles used with these systems are reported in Table 6-3.

Hoist-Truck Systems   In the past, hoist trucks were used widely at military installations (see Fig. 6-6). With the advent of self-loading collection vehicles, however, this system appears to be applicable in only a limited number of cases, the most important of which are as follows:

1. For the collection of wastes by a collector who has a small operation and collects from only a few pickup points at which a considerable amount of wastes are generated. Generally, for such operations the purchase of newer and more efficient collection equipment cannot be justified economically.
2. For the collection of bulky items and industrial rubbish not suitable for collection with compaction vehicles.

Tilt-Frame Container Systems   Systems that use tilt-frame–loaded vehicles

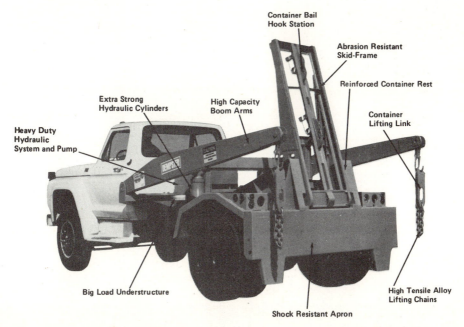

FIG. 6-6 Hoist-truck mechanism mounted on truck frame. (Dempster Dumpster Systems.)

TABLE 6-3 TYPICAL DATA ON VEHICLES USED FOR THE COLLECTION OF SOLID WASTES

Collection vehicle			Typical overall collection vehicle dimensions				
Type	Available container or truck body capacities*, yd³	Number of axles	With indicated container or truck body capacity†, yd³	Width, in	Height, in	Length‡, in	Unloading method
Hauled container systems							
Hoist truck	6–12	2	10	94	80–100	110–150	Gravity, bottom opening
Tilt-frame	12–50	3	30	96	80–90	220–300	Gravity, inclined tipping
Truck-tractor trash-trailer	15–40	3	40	96	90–150	220–450	Gravity, inclined tipping
Stationary container system							
Compactor (mechanically loaded)							
Front loading	20–45	3	30	96	140–150	240–290	Hydraulic ejector panel
Side loading	10–36	3	30	96	132–150	220–260	Hydraulic ejector panel
Rear loading	10–30	2	20	96	125–135	210–230	Hydraulic ejector panel
Compactor (manually loaded)							
Side loading	10–37	3	37	96	132–150	240–300	Hydraulic ejector panel
Rear loading	10–30	2	20	96	125–135	210–230	Hydraulic ejector panel

* See Table 6-2 and Table 5-2.
† See Table 5-2 for dimensions of typical containers.
‡ From front of truck to rear of container or truck body.
Note: yd³ × 0.7646 = m³
  in × 0.0254 = m

FIG. 6-7 Truck with tilt-frame loading mechanism. (Dempster Dumpster Systems.)

(see Fig. 6-7) and large containers, often called "drop boxes," are ideally suited for the collection of all types of solid waste and rubbish from locations where the generation rate warrants the use of large containers (see Fig. 6-8). As noted in Table 6-2, various types of large containers are available for use with tilt-frame collection vehicles. Open-top containers are used routinely at warehouses and construction sites. Large containers used in conjunction with stationary compactors are common at apartment complexes, commercial services, and transfer stations (see Chap. 7). Because of the large volume that can be hauled, the use of the tilt-frame hauled container system has become widespread, especially among private collectors servicing commercial accounts.

Trash-Trailer Systems   The application of trash-trailers is similar to that for tilt-frame container systems. Trash-trailers are better for the collection of especially heavy rubbish, such as sand, timber, and metal scrap, and often are used for the collection of demolition wastes at construction sites (see Fig. 6-9).

Stationary Container Systems and Equipment

Stationary container systems may be used for the collection of all types of wastes. The systems vary according to the type and quantity of wastes to be handled, as well as the number of generation points. There are two main types: (1) systems in which self-loading compactors are used, and (2) systems in which manually loaded vehicles are used. Some typical

FIG. 6-8   Contents of large tilt-frame loaded container being emptied at landfill.

FIG. 6-9   Contents of trash-trailer used for demolition wastes being unloaded at landfill.

examples of the vehicles and containers used are shown in Figs. 6-2, 6-3, 6-5, 6-10, and 6-11. Data on the collection vehicles used in this system are reported in Table 6-3.

Systems with Self-loading Compactors   Container size and utilization are not as critical in stationary container systems using collection vehicles equipped with a compaction mechanism as they are in hoist-truck systems. Trips to the disposal site, transfer station, or processing station are made after the contents of a number of containers have been collected (see Fig. 6-10) and compacted and the collection vehicle is full. For this reason, use of the driver in terms of the quantities of wastes hauled is considerably greater for these systems than for hauled container systems.

   A variety of container sizes are available for use with these systems (see Table 6-2 and Figs 5-5, 6-5, and 6-10). They vary from relatively small sizes (1 yd³) to sizes comparable to those handled with a hoist truck (see Table 6-2). The use of smaller containers offers greater flexibility in terms of shape, ease of loading, and special features available. By using small, easier-to-load containers, utilization of containers can be increased considerably. These systems can also be used for the collection of residential wastes where one large container can be substituted for a number of small containers.

   Because truck bodies are difficult to maintain and because of the weight involved, these systems are not ideally suited for the collection of heavy industrial wastes and bulk rubbish, such as that produced at construction and demolition sites. Locations where high volumes of rubbish are produced are also difficult to service because of the space requirements for the large number of containers.

Systems with Manually Loaded Vehicles   The major application of manual transfer and loading methods is in the collection of residential wastes and litter (see Fig. 6-11). Manual loading can compete effectively with mechanical loading in residential areas because the quantity picked up at each location is small and the loading time is short. In addition, manual methods are used for residential collection because many individual pickup points are inaccessible to the collection vehicle.

Transfer Operations

Transfer operations, in which the wastes, containers, or collection vehicle bodies holding the wastes are transferred from a collection vehicle to a transfer or haul vehicle, are used primarily because of economic considerations. Transfer operations may prove economical when (1) relatively small, manually loaded collection vehicles are used for the collection of residential wastes and long haul distances are involved, (2) extremely large quantities of wastes must be hauled over long distances, and (3) one

(a)

(b)

FIG. 6-10   Loading cycle for front self-loading compaction-type collection vehicle. (a) Loading forks being engaged in container sleeves. (b) Container being raised. (c) Container beginning to tilt. (d) Contents of container being emptied into collection vehicle.

(c)

(d)

Figure 6-11 Collector manually emptying the contents of a tote container into a rear-loaded compaction-type collection vehicle. (This type of vehicle is commonly used with two- and three-person crews for the collection of residential wastes throughout the United States.)

transfer station can be used by a number of collection vehicles. Transfer and transport operations are considered in detail in Chap. 7.

Labor Requirements

Labor requirements for the collection of solid wastes vary with both the type of service provided and the type of collection system used.

Hauled Container Systems   In most hauled container systems a single collector is used. The collector is responsible for driving the vehicle, loading full containers and unloading empty containers, and emptying the contents of the container at the disposal site. In some cases, for safety reasons, both a driver and helper are used. The helper usually is responsible for attaching and detaching any chains or cables used in loading and unloading containers on and off the collection vehicle; the driver is responsible for the operation of the vehicle. A driver and helper should always be used where hazardous wastes are to be handled.

Stationary Container Systems (Mechanically Loaded)   Labor requirements for mechanically loaded stationary container systems are essentially the same as for hauled container systems. Where a helper is used, the driver often assists the helper in bringing loaded containers mounted on rollers to the collection vehicle and returning the empty containers (see Fig. 6-5).

Occasionally, a driver and two helpers are used where the containers to be emptied must be rolled (transferred) to the collection vehicle from inaccessible locations, such as in congested downtown commercial areas.

Stationary Container Systems (Manually Loaded) In stationary container systems where the collection vehicle is loaded manually, the number of collectors varies from one to three, in most cases, depending on the type of service and the collection equipment. Typically, a single collector is used for curb and alley service, and a multiperson crew is used for backyard carry service (see Table 6-1). A single collector-driver is also used with most satellite-vehicle systems for curb collection (see Fig. 6-3). While the aforementioned crew sizes are representative of current practices, there are many exceptions. In many cities multiperson crews are used for curb service as well as for backyard carry service.

Special attention must be given to the design of the collection vehicle intended for use with a single collector. At present, it appears that a side-loaded compactor, such as the one shown in Fig. 6-12 equipped with standup right-hand drive, is best suited for curb and alley collection.

6-3 ANALYSIS OF COLLECTION SYSTEMS

To establish vehicle and labor requirements for the various collection systems and methods, the unit time required to perform each task must be

FIG. 6-12 Standup right-hand-drive side-loaded collection vehicle.

determined. By separating the collection activities into unit operations, it is possible (1) to develop design data and relationships that can be used universally and (2) to evaluate both the variables associated with collection activities and the variables related to, or controlled by, the particular location.

Definition of Terms

Before the relationships for collection systems can be modeled effectively, the component tasks must be delineated. The operational tasks for the hauled container and stationary container systems are shown schematically in Figs. 6-13 and 6-14, respectively. On the basis of previous work [14, 16], the activities involved in the collection of solid wastes can be resolved into four unit operations: (1) pickup, (2) haul, (3) at-site, and (4) off-route.

Pickup   The definition of the term pickup depends on the type of collection system used [14].

1. For hauled container systems operated in the conventional mode (see Fig. 6-13a), pickup (P_{hcs}) refers to the time spent driving to the next container after an empty container has been deposited, the time spent picking up the loaded container, and the time required to redeposit the container after its contents have been emptied. For hauled container systems operated in the exchange-container mode (see Fig. 6-13b), pickup includes the time required to pick up a loaded container and to redeposit the container at the next location after its contents have been emptied.

2. For stationary container systems (see Fig. 6-14), pickup (P_{scs}) refers to the time spent loading the collection vehicle, beginning with the stopping of the vehicle prior to loading the contents of the first container and ending when the contents of the last container to be emptied have been loaded. The specific tasks in the pickup operation depend on the type of collection vehicle as well as the collection methods used.

Haul   The definition of the term *haul* (h) also depends on the type of collection system used [14].

1. For hauled container systems, haul represents the time required to reach the disposal site, starting after a container whose contents are to be emptied has been loaded on the truck, plus the time after leaving the disposal site until the truck arrives at the location where the empty container is to be redeposited. It does not include any time spent at the disposal site.

2. For stationary container systems, haul refers to the time required to

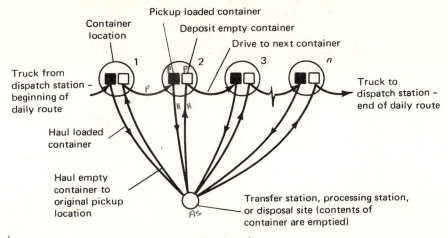

(a) **Conventional mode**

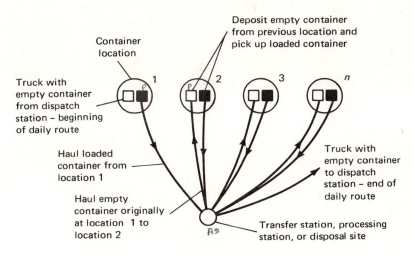

(b) **Exchange container mode**

FIG. 6-13 Schematic of operational sequence for hauled container system.

reach the disposal site, starting after the last container on the route has been emptied or the collection vehicle is filled, plus the time after leaving the disposal site until the truck arrives at the location of the first container to be emptied on the next collection route. It does not include the time spent at the disposal site.

At-Site   The unit operation at-site (s) refers to the time spent at the disposal site and includes the time spent waiting to unload as well as the time spent unloading [14].

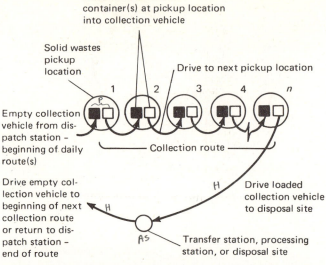

FIG. 6-14   Schematic of operational sequence for stationary container system.

Off-Route   The unit operation off-route (W) includes all time spent on activities that are nonproductive from the point of view of the overall collection operation. Many of the activities associated with off-route times are sometimes necessary or inherent in the operation. Therefore, the time spent on off-route activities may be subdivided into two categories: necessary and unnecessary. In practice, however, both necessary and unnecessary off-route times are considered together because they must be distributed equally over the entire operation [14].

Necessary off-route time includes (1) time spent checking in and out in the morning and at the end of the day, (2) time spent driving to the first pickup point and/or from the approximate location of the last pickup point to the dispatch station at the end of the day (the term *approximate location* is used because, in the stationary container system, the collection vehicle is normally driven directly back to the dispatch station after the wastes collected on the last route have been emptied), (3) time lost due to unavoidable congestion, and (4) time spent on equipment repairs and maintenance, etc. Unnecessary off-route time includes time spent for lunch in excess of the stated lunch period and time spent on taking unauthorized coffee breaks, talking to friends, etc.

Hauled Container Systems

The time required per trip, which also corresponds to the time required per container, is equal to the sum of the pickup, at-site, and haul times, times a factor accounting for off-route activities, and is given by the following equation:

$$T_{hcs} = (P_{hcs} + s + h)/(1 - W)$$ (6-1)

where T_{hcs} = time per trip for hauled container system, h/trip
  P_{hcs} = pickup time per trip for hauled container system, h/trip
    s = at-site time per trip, h/trip
    h = haul time per trip, h/trip
    W = off-route factor, expressed as a fraction

While the pickup and at-site times for hauled container systems are relatively constant, the haul time depends on both haul speed and distance. From an analysis of a considerable amount of haul data for various collection vehicles (see Fig. 6-15), it has been found that the haul time h may be approximated by the following expression:

$$h = a + bx$$ (6-2)

where h = total haul time, h/trip
    a = empirical constant, h/trip
    b = empirical constant, h/mi
    x = round-trip-haul distance, mi/trip

The analysis of haul time and mileage data, using Eq. 6-2 and the relationship presented in Fig. 6-15, is illustrated in Example 6-1, presented at the end of the discussion dealing with hauled container systems.

Substituting in Eq. 6-1 the expression for h given in Eq. 6-2, the time per trip can be expressed as follows:

$$T_{hcs} = (P_{hcs} + s + a + bx)/(1 - W)$$ (6-3)

The pickup time per trip P_{hcs} for the hauled container system is then

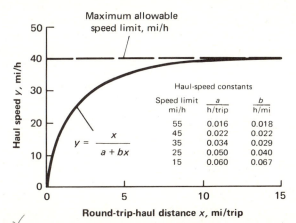

FIG. 6-15 Correlation between average haul speed and round-trip-haul distance.

equal to

$$P_{hcs} = pc + uc + dbc \tag{6-4}$$

where P_{hcs} = pickup time per trip, h/trip
   pc = time required to pick up loaded container, h/trip
   uc = time required to unload empty container, h/trip
   dbc = time required to drive between container locations, h/trip

In computing the pickup time per trip, if the average time required to drive between containers is unknown, the time can be estimated by using Eq. 6-2, where the distance between containers is substituted for the round-trip-haul distance.

The number of trips that can be made per vehicle per day with a hauled container system can be determined by using Eq. 6-5:

$$N_d = (1 - W)H/(P_{hcs} + s + a + bx) \tag{6-5}$$

where N_d = number of trips per day, trips/day
   H = length of work day, h/day
other terms = as defined previously

Data that can be used in the solution of Eq. 6-5 for various types of hauled container systems are given in Fig. 6-15 and Table 6-4. The off-route factor

TABLE 6-4   TYPICAL DATA FOR COMPUTING EQUIPMENT AND LABOR REQUIREMENTS FOR VARIOUS COLLECTION SYSTEMS*

Collection			Time required to pick up loaded container and to deposit empty container, h/trip	Time required to empty contents of loaded container, h/container	At-site time, h/trip
Vehicle	Loading method	Compaction ratio, r			
Hauled container systems					
Hoist truck	Mechanical	—	0.067		0.053
Tilt-frame	Mechanical	—	0.40		0.127
Tilt-frame	Mechanical	2.0–4.0†	0.40		0.133
Stationary container systems					
Compactor	Mechanical	2.0–2.5		0.050	0.10
Compactor	Manual	2.0–2.5		—	0.10

* Adapted from Ref. 14.
† Containers used in conjunction with stationary compactor.

in Eq. 6-5 varies from 0.10 to 0.25; a factor of 0.15 is representative for most operations.

In some cases where especially long distances are involved, the time spent driving from and to the dispatch station at the beginning and end of the day is subtracted from the length of the workday in Eq. 6-5. If this is done, it is also important to remember to adjust the off-route factor.

Assuming that the number of containers to be emptied per week is known, the time required per week can be computed by using Eq. 6-6.

$$D_w = t_w(P_{hcs} + s + a + bx)/[(1 - W)H] \qquad (6\text{-}6)$$

where D_w = time required per week, days/wk
        t_w = integer number of trips per week, trips/wk
other terms = as defined previously

If the weekly number of trips is unknown, it can be estimated by using the following expression:

$$N_w = V_w/(cf) \qquad (6\text{-}7)$$

where N_w = number of trips per week, trips/wk
        V_w = weekly waste generation rate, yd³/wk
        c = average container size, yd³/trip
        f = weighted average container utilization factor

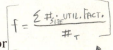

$$f = \frac{\sum \#_{size} \cdot \text{UTIL. FACT.}}{\#_T}$$

As noted previously, the *container utilization factor* may be defined as the fraction of the container volume occupied by solid wastes. Because this factor will vary with the size of the container, a weighted factor should be used in Eq. 6-7. The weighted factor is found by dividing the sum of the values obtained by multiplying the number of containers in each size by their corresponding utilization factor by the total number of containers.

The integer number of trips to be used in computing the time required per week in Eq. 6-6 is obtained by appropriately rounding off the value of N_w obtained using Eq. 6-7 to an integer value. In terms of the system, if the value of N_w is rounded off to a lower integer, the physical significance is that one or more of the containers will be fuller than usual. If the value of N_w is rounded off to a higher integer, one or more of the containers will not be as full as usual.

The weekly labor requirement in collector-days per week is obtained by multiplying the time required per week by the number of collectors. Finally, the required number of collection vehicles can be determined by dividing the weekly time requirement, expressed in days per week, by the number of workdays per week and rounding off the result to the next highest whole integer. Thus, for $D_w/5$ values of 0.7, 1.2, and 3.7, the number of collection vehicles would be equal to 1, 2, and 4, respectively. To improve the efficiency of the operation where fractional equipment and labor requirements are obtained, the use of larger containers and reduced collection frequency should be investigated.

TABLE 6-5   AVERAGE HAUL SPEEDS FOR EXAMPLE 6-1

Round-trip distance x, mi/trip	Total time, h	Average haul speed y, mi/h
2	0.12	17
5	0.18	28
8	0.25	32
12	0.33	36
16	0.40	40
20	0.48	42
25	0.56	45

EXAMPLE 6-1   *Haul-Speed Analysis*

The following average speeds were obtained for various round-trip distances to a disposal site. Find the haul-speed constants *a* and *b* and the round-trip-haul time for a site that is located 11.0 mi away. See Table 6.5.

Solution

1. Linearize the haul-speed equation given in Fig. 6-15. The basic haul-speed equation is

$$y = \frac{x}{a + bx}$$

The linearized form of equation is

$$\frac{x}{y} = h = a + bx$$

2. Plot x/y, which is the travel time versus the round-trip distance x (see Fig. 6-16).

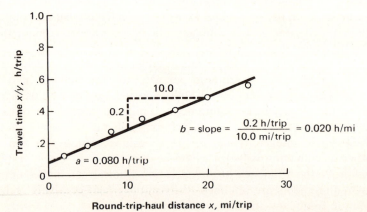

FIG. 6-16   Travel time versus round-trip-haul distance for Example 6-1.

3. Determine the constants a and b using Fig. 6-16. When $x = 0$, $a =$ intercept value $= 0.080$ h/trip, $b =$ slope of line $= (0.2$ h$)/(10$ mi$) = 0.020$ h/mi $(0.012$ h/km$)$.

4. Find the round-trip-haul time for a site that is located 11.0 mi away.

$$\text{Round-trip distance} = 2(11.0 \text{ mi/trip}) = 22 \text{ mi/trip}$$

$$
\begin{aligned}
\text{Haul time } h = a + bx \\
= 0.080 \text{ h/trip} + (0.020 \text{ h/mi})(22 \text{ mi/trip}) \\
= 0.52 \text{ h/trip}
\end{aligned}
$$

Stationary Container Systems

Because of differences in the loading process, mechanically and manually loaded stationary container systems are considered separately in the following discussion.

Mechanically Loaded Vehicles   For systems using self-loading compactors, the time per trip is expressed as

$$T_{scs} = (P_{scs} + s + a + bx)/(1 - W) \tag{6-8}$$

where T_{scs} = time per trip for stationary container system, h/trip
$\quad\quad P_{scs}$ = pickup time per trip for stationary container system, h/trip
$\quad\quad s$ = at-site time per trip, h/trip
$\quad\quad a$ = empirical constant, h/trip
$\quad\quad b$ = empirical constant, h/mi
$\quad\quad x$ = round-trip-haul distance, mi/trip
$\quad\quad W$ = off-route factor, expressed as a fraction

The only difference between Eq. 6-8 and Eq. 6-3 for hauled container systems is the pickup term. For the stationary container system, the pickup time is given by

$$P_{scs} = C_t(uc) + (n_p - 1)(dbc) \tag{6-9}$$

where $\quad P_{scs}$ = pickup time per trip for stationary container systems, h/trip
$\quad\quad C_t$ = number of containers emptied per trip, containers/trip
$\quad\quad uc$ = average unloading time per container for stationary container systems, h/container
$\quad\quad n_p$ = number of container pickup locations per trip, locations/trip
$\quad\quad dbc$ = average time spent driving between container locations, h/location

The term $(n_p - 1)$ accounts for the fact that the number of times the collection vehicle will have to be driven between container locations is equal to the number of containers less 1. As in the case of the hauled container system, if the time spent driving between container locations is unknown, it can be estimated by using Eq. 6-2 where the distance between containers is substituted for the round-trip-haul distance.

The number of containers that can be emptied per collection trip is related directly to the volume of the collection vehicle and the compaction ratio that can be achieved. This number is given by

$$C_t = vr/(cf) \tag{6-10}$$

where C_t = number of containers emptied per trip, containers/trip
 v = volume of collection vehicle, yd^3/trip
 r = compaction ratio
 c = container volume, yd^3/container
 f = weighted container utilization factor

The number of trips required per week can be estimated by using the following equation:

$$N_w = V_w/(vr) \tag{6-11}$$

where N_w = number of collection trips required per week, trips/wk
 V_w = weekly waste generation rate, yd^3/wk
other terms = as defined previously

The time required per week can be expressed as follows:

$$* \quad D_w = [(N_w)P_{scs} + t_w(s + a + bx)]/[(1 - W)H] \tag{6-12}$$

where D_w = time required per week, days/wk
 t_w = the value of (N_w) rounded off to the next highest integer, which accounts for the fact that even though the truck may be only partially loaded on the last trip, a full trip to the disposal site is still required
 H = length of workday, h/day
other terms = as defined previously

In applications in which an integer number of trips do not have to be made and partial loads do not have to be emptied at the end of the day, the truck size to be used can be determined as follows. Assume two or three available truck sizes and compute the time requirements for each size, using Eq. 6-12. The truck size that results in minimum labor use should then be selected. For example, if the labor requirements with one collector are 2.2 collector-days/wk using a 24-yd^3 compactor and 2.0 collector-days/wk using a 30-yd^3 compactor, the larger one should be selected, because in most operations it will be difficult to use a collector for part of the day. Thus, by calculating the time and labor requirements for various truck sizes, the optimum vehicle size can be selected.

Where an integer number of trips are to be made each day, the proper combination of trips per day and the size of the vehicle can be determined by using Eq. 6-13 in conjunction with an economic analysis:

$$H = N_d(P_{scs} + s + a + bx)/(1 - W) \tag{6-13}$$

where N_d = number of collection trips per day, trips/day
other terms = as defined previously

To determine the required truck volume, substitute two or three different values for N_d in Eq. 6-13 and determine the available pickup times per trip. Then, by trial and error, using Eqs. 6-9 and 6-10, determine the truck volume required for each value of N_d. From the available truck sizes, select the ones that most nearly correspond to the computed values. If available truck sizes are smaller than the required values, compute the actual time per day that will be required using these sizes.

Once the labor requirements for each combination of truck size and number of trips per day have been determined, the most cost-effective combination can be selected. For example, where long haul distances are involved, it may be more economical to use a large collection vehicle and make 2 trips/day (even though some time at the end of the day may not be used) than to use a smaller vehicle and make 3 trips/day by using all the available time.

When the truck size is fixed and an integral number of trips must be made each day, the length of the required workday can be estimated using Eqs. 6-9, 6-10, and 6-11. The analysis and comparison of a hauled and stationary container system are illustrated in Example 6-2.

EXAMPLE 6-2 *Analysis of Solid Waste Collection Systems*

A private solid waste collector wishes to locate a disposal site near a commercial area. The collector would like to use a hauled container system but fears that the haul costs might be prohibitive. What is the maximum distance away from the commercial area that the disposal site can be located so that the weekly costs of the hauled container system do not exceed those of a stationary container system? Assume that one collector-driver will be used with each system and that the following data are applicable.

1. Hauled container system
   - *a*. Quantity of solid wastes = 300 yd³/wk
   - *b*. Container size = 8 yd³/trip
   - *c*. Container utilization factor = 0.67
   - *d*. Container pickup time = 0.033 h/trip
   - *e*. Container unloading time = 0.033h/trip
   - *f*. At-site time = 0.053 h/trip
   - *g*. Overhead costs = $400/wk
   - *h*. Operational costs = $15/h of operation

2. Stationary container system
   - *a*. Quantity of solid wastes = 300 yd³/wk
   - *b*. Container size = 8 yd³/location
   - *c*. Container utilization factor = 0.67
   - *d*. Collection vehicle capacity = 30 yd³/trip
   - *e*. Collection vehicle compaction ratio = 2

    f. Container unloading time = 0.05 h/container
    g. Overhead costs = \$750/wk
    h. Operational costs = \$15/h of operation
    i. At-site time = 0.10 h/trip

3. Location characteristics
    a. Average distance between container locations = 0.1 mi
    b. Constants for estimating driving time between container loca-
        tions for hauled container system: $a' = 0.060$ h/trip and $b' = 0.067$ h/mi
    c. Constants for estimating driving time between container loca-
        tions for stationary container system: $a' = 0.060$ h/location and
        $b' = 0.067$ h/mi
    d. Constants for estimating haul time: $a = 0.022$ h/trip and $b = 0.022$ h/mi

Solution

1. Hauled container system
    a. Determine the number of trips per week, using Eq. 6-7.

$$N_w = v_w/(cf) = (300 \text{ yd}^3/\text{wk})/(8 \text{ yd}^3/\text{trip})(0.67)$$
$$= 56.0 \text{ trips/wk}$$

    b. Determine the integer number of trips to use in computing the
        time required per week, using Eq. 6-6.

$$N_w = 56 \text{ trips/wk}$$
$$\therefore t_w = 56 \text{ trips/wk}$$

    c. Estimate the pickup time for the hauled container system, using
        Eq. 6-4.

$$P_{hcs} = pc + uc + dbc = pc + uc + a' + b'x'$$

$$= 0.033 \text{ h/trip} + 0.033 \text{ h/trip}$$

$$+ 0.060 \text{ h/trip}$$

$$+ (0.067 \text{ h/mi})(0.1 \text{ mi/trip})$$

$$= 0.133 \text{ h/trip}$$

    d. Estimate the time required per week as a function of the round-
        trip-haul distance, using Eq. 6-6.

$$D_w = t_w(P_{hcs} + s + a + bx)/[(1 - W)H]$$

$$D_w = (56 \text{ trips/wk})[0.133 \text{ h/trip}$$

$$+ 0.053 \text{ h/trip} + 0.022 \text{ h/trip}$$

$$+ (0.022 \text{ h/mi})(x)]/[(1 - 0.15)(8 \text{ h/day})]$$

$$= [1.70 + (0.181/\text{mi})(x)] \text{ days/wk}$$

e. Determine the weekly operational cost as a function of the round-trip-haul distance.

Operational cost = ($15/h)(8 h/day)

$$[1.70 + (0.181/\text{mi})(x)] \text{ days/wk}$$

$$= [204 + (21.7/\text{mi})(x)] \text{ \$/wk} \checkmark$$

2. Stationary container system

a. Determine the number of containers emptied per trip, using Eq. 6-10.

$\checkmark\ C_t = vr/(cf) = (30 \text{ yd}^3/\text{trip})(2)/(8 \text{ yd}^3/\text{container})(0.67)$

$\quad = 11.19 \text{ containers/trip} = 11 \text{ containers/trip}$

b. Estimate the pickup time per container by using Eq. 6-9.

$\checkmark P_{scs} = C_t(uc) + (n_p - 1)(dbc)$

$$= C_t(uc) + (n_p - 1)(a' + b'x')$$

$$= (11 \text{ containers/trip})(0.050 \text{ h/container})$$

$$+ (11 - 1 \text{ locations/trip})[(0.06 \text{ h/location})$$

$$+ (0.067 \text{ h/mi})(0.1 \text{ mi/location})]$$

$$= 1.22 \text{ h/trip}$$

c. Determine the number of trips required per week by using Eq. 6-11.

$N_w = V_w/(vr) = (300 \text{ yd}^3/\text{wk})/(30 \text{ yd}^3/\text{trip})(2)$

$\checkmark$

$\quad = 5 \text{ trips/wk}$

d. Determine the time required per week as a function of the round-trip-haul distance using Eq. 6-12.

$\checkmark D_w = [(N_w)P_{scs} + t_w(s + a + bx)]/[(1 - W)H]$

$$= \{(5 \text{ trips/wk})(1.22 \text{ h/trip})$$

$$+ (5 \text{ trips/wk})$$

$$\times [0.10 \text{ h/trip} + 0.022 \text{ h/trip}$$

$$+ (0.022 \text{ h/mi})(x)]\}/[(1 - 0.15)(8 \text{ h/day})]$$

$$= [0.99 + (0.016/\text{mi})(x)] \text{ days/wk}$$

e. Determine the weekly operational costs as a function of the round-trip-haul distance.

Operational cost = ($15/h)(8 h/day)

$$\times [0.99 + (0.016/\text{mi})(x)] \text{ days/wk}$$

$$\checkmark\quad = [118.8 + (1.92/\text{mi})(x)] \text{ \$/wk}$$

3. Comparison of systems
    a. Determine the maximum round-trip-haul distance at which the cost for hauled container systems equals the cost for the stationary container systems by equating the total costs for the two systems and solving for x.

$$\$400/wk + [204 + (21.7/mi)(x)]\ \$/wk$$

$$= \$750/wk + [118.8 + (1.92/mi)(x)]\ \$/wk$$

$$(19.8/mi)(x) = 264.8$$

$$x = 13.4\ \text{mi (one-way distance} = 6.7\ \text{mi)}$$

$$= 21.6\ \text{km (one-way distance} = 10.8\ \text{km)}$$

   b. Plot the weekly cost versus round-trip-haul distance for each system (see Fig. 6-17).

   Comment   The curves shown in Fig. 6-17 are characteristic of those obtained when hauled container systems are compared to stationary container systems. In most cases the round-trip-haul distance at which hauled container systems are no longer competitive is much shorter.

Manually Loaded Vehicles   The analysis and design of residential collection systems using manually loaded vehicles may be outlined as follows. If H hours are worked per day and the number of trips to be made per day is known or fixed, the time available for the pickup operation can be computed by using Eq. 6-13 because either all the factors are known or they can be assumed. Once the pickup time per trip is known, the number of pickup locations from which wastes can be collected per trip can be

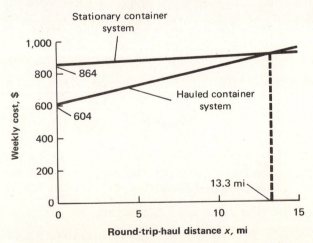

FIG. 6-17 Weekly cost versus round-trip-haul distance for Example 6-2.

estimated as follows:

$$✶ \sqrt{} \quad N_p = 60 P_{scs} n / t_p \tag{6-14}$$

where N_p = number of pickup locations per trip, locations/trip
 60 = conversion factor from hours to minutes, 60 min/h
 P_{scs} = pickup time per trip, h/trip
 n = number of collectors, collectors
 t_p = pickup time per pickup location, collector-min/location

Pickup-time data derived from field observations for a two-person collection crew are shown in Fig. 6-18. The pickup time t_p per location was found to be related to the number of containers per pickup location and the percent of rear-of-house pickup points. The corresponding relationship is

$$✶ \quad t_p = 0.72 + 0.18(C_n) + 0.014(\text{PRH}) \tag{6-15}$$

where t_p = average pickup time per pickup location, collector-min/
 location
 C_n = average number of containers at each pickup location
 PRH = rear-of-house pickup locations, percent

Equation 6-15 is typical of the types of equations derived from field observations for the pickup time per location. Usually, the first term in such equations represents the time spent driving between pickup locations. This value will, of course, depend on the characteristics of the residential area. The use of Eq. 6-15 is illustrated in Examples 6-3 and 6-4.

If curb collection is made once per week, the data in Table 6-6 may be used to estimate the labor requirements. These data were observed for operations using one collector and a side-loaded vehicle equipped with a standup drive [14] (see Fig. 6-12). If conventional trucks are used for curb collection, the pickup time per service reported in Table 6-6 should be increased by 5 to 10 percent.

The difference between the times obtained by comparing Eq. 6-15 to

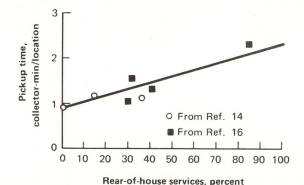

FIG. 6-18 Relationship between time requirements for pickup and percent of rear-of-house services for a two-person crew [14].

TABLE 6-6   LABOR REQUIREMENTS FOR CURB COLLECTION*

Average number of containers and/or boxes per pickup location	Pickup time, collector-min/location
1 to 2	0.50–0.60
3 or more, or unlimited service	0.92

— ONE COLLECTOR
— SIDE-LOADED VEHICLE

* Adapted from Ref. 14.

the data in Table 6-6 is accounted for by the fact that, where two- or three-person crews are used, the size of the lots tends to be larger. Although Eq. 6-15 and the data in Table 6-6 can be used to estimate the time per pickup location, it is recommended that field measurements be made whenever possible because residential collection operations are so variable.

EXAMPLE 6-3   *Analysis of Collection Operation*

The agency responsible for the collection of solid wastes presently allows two containers per service, picked up at the backyard. Consideration is being given to limiting backyard service to one container only; the remaining services would be allowed two containers at curbside. About 10 percent of all services would be expected to ask for the backyard service. How many additional containers can be collected per day? At present there are 300 collection stops per day. Assume that the average pickup time per service can be estimated using Eq. 6-15.

Solution

1. Determine the collection time for the present operation using Eq. 6-15.

    Collection time

    $$= [(0.72 + 0.18(2) + 0.014(100)] \text{ min/service} \times (300 \text{ services})$$
    $$= (0.72 + 0.36 + 1.40)(300)$$
    $$\checkmark = 744 \text{ min}$$

2. Determine the total number of pickup locations Tp that can be picked up if the proposed new service is instituted.

    Collection time

    $$= [0.72 + 0.18(2) + 0.014(0)] \text{ min/service} \times (0.9Tp)$$
    $$+ [0.72 + 0.18(1) + 0.014(100)] \text{ min/service} \times (0.1Tp)$$

    $$744 = (0.72 + 0.36)(0.9Tp) + (0.72 + 0.18 + 1.40)(0.1)Tp$$
    $$744 \text{ min} = 1.20Tp$$
    $$\checkmark \quad Tp = \frac{744}{1.20} = 620 \text{ services}$$

3. Determine number of additional containers that can be collected.

Containers collected at present
$$= (2 \text{ containers/service})(300 \text{ services})$$
$$= 600 \text{ containers}$$

Containers collected, proposed
$$= (2 \text{ containers/service})(0.90)(620 \text{ services})$$
$$+ (1 \text{ container/service})(0.10)(620 \text{ services})$$
$$= 1{,}116 + 62 = 1{,}178 \text{ containers}$$
Additional collected per day $= 1{,}178 - 600 = 578$ containers

Comment It should be noted that the computed collection time (744 min) exceeds the time one crew has available in a typical 8-h day. In practice, either a second crew would be used or the collection operation would be completed on the following day.

Once the number of pickup locations per trip is known, the proper size of collection vehicle can then be estimated as follows:

$$v = V_p N_p / r \tag{6-16}$$

where v = volume of collection vehicle, yd³/trip
V_p = volume of solid wastes collected per pickup location, yd³/location
N_p = number of pickup locations per trip, locations/trip
r = compaction ratio

When the number of pickup locations per trip is known, the weekly labor requirements can be computed, using Eq. 6-12, by multiplying the right-hand side of the equation by n, the number of collectors. The number of trips per week is computed using the following expression:

$$N_w = T_p F / N_p \tag{6-17}$$

where N_w = number of collection trips per week, trips/wk
T_p = total number of pickup locations, locations
F = collection frequency per week, times/wk
N_p = number of pickup locations per trip, locations/trip

In many housing areas, the collection frequency is twice per week. The effect of twice-per-week collection on the amount of wastes collected is discussed in Chap. 4. In terms of labor requirements, it has been found that the requirements for the second weekly collection are about 0.9 and 0.95 times those for the first weekly collection. In general, the labor requirements are not significantly different because container handling time is about the same for both full and partially full containers. Often this difference is neglected in computing the labor requirements.

The number of collection vehicles required can be computed by dividing D_w, the labor requirement, by n, the number of collectors per truck, and by the number of workdays per week. For fractional values it may be necessary to adjust the routes to achieve optimum cost effectiveness.

EXAMPLE 6-4 *Design of Residential Collection System*

Design a solid waste collection system to service a residential area with 1,000 single-family dwellings. Assume that a two-person collection crew will be used and that the following data are applicable.

1. Average number of residents per service = 3.5
2. Solid waste generation rate per capita = 2.0 lb/capita/day
3. Density of solid wastes (at containers) = 200 lb/yd^3
4. Containers per service = two 32-gal containers
5. Type of service = 50 percent rear-of-house, 50 percent alley
6. Collection frequency = once per week
7. Collection vehicle = rear-loaded compactor, compaction ratio = 2
8. Round-trip-haul distance = 15 mi
9. Length of workday = 8 h
10. Trips per day = 2
11. Off-route factor = 0.15
12. Constants for estimating haul time: a = 0.016 h/trip and b = 0.018 mi/h
13. Assume at-site time per trip = 0.10 h/trip
14. Assume Eq. 6-15 can be used to estimate the pickup time per pickup location.

Solution

1. Determine the time available for the pickup operation using Eq. 6-13.

$$H = N_d(P_{scs} + s + a + bx)/(1 - W)$$

$$P_{scs} = (1 - W)H/N_d - (s + a + bx)$$

$$= (1 - 0.15)(8 \text{ h/day})/(2 \text{ trips/day})$$

$$- [0.10 \text{ h/trip} + 0.016 \text{ h/trip} + (0.018 \text{ h/mi})(15 \text{ mi/trip})]$$

$$= (3.40 - 0.39) \text{ h/trip}$$

$$= 3.01 \text{ h/trip}$$

2. Determine the pickup time required per pickup location using Eq. 6-15.

$$t_p = 0.72 + 0.18(C_n) + 0.014(PRH)$$

$$= 0.72 + 0.18(2) + 0.014(50)$$

$$\checkmark = 1.76 \text{ collector-min/location}$$

3. Determine the number of pickup locations from which wastes can be collected using Eq. 6-14.

$$N_p = 60P_{scs}\,n/t_p$$

$$= (60 \text{ min/h})(3.01 \text{ h/trip})(2 \text{ collectors})/(1.76 \text{ collector-min/location})$$

$$\checkmark = 205 \text{ locations/trip}$$

4. Determine the volume of wastes generated per pickup location per week.

$$\text{Volume per week} = (2.0 \text{ lb/person/day})(3.5 \text{ persons/pickup location})(7 \text{ days/wk})/(200 \text{ lb/yd}^3)(1/\text{wk})$$

$$\checkmark = 0.245 \text{ yd}^3/\text{location}$$

5. Determine the required truck volume using Eq. 6-16.

$$v = V_p N_p/r$$

$$= (0.245 \text{ yd}^3/\text{location})(205 \text{ locations/trip})/2$$

$$\checkmark = 20.1 \text{ yd}^3/\text{trip}$$

6. Determine the number of trips required per week using Eq. 6-17.

$$N_w = T_p F/N_p$$

$$= (1{,}000 \text{ locations})(1/\text{wk})/(205 \text{ locations/trip})$$

$$\checkmark = 4.88 \text{ trips/wk}$$

7. Determine the time requirements using Eq. 6-12 by multiplying the right-hand side of the equation by n, the number of collectors.

$$D_w = \overset{\text{COLLECTORS}}{n}[(N_w)P_{scs} + t_w(s + a + bx)]/[(1 - w)H]$$

$$= 2\{(4.88 \text{ trips/wk})(3.01 \text{ h/trip})$$

$$+ (5 \text{ trips/wk})[0.10 \text{ h/trip} + 0.016 \text{ h/trip}$$

$$+ (0.018 \text{ h/mi})(15 \text{ mi/trip})]\}/(1 - 0.15)(8 \text{ h/day})$$

$$\checkmark = 4.89 \text{ collector-days/wk}$$

Because two collectors are used, 2.45 days/wk will be required.

Comment In most cases, most of the data used in this example would be known to the designer, and missing information could be either determined easily from limited field studies or estimated from data presented in this chapter. It should also be noted that a key assumption in this example is that two trips will be made each day.

Discussion

The preceding analysis of hauled and stationary container systems illustrates the types of interrelationships that exist among the various components of solid waste collection systems. This presentation is intended not to be a compendium of such relationships, but rather to indicate the types of data that should be collected to evaluate such systems properly. The application of these relationships has been illustrated in Examples 6-1 through 6-4. One additional example in which these relationships are applied is presented at the end of the following section dealing with collection routes.

6-4 COLLECTION ROUTES

Once the equipment and labor requirements have been determined, collection routes must be laid out so that both workforce and equipment are used effectively. In general, the layout of collection routes is a trial-and-error process. There are no fixed rules that can be applied to all situations.
Some of the factors that should be taken into consideration when laying out routes are as follows:

1. Existing policies and regulations related to such items as the point of collection and frequency of collection must be identified.

2. Existing system conditions such as crew size and vehicle types must be coordinated.

3. Wherever possible, routes should be laid out so that they begin and end near arterial streets, using topographical and physical barriers as route boundaries.

4. In hilly areas, routes should start at the top of the grade and proceed downhill as the vehicle becomes loaded.

5. Routes should be laid out so that the last container to be collected on the route is located nearest to the disposal site.

6. Wastes generated at traffic-congested locations should be collected as early in the day as possible.

7. Sources at which extremely large quantities of wastes are generated should be serviced during the first part of the day.

8. Scattered pickup points where small quantities of solid wastes are

generated that receive the same collection frequency should, if possible, be serviced during one trip or on the same day.

Layout of Routes

The general steps involved in establishing collection routes include (1) preparation of location maps showing pertinent data and information concerning the waste generation sources, (2) data analysis and, as required, preparation of information summary tables, (3) preliminary layout of routes, and (4) comparison of preliminary routes and the development of balanced routes by trial and error. Step 1 is discussed below. Because the application of steps 2, 3, and 4 is different for the hauled and stationary container systems, each will be discussed separately. Following a brief discussion of the preparation of route schedules, the layout of collection routes for both the hauled and stationary container systems is illustrated in detail in Example 6-5. Additional details on the layout of collection routes may be found in Refs. 7 and 13.

Step 1   On a relatively large-scale map of the commercial, industrial, or housing area, the following data should be plotted for each solid waste pickup point: location, number of containers, collection frequency, and, if a stationary container system with self-loading compactors is used, the estimated quantity of wastes to be collected at each pickup. To aid in problem analysis, the following symbols may be used.

	Stationary container system	
Hauled container system	Self-loading compactors	Manually loaded vehicles
$\dfrac{F}{N}$	$\dfrac{SW}{N \mid F}$	$\frac{1}{1}$ ○ ● $\frac{2}{1}$ $\frac{2}{1}$ □ ■ $\frac{2}{2}$ $\frac{u}{1}$ △ ▲ $\frac{u}{2}$

where F = collection frequency
N = number of containers
SW = amount of solid wastes collected, yd^3/trip
○ = one container, once per week
□ = two containers, once per week
△ = unlimited service, once per week
● = one container, twice per week
■ = two containers, twice per week
▲ = unlimited service, twice per week

For hauled container systems, the rate of waste generation at each pickup point is not important because it usually has no direct effect on the layout of collection routes. For stationary container systems, however, the rate of waste generation at each pickup point determines the number of containers that can be emptied per trip. For residential sources it is generally assumed that the same quantity will be collected from each source.

Because a trial-and-error method is used for laying out routes, tracing paper should be used once the basic data have been entered on the work map. Depending on the size of the area and the number of pickup points, the area should be subdivided into rectangular and square areas corresponding roughly to functional land-use areas. For locations with less than 20 to 30 pickup points, this step is usually not necessary. For larger areas it may be necessary to subdivide further each of the functional-use areas into smaller areas, taking into account factors such as waste generation rates and collection frequency.

Steps 2, 3, and 4 for Hauled Container Systems Assume that collection routes are to be established for the hypothetical functional-use area shown in Fig. 6-19, and that the following data are known in addition to the data shown on the map (prepared in step 1).

Collection vehicle: hoist truck

Collection operation: 5 days/wk

Average number of trips per day: 9

Step 2. First summarize the number of pickup locations, each of which receives the same collection frequency, as shown in columns 1 and 2 of Table 6-7. Next determine the number of containers receiving the same collection frequency to be collected each day, as shown in columns 4

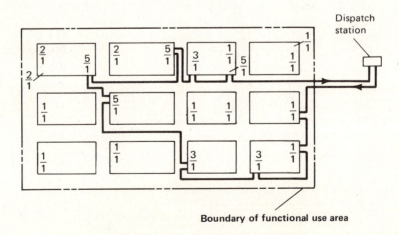

$\dfrac{F}{N}$ { Collection frequency, times/wk
      Number of containers
━━ Typical collection route for Monday

FIG. 6-19 Plan of a typical functional use area.

TABLE 6-7 SUMMARY DATA ON A TYPICAL FUNCTIONAL-USE AREA*

Collections/wk (1)	Number of pickup points (2)	Trips/wk (1) × (2) (3)	Number of containers (receiving the same collection frequency) emptied per day				
			Mon. (4)	Tues. (5)	Wed. (6)	Thurs. (7)	Fri. (8)
1	10	10	2	2	2	2	2
2	3	6	0	3	0	3	0
3	3	9	3	0	3	0	3
4	0	0	0	0	0	0	0
5	4	20	4	4	4	4	4
		—	—	—	—	—	—
Total		45	9	9	9	9	9

* See Fig. 6-19 for plan of typical functional-use area.

through 8 of Table 6-7. With this information the preliminary collection routes can be laid out.

Step 3. Starting from the dispatch station or where the vehicles are parked, lay out the collection routes for each day so that they begin and end near the dispatch station. A typical route for Monday is shown in Fig. 6-19. The collection operation should proceed in a logical manner, taking into account factors such as traffic conditions, type of activity, etc.

Step 4. When five preliminary routes have been laid out, the average distance to be traveled between containers should be computed. If the routes are unbalanced, they should be redesigned so that each route covers approximately the same distance. In general, a number of collection routes must be tried before the final ones are selected. When more than one collection vehicle is required, collection routes for each functional-use area must be laid out, and the work loads for each driver must be balanced.

The layout of routes will not always be as orderly and efficient as that shown in Fig. 6-19. The greatest problem is with private haulers who meet open competition for accounts. However, even in these cases, layout of functional-use areas will be useful. Functional-use boundaries must be adjusted to reflect added or lost accounts.

Steps 2, 3, and 4 for Stationary Container Systems (with Self-loading Compactors) Assume that collection routes are to be laid out for the area shown in Fig. 6-20 and that the following data are known in addition to the data shown on the map (prepared in step 1):

Collection vehicle: 30-yd³ self-loading compactor
Compaction ratio: 3

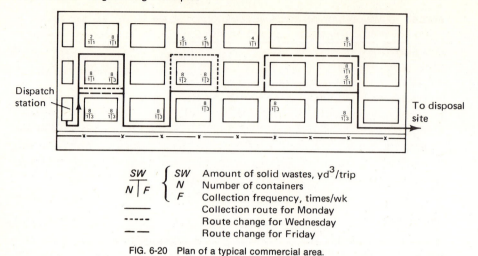

$$\frac{SW}{N\,|\,F} \quad \left\{ \begin{array}{ll} SW & \text{Amount of solid wastes, yd}^3\text{/trip} \\ N & \text{Number of containers} \\ F & \text{Collection frequency, times/wk} \end{array} \right.$$

———— Collection route for Monday
-----   Route change for Wednesday
— —   Route change for Friday

FIG. 6-20   Plan of a typical commercial area.

Number of days per week collection operation will be conducted: 3 (Monday, Wednesday, and Friday)

Number of trips per day: 1

Step 2. First, estimate the quantity of wastes collected from pickup locations serviced each day that the collection operation is conducted. From Fig. 6-20 it can be seen that there are eight locations to be serviced during each collection day, and that the quantity of wastes to be collected is 64 yd³.

Because the effective volume of the collection vehicle is 90 yd³ (30 yd³ × 3), an additional 26 yd³ of wastes can be picked up from the remaining locations to fill the collection vehicle. Further, inspection of Fig. 6-20 reveals that two locations produce a total of 16 yd³ that must be collected twice per week. If it is assumed that these locations will be serviced on Monday and Wednesday, the additional wastes that should be collected from other locations on Monday, Wednesday, and Friday will be 10, 10, and 26 yd³, respectively. These amounts must be collected from locations that are serviced once per week.

Step 3. Once the foregoing information is known, the layout of collection routes can proceed as follows. Starting from the dispatch station or where the collection vehicles are parked, a route should be laid out that connects all the pickup points to be serviced during each collection day. This route should be laid out so that the last of these locations is nearest the disposal site. The solid line shown in Fig. 6-20, with the exception of that portion that includes the two pickup locations serviced once per week, is the basic route for this simplified example.

The next step is to modify the basic route to include the additional pickup locations that will have to be serviced before the load can be completed. These modifications should be made so that a certain area or portion of the area is serviced with each collection route, as shown in Fig. 6-20. For large areas that have been subdivided and that are serviced daily, it will be necessary to establish basic routes in each subdivided area; in some cases, between them, depending upon the number of trips to be made per day.

Step 4. When the collection routes have been laid out, the actual container density and haul distance for each route should be determined. Using these data, the labor requirements per day should be checked against the available work time per day. In some cases it may be necessary to readjust the collection routes to balance the work load. After the routes have been established, they should be drawn on the master map.

Schedules

A master schedule for each collection route should be prepared for use by the engineering department and the transportation dispatcher. A schedule for each route, on which can be found the location and order of each pickup point to be serviced, should be prepared for the driver. In addition a route book should be maintained by each truck driver. The driver uses the route book to check the location and status of accounts. It is also a convenient place in which to record any problems with the accounts.

EXAMPLE 6-5   *Layout of Collection Routes and Analysis of Collection Systems*

Lay out collection routes for both a hauled container and stationary container collection system for the service area shown in Fig. 6-21 (a map such as shown in Fig. 6-21 would be prepared as the first step in the layout of collection routes). For both systems determine the maximum distance away from point *B* that the disposal site can be located so that the specified amount of work can be accomplished in an 8-h day. Assume that the following conditions apply:

1. Containers with a collection frequency of twice per week must be picked up on Tuesday and Friday.

2. Containers with a collection frequency of three times per week must be picked up on Monday, Wednesday, and Friday.

3. Containers may be picked up from any side of the intersection where they are stationed.

4. Start and finish each day at the dispatch station (point *A*).

5. For the hauled container system, collection will be provided Monday through Friday.

6. Hauled containers are exchanged rather than returned to the original location (see Fig. 6-13*b*).

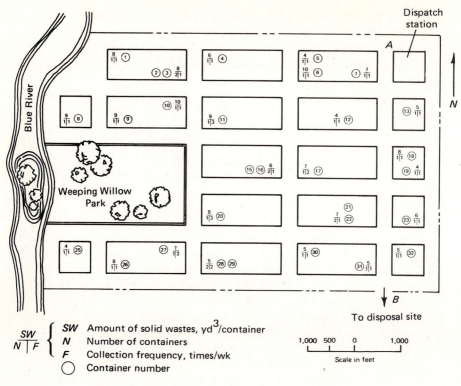

FIG. 6-21   Solid waste collection service area for Example 6-5.

7. Operational data for the hauled container system are as follows:
   a. Container pickup and unloading time = 0.033 h/trip
   b. At-site time = 0.053 h/trip   [SEPARATE]

8. For the stationary container system, collection will be provided only 4 days/wk (Monday, Tuesday, Wednesday, and Friday) with only 1 trip/day.

9. For the stationary container system, the collection vehicle will be a self-loading compactor with a capacity of 35 yd³ and a compaction ratio of 2.

10. Operational data for the stationary container system are as follows:
    a. Container unloading time = 0.050 h/container
    b. At-site time = 0.10 h/trip
    c. Constants for estimating driving time between container locations: $a' = 0.060$ h/trip and $b' = 0.067$ h/mi

11. Determine the haul time for both systems using the following haul constants: $a = 0.080$ h/trip, $b = 0.025$ h/mi.

12. Off-route factor for both systems = 0.15

Solution

1. Hauled container system

   a. Set up a summary table for the data reported in Fig. 6-21 (step 2 for route layout), as follows:

			Number of containers (receiving the same collection frequency) emptied per day				
Collections/wk	Number of pickup points	Number of trips/wk	Mon.	Tues.	Wed.	Thurs.	Fri.
1	26	26	6	4	6	8	2
2	4	8	—	4	—	—	4
3	2	6	2	—	2	—	2
	—	—	—	—	—	—	—
Total	32	40	8	8	8	8	8

   The routes will vary from one solution to another, but containers 11 and 20 must be picked up on Monday, Wednesday, and Friday, and containers 17, 27, 28, and 29 must be picked up on Tuesday and Friday. The optimum solution will be to have an equal number of containers picked up on each day as well as equal distances driven on each day. If one day has more pickups to be made or a longer distance to be driven, then that day's route will take more time and will therefore limit the maximum distance away that the disposal site can be located.

   b. Assuming that the containers are exchanged from location to location, lay out balanced collection routes for each day of the week by trial and error (steps 3 and 4).

      i. The resulting weekly routes and distance computations are shown on the following tabulations.

      ii. The limiting routes for the hauled container system are on Tuesday, Wednesday, Thursday, and Friday, with each route having eight containers to be picked up and about 86,000 ft to be driven between points A and B.

   c. Determine the maximum distance from point B to the disposal site.

      i. Determine the pickup time per trip. Because the exchange container system is used, the pickup time per trip consists of the time required to pick up each container.

$$P_{hcs} = pc + uc$$

$$= (0.033 \text{ h/trip}) + (0.033 \text{ h/trip})$$

$$= 0.066 \text{ h/trip}$$

      ii. Weekly route and distance tabulations for hauled container system, Example 6-5. See Table 6-8.

TABLE 6-8   PICKUP SCHEDULE FOR HAULED CONTAINER SYSTEM IN EXAMPLE 6-5

Container picked up	Between what points driven	Distance driven, 1,000 ft	Container picked up	Between what points driven	Distance driven, 1,000 ft
	Monday			Thursday	
	A to 1	6		A to 2	4
1	1 to B	11	2	2 to B	9
9	B to 9 to B	18	6	B to 6 to B	12
11	B to 11 to B	14	18	B to 18 to B	6
20	B to 20 to B	10	15	B to 15 to B	8
22	B to 22 to B	4	16	B to 16 to B	8
30	B to 30 to B	6	24	B to 24 to B	16
19	B to 19 to B	6	25	B to 25 to B	16
23	B to 23 to B	4	32	B to 32 to B	2
	B to A	5		B to A	5
		—			—
Total		84	Total		86
	Tuesday			Friday	
	A to 7	1		A to 13	2
7	7 to B	4	13	13 to B	5
10	B to 10 to B	16	5	B to 5 to B	16
14	B to 14 to B	14	11	B to 11 to B	14
17	B to 17 to B	8	17	B to 17 to B	8
26	B to 26 to B	12	20	B to 20 to B	10
27	B to 27 to B	10	27	B to 27 to B	10
28	B to 28 to B	8	28	B to 28 to B	8
29	B to 29 to B	8	29	B to 29 to B	8
	B to A	5		B to A	5
		—			—
Total		86	Total		86
	Wednesday				
	A to 3	2			
3	3 to B	7			
8	B to 8 to B	20			
4	B to 4 to B	16			
11	B to 11 to B	14			
12	B to 12 to B	8			
20	B to 20 to B	10			
21	B to 21 to B	4			
31	B to 31 to B	0			
	B to A	5			
		—			
Total		86			

iii. Determine the round-trip-haul distance by using Eq. 6-5.

$$H = N_d(P_{hcs} + s + a + bx)/(1 - W)$$

8 h/day = (8 trips/day)[0.066 h/trip

+ 0.053 h/trip + 0.080 h/trip

+ (0.025 h/mi)(x)]/(1 − 0.15)

[(1 − 0.15) − 0.20]h/day = [(0.025 h/mi)(x)] trip/day

$$x = 26 \text{ mi/trip}$$

iv. Determine the distance from point *B* to the disposal site.

Round-trip distance from *B*

= (26.0 mi/trip) − (86,000 ft/day)/(5,280 ft/mi)

× (8 trips/day)

= 24 mi/trip (38.6 km/trip)

Distance from *B* to disposal site
= (24 mi/trip)/2 = 12 mi/trip (19.3 km/trip)

2. Stationary container system

  a. Set up a summary table for the data reported on Fig. 6-21 (step 2 for route layout).

	Total wastes, yd³	Quantity of wastes collected per day, yd³				
Collections/wk		Mon.	Tues.	Wed.	Thurs.	Fri.
1	178	53	45	52	0	28
2	48	—	24	—	0	24
3	51	17	—	17	0	17
Total	277	70	69	69	0	69

Collection routes for the stationary container system will vary, but they must have containers 11 and 20 picked up on Monday, Wednesday, and Friday, and containers 17, 27, 28, and 29 picked up on Tuesday and Friday. Again, the optimum solution will be to have an equal number of containers picked up on each day as well as equal distances driven on each day.

  b. Lay out balanced collection routes by trial and error, in terms of quantity of wastes collected (steps 3 and 4). Note that the maximum quantity of wastes that can be collected per day is 70 yd³ (35 yd³ × 2). The resulting routes and distance computation are shown on the following tabulations. Because the same number of containers is picked up on all the days, the limiting day will be Tuesday with about 28,000 ft driven between points A and B.

i. Weekly routes and distance tabulations for stationary container system, Example 6-5; see Table 6-9.

ii. Distance traveled between points A and B

Day	Distance, ft
Monday	26,000
✓ Tuesday	28,000
Wednesday	26,000
Friday	22,000

c. Assume that the average distance driven between points A and

TABLE 6-9 PICKUP SCHEDULE FOR STATIONARY CONTAINER SYSTEM IN EXAMPLE 6-5

Order of container pickup	Quantity of wastes, yd³	Order of container pickup	Quantity of wastes, yd³
Monday		**Wednesday**	
13	5	18	8
7	7	12	4
6	10	11	9
4	8	20	8
5	8	24	9
11	9	25	4
20	8	26	8
19	4	30	5
23	6	21	7
32	5	22	7
Total	70	Total	69
Tuesday		**Friday**	
2	6	3	4
1	8	10	10
8	9	11	9
9	9	14	10
15	6	17	7
16	6	20	8
17	7	27	7
27	7	28	5
28	5	29	5
29	5	31	5
Total	68	Total	70

B is equal to 25,500 ft and that the average distance between containers is equal to 2,550 ft (25,500/10) = 0.48 mi. Determine the pickup time per container using a modified form of Eq. 6-9.

$$P_{scs} = C_t(uc + dbc) = C_t[uc + (a' + b'x')]$$

$$= (10 \text{ containers/trip})[0.050 \text{ h/container}$$

$$+ 0.060 \text{ h/container} + 0.067 \text{ h/mi})(0.48 \text{ mi/container})]$$

$$= 1.42 \text{ h/trip}$$

d. Determine the maximum round-trip-haul distance from point B to the disposal site, using Eq. 6-13.

$$H = N_d(P_{scs} + s + a + bx)/(1 - W)$$

$$8 \text{ h/day} = (1 \text{ trip/day})[1.42 \text{ h/trip}$$

$$+ 0.10 \text{ h/trip} + 0.080 \text{ h/trip}$$

$$+ (0.025 \text{ h/mi})(x)]/(1 - 0.15)$$

$$[8(1 - 0.15) - 1.6] \text{ h/day} = [(0.025 \text{ h/mi})(x)] \text{ trip/day}$$

$$x = 5.2/0.025 = 208 \text{ mi/trip (335 km/trip)}$$

e. Determine the distance from point B to the disposal site.

Distance from point B to disposal site

$$= (208 \text{ mi/trip})/2$$

$$= 104 \text{ mi/trip (167 km/trip)}$$

Comment   Here again, as noted previously in Example 6-2, the competitive advantage of the stationary container system is clear when the round-trip-haul distances are compared.

6-5 ADVANCED TECHNIQUES OF ANALYSIS

Interest in the analysis of solid waste collection systems arises from the need to improve (optimize) the operation of existing systems and to develop data and techniques that can be used to design or evaluate new or future systems. In the past, the design and operation of solid waste collection systems were based largely on experience and intuition. As collection systems and operations have increased in size and complexity, this method has proved to be less reliable. The main reason is that, because of the many variables, operations, and interrelationships that must be considered (in large systems), management has become structured to the extent that no one person can know or understand the entire system.

In an effort to operate existing systems and design new ones more efficiently, techniques and tools that were developed in related areas are now being applied to problems of waste collection. Terms such as systems analysis, operations research, system simulation, and systems and operations modeling are becoming part of the vocabulary in this field, and it is anticipated that systems engineering will find even wider application in the analysis of solid waste collection systems. Therefore, the purpose of this section is to discuss briefly some of the advanced techniques that have been applied to the analysis of solid waste collection systems.

Systems Analysis

Systems analysis or engineering is concerned with the selection of appropriate relationships, procedures, and elements to achieve a specific purpose. This definition is quite general and may be used in a variety of contexts. For example, systems analysis can be applied to the design of solid waste collection systems or to the selection of the equipment combinations required for the rail haul of solid wastes for disposal. Depending on the nature of the system under investigation, systems analysis techniques, such as operations research and simulation, have found wide application. Some texts make a clear-cut distinction that operations research deals with operation of existing systems, whereas simulation deals with the study and design of new or proposed systems. Both these techniques are included within the framework of systems engineering. As a matter of fact, from the development of operations research and other operational techniques originated the term *systems engineering,* which is dependent on the use of such techniques.

Operations Research

The field of operations research was first developed in a military context in England in the early 1940s [2]. In a general sense it may be defined as the scientific approach to decision-making that involves the operations of any organized system. The meaning of this definition can be more clearly understood in terms of the steps or operations that characterize an operations research study [3]:

1. Formulating the problem
2. Constructing a mathematical model to represent the system under study
3. Deriving a solution from the model
4. Testing the model and the solution derived from it
5. Establishing controls over the solution
6. Putting the solution to work: implementation

With some generalizations, this same list can also be used to describe the steps involved in a systems engineering study. In such a study, the model-construction phase may involve the use of a variety of different models. The behavior of proposed systems may also be studied using simulation models (see following discussion). For further details, Refs. 2 and 12 may be consulted.

Simulation

Simulation may be defined as the conduct of experiments involving physical, analog, or symbolic models which are used to describe the response of the system (or its components) under study. While operations research is concerned with the formulation and solution of mathematical models that represent real systems, simulation is used to describe the operation of both real and proposed systems in terms of their individual components. Experiments are conducted with models because experiments on any actual solid waste collection system may be impossible to conduct [8, 10]. By simulating the operation of the proposed system it is possible to make predictions about it and to study the effects of changes in vehicles, routes, and transfer and disposal sites.

A model may be defined as a representation of some subject of investigation (such as objects, events, processes, systems). Usually, such a model is to be used for the purpose of operation control or prediction. The general types of simulation models that have been used for representational purposes are iconic, analog, and symbolic [3].

Models that look like what they represent are said to be *iconic*. Examples of iconic models are laboratory-scale models of a hydraulic channel, bridge, or structure. Models in which one set of properties is used to model another set are said to be *analog*. The modeling of a pipe network using electrical components is a familiar analog modeling method used in the waterworks field. Models in which symbols or groups of symbols are used to represent the component or system under study are said to be *symbolic*.

Symbolic models are used most commonly in the analysis of solid waste collection systems because they are the most general and they can be altered easily. This is important because of the size and complexity of the systems that must be modeled. In this context, models used for solid waste collection operations, such as those described earlier in this chapter, should symbolically represent what takes place in the field to the extent the process is identifiable and can be quantified.

Applications

The techniques mentioned can be applied to (1) the evaluation of existing systems, (2) the design of new components within existing systems, and (3) the design of new or proposed systems.

Evaluation of Existing Systems   The purpose here is to evaluate the performance of existing collection operations and equipment and to improve the operation of existing systems [7, 10]. Evaluations of existing systems are usually based on economic comparisons, in which costs are compared between the existing operation and recommended alternative methods of operation or systems. A classic example of such an analysis occurred in the United States at most military installations when it became apparent that the cost of using a hauled container system (see Fig. 6-6) was significantly higher than the cost of using a stationary container system with mechanically self-loading compactors (see Fig. 6-10).

In many collection systems, savings can result from analysis and reorganization of the collection routes. In this case the component relationships must be coupled to a routing model to find the optimum combination of routes subject to the given system constraints. The problem of finding the optimum collection route has a direct analogy in the field of operations research where a similar problem is encountered in routing a traveling salesperson. In operations research texts, it is referred to as "the traveling salesperson problem" [2, 3].

Modification of Existing Systems Another extremely valuable application of these techniques is in the modification of existing systems in light of technological and operational changes. For example, if a group of cities in a metropolitan area were to adopt rail haul as a suitable means of disposal, many of the existing collection routes would have to be modified to minimize collection costs.

Design of Proposed Systems Problems posed in the design of new systems usually are related to the lack of specific data, such as the quantities of solid wastes that will be generated and the characteristics of the collection system(s) to be used. The quantity of solid wastes to be generated becomes an even more important consideration in industrial or commercial collection operations because the generation rates are usually not fixed, but follow some statistical frequency pattern. It is in these situations that simulation techniques can be used most effectively.

For example, the quantity of solid wastes to be expected can be estimated using various simulation techniques [9, 10]. In most cases, data from existing operations can be used in the development and verification of these models. The development of simulation models is discussed in Refs. 5, 8, 10, and 15. The design of proposed collection systems also involves the same sort of procedure, using (1) simulated solid waste generation data, (2) generation distribution patterns (i.e., proposed collection points), and (3) disposal alternatives. On the basis of this information, a number of alternative systems are investigated by using a system simulation model. The alternative solutions must then be evaluated in terms of engineering judgment and consideration of intangible factors.

Discussion

The effective application of these advanced techniques requires that the problem under investigation be *well defined*. It has been said that "a problem well defined is half solved." Unfortunately, the truth of this statement is often appreciated only after the fact or when it is too late to change course. The application of these techniques to a poorly defined problem can lead only to frustration and a mistrust of the techniques which, if properly applied, could materially aid in improving the operation of all types of solid waste collection systems.

6-6 DISCUSSION TOPICS AND PROBLEMS

6-1. Drive around your community and identify the principal types of systems and equipment used for the collection of residential and commercial solid wastes. Select two of the more common systems, and time the various activities associated with the collection of wastes. How do your values compare with those given in this chapter? If your values are significantly different, explain why.

6-2. Determine the haul equation constants a and b for the following data.

Average haul speed y, mi/h	Round-trip distance x, mi/trip
10	0.8
21	2.5
30	5.0
34	7.5
37.5	10.0
38	12.5
39.5	15.0
39.8	17.5
40	20.0

6-3. You are the city engineer in a medium-sized rural town. During a council meeting you are asked to compare the satellite method of collection with the more traditional curb-and-alley collection service that the city is currently providing. Startled, because you fell asleep during the preceding 4-h debate concerning the merits of the city slogan, you try to collect your thoughts. What are some of the important considerations that must be brought out in your discussion?

6-4. Develop an equation similar to those presented in this chapter that can be used to determine the labor requirements for a stationary container system employing satellite collection vehicles (see Fig. 6-3 and Ref. 4).

6-5. Because of a difference of opinion among city staff members, you have been retained as an outside consultant to evaluate the collection operation of the city of Davisville. The basic question centers around the amount of time spent on off-route activities by the collectors. The collectors say that they spend less than 15 percent of each 8-h workday on off-route activities; management claims that the amount of time spent is more than 15 percent. You are given the following information that has been verified by both the collectors and management:

1. A hauled container system, without container exchange, is used.

2. The average time spent driving from the corporation yard to the first container is 20 min, and no off-route activities occur.

3. The average pickup time per container is 6 min.

4. The average time taken to drive between containers is 6 min.

5. The average time required to empty the container at the disposal site is 6 min.

6. The average round-trip distance to the disposal site is 10 mi/trip, and the haul equation $(a + bx)$ constants are $a = 0.004$ h/trip and $b = 0.02$ h/mi.

7. The time required to redeposit a container after it has been emptied is 6 min.

8. The average time spent driving from the last container to the corporation yard is 15 min, and no off-route activities occur.

9. The number of containers emptied per day is 10.

From this information, determine whether the truth is on the side of the drivers or the management.

6-6. The amount of solid wastes generated per week in a large residential complex is about 600 yd³. There are two containers, each with a capacity of 40 gal, at the rear of every house. The solid wastes are collected by a two-person crew using a 35-yd³ manually loaded compactor once a week.

Determine the time per trip and the weekly labor requirement in person-days. The disposal site is located 15 mi away; haul-speed constants a and b are 0.08 h/trip and 0.025 h/mi, respectively; at-site time is 0.10 h/trip; the off-route factor is 0.15; the container utilization factor is 0.7; and the compaction ratio is 2. Assume that collection is based on an 8-h day.

6-7. A city desires to determine the impact of a new subdivision on solid waste collection services. The subdivision will add 150 new houses. A two-person crew will collect the wastes twice a week, using a 24-m³ manually loaded compactor. The allowable container size is 0.14 m³. It is estimated that there will be 3.2 persons per household and that each person will dispose of 2.5 kg of waste daily. Determine the number of containers that will be needed per household, the average container utilization factor, and the weekly labor requirement in person-days. The compaction ratio for the collection vehicle is 2.5, the average density of the solid wastes in the containers is 120 kg/m³, the disposal site is located 25 km away, and haul-speed constants a and b are 0.08 h/trip and 0.015 h/km, respectively. Collection is during an 8-h day. Collection is at curbside except for elderly persons (about 5 percent) who receive backyard service.

6-8. A new residential area composed of 800 low-rise detached dwellings is about to be occupied. Assuming that either two or three trips per day will be made to the disposal site, design the collection system and compare the two alternatives. The following data are applicable:

1. Solid waste generation rate = 0.032 yd³/day/home

2. Containers per service = 2

3. Type of service = 75 percent curbside and 25 percent rear of house

4. Collection frequency = once per week

5. Collection vehicle is a rear-loaded compactor with a compaction ratio of 2.5

6. Length of workday = 8 h

7. Collection crew = two persons

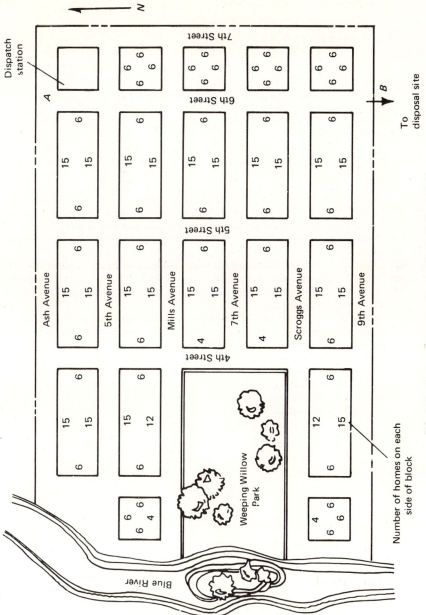

FIG. 6-22 Residential service area for Probs. 6-12 and 6-13.

8. Round-trip-haul distance = 20 mi

9. Haul constants: $a = 0.08$ h/trip and $b = 0.025$ h/mi

10. At-site time per trip = 0.083 h/trip

6-9. TT&E Corporation has four business locations that are each conveniently located 5 mi apart and 5 mi from the disposal site. TT&E presently uses a conventional hauled container system with large open-top containers. It has been suggested to TT&E that money could be saved by renting a fifth container from the waste collection company at a cost of $120/month and switching the operation to the container-exchange mode (see Fig. 6-13b). Each location will be serviced 8 times per month. The extra container will be stored at the collection company's dispatch station. Assuming that the operating costs are $20/h, compute the costs for both systems. Is it a wise decision for TT&E to rent the fifth container? Assume that $a = 0.034$ h/trip and $b = 0.029$ h/mi for all cases. Clearly state any additional assumptions.

6-10. Your friend and her friend are looking for some part-time work. You live in a small rural community that does not receive regular waste collection service. Your friend thinks it would be a good idea to provide waste collection service using your new $\frac{3}{4}$-ton four-wheel-drive pickup truck. There are 30 houses, and each one uses two 32-gal containers. All the houses would receive backyard carry service once per week. The haul constants are 0.08 h/trip and 0.025 h/mi. Assume that the at-site time is equal to 0.5 h. The round-trip-haul distance to the disposal site is 32 mi. The size of the pickup truck bed is $6 \times 8 \times 3$ ft.^{3} Assuming that your friend and her friend can devote 10 h/wk to this project, can they do it?

6-11. You have been called in to submit a proposal to evaluate your university's solid waste collection operation. Prepare a proposal, in outline form, to be submitted to the university. Note clearly the major divisions or tasks into which the work effort would be divided. Based on your knowledge to date, estimate the person-months of effort that would be required to do the work outlined in your proposal. Use an outline format in answering this question.

6-12. Lay out collection routes for the residential area shown in Fig. 6-22. Assume that the following data are applicable.

1. Occupants per residence = 4.0

2. Solid waste generation rate = 3.5 lb/capita/day

3. Collection frequency = once per week

4. Type of collection service = curb

5. Collection crew size = one person

6. Collection vehicle capacity = 30 yd^3

7. Compacted density of solid wastes in collection vehicle is equal to 552 lb/yd^3.

6-13. Lay out collection routes for the area shown in Fig. 6-22 using the data given in Problem 6-12 assuming that 4th and 6th Streets are one-way running from south to north and that 5th and 7th are one-way running from north to south.

6-14. Prepare a one-page abstract of Ref. 10. Do you feel the methods and techniques discussed in the article are applicable in your community?

6-7 REFERENCES

1. American Public Works Association: "Municipal Refuse Disposal," 3d ed., Public Administration Service, Chicago, 1970.

2. American Public Works Association, Institute for Solid Wastes: "Solid Waste Collection Practice," 4th ed., American Public Works Association, Chicago, 1975.

3. Churchman, C. W., R. L. Ackoff, and L. D. Arnoff: "Introduction to Operations Research," Wiley, New York, 1957.

4. Delaney, J. E.: *Satellite Vehicle Waste Collection Systems,* U.S. Environmental Protection Agency, Publication SW-82ts.1, Washington, D.C., 1972.

5. Golueke, C. G. and P. H. McGauhey: *Comprehensive Studies of Solid Wastes Management,* First Annual Report, Sanitary Engineering Research Laboratory, SERL Report 67-7, University of California, Berkeley, 1967.

6. *Guidelines for Local Governments on Solid Waste Management,* U.S. Environmental Protection Agency, Publication SW-17c, Washington, D.C., 1971.

7. Liebman, J. C.: *Routing of Solid Waste Collection Vehicles,* Final Report on Project 801289, Office of Research and Monitoring, U.S. Environmental Protection Agency, Washington, D.C., 1973.

8. Marks, D. H. and J. C. Liebman: *Mathematical Analysis of Solid Waste Collection,* U.S. Department of Health, Education, and Welfare, Public Health Service, Publication 2104, Washington, D.C., 1970.

9. Naylor, T. H., J. L. Balintfy, D. S. Burdick, and K. Chu: "Computer Simulation Techniques," Wiley, New York, 1966.

10. Quon, J. E., A. Charnes, and S. J. Wenson: Simulation and Analysis of a Refuse Collection System, *Proceedings ASCE, Journal of the Sanitary Engineering Division,* vol. 91, no. SA5, 1965.

11. Stone, Ralph, and Company, Inc.: *A Study of Solid Waste Collection Systems Comparing One-Man with Multi-Man Crews,* U.S. Department of Health, Education, and Welfare, Public Health Service, Publication 1892, Washington, D.C., 1969.

12. Sasieni, M., A. Yaspan, and L. Friedman: "Operations Research Methods and Problems," Wiley, New York, 1959.

13. Shuster, K. A. and D. A. Schur: *Heuristic Routing for Solid Waste Collection Vehicles,* U.S. Environmental Protection Agency, Publication SW-113, Washington, D.C., 1974.

14. Tchobanoglous, G. and G. Klein: *An Engineering Evaluation of Refuse Collection Systems Applicable to the Shore Establishment of the U.S. Navy,* Sanitary Engineering Research Laboratory, University of California, Berkeley, 1962.

15. Truitt, M. M., J. C. Liebman, and C. W. Kruse: *Mathematical Modeling of Solid Waste Collection Policies,* vols. 1 and 2, U.S. Department of Health, Education, and Welfare, Public Health Service, Publication 2030, Washington, D.C., 1970.

16. University of California: *An Analysis of Refuse Collection and Sanitary Landfill Disposal,* Technical Bulletin 8, Series 37, University of California Press, Berkeley, 1952.

7

Transfer and Transport

In the field of solid wastes, the functional element of *transfer and transport* refers to the means, facilities, and appurtenances used to effect the transfer of wastes from relatively small collection vehicles to larger vehicles and to transport them over extended distances to either processing centers or disposal sites. Thus, the discussion in this chapter focuses on (1) the need for transfer operations, (2) descriptions of the principal types of transfer stations, (3) alternative transport means and methods, and (4) the location of transfer stations.

It is noted that transfer stations may be either in the same location as processing stations or in entirely separate locations. (Processing stations, in turn, may be in the same location as transfer stations, in separate locations, or in the same location as disposal sites.) It is further noted that, for the sake of simplicity, the disposal site is generally mentioned in this chapter as the destination point of the transport vehicles. Where applicable, however, the processing center is an alternative (intermediate) destination.

7-1 THE NEED FOR TRANSFER OPERATIONS

Transfer operations can be used successfully with almost any type of collection system. Factors that tend to make the use of transfer operations attractive include (1) the presence of illegal dumps and large amounts of litter, (2) the location of disposal sites relatively far from collection routes (typically more than 10 mi), (3) the use of small-capacity collection trucks (generally under 20 yd³), (4) the existence of low-density residential areas (lots of 1 acre or larger with long driveways), (5) the widespread use of

medium-sized containers for the collection of wastes from commercial
sources, and (6) the use of hydraulic or pneumatic collection systems.

Transfer and transport operations become a necessity when haul
distances to available disposal sites or processing centers increase to the
point that direct hauling is no longer economically feasible. They also
become a necessity when disposal sites or processing centers are in remote
locations and cannot be reached directly by highway. Both these situations
are discussed in this section.

Excessive Haul Distances

In the early days when horse-drawn carts were used for the collection of
solid wastes, it was common practice to dump the contents of the loaded
carts into some auxiliary vehicle for transport to the disposal site or some
intermediate point for processing [8]. However, with the advent of the
modern motor truck and the availability of low-cost fuel, transfer operations in most cities were abandoned and direct hauling was adopted.
Today, with rising labor, operating, and fuel costs, the trend is reversing,
and transfer stations are again becoming common.

Typically, the decision to use a transfer operation is based on
economics. For example, in Chap. 6 (Examples 6-2 and 6-5), the time and
economic advantages of the stationary container system over the hauled
container system are demonstrated clearly. Simply stated, it is cheaper to
haul a large volume of wastes in large increments over a long distance than
it is to haul a large volume of wastes in small increments over a long
distance. This is illustrated in Example 7-1. Management aspects concerning the use of transfer operations are considered further in Chap. 15.

EXAMPLE 7-1 *Economic Comparison of Transport Alternatives*

Determine the break-even time for a hauled and a stationary container
system as compared to a system using transfer and transport operations
for transporting wastes collected from a metropolitan area to a landfill
disposal site. Assume that the following cost data are applicable:

1. Transportation costs
   a. Hauled container system using a hoist truck with an 8-yd³
      container = $8/h
   b. Stationary container system using a 20-yd³ compactor = $12/h
   c. Tractor-trailer–trailer transport unit with a capacity of 120 yd³ =
      $16/h
2. Other costs
   a. Transfer station operation cost including amortization = $0.30/
      yd³

    b. Cost of extra unloading time for tractor-trailer–trailer transport unit as compared to the cost for other vehicles = \$0.05/yd³

Solution

1. Convert the haul cost data to units of dollars per cubic yard per minute (see comment at end of this example).
   a. Hoist truck = \$0.0167/yd³/min $\left(\$8/hr / 8 c.y. / 60 min/hr \right)$
   b. Compactor = \$0.0100/yd³/min $\left(\$12/hr / 20 c.y. / 60 \right)$
   c. Transfer equipment = \$0.0022/yd³/min $\left(16 / 120 / 60 \right)$
2. Prepare a plot, as shown in Fig. 7-1, of the cost per cubic yard versus the round-trip driving time expressed in minutes for the three alternatives.
3. Determine break-even times for hauled and stationary container systems (see Fig. 7-1).
   a. Hauled container system = 23 min
   b. Stationary container system = 46 min
   Thus, for example, if a stationary container system is used and the round-trip driving time to the disposal site is more than 46 min, the use of a transfer station should be investigated.

Comment   In most articles and reference books dealing with the long-distance hauling of solid wastes, cost data are expressed in terms

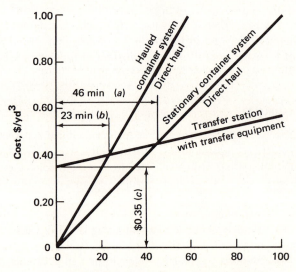

(a) Break-even time for stationary container system
(b) Break-even time for hauled container system
(c) Transfer station operating cost

FIG. 7-1   Economic evaluation of alternative means of transport for Example 7-1.

of dollars per ton per minute or dollars per ton per mile. This practice is widely accepted for transfer station analyses because weight is the most critical measure for efficient highway or rail movement. Such cost data can be misleading, however, when the densities of solid wastes vary significantly from location to location or container to container. For example, if the density of the wastes in two hoist-truck containers varies by a factor of 3, then comparing the costs of hauling two containers of the same size on a per-ton basis would tend to be misleading because the actual cost is the same for both. On the other hand, a comparison based on dollars per cubic yard per minute or dollars per minute would be valuable in comparing the two operations.

Remote Disposal Sites or Processing Centers

Transfer operations must be used when the disposal site or processing station is in such a remote location that conventional highway transportation alone is not feasible. For example, transfer stations are required when rail cars or ocean-going barges must be used to transport wastes to the final point of disposition. If solid wastes are transported by pipeline, a combination transfer-processing station is usually necessary. These subjects are considered further in the following sections.

7-2 TRANSFER STATIONS

Transfer stations, as mentioned, are used to accomplish the removal and transfer of solid wastes from collection and other small vehicles to larger transport equipment. Transfer stations may be classified with respect to capacity as follows: small, less than 100 tons/day; medium, between 100 and 500 tons/day; and large, more than 500 tons/day. Although specific details vary with size, important factors that must be considered in the design of transfer stations include (1) the type of transfer operation to be used, (2) capacity requirements, (3) equipment and accessory requirements, and (4) sanitation requirements.

Types of Transfer Stations

Depending on the method used to load the transport vehicles, transfer stations may be classified into three types: (1) direct discharge, (2) storage discharge, and (3) combined direct and storage discharge.

Direct Discharge (Large) In a large-capacity direct-discharge transfer station, the wastes in the collection vehicles usually are emptied directly into the vehicle to be used to transport them to a place of final disposition. To accomplish this, these transfer stations usually are constructed in a two-level arrangement. The unloading dock or platform from which wastes

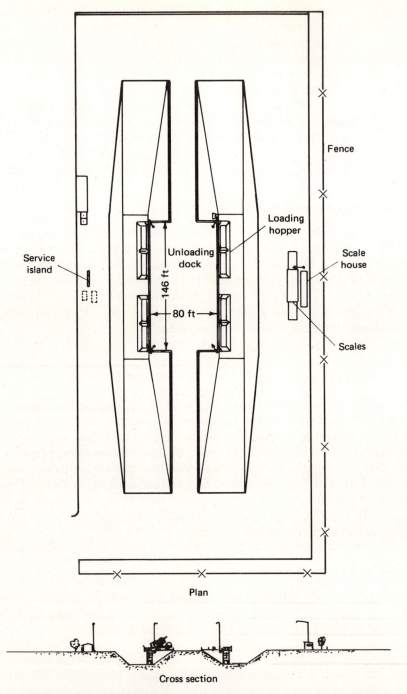

Plan

Cross section

FIG. 7-2 Typical direct-discharge transfer station with trailers located in depressed ramps. (Orange County, California [12].)

(a)

(b)

FIG. 7-3   Facilities and equipment used at transfer station shown in Fig. 7-2, Orange County, California. (a) End view showing trailers positioned under loading hopper in depressed ramp and clamshell mounted on rubber-tired tractor on ground-level unloading platform. (b) Clamshell is used to distribute and compact wastes in trailers and to pick up wastes spilled on unloading platform.

from collection vehicles are discharged into the transport trailers can be elevated, or the transport trailers can be located in a depressed ramp (see Fig. 7-2). Photographs of the facility shown in Fig. 7-2 and some of the equipment used are shown in Fig. 7-3. In some direct-discharge transfer stations, the contents of the collection vehicles can be emptied onto the unloading platform if the trailers are filled. The wastes are then pushed into the transport trailers.

Briefly, the operation of the direct-discharge transfer station such as shown in Fig. 7-2 may be summarized as follows. Upon arrival at the transfer station, all vehicles hauling wastes are weighed by the weighmaster who then indicates where the wastes should be dumped by giving the driver an appropriate stall number. After the collection vehicles have been unloaded, they are reweighed and the dumping fee is determined. Commercial vehicles that regularly use the transfer station are issued credit cards showing the firm name and the truck tare weight, thereby eliminating the second weighing for these vehicles.

As the trailers become loaded, the wastes in the trailer are shifted and compacted with a clamshell mounted on a rubber-tired tractor (see Fig. 7-3b). When the trailers are full or the maximum allowable tonnage has been placed in them, as indicated by the weighmaster, they are removed and prepared for the haul operation.

Direct Discharge (Medium and Small)   The layout of medium- and small-capacity transfer stations depends on the specific application and the site conditions. An example of an enclosed medium-capacity direct-discharge transfer station employing stationary compactors is shown in Fig. 7-4. The decision to enclose a transfer station usually depends on local weather conditions and environmental concerns.

Two small-capacity direct-discharge transfer stations are shown here in Figs. 7-5 and 7-6. In the station shown in Fig. 7-5, both a stationary compactor and an open-top container are used. Compressible items are discharged into the stationary compactor hopper, and bulky items, such as refrigerators, are discharged into the open-top container. Such a transfer station would be used by the general public (it must be attended when in use) as an alternative to driving to some distant disposal site. Transfer stations of this type are often called "public convenience centers."

Used in rural and recreational areas, the small-capacity direct-discharge transfer station shown in Fig. 7-6 is designed so that the loaded containers are emptied into a stationary container system for transport to the disposal site. In the design and layout of such stations, which are usually unattended, the key consideration should be simplicity. This is no place for complex mechanical systems. The number of containers used depends on the area served and the collection frequency that can be provided. To facilitate unloading, the tops of the containers may be set about 3 ft above the top of the unloading-area platform. Alternatively, the tops of the containers may be set level with the unloading area (see Fig.

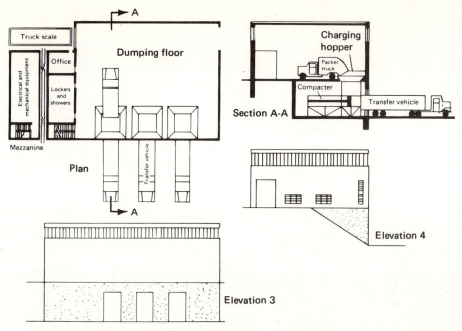

FIG. 7-4 Enclosed medium-capacity direct-discharge transfer station equipped with stationary compactors [7].

7-7), and the area behind the containers can be excavated to provide space for maneuvering the collection vehicles when the contents of the containers are emptied [6].

Storage Discharge In the storage discharge transfer station, wastes are emptied either into a storage pit or onto a platform from which they are loaded into transport vehicles by various types of auxiliary equipment. Perhaps the most well-known example of this type of transfer station is the San Francisco facility, shown schematically in Fig. 7-8 and pictorially in Fig. 7-9.

In this station, all incoming collection trucks are routed to a computerized weigh station where each truck is weighed. In addition, the weighmaster records the name of the unloading company, the identification of the particular truck, and the time it entered. Then the weighmaster directs the driver to either the east or west side of the main entrance of the enclosed transfer station. Once inside, the driver backs up the collection vehicle at a 50° angle to the edge of a depressed central waste storage pit. The contents of the vehicle are emptied into the pit (see Fig. 7-9a), and the empty vehicle is driven out of the transfer station.

Within the pit area, two bulldozers are used to break up the wastes and to push them into loading hoppers that are located at one end of the pit (see Fig. 7-9b). Two articulated bucket-type hoists, located on the other

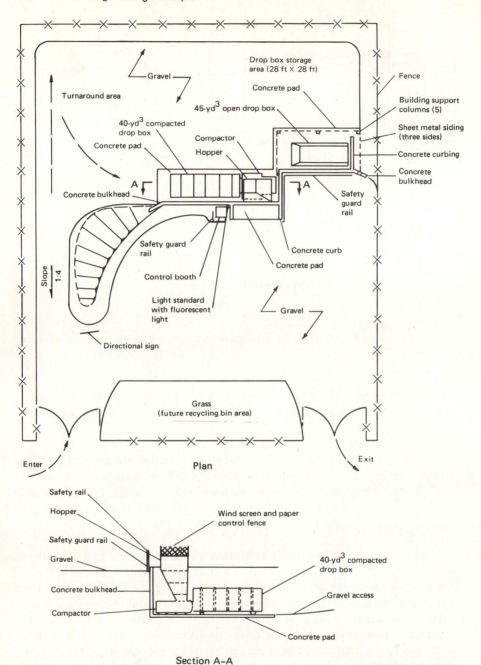

Plan

Section A-A

FIG. 7-5 Small-capacity direct-discharge transfer station with stationary compactor and open-top container [3].

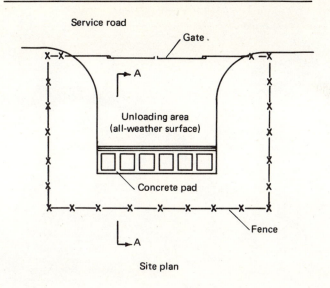

Site plan

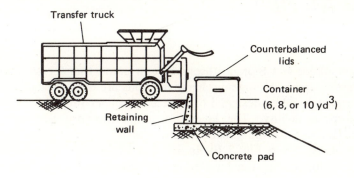

Section A-A

FIG. 7-6   Small-capacity direct-discharge transfer station for rural or recreational areas [6].

side of the hoppers, are used to remove any wastes that could damage the transport trailers. The wastes fall through the hoppers into trailers located on scales on a lower level (see Fig.7-9c). When the allowable weight limit has been reached, the hoist operator signals the truck driver. The loaded trailers are then driven out of the loading area, and wire screens are placed over the open trailer tops to prevent any papers or other solid wastes from blowing away during transport.

FIG. 7-7 Small rural public convenience transfer station with open-top containers placed against retaining wall at same level as unloading platform. (California Department of Public Health, Bureau of Vector Control.)

Combined Direct and Storage Discharge   In some transfer stations, both direct-discharge and storage discharge methods are used. Usually these are multipurpose facilities designed to service a broader range of users than a single-purpose facility. The layout of a multipurpose transfer station, designed for use by the general public and by various waste collection agencies, is shown in Fig. 7-10. In addition to serving a broader range of users, a multipurpose transfer station can also house a materials salvage operation.

The operation may be described as follows. All waste haulers (general public as well as commercial haulers) wishing to use the transfer station must check in at the scale house. Large commercial collection vehicles are weighed, and a commercial customer ticket is stamped and given to the vehicle driver. The driver then proceeds to the direct trailer-loading hoppers on the south side of the transfer station, backs up the vehicle to the hopper, and empties the wastes directly into the transport trailer. From there the driver returns the vehicle to the scale house for reweighing and turns in his customer ticket. The empty-vehicle weight is recorded while a discharge fee is calculated. In the community where this transfer station is used, all food and putrescible wastes are collected by waste collection agencies. Therefore, all putrescible wastes are discharged directly into transport trailers for daily delivery to a disposal site.

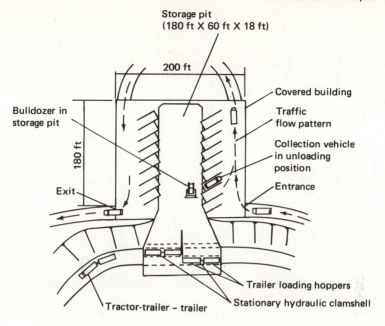

(a) **Plan view of transfer station**

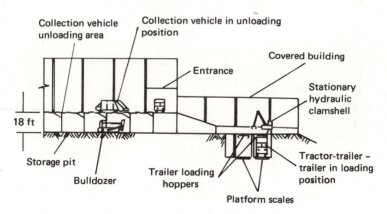

(b) **Section through portion of transfer station**

FIG. 7-8 Enclosed large-capacity (2,000 tons/day) storage discharge transfer station, San Francisco. (Solid Waste Engineering and Transfer Systems, Inc.)

Residents in this community and small independent haulers also haul significant quantities of yard wastes, tree trimmings, and bulky wastes (stoves, lawn mowers, refrigerators, etc.) to the transfer station. All automobiles, automobiles pulling trailers, and pickup trucks containing wastes must be checked in at the scale house. These vehicles are not weighed, but users do pay a discharge fee that is collected at the

(a)

(b)

FIG. 7-9 Operation details for storage discharge transfer station shown in Fig. 7-8, San Francisco. (a) Inside transfer station, contents of collection vehicles are emptied into storage pit. Two crawler tractors are used to break up wastes and push them to the hoppers used to load the trailers and to processing facilities (see Fig. 7-9d). (b) Solid wastes are pushed into trailer loading hoppers where they fall by gravity into trailers parked on platform scales located on lower level (see Fig. 7-8).

(c)

(d)

Stationary hydraulic clamshell is used to assist in the loading operation. (c) Exterior view showing tractor-trailer–trailer combination positioned under loading hopper and collection vehicle exiting from transfer station at upper level. (d) Solid wastes being pushed to feed conveyor leading to ferrous metal separation facilities. Constructed after the transfer station was built, the separation facilities are designed to process about 1,000 tons/day of solid wastes.

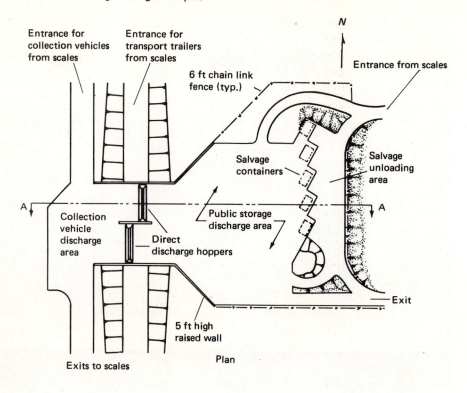

Entrance for
collection vehicles
from scales

Entrance for
transport trailers
from scales

N

Entrance from scales

6 ft chain link
fence (typ.)

Salvage
containers

Salvage
unloading
area

A

A

Collection
vehicle
discharge
area

Public storage
discharge area

Direct
discharge hoppers

Exit

5 ft high
raised wall

Exits to scales

Plan

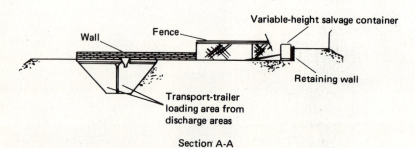

Variable-height salvage container

Wall

Fence

Retaining wall

Transport-trailer
loading area from
discharge areas

Section A-A

FIG. 7-10 Combination direct- and storage discharge transfer station with salvage operation.
(Division of Solid Waste Management, Sacramento County.)

scalehouse by the attendant who gives the user a cash receipt. The scale attendant visually checks the waste load to determine if it contains any salvageable metals. If it does, the attendant instructs the driver to deposit the metals at the salvage discharge area before proceeding to the public storage discharge area.

In reference to Fig. 7-10, once past the scales, noncommercial vehicles move either to the salvage unloading area or directly to the storage discharge area if there are no salvageable materials. A transfer station employee assists in unloading all salvageable materials. If the waste load contains a predetermined amount of salvageable materials, the driver is given a free pass for a vehicle of the type in which the wastes were delivered for future use. The driver then proceeds to the storage discharge pad and unloads any remaining wastes.

If there are no salvageable materials, the driver proceeds directly to the public storage discharge area. This area is separated from the direct discharge area used by commercial vehicles, by the two 40-ft trailer-loading hopper openings. Bulky wastes and trimmings that accumulate in the storage discharge area are periodically pushed into one of the loading hoppers by a rubber-tired loader.

Caution must be used in selecting and designing such transfer stations, for the cost of adding multipurpose facilities is often not justified in terms of the benefits achieved. Station users and discharge methods should be separated to prevent interferences and accidents between the large collection trucks and the smaller private vehicles. The physical separation of the discharge areas usually is the only positive way to maintain system efficiency.

Capacity Requirements

Both the operational and storage capacity requirements must be evaluated carefully in the planning and designing of transfer facilities. The operational capacity of a transfer station must be such that the collection vehicles do not have to wait too long to unload. In most cases, it will not be cost-effective to design the station to handle the ultimate peak number of hourly loads. Ideally, an economic tradeoff analysis should be made. For example, for both types of transfer stations, the annual cost of the time spent by the collection vehicles waiting to unload must be traded off against the incremental annual cost of a larger transfer station and/or the use of more transport equipment.

Because of the increased cost of transport equipment, a tradeoff analysis must also be made between the capacity of the transfer station and the cost of the transport operation, including both equipment and labor components. For example, in a given situation, it may be more cost-effective to increase the capacity of a transfer station and to operate with fewer transport vehicles by increasing the working hours than to use a smaller transfer station and purchase more transport vehicles. In a storage discharge transfer station, the equivalent storage capacity varies from about one-half to one day's volume of wastes. The capacity also varies with the type of auxiliary equipment used to load the transport vehicles. Seldom will the nominal storage capacity exceed two days' volume of wastes.

Equipment and Accessory Requirements

The equipment and accessories used in conjunction with a transfer station depend on the station's function in the waste management system. In a direct-discharge transfer station, some sort of rig, usually rubber-tired, is required to push the wastes into the transfer vehicles. Another rig is required to distribute the wastes and to equalize the load in the transfer vehicles. The types and amounts of equipment required vary with the capacity of the station. In a storage discharge transfer station, one or more tractors are required to break up the wastes and to push them into the loading hopper. Additional equipment is required to distribute the wastes and to equalize the loads. In some installations an overhead clamshell crane has been used successfully for both purposes.

Scales should be provided at all medium and large transfer stations both to monitor the operation and to develop meaningful management and engineering data. Scales are also necessary when the transfer station is to be used by the public and the charges are to be based on weight. If scales are to be used, it will usually be necessary to provide an enclosure for them. The scale house, as it is commonly called, should also have an office equipped with a telephone and a two-way speaker system so that the weighmaster can talk with the drivers.

If the transfer station is to be used as a dispatch center or district headquarters for a solid waste collection operation, a more complete facility should be constructed. For a headquarters facility, a lunch room, meeting rooms, offices, locker rooms, showers, and toilets should be provided.

Sanitation Requirements

By proper construction and operation, the objectionable features of transfer stations can be minimized. Most of the modern, large transfer stations are enclosed and are constructed of materials that can be maintained and cleaned easily. In most cases, fireproof construction is used. For direct-discharge transfer stations with open loading areas, special attention must be given to the problem of blowing papers. Wind screens or other barriers are commonly used. Regardless of the type of station, the design and construction should be such that all accessible areas where rubbish or paper can accumulate are eliminated [2].

The best way to maintain the overall sanitation of a transfer station is to monitor the operation continually. Spilled solid wastes should be picked up immediately or in any case should not be allowed to accumulate for more than 1 or 2 h. Overhead water sprays are often used to keep the dust down in the storage area of a storage discharge transfer station. To prevent dust inhalation, workers should wear dust masks. In the San Francisco transfer station, tractors in the pit area have enclosed cabs equipped with air conditioning and dust filtering units.

7-3 TRANSPORT MEANS AND METHODS

Motor vehicles, railroads, and ocean-going vessels are the principal means now used to transport solid wastes. Pneumatic and hydraulic systems have also been used. Still other systems have been suggested, but most have not been tested.

Motor Vehicles Transport

Where the point of final disposition can be reached by motor vehicles, the most common means used to transport solid wastes from transfer stations are trailers, semitrailers, and compactors. All types of vehicles can be used in conjunction with either type of transfer station. In general, vehicles used for hauling on highways should satisfy the following requirements: (1) wastes must be transported at minimum cost, (2) wastes must be covered during the haul operation, (3) vehicles must be designed for highway traffic, (4) vehicle capacity must be such that allowable weight limits are not exceeded, and (5) methods used for unloading must be simple and dependable.

Trailers and Semitrailers   In recent years, because of their simplicity and dependability, open-top trailers and semitrailers have found wide acceptance (see Fig. 7-11). When equipped for use with a dolly supporting the front end, semitrailers can be used interchangeably as the first or second trailer in tractor-trailer–trailer combinations, thus lending flexibility to the operation.

The maximum volume that can be hauled in highway transport vehicles depends on the regulations in force in the state in which they are operated. These regulations usually limit the outside dimensions of the vehicles or combinations of vehicles, as well as the weight per axle and the total weight. To maximize the payload, transport trailers are often designed so that they are higher than the legal limit when empty and lower when full. Typical data on transport trailers are summarized in Table 7-1.

Methods used to unload the transport trailers may be classified as (1) self-emptying and (2) requiring the aid of auxiliary equipment. Self-emptying transport trailers are mechanisms such as hydraulic dump beds, powered diaphragms, and moving floors that are part of the vehicle. Moving floors are an adaptation of equipment used in the construction industry for unloading trailers that carry gravel and asphalt. The operation of two different types of moving-floor systems is shown schematically in Fig. 7-12. The moving floor usually has two or more sections extending across the width of the trailer. Thus, if one section becomes inoperable, the moving floor does not prevent unloading because the system will function with the remaining operable section(s). This feature is important in terms of system reliability. Another advantage of the moving-floor trailer is the rapid turnaround time (typically 6 to 10 min) achieved at the disposal

(a)

(b)

(c)

FIG. 7-11 Typical transport vehicles used in conjunction with transfer facilities. (a) 96-yd^3 open-top trailer with moving-floor unloading mechanism (see Fig. 7-12a). (b) 75-yd^3 enclosed trailer used with stationary compactor (see Fig. 8-3). Trailer is unloaded with self-contained ejector plate. (c) 70- and 75-yd^3 open-top trailers unloaded with hydraulic tipping ramps (see Fig. 7-13).

TABLE 7-1 DATA ON HAUL VEHICLES USED AT LARGE- AND MEDIUM-CAPACITY TRANSFER STATIONS

| Station | | Capacity per trailer | | Dimensions of single trailer | | | | | |
Location	Capacity tons/day	Number of trailers	yd³	tons	Width, ft	Length, ft	Approx. height, empty, ft	Lenth of tractor and trailer units, ft	Method used for covering wastes	Method used for unloading trailers
Station No. 3, Orange County, California	960	23 tractor-trailer–trailer	70	12.5	8	27	13.5	65	Canvas tarp	Wire-cable
San Francisco, California	2,000	19 tractor-trailers	70	12	8	33	13.5	61	Wire-screen hinged cover	Tilting ramp at disposal site
		19 trailer units	75	14	8	28	13.5			
Seattle, Washington	2,000	10 single units	96	19	8	40	13.5	60	Nylon-mesh hinged cover	Chain-driven movable floor
		27 single units	96	19	8	40	13.5	60	Neoprene-canvas hinged cover	Wire-cable sling

Note: yd³/day × 0.7646 = m³/day
yd³ × 0.7646 = m³
ft × 0.3048 = m

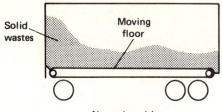

Normal position

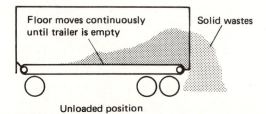

Unloaded position

(a) Continuous-belt moving floor

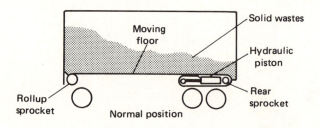

Normal position

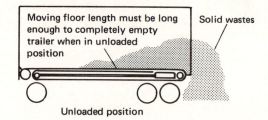

Unloaded position

(b) Rollup moving floor

FIG. 7-12　Moving-floor unloading systems for transport trailers. (a) Continuous-belt moving floor. (b) Rollup moving floor.

FIG. 7-13 Unloading operations using hydraulically operated tipping ramps. (*a*) Rear trailer being backed onto tipping ramp for unloading. (*b*) As tipping ramp is elevated, solid wastes fall out by gravity. (*c*) Tipping ramp in fully elevated position.

(a)

(b)

(c)

site without the need for auxiliary equipment. In some designs the rear of the trailer is made larger to facilitate the unloading operation. Trailers such as those shown in Fig. 7-12 are equipped with sumps to collect any liquids that accumulate from the solid wastes. The sumps are equipped with drains so that they can be emptied at the disposal site.

Unloading systems that require auxiliary equipment are usually of the "pull-off" type, in which the wastes are pulled out of the truck by either a movable bulkhead or wire-cable slings placed forward of the load. The disadvantage of requiring auxiliary equipment and workforce to unload at the disposal site is relatively minor in view of the simplicity and reliability of the method. An additional disadvantage, however, is the unavoidable waiting time during which the haul vehicle remains idle at the disposal site until the auxiliary equipment can be placed in the required position.

Another auxiliary unloading system that has proved very effective and efficient involves the use of movable, hydraulically operated tipping ramps located at the disposal site. Operationally, the semitrailer of a tractor-trailer–trailer combination is backed up onto a tipping ramp and uncoupled from the tractor-trailer (see Fig. 7-13). Once uncoupled, the tractor-trailer combination is backed up onto a second tipping ramp. The backs of the trailers are opened, and the units are then tilted up until the wastes fall by gravity out of the backs of the trailers into the disposal area. After being emptied, the truck-trailer and semitrailer are returned to their original positions. The tractor-trailer is driven off the ramp and is backed up to the ramp used for the semitrailer. The semitrailer is reattached, and the transfer rig is returned to the transfer station. The time required for the entire unloading operation typically is about 6 min/trip.

Compactors   Large-capacity containers in conjunction with stationary compactors are also used in a number of transfer stations (see Fig. 7-5). In some cases, the compaction mechanism is an integral part of the container. Representative data for such units are reported in Table 7-2. When containers are equipped with a self-contained compaction mechanism, the movable bulkhead used to compress the wastes is also used to discharge the compacted wastes. The contents of containers used with stationary compactors usually are unloaded by tilting the container and allowing the contents to fall out by gravity. If the wastes are compressed too tightly, unloading can be a problem. Various ejection devices also are available to empty the contents of the containers. The most common device is the movable bulkhead that is pulled out by cables.

Other Vehicles   Almost every imaginable type of vehicle has been used at one time or another for the transport of solid wastes. Because a complete discussion of alternative vehicle types is beyond the scope of this text, Refs. 1, 2, 5, and 11 are recommended.

TABLE 7-2 TYPICAL DATA ON CONTAINERS USED WITH STATIONARY COMPACTORS AND CONTAINER-COMPACTOR UNITS FOR MEDIUM AND SMALL TRANSFER STATIONS

Type	Rated capacity, yd³	Dimensions			Approx. weight, lb	Remarks
		Width, ft	Length, ft	Height, ft		
Container						
Small	20	8	14	6	8,000	Door openings where containers are attached to stationary compactor usually are reinforced.
Medium	30	8	18	6	9,000	
Large	45	8	22	9	10,000	
Container-compactor						
Small	3.7	6.5	6.5	4.5	1,500	Available with watertight sumps and leakproof doors. Other features on request.
Medium	15	7.5	15	6	6,000	
Large	30	8	22	8	9,000	

Note: yd³ × 0.7646 = m³
ft × 0.3048 = m
lb × 0.4536 = kg

Railroad Transport

Although railroads were commonly used for the transport of solid wastes in the past [4], they are now used by only a few communities. However, renewed interest is again developing in the use of railroads for hauling solid wastes, especially to remote areas where highway travel is difficult and railroad lines now exist, and where railroads own property or adjacent land for filling is available. If the use of railroad transport is being considered, Refs. 1, 2, and 8 are recommended.

Water Transport

Barges, scows, and special boats have been used in the past to transport solid wastes to processing locations and to seaside and ocean disposal sites, but ocean disposal is no longer practiced by the United States [2]. Although some self-propelled vessels (such as United States Navy garbage scows and other special boats) have been used, the most common practice is to use vessels towed by tugs or other special boats.

One of the major problems encountered when ocean vessels are used for the transport of solid wastes is that it is often impossible to move the barges and boats during storms or during times of heavy seas. In such cases, the wastes must be stored and the construction of costly storage facilities may be necessary.

Pneumatic, Hydraulic, and Other Systems of Transport

Both low-pressure air and vacuum conduit transport systems have been used to transport solid wastes. The most common application is the transport of wastes from high-density apartments or commercial activities to a central location for processing or for loading into transport vehicles. The largest pneumatic system in the United States is now used at the Walt Disney World amusement park in Orlando, Florida. The layout of this system is shown schematically in Fig. 7-14.

From a design and operational standpoint, pneumatic systems are more complex than hydraulic systems because of the complex control valves and ancillary mechanisms that are required. The necessity to use blowers or high-speed turbines further complicates the installation from a maintenance standpoint. Because installation costs for such systems are quite high, they are most cost-effective when used in new facilities.

The concept of using water to transport wastes is not new. Hydraulic transport is now commonly used for the transport of a portion of food wastes (where home grinders are used). One of the major problems with this method is that ultimately the water or waste water used for transporting the wastes must be treated. As a result of solubilization, the organic strength of this waste water is considerably greater than that of other domestic waste water. Hydraulic systems may be practical in areas where

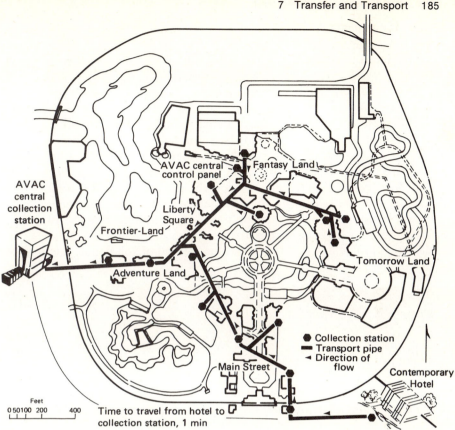

FIG. 7-14 Pneumatic solid waste collection system for Walt Disney World, Florida. (AVAC Systems, Inc.)

proper preprocessing and postprocessing facilities are incorporated into the treatment system. Usually, such applications are limited to areas with high population densities.

Other systems that have been suggested for the transport of solid wastes include various types of conveyors, air-cushion and rubber-tired trolleys, and underground conduits with magnetically transported gondolas, but these systems have never been put into operation.

7-4 LOCATION OF TRANSFER STATIONS

Whenever possible, transfer stations should be located (1) as near as possible to the weighted center of the individual solid waste production areas to be served, (2) within easy access to major arterial highway routes as well as near secondary or supplemental means of transportation, (3) where there will be a minimum of public and environmental objection to the transfer operations, and (4) where construction and operation will be most economical [2]. Additionally, if the transfer station site is to be used for processing operations involving materials recovery and/or energy production, the requirements for those operations must also be assessed. In some cases, these latter requirements may be controlling.

Because all the above considerations can seldom, if ever, be satisfied simultaneously, it is usually necessary to perform a tradeoff analysis among these factors. An approximate method for making an economic tradeoff among different locations based on haul cost is described in detail in this section. This method is applicable not only when a selection must be made from among several potential transfer station locations, but also in the more complex situation when two or more transfer stations and disposal sites are to be used. In the latter case, the basic question that must be answered is: What is the optimum allocation of wastes from each transfer station to each disposal site? In the discussion that follows, this allocation problem is described, and the method of solution is illustrated in an example.

Waste Allocation Problem

The waste allocation problem can be analyzed as follows. Assume that a determination must be made of the amount of solid wastes that should be hauled to each of three disposal sites from each of three transfer stations, so that the total haul cost will be the minimum possible value. A definition sketch for this situation is shown in Fig. 7-15. Also assume (1) that the

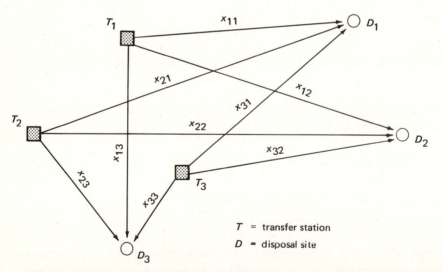

FIG. 7-15 Definition sketch for allocation of solid wastes from three transfer stations to three disposal sites.

total amount of wastes hauled to all the disposal sites must be equal to the amount delivered to the transfer station (materials-balance requirement), (2) that only specified amounts of wastes can be accepted at each disposal site (this constraint could arise as a result of limited highway access to a given disposal site), and (3) that the amount of wastes hauled from each transfer station is equal to or greater than zero. In the symbolic form, the allocation problem is set up as follows:

1. Let the transfer station sites be designated by i.
2. Let the disposal sites be designated by j.
3. Then let X_{ij} = the amount of wastes hauled from transfer station i to disposal site j.
4. Let C_{ij} = the cost of hauling wastes from transfer station i to disposal site j.
5. Let R_i = the total amount of wastes delivered to transfer station i.
6. Let D_j = the total amount of wastes that can be accepted at disposal site j.
7. If the total haul costs are to be minimized, then an objective function, which is defined as the sum of the following terms, must be minimized subject to the problem constraints:

$$X_{11}C_{11} + X_{12}C_{12} + X_{13}C_{13} + X_{21}C_{21} + X_{22}C_{22} + X_{23}C_{23} + X_{31}C_{31} + X_{32}C_{32} + X_{33}C_{33} = \text{objective function}$$

8. Expressed in mathematical summation form, the problem is to minimize the function

$$\text{Objective function} = \sum_{j=1}^{3}\sum_{i=1}^{3} X_{ij}C_{ij} \tag{7-1}$$

subject to the following constraints:

$$\sum_{j=1}^{3} X_{ij} = R_i \qquad i = 1 \text{ to } 3 \tag{7-2}$$

$$\sum_{i=1}^{3} X_{ij} \le D_j \qquad j = 1 \text{ to } 3 \tag{7-3}$$

$$X_{ij} \ge 0 \tag{7-4}$$

Solution to Waste Allocation Problem

The problem as set up in step 8 is commonly known as the "transportation problem" in the field of operations research (see Refs. 3 and 12 in Chap. 6). At present, a number of solution methods are available; the most common one is the simplex method [4, 9, 10]. However, most of the methods require the aid of modern computers. As an alternative, VAM (Vogel's approximation method) is a manual technique that can be used to find an initial solution of an allocation matrix associated with the transfer of material from one location to another (i.e., the transportation problem) [9]. Because the solution obtained will be close to the optimal solution (within 10 percent), it is sufficiently accurate for most practical applications in the field of solid waste management. Moreover, VAM is quick; by comparison, the time required to set up a computer program would be significantly greater. The optimal solution may be obtained by a second method, as discussed in Ref. 10. VAM is presented and illustrated in Example 7-2.

EXAMPLE 7-2   *Approximate Allocation of Solid Wastes among Multiple Transfer Stations and Disposal Sites Using VAM*

Determine the number of units of solid wastes to be hauled to each of four disposal sites (D_j) from four different transfer stations (T_i) to minimize the total haul cost. Use the following data.

1. Quantity of solid wastes to be disposed of from each transfer station:

Transfer station	Wastes, units/day
1	2
2	4
3	3
4	2
	—
Total	11

2. Disposal site capacity:

Disposal site	Capacity, units/day
1	5
2	5
3	6
4	6
	—
Total	22

3. Round-trip-haul distance in miles from each transfer station to each disposal site:

Transfer station	Haul distance to disposal site, mi			
	Site 1	Site 2	Site 3	Site 4
1	40	16	12	18
2	30	30	10	28
3	40	24	40	24
4	20	40	30	36

4. Haul time can be computed by using the following expression:

$$\text{Haul time, h/trip} = 0.08 \text{ h/trip} + 0.025 \text{ h/mi } (x)$$

where x = round-trip-haul distance, mi/trip.

5. Assume the haul cost = \$20/h

Solution

1. Because the solution depends on the cost of hauling wastes from each transfer station to each disposal site, develop a matrix of haul costs. The cost matrix that follows relates the cost of transporting one unit of solid wastes from each transfer station (T_i) to each disposal site (D_j) using the given data.

Transfer station	Haul cost to disposal site, \$/unit of wastes			
	D_1	D_2	D_3	D_4
T_1	21.60	9.60	7.60	10.60
T_2	16.60	16.60	6.60	15.60
T_3	21.60	13.60	21.60	13.60
T_4	11.60	21.60	16.60	19.60

2. Develop an allocation matrix, as shown in Table 7-3. The entries in the various rows and columns in the matrix are as follows:

   a. Haul cost. The cost of hauling one unit of solid wastes from a transfer station to a given disposal site is shown above and to the left of each diagonal line in the cost matrix.

   b. Haul units. The number of units to be hauled from a transfer station to a given disposal site is to be shown below and to the right of each diagonal line.

   c. Penalty cost. The difference between the lowest two-unit costs of a column or row is the penalty cost. This cost is shown in the "penalty" column and row.

TABLE 7-3   COST ALLOCATION MATRIX FOR EXAMPLE 7-2

Transfer station	Disposal site				Penalty	Rim requirements
	D_1	D_2	D_3	D_4		
T_1	21.60	9.60	7.60	10.60	2.00	2
T_2	16.60	16.60	6.60	15.60	9.00	4
T_3	21.60	13.60	21.60	13.60	0	3
T_4	11.60	21.60	16.60	19.60	5.00	2
T_d	100.00	100.00	100.00	100.00	0	22 − 11 = 11
Penalty	5.00	4.00	1.00	3.00		
Rim requirements	5	5	6	6		22 / 22

d. Rim. The rim requirements are the maximum number of units of solid wastes that have to be disposed of from each transfer station that each disposal site can accept. These are the boundaries of the problem that must be known before a solution can be obtained.

e. Dummy transfer station, T_d. The sum of the transfer-station rim requirements must equal the sum of the disposal-site requirements to make the problem mathematically correct. To do this, a dummy transfer station is added so that its rim requirement equalizes the sum of the disposal-site and transfer-station requirements.

    f. Haul cost from dummy transfer station. Dummy variables are always given identical haul costs in the matrix. This makes each dummy square of the matrix equivalent when the number of units to be hauled are being allocated. The dummy haul cost is set much higher than the other haul costs. This ensures that the allocation to the dummy squares will be made only after the most economical allocation has been made within the rest of the matrix.

3. The solution procedure is as follows:

    a. Subtract the lowest cost from the next lowest cost of each row and column, and enter it in the penalty box.

    b. Find the row or column with the greatest penalty.

    c. Allocate a maximum number of units to the cell with the lowest cost for the row or column selected in accordance with the rim requirements of that cell.

    d. Eliminate the row or column whose rim requirements have been exhausted.

    e. Repeat steps 1 through 5.

    f. The solution is found when all rim requirements are satisfied.

4. The solution to the problem is as follows:

    a. Iteration 1 (see Table 7-4)

        i. Compute the penalty cost for each row and column.

        ii. As shown, T_2 has the highest penalty.

        iii. Because the combination T_2 to D_3 has the lowest haul cost, a maximum of 4 units will be assigned consistant with the rim requirements on T_2. As a result, 4 units must be subtracted from the T_2 and D_3 rim requirements.

        iv. Row T_2 is eliminated because its rim requirement is now zero.

    b. Iteration 2 (see Table 7-5)

        i. Reevaluate all penalty costs.

        ii. The highest penalty cost, 10, is found to be column D_1.

        iii. Because T_4 to D_1 has the lowest haul cost, a maximum of 2 units is assigned to that square, thus satisfying the rim requirement of T_4. Two units are subtracted from the rim requirements of T_4 and D_1.

        iv. Row T_4 is now eliminated.

TABLE 7-4   SOLUTION PROCEDURE FOR EXAMPLE 7-2, ITERATION 1

Transfer station	Disposal site				Penalty	Rim requirements
	D_1	D_2	D_3	D_4		
T_1	21.60	9.60	7.60	10.60	2.00	2
T_2	16.60	16.60	† 6.60 ④	15.60	* 9.00	⁴ 0
T_3	21.60	13.60	21.60	13.60	0	3
T_4	11.60	21.60	16.60	19.60	5.00	2
T_d	100.00	100.00	100.00	100.00	0	22 − 11 = 11
Penalty	5.00	4.00	1.00	3.00		
Rim requirements	5	5	⁶ 2	6		22 / 22

* Highest penalty.
† Lowest cost.

    c. Iteration 3 (see Table 7-6)

      i. Revaluate all penalty costs.

      ii. The highest penalty, 14, is found to be in column D_3.

      iii. Because T_1 to D_3 has the lowest haul cost, 2 units are assigned to that square. Then 2 units are subtracted from rim

TABLE 7-5 SOLUTION PROCEDURE FOR EXAMPLE 7-2, ITERATION 2

Transfer station	Disposal site				Penalty	Rim requirements
	D_1	D_2	D_3	D_4		
T_1	21.60	9.60	7.60	10.60	2	2
T_2			④			
T_3	21.60	13.60	21.60	13.60	0	3
T_4	† 11.60 ②	21.60	16.60	19.60	5	2̶ 0
T_d	100.00	100.00	100.00	100.00	0	11
Penalty	* 10	4	9	3		
Rim requirements	5̶ 3	5	2	6		

* Highest penalty.
† Lowest cost.

requirements. Note that both rim requirements T_1 and D_3 have been satisfied.

    iv. Thus, both row T_1 and column D_3 have been eliminated.

  d. Iteration 4 (see Table 7-7)

    i. Reevaluate all penalty costs.

TABLE 7-6　SOLUTION PROCEDURE FOR EXAMPLE 7-2, ITERATION 3

Transfer station	Disposal site				Penalty	Rim requirements
	D_1	D_2	D_3	D_4		
T_1	21.60	9.60	† 7.60 ②	10.60	2	2̸ 0
T_2			④			
T_3	21.60	13.60	21.60	13.60	0	3
T_4	②					
T_d	100.00	100.00	100.00	100.00	0	11
Penalty	0	4	* 14	3		
Rim requirements	3	5	2̸ 0	6		

* Highest penalty.
† Lowest cost.

    ii. Because both D_2 and D_4 have penalty costs of \$86.40, an arbitrary decision must be made as to which one to use. For this problem, choose D_4.

    iii. Because T_3 to D_4 has the lowest haul cost, 3 units (note rim requirement) are assigned to the square.

    iv. Row T_3 has now been eliminated. Because all the real

TABLE 7-7 SOLUTION PROCEDURE FOR EXAMPLE 7-2, ITERATION 4

Transfer station	Disposal site				Penalty	Rim requirements
	D_1	D_2	D_3	D_4		
T_1			②			
T_2			④			
T_3	21.60	13.60		† 13.60 ③	0	3̶ 0
T_4	②					
T_d	100.00	100.00		100.00	0	11
Penalty	78.40	86.40		* 86.40		
Rim requirements	3	5		6̶ 3		

* Highest penalty (arbitrarily chosen).
† Lowest cost (arbitrarily chosen).

transfer stations have been eliminated, the required solution has been reached. The final problem solution is then only of academic interest.

  e. Iteration 5 (see Table 7-8)
  Forced solution. There is only one row left, T_d. Therefore, the time requirements determine the allocation in that row.

TABLE 7-8   SOLUTION PROCEDURE FOR EXAMPLE 7-2, ITERATION 5

Transfer station	Disposal site				Penalty	Rim requirements
	D_1	D_2	D_3	D_4		
T_1			②			
T_2			④			
T_3				③		
T_4	②					
T_d	100.00 ③	100.00 ⑤		100.00 ③	0	1̶ 0
Penalty	100.00	100.00		100.00		
Rim requirements	3̶ 0	5̶ 0		3̶ 0		

  f. Iteration 6
    Because all rim requirements have been satisfied, the final solution has been reached.
5. The problem summary is as follows:
  a. Solution matrix showing allocations of wastes (in units per day)

from transfer station to disposal site:

Transfer station	Disposal site			
	D_1	D_2	D_3	D_4
T_1	0	0	2	0
T_2	0	0	4	0
T_3	0	0	0	3
T_4	2	0	0	0

b. Cost summary.

Haul operation			Haul cost, $	
From	to	Units	Per unit	Total
T_1	D_3	2	7.60	15.20
T_2	D_3	4	6.60	26.40
T_3	D_4	3	13.60	40.80
T_4	D_1	2	11.60	23.20
Total				105.60

Comment   The dummy variable used in the solution was added for mathematical reasons and is not a part of the solution. Dummy variables can take on greater importance in more complicated problems.

This solution happens to be the optimal solution. A less-than-optimal solution results when a tie in penalty costs is broken in the wrong direction. In such a case, experience must be relied upon to help solve the problem. However, if the matrix is not large, both solutions can be computed and checked.

7-5 DISCUSSION TOPICS AND PROBLEMS

7-1.   Why was a cost for extra unloading time for the tractor-trailer–trailer transport units included in the cost analysis prepared in Example 7-1?

7-2.   Given the following data, determine the break-even times for the two stationary container systems versus the use of a transfer and transport system. Base your computations on dollars per ton per minute.

Transportation costs
 Stationary container systems
  1-ton capacity at $6.00/h
  8-ton capacity at $12.00/h
 Transport trailer
  20-ton capacity at $20.00/h

Transfer station costs
    Transfer station = $1.25/ton
    Extra unloading time at the disposal site = $0.25/ton

7-3.   Determine the round-trip break-even time for a solid waste collection system in which the 30-yd³ self-loading compactors used for collection are driven to the disposal site as compared to using a transfer and transport system. Assume that the following data are applicable.
   1. Density of wastes in self-loading compactor = 600 lb/yd³
   2. Density of wastes in transport trailers = 325 lb/yd³
   3. Volume of tractor-trailer–trailer transport unit = 120 yd³
   4. Operational cost for self-loading compactor = $20/h
   5. Operational cost for tractor-trailer–trailer transport unit = $30/h
   6. Transfer station operational costs including amortization = $2.10/ton
   7. Extra unloading time cost for transport units as compared to compactors = $0.40/ton

7-4.   Do either Problem 7-4 or 7-5, depending on whether your community has a transfer station. If your community does not have a transfer station, estimate the break-even time at which a transfer station operation would become feasible. How does this time

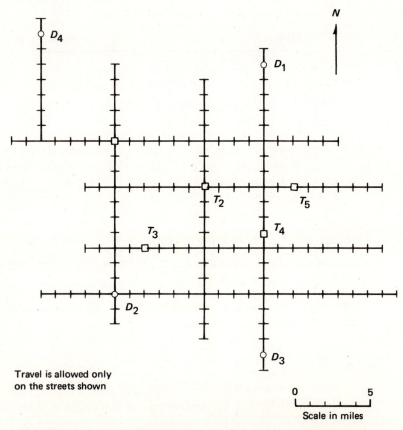

Travel is allowed only
on the streets shown

FIG. 7-16   Location map for disposal sites and transfer stations for Prob. 7-8.

compare to the actual time now spent by the collection vehicles in the haul operation? State clearly all your assumptions.

7-5. If your community has a transfer station, determine what the break-even time would be for a direct-haul operation. How does this time compare to the actual time now spent by the transport units in transport operation? State clearly all your assumptions.

7-6. A 1,000 ton/day transfer station is to be constructed. Consideration is being given to both a direct-discharge transfer station employing stationary compactors such as shown in Fig. 7-4 and a storage discharge type such as shown in Fig. 7-8. Identify and discuss the important factors that must be considered in selecting one of these two choices.

7-7. Given the following information, determine by the long-hand method of evaluating every possibility the most economical allocation of wastes from each of two disposal sites on the basis of transportation cost only.

Transfer station	Waste, units/day	Disposal site	Capacity, units/day
1	4	1	4
2	2	2	4

The round-trip-haul distance from transfer station 1 to disposal sites 1 and 2 is 10 and 20 mi, respectively. The distances from transfer station 2 to disposal sites 1 and 2 are 30 and 40 mi, respectively. Assume that the transport time in hours per trip is given by the expression $[0.08 \text{ h/trip} + 0.025 \text{ h/mi } (x)]$, where x is the round-trip-haul distance in miles per trip, and that the transportation cost is \$35/h.

7-8. The city shown in Fig. 7-16 has four disposal sites D_1, D_2, D_3, and D_4 and needs four transfer stations to handle the solid waste. The locations of transfer sites T_1, T_2, and T_3 have already been selected. The fourth site has been narrowed to two possibilities, T_4 and T_5 as shown. The following disposal site and transfer station data were collected for the city.

Disposal site	Capacity, units/day	Transfer station	Waste, units/day
D_1	4	T_1	3
D_2	10	T_2	3
D_3	3	T_3	5
D_4	8	T_4 or T_5	2

On the basis of transport cost alone, determine the more economical location for transfer station 4 (T_4 or T_5). Assume that the transport time in hours per trip is given by the expression $[0.08 \text{ h/trip} + 0.025 \text{ h/mi } (x)]$, where x is the round-trip-haul distance in miles per trip, and that the transport cost is \$35/h.

7-6 REFERENCES

1. American Public Works Association: *Rail Transport of Solid Wastes,* U.S. Environmental Protection Agency, NTIS Publication PB-222-709, Springfield, Va., 1973.
2. American Public Works Association, Institute for Solid Wastes: "Solid Waste Collection Practice," 4th ed., American Public Works Association, Chicago, 1975.

3. COR-MET: *Metropolitan Service District Solid Waste Management Action Plan,* vol. 1, Portland, Oreg., 1974.

4. Hadley, G.: "Linear Programming," Addison-Wesley, Reading, Mass., 1962.

5. Hegdahl, T. A.: *Solid Waste Transfer Stations: A State-of-the-Art Report on Systems Incorporating Highway Transportation,* U.S. Environmental Protection Agency, NTIS Publication PB-213-511, Springfield, Va., 1972.

6. Little, H. R.: *Design Criteria for Solid Waste Management in Recreational Areas,* U.S. Environmental Protection Agency, Publication SW-91ts, Washington, D.C., 1972.

7. Metcalf & Eddy, Inc: *Greater Bridgeport Regional Solid Wastes Management Study,* Boston, 1972.

8. Parsons, H. de B.: "The Disposal of Municipal Refuse," 1st ed., Wiley, New York, 1906.

9. Reinfeld, N. V. and W. R. Vogel: "Mathematical Programming," Prentice-Hall, Englewood Cliffs, N.J., 1958.

10. Riggs, J. L.: "Economic Decision Models," McGraw-Hill, New York, 1968.

11. Tchobanoglous, G. and G. Klein: *An Engineering Evaluation of Refuse Collection Systems Applicable to the Shore Establishment of the U.S. Navy,* Sanitary Engineering Research Laboratory, University of California, Berkeley, 1962.

12. *The Orange County Refuse Disposal Program,* The Orange County Road Department, Santa Ana, Calif., 1965.

8

Processing Techniques and Equipment

Processing techniques are used in solid waste management systems to improve the efficiency of operations, to recover resources (usable materials), and to recover conversion products and energy. The purpose of this chapter is to describe the more important techniques used for processing solid wastes. Because many of the techniques, especially those associated with the recovery of materials and energy, are in a state of flux with respect to design criteria, the objective here is only to introduce them to the reader. When adequate engineering information is available, it is presented. When known, factors that should be considered in the selection of equipment, other than cost, are also mentioned. It is strongly emphasized, however, that if these techniques are to be considered in the development of waste management systems, current engineering design and performance data must be obtained from the records of operating installations, from field tests, from equipment manufacturers, and from the literature.

Following a brief discussion of the main purposes of processing, five techniques and the equipment involved in each one are described. These techniques are: (1) mechanical volume reduction (compaction), (2) chemical volume reduction (incineration), (3) mechanical size reduction (shredding), (4) component separation (manual and mechanical), and (5) drying and dewatering (moisture content reduction). Of these, the first two have been used for the processing of solid wastes since the turn of the century. Although used extensively in other fields, the latter three techniques do not have a long history of application in the processing of solid wastes.

Flowsheets using many of these techniques are presented and discussed in Chap 9.

8-1   PURPOSES OF PROCESSING

The selection of specific processing techniques for a solid waste management system depends on the purposes to be achieved. As previously mentioned, the three main purposes of processing are to improve the efficiency of solid waste management systems, to recover usable materials, and to recover conversion products and energy.

Improving Efficiency of Solid Waste Management Systems

Various processing techniques are available to improve the efficiency of solid waste management systems. For example, to reduce storage requirements at medium- and high-rise apartment buildings, both incineration and baling are used (see Chap. 5). Before wastepaper is reused, it usually is baled to reduce shipping and storage volume requirements. In some cases, wastes are baled to reduce haul costs to the disposal site. At the disposal site, solid wastes are compacted to use the available land effectively. If solid wastes are to be transported hydraulically or pneumatically, some form of shredding is required. Shredding is also used to improve the efficiency of disposal sites. The selection of processing techniques for these purposes depends on the components of the overall waste management system and, in most cases, is situation-specific.

Recovery of Materials for Reuse

The principal components of residential solid wastes are reported in Chap. 4. As a practical matter, components that are most amenable to recovery are those for which markets exist and which are present in the wastes in sufficient quantity to justify their separation.

Materials that have been recovered from solid wastes include paper, cardboard, plastic, glass, ferrous metal, aluminum, and other residual nonferrous metal. Because all these materials may be of sufficient economic value to warrant their separation (depending on market conditions), a variety of techniques have been developed to accomplish this for each component. Some of the more established techniques are discussed later in this chapter.

Recovery of Conversion Products and Energy

Combustible organic materials can be converted to intermediate products and ultimately to energy in a number of ways, including (1) incineration or direct combustion in power boilers to produce steam, (2) pyrolysis to produce a synthetic gas or liquid fuel, and (3) biodigestion with and without sewage sludge to generate methane. These topics are considered further in Chap. 9. What is important with respect to this chapter is to point out that, with a few exceptions, the combustible organic materials must be separated from the other solid waste components as a first step. Once they are separated, further processing is usually necessary before the materials can be used for the production of power. Typically, they must be shredded and dried before use. These and other techniques are considered in the remainder of this chapter. Complete energy recovery systems are discussed in Chap. 9.

8-2 MECHANICAL VOLUME REDUCTION

Volume reduction is an important factor in the development and operation of almost all solid waste management systems. In most cities, vehicles equipped with compaction mechanisms are used for the collection of solid wastes. To increase the useful life of landfills, wastes usually are compacted before being covered. Paper for recycling is baled for shipping to processing centers. Recently, high-pressure compaction systems have been developed to reduce landfill requirements and to produce materials suitable for various alternative uses. These and other topics related to volume reduction achieved by compaction techniques are discussed in this section. Weight reduction of solid wastes is considered later in this chapter (see Sec. 8-6).

Compaction Equipment

The types of compaction equipment used in solid waste operations may be classified as stationary and movable. Where wastes are brought to and loaded into the compactor either manually or mechanically, the compactor is stationary. Using this definition, the compaction mechanism used to compress wastes in a collection vehicle is, in fact, a stationary compactor. By contrast, the wheeled and tracked equipment used to place and compact solid wastes in a sanitary landfill is classified as movable. The types and applications of compaction equipment used routinely are reported in Table 8-1.

Typically, stationary compactors may be described according to their application as (1) light duty, such as those used in residential areas; (2) commercial or light industrial; (3) heavy industrial; and (4) transfer station. Compactors used at transfer stations may further be divided according to the compaction pressure: low pressure, less than 100 lb/in²; high pressure, more than 100 lb/in². In general, all compactors in the other applications would also be classified as low-pressure units.

Where large stationary compactors are used, the wastes can be compressed (1) directly into the transport vehicle (see Chap. 7), (2) into steel containers that can be subsequently moved manually or mechanically, (3) into specifically designed steel chambers where the compressed

TABLE 8-1 COMPACTION EQUIPMENT USED FOR VOLUME REDUCTION

Location or operation	Type of compactor	Remarks
Solid waste generation points	Stationary/residential	
	Vertical	Vertical compaction ram; may be mechanically or hydraulically operated; usually hand-fed; wastes compacted into corrugated box containers or paper or plastic bags; used in medium- and high-rise apartments.
	Rotary	Ram mechanism used to compact wastes into paper or plastic bags on rotating platform, platform rotates as containers are filled; used in medium- and high-rise apartments.
	Bag or extruder	Compactor can be chute-fed; either vertical or horizontal rams; single or continuous multibags; single bags must be replaced and continuous bags must be tied off and replaced; used in medium- and high-rise apartments.
	Undercounter	Small compactors used in individual residences and apartment units; wastes compacted into special paper bags; after wastes are dropped through a panel door into bag and door is closed, they are sprayed for odor control; button is pushed to activate compaction mechanism.
	Stationary/commercial	Compactor with vertical or horizontal ram; waste compressed into steel container; compressed wastes are manually tied and removed; used in low-, medium-, and high-rise apartments, commercial and industrial facilities.
Collection	Stationary/packer	Collection vehicles equipped with compaction mechanisms (see Chap. 6).
Transfer and/or processing station	Stationary/transfer trailer	Transport trailer, usually enclosed, equipped with self-contained compaction mechanism.
	Stationary	
	Low pressure	Wastes are compacted into large containers.
	High pressure	Wastes are compacted into dense bales or other forms.
Disposal site	Movable wheeled or tracted equipment	Specially designed equipment to achieve maximum compaction of wastes.
	Stationary/track-mounted	High-pressure movable stationary compactors used for volume reduction at disposal sites.

TABLE 8-2 IMPORTANT DESIGN FACTORS IN THE SELECTION OF CONVENTIONAL COMPACTION EQUIPMENT*

| Factor | Value | | Remarks |
	Unit	Range	
Size of loading chamber	yd³	<1–11	Fixes the maximum size of wastes that can be placed in the unit.
Cycle time	s	20–60	The time required for the face of the compaction ram, starting in the fully retracted position, to pack wastes in the loading chamber into the receiving container and return to the starting position.
Machine volume displacement	yd³/h	30–1,500	The volume of wastes that can be displaced by the ram in 1 h.
Compaction pressure	lb/in²	15–50	The pressure on the face of the ram.
Ram penetration	in	4–26	The distance that the compaction ram penetrates into the receiving container during the compaction cycle. The further the distance, the less chance there is for wastes to fall back into the charging chamber and the greater the degree of compaction that can be achieved.
Compaction ratio		2:1–8:1	The initial volume divided by the final volume after compaction. Ratio varies significantly with waste composition.
Physical dimensions of unit	variable	variable	Affects the design of service areas in new buildings and provision of service to existing facilities.

* Adapted in part from Ref. 2.
Note: yd³ × 0.7646 = m³
yd³/h × 0.7646 = m³/h
lb/in² × 0.0703 = kg/cm²
in × 2.54 = cm

block is banded or tied by some means before being removed, or (4) into chambers where they are compressed into a block and then released and hauled away untied.

Low-Pressure Compaction   Typically, low-pressure compactors include those used at apartments and commercial establishments (see Fig. 8-1), baling equipment used for wastepaper and cardboard (see Fig. 8-2), and stationary compactors used at transfer stations (see Fig. 8-3). Portable stationary compactors are being used increasingly by a number of industries in conjunction with materials recovery operations, especially for wastepaper and cardboard.

High-Pressure Compaction   Recently, a number of high-pressure (up to 5,000 lb/in^2) compaction systems have been developed. In most of these systems, specialized compaction equipment is used to produce compressed solid wastes in blocks or bales of various sizes. In one system the size of the completed block is about 4 ft $\times$ 4 ft $\times$ 16 in, and the density is about 1,600 to 1,850 lb/yd^3. In another system, pulverized wastes are extruded

(a)

FIG. 8-1 Low-pressure compactors used at apartments and commercial establishments. (a) Compactor used with small container. Contents of loaded containers are emptied with front self-loading compactor (see Fig. 6-10). (b) Compactor used with large containers. Loaded container is hauled to the disposal site, emptied, and returned using a truck equipped with a tilt-frame loading mechanism (see Fig. 6-7).

after compaction in the form of logs approximately 9 in in diameter. Final densities achieved with this process range from 1,600 to 1,700 lb/yd³. The volume reduction achieved with these high-pressure compaction systems varies with the characteristics of the wastes. Typically, the reduction ranges from about 3 to 1 through 8 to 1.

Selection of Compaction Equipment

Factors that must be considered in the selection of compaction equipment include:

1. Characteristics of the wastes to be compacted, including size, composition, moisture content, and bulk density
2. Method of transferring and feeding wastes to the compactor
3. Handling methods and uses for compacted waste materials
4. Compactor design characteristics (see Table 8-2)
5. Operational characteristics including energy requirements, routine and specialized maintenance requirements, simplicity of operation, proved performance and reliability, noise level, and air and water pollution control requirements

(b)

FIG. 8-2   Baler used for shredded cardboard.

6. Site considerations including space and height, access, noise, and related environmental limitations

Additional details and factors that must be considered in various specific applications may be found in Refs. 1, 3, and 7. Because much confusion exists concerning the use and application of compaction ratio data, this subject is considered further below.

When wastes are compressed, their volume is reduced. The reduction in volume expressed in percent is given by the following:

$$\text{Volume reduction (\%)} = \left(\frac{V_i - V_f}{V_i}\right)100 \tag{8-1}$$

where V_i = initial volume of wastes before compaction, yd³
V_f = final volume of wastes after compaction, yd³

The compaction ratio is defined as

$$\text{Compaction ratio} = \frac{V_i}{V_f} \tag{8-2}$$

where V_i, V_f = as defined in Eq. 8-1.

The relationship between the compaction ratio and the percent of volume reduction is shown graphically in Fig. 8-4. Because of the nature of the relationship, it can be seen that to achieve more than about 80 percent

FIG. 8-3 Horizontal-ram stationary compactor used in conjunction with enclosed transfer trailer (see Fig. 7-11b).

reduction requires a disproportionate increase in the compaction ratio. For example, to achieve an increase from 80 to 90 percent requires an increase in the compaction ratio from 5 to 10. This relationship is important in making a tradeoff analysis between compaction ratio and overall compactor cost [8].

Another important factor that must be considered is the final density of the wastes after compaction. Some typical curves for unprocessed municipal solid wastes are presented in Fig. 8-5. The asymptotic value used in developing these curves is 1,800 lb/yd³, which is consistent with the values obtained by using high-pressure compactors. When shredded wastes are compacted under the same conditions, the density may be up to 35

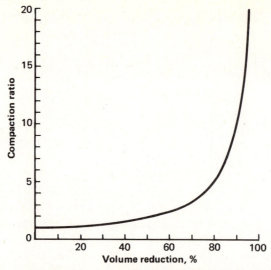

FIG. 8-4   Compaction ratio versus percent volume reduction.

percent greater than that of unprocessed wastes, up to an applied pressure of 100 lb/in² [15]. The maximum density reached under the application of very high pressure is not affected significantly by shredding.

Perhaps the most significant fact to be noted in Fig. 8-5 is that the initial density increase brought about by the application of pressure is highly dependent on the initial density of the wastes to be compacted. This fact is especially important when considering the claims made by manufacturers of compaction equipment. The moisture content, which varies with location, is another variable that has a major effect on the degree of

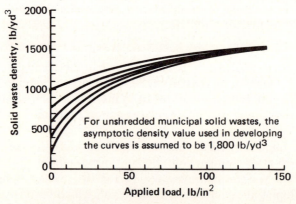

For unshredded municipal solid wastes, the asymptotic density value used in developing the curves is assumed to be 1,800 lb/yd³

FIG. 8-5   Density of solid wastes versus applied pressure. (Derived in part from Refs. 8 and 15.)

compaction achieved. In some stationary compactors, provision is made to add moisture, usually in the form of water, during the compaction process.

8-3 CHEMICAL VOLUME REDUCTION

In addition to mechanical volume reduction, various chemical processes have been used to reduce the volume of solid wastes. As noted in Chap. 2, until the early 1970s open burning was a common practice at many disposal sites. In some parts of the country this method is still used. In the early part of this century, chemical reduction was used to recover grease from food wastes, and in the process the volume was reduced. Since the turn of the century, incineration has been the method most commonly used to reduce the volume of wastes chemically. Although other chemical processes such as pyrolysis, hydrolysis, and chemical conversion are also effective in reducing the volume of solid wastes, they are not considered in this section because their use is primarily for the recovery of conversion products. Chemical conversion processes are considered in detail in Chap. 9.

Because incineration is now used both for volume reduction and for power production, the discussion of incineration in this chapter is limited to its application for volume reduction. The incineration process in power production is considered in detail in Chap. 9. The computations necessary to determine the amount of steam that can be produced from the incineration of solid wastes are also delineated in Chap. 9. Because the design and operation of modern municipal incinerators are a highly specialized undertaking, the following discussion is only intended to serve as an introduction to this subject. Topics to be considered include (1) discussion of the incineration of municipal wastes, (2) description of the incineration process for municipal solid wastes, (3) discussion of air pollution control facilities and equipment, and (4) some of the important design and performance considerations.

Incineration of Municipal Wastes

One of the most attractive features of the incineration process is that it can be used to reduce the original volume of combustible solid wastes by 80 to 90 percent. In some of the newer incinerators designed to operate at temperatures high enough to produce a molten material before cooling, it may be possible to reduce the volume to about 5 percent or less. Although the technology of incineration has advanced in the past two decades, air pollution control remains a major problem in implementation. Even if stricter air pollution control requirements can be met through the use of existing and developing technology, the question of economy remains more of a problem than with other alternatives.

In addition to the use of large municipal incinerators, onsite incineration is also used at individual residences, apartments, stores, industries, hospitals, and other institutions. Design of onsite incinerators varies with the type of service and local air pollution control requirements. Because most large cities in the United States have adopted some type of air pollution control ordinance, it is anticipated that, in the future, continued use of onsite incinerators will be limited to specially designed units that can meet local air pollution control requirements. For this reason a detailed discussion of onsite incineration is not included in this section. Some of the different types of onsite incinerators are described in Chap. 5, and additional details may be found in Refs. 1 and 4.

Description of Incineration Process

The basic operations involved in the incineration of solid wastes are identified in Fig. 8-6. The operation begins with the unloading of solid wastes from collection trucks (1) into a storage bin (2). The length of the unloading platform and storage bin is a function of the number of trucks that must unload simultaneously. The depth and width of the storage bin are determined by both the rate at which waste loads are received and the rate of burning. Storage capacity usually averages about the volume of 1 day. The overhead crane (3) is used to batch load wastes into the charging hopper (4). The crane operator can select the mix of wastes to achieve a

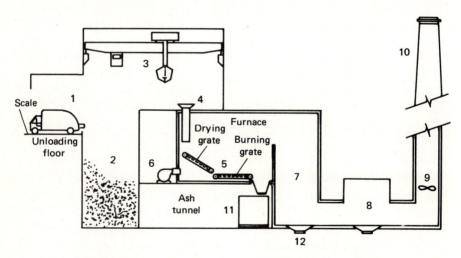

1. Collection truck	7. Combustion chamber
2. Storage bin	8. Gas cleaning equipment
3. Overhead crane	9. Induced draft fan
4. Charging hopper	10. Stack
5. Traveling grate stokers	11. Residue hopper
6. Forced draft fan	12. Flyash sluiceway

FIG. 8-6   Section through a typical continuous-feed mass-fired municipal incinerator.

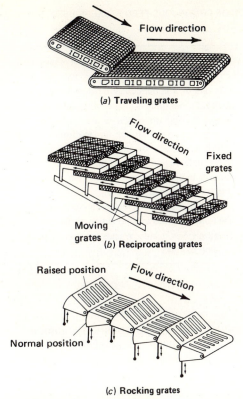

(a) **Traveling grates**

(b) **Reciprocating grates**

(c) **Rocking grates**

FIG. 8-7 Typical stokers used in mass-fired incinerators. (Adapted in part from Ref. 6.)

fairly even moisture content in the charge. Large or incombustible items
are also removed from the wastes. Solid wastes from the charging hopper
fall onto the stokers (5) where they are mass-fired. Several different types
of mechanical stokers are commonly used. Their characteristics are
described in Table 8-3, and some representative stokers are shown
pictorially in Fig. 8-7. Other methods of firing and stokers used with
processed solid wastes are discussed in Chap. 9.

Air may be introduced from the bottom of the grates (under-fire air) by
means of a forced-draft fan (6) or above the grates (over-fire air) to control
burning rates and furnace temperature. The hottest part of the fire is above
the burning grate. The heated air rises over the incoming high-moisture
wastes at the top of the drying grate and thus drives off the moisture to
permit burning as the wastes travel down the grate. Because most organic
wastes are thermally unstable, various gases are driven off in the combustion process taking place in the furnace, where the temperature is about
1400°F. These gases and small organic particles pass into a secondary

TABLE 8-3   OPERATION OF CONTINUOUS-FEED STOKERS USED IN MUNICIPAL SOLID WASTE INCINERATORS*

Stoker type	Description of operation
Traveling grate†	Consists of a continuously moving feeder grate and one or more burner grates. Feeder grate is located directly under a charging hopper from which wastes fall onto grate. Wastes are partially dried while on feeder grate.
Reciprocating grate†	Wastes are moved through furnace from the hopper while grate is actually stationary, except for alternating reciprocating movements of component stoker bars. Action of stoker bars turns wastes over and then tumbles them forward to next successive stoker bar. Burning rate is adjusted by controlling the speed of stoker bars.
Rocking grate†	Operation is similar to reciprocating grate, but wastes are moved through the furnace by the rocking action of the grates.
Barrel grate	Relatively new design. Wastes are burned as they are moved by a series of rotating barrels.

* Adapted from Ref. 18.
† See Fig. 8-7.

chamber, commonly called a "combustion chamber" (7), and burn at temperatures in excess of 1600°F. Odor-producing compounds usually are destroyed at temperatures above about 1400 to 1600°F.

Some fly ash and other particulates may be carried through the combustion chamber. To meet local air pollution control regulations, space must be provided for air-cleaning equipment (8). To secure adequate air flows to provide for head losses through air-cleaning equipment, as well as to supply air to the incinerator itself, an induced-draft fan (9) may be needed. It may also be done with the forced-draft fan.

The end products of incineration are the cleaned gases that are discharged to the stack (10). Ashes and unburned materials from the grates fall into a residue hopper (11) located below the grates where they are quenched with water. Fly ash which settles in the combustion chamber is removed by means of a fly ash sluiceway (12). Residue from the storage hopper may be taken to a sanitary landfill or to a resource recovery plant. Fly ash from the sluiceway and wastes from the air-cleaning equipment are taken to a sanitary landfill.

Air Pollution Control

With most incinerators, the primary concern in air pollution control is with particulate emissions rather than with gases and odors [18]. Typically, the size of particles that make up the emissions from most incinerators varies from less than 5 μm to about 120 μm. About one-third of the particles have a diameter of less than 10 μm [18]. In terms of size, these particles would be classified as fine dust, as shown in Fig. 8-8.

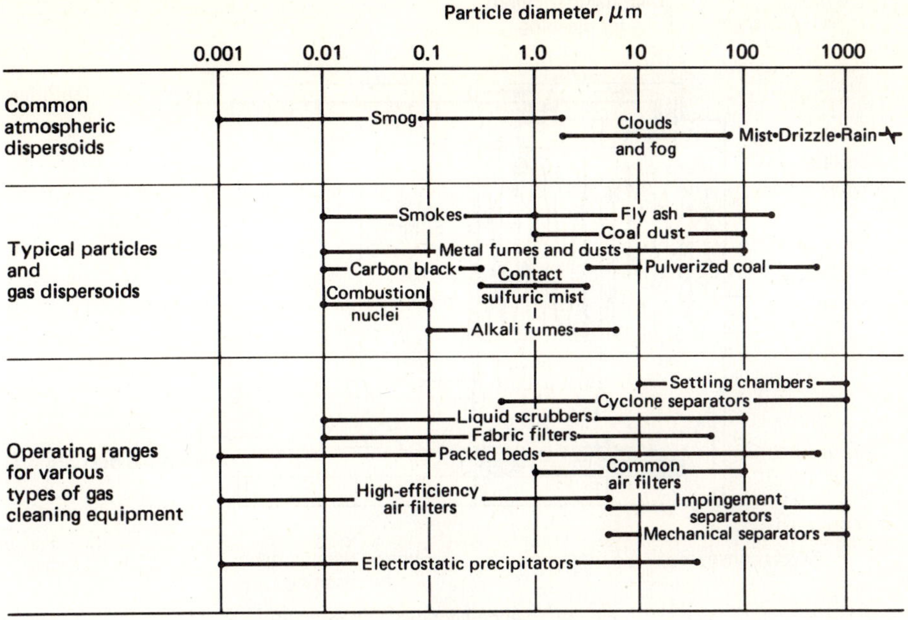

Particle diameter, μm

FIG. 8-8 Particle classification chart.

TABLE 8-4 EMISSION CONTROL FACILITIES AND EQUIPMENT FOR MUNICIPAL INCINERATORS*

Item	Description
Settling chamber	A large chamber usually located immediately after combustion chamber (see Fig. 8-6) for removal of large fly ash particles and as pretreatment operation for subsequent removal processes.
Baffled collectors	Baffles constructed of brick or metal that can be operated in wet or dry mode. Usually located after combustion chamber. Particles 50 μm or larger can be removed by impingement, velocity reduction, or centrifugal action. Efficiency depends on design and placement.
Scrubber	Fly ash is impacted on water droplets and subsequently removed. Method of removing wetted fly ash depends on equipment to be used and design of incinerator.
Cyclone separator	Dry separation of fly ash particles by means of centrifugal action in which particles are thrown or impinged against walls of collector.
Electrostatic precipitator	Fly ash particles are charged by means of an electrode. Charged particles are removed on collecting surfaces placed in an electrical field of high intensity. Once on the collecting surface, particles lose charge and adhere lightly. Can be moved by light tapping.
Fabric filter	Combustion gases are filtered through filter bags made of various materials.

* Adapted in part from Ref. 18.

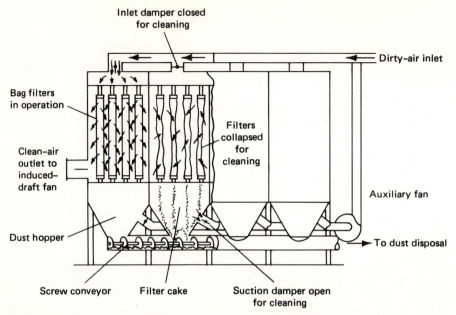

FIG. 8-9   Fabric-filter dust collector [18].

To control these particulate emissions, various design techniques and equipment have been used. The characteristics of some representative emission control facilities and equipment are reported in Table 8-4. The operative range of the facilities and equipment reported in Table 8-4 is also shown in Fig. 8-8. A typical fabric-filter dust collector and an electrostatic precipitator are shown in Figs. 8-9 and 8-10, respectively. The efficiency of various control methods is summarized in Fig. 8-11. Comparative air pollution control data for municipal incinerators are reported in Table 8-5.

Design and Performance Considerations

The principal elements that must be considered in the mechanical design of an incinerator are summarized in Table 8-6. Because the design of large, modern incinerators has become very complex, engineering firms and specialized groups within large engineering firms have been formed to design such facilities. Additional details on the design of incinerators may be found in Refs. 4, 14, 16, and 18.

Among the factors that must be considered in assessing incinerator performance are the amount of residue remaining after incineration and whether auxiliary fuel will be required when heat recovery is not of primary concern. The need for auxiliary fuel is considered in Chap. 9. The amount of residue depends on the nature of the wastes to be incinerated. Typical data on the residue from various solid waste components are

TABLE 8-5 COMPARATIVE AIR POLLUTION CONTROL DATA FOR MUNICIPAL INCINERATORS*

Collector	Relative capital cost factor, FOB	Relative space, percent	Collection efficiency, percent	Water to collector, gal/min/1,000 ft³/min	Pressure drop, of water	Relative operating-cost factor
Settling chamber	Not applicable	60	0–30	2–3	0.5–1	0.25
Multicyclone	1	20	30–80	None	3–4	1.0
Cyclones to 60-in diameter	1.5	30	30–70	None	1–2	0.5
Scrubber*	3	30	80–96	4–8	6–8	2.5
Electrostatic precipitator	6	100	90–97	None†	0.5–1	0.75
Fabric filter	6	100	97–99.9	None	5–7	2.5

* From Ref. 18.
† Gases usually are cooled with water-spray scrubber before electrostatic precipitator.
Note: gal/min × 0.0631 = 1/s
ft³/min × 0.028 = m³/min
in × 2.54 = cm

TABLE 8-6   PRINCIPAL COMPONENTS IN THE DESIGN OF LARGE MUNICIPAL INCINERATORS*

Component	Purpose/Description
Scales	Required to maintain accurate records of the amount of wastes processed.
Storage pits	Design of pits depends on furnace capacity, storage requirements (approximately 1-day capacity), collection schedules, and truck-discharge methods.
Cranes	Used to transfer wastes from storage pit to charging hoppers to mix and redistribute wastes in storage pit.
Charging hoppers	Constructed of metal or concrete, used to introduce wastes to furnace grates.
Furnace grates	Used to move wastes through furnace. Traveling, reciprocating, rocker arm, and barrel grates have been used successfully. Burning rate of 60 to 65 lb/ft²/h has been adopted as a "generally allowable" standard for mass firing.
Combustion chamber	Water-walled and refractory chambers are used.
Heat-recovery system	Types of systems vary. Typically, two boiler sections are used: convection and economizer (see Chap. 9).
Auxiliary heat	Need depends on moisture content of wastes as delivered.
Air pollution control facilities	Used to control particulate emissions (see Table 8-4).
Auxiliary facilities and equipment	Normally includes residue handling facilities, air supply and exhaust fans, incinerator stacks, control building, etc.

* Adapted in part from Ref. 18.

reported in Table 4-9. The composition of the residue from incinerators is reported in Table 8-7. Computations required to assess the quantity and composition of residue after incineration are illustrated in Example 8-1.

EXAMPLE 8-1   *Computations of Incinerator Residue*

Determine the quantity and composition of the residue from an incinerator used for municipal solid wastes with the average composition given in Table 4-9. Estimate the reduction in volume if it is assumed that the density weight of the residue is 1,000 lb/yd³.

Solution

1. Set up a computation table to determine the amount of residue and its percentage distribution by weight. The completed computation table is presented as Table 8-8.

2. Estimate the original and final volumes before and after burning. To estimate the approximate initial volume, assume that the

average density of the solid wastes in the incinerator storage pit is about 375 lb/ yd³.

$$\text{Original volume} = \frac{1,000 \text{ lb}}{375 \text{ lb/yd}^3} = 2.67 \text{ yd}^3$$

$$\text{Residue volume} = \frac{238.1 \text{ lb}}{1,000 \text{ lb/yd}^3} = 0.24 \text{ yd}^3 \ (0.18 \text{ m}^3)$$

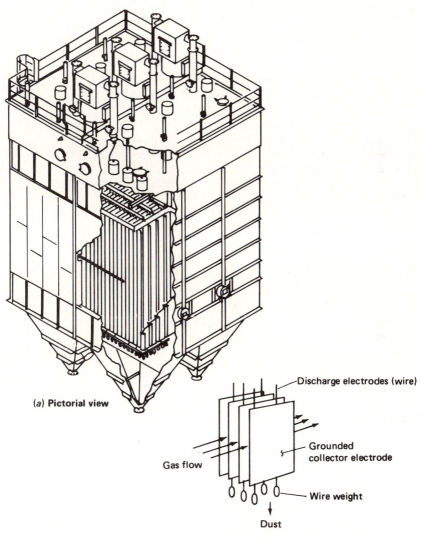

(a) Pictorial view

Discharge electrodes (wire)

Grounded collector electrode

Gas flow

Wire weight

Dust

(b) Definition sketch

FIG. 8-10 Electrostatic precipitator. (a) Pictorial view. (Research-Cotrell, Inc.) (b) Definition sketch.

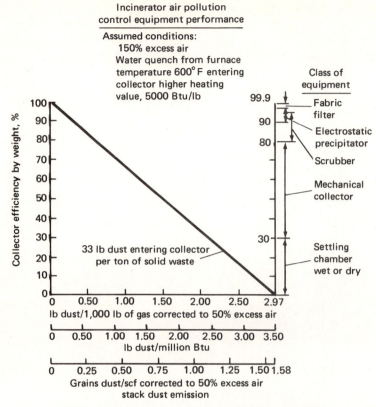

FIG. 8-11 Collector efficiency versus stack-dust emissions [18].

TABLE 8-7 COMPOSITION OF RESIDUE FROM INCINERATION OF MUNICIPAL SOLID WASTES

Component	Percent by weight	
	Range	Typical
Partially burned or unburned organic matter	3–10	5
Tin cans	10–25	18
Other iron and steel	6–15	10
Other metals	1–4	2
Glass	30–50	35
Ceramics, stones, bricks	2–8	5
Ash	10–35	25
Total		100

TABLE 8-8 INCINERATION RESIDUE COMPUTATION FOR EXAMPLE 8-1

Component	Solid wastes*, lb	Inert residue†, percent	Residue lb	Residue percent
Food wastes	150	5	7.5	3.2
Paper	400	6	24	10.1
Cardboard	40	5	2	0.8
Plastics	30	10	3	1.3
Textiles	20	2.5	0.5	0.2
Rubber	5	10	0.5	0.2
Leather	5	10	0.5	0.2
Garden trimmings	120	4.5	5.4	2.3
Wood	20	1.5	0.3	0.1
Glass	80	98	78.4	32.9
Tin cans	60	98	58.8	24.7
Nonferrous metals	10	96	9.6	4.0
Ferrous metals	20	98	19.6	8.2
Dirt, ashes, brick, etc.	40	70	28.0	11.8
Total	1,000		238.1	100

* Based on 1,000 lb of solid wastes (See Table 4-4).
† From Table 4-9.
Note: lb × 0.4536 = kg

3. Estimate the volume reduction by using Eq. 8-1.

$$\text{Volume reduction} = \left(\frac{2.67 - 0.24}{2.67}\right)100 = 91 \text{ percent}$$

8-4 MECHANICAL SIZE REDUCTION

Size reduction is the term applied to the conversion of solid wastes as they are collected into smaller pieces. The objective of size reduction is to obtain a final product that is reasonably uniform and considerably reduced in size in comparison to its original form. It is important to note that size reduction does not necessarily imply volume reduction. In some situations, the total volume of the material after size reduction may be greater than that of the original volume. In practice, the terms shredding, grinding, and milling are used interchangeably to describe mechanical size-reduction operations. The principal types of equipment and the important design factors are discussed in this section.

Size reduction is an important factor not only in the design and operation of solid waste management systems, but also in the recovery of materials for reuse and for conversion to energy. For example, some form of size reduction is necessary for the liquid transport of solid wastes.

Central grinding stations are described in Ref. 1. To achieve a higher density at a lower compaction pressure, wastes are shredded before they are baled. The disposal of shredded wastes in landfills without the use of daily cover is another important application of size reduction. This subject is considered further in Chap. 10 and Ref. 15.

Shredding is commonly used in systems designed to recover materials and energy from solid wastes. Because of varying particle size, moisture content, chemical composition, and physical characteristics, municipal solid wastes are not an ideal fuel. However, by dry (as received) or wet shredding, followed by separation, the organic materials in raw solid wastes can be transformed into a relatively homogeneous mixture with uniform size, heating value, and moisture content. Components from the remaining inorganic materials can also be recovered more easily because of their reduced size. This subject is considered further in a subsequent section of this chapter (see Sec. 8-6).

Size-Reduction Equipment

The types of equipment that have been used for reducing the size of and for homogenizing solid wastes include small grinders, chippers, large grinders, jaw crushers, rasp mills, shredders, hammer mills, and hydropulpers. The modes of action and principal applications of this equipment are listed in Table 8-9; some of the more commonly used types for solid wastes are considered further in the following discussion.

TABLE 8-9   TYPES, MODE OF ACTION, AND APPLICATIONS OF EQUIPMENT USED FOR MECHANICAL SIZE REDUCTION

Type	Mode of action	Application
Small grinders	Grinding, mashing	Organic residential solid wastes.
Chippers	Cutting, slicing	Paper, cardboard, tree trimmings, yard wastes, wood, plastics.
Large grinders	Grinding, mashing	Brittle and friable materials. Used mostly in industrial operations.
Jaw crushers	Crushing, breaking	Large solids.
Rasp mills	Shredding, tearing	Moistened solid wastes. Most commonly used in Europe.
Shredders,	Shearing, tearing	All types of municipal wastes.
cutters, clippers	Shearing, tearing	All types of municipal wastes.
Hammer mills	Breaking, tearing, cutting, crushing	All types of municipal wastes. Most commonly used equipment for reducing size and homogenizing composition of wastes.
Hydropulper	Shearing, tearing	Ideally suited for use with pulpable wastes, including paper, wood chips. Used primarily in the papermaking industry. Also used to destroy paper records.

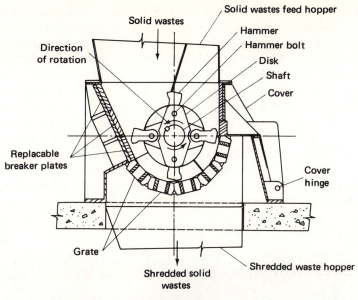

Solid wastes

Solid wastes feed hopper

Hammer

Hammer bolt

Direction
of rotation

Disk

Shaft

Cover

Replacable
breaker plates

Cover
hinge

Grate

Shredded solid
wastes

Shredded waste hopper

(a) One-way type

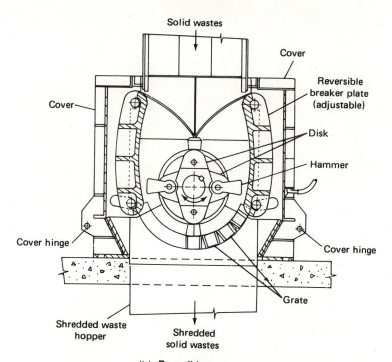

Solid wastes

Cover

Reversible
breaker plate
(adjustable)

Cover

Disk

Hammer

Cover hinge

Cover hinge

Grate

Shredded waste
hopper

Shredded
solid wastes

(b) Reversible type

FIG. 8-12 Hammermills used for the size reduction of solid wastes. (a) One-way
type. (b) Reversible type. (Williams Patent Crusher and Pulverizer Company, Inc.)

Hammer Mills (Horizontal Shaft)   Of the equipment reported in Table 8-9, horizontal-shaft hammer mills of the types shown in Fig. 8-12*a* and 8-12*b* are used most often in large commercial operations for reducing the size of solid wastes (see Fig. 8-13). Operationally, a hammer mill is an impact device in which a number of hammers are fastened flexibly to an inner shaft or disk(s) that is rotated at a high speed (see Fig. 8-12). Owing to centrifugal force the hammers extend radially from the center shaft. As solid wastes enter the mill, they are hit with sufficient force to crush or tear them, and with a velocity such that they do not adhere to the hammers. Wastes are further reduced in size by being struck against breaker plates and/or cutting bars fixed around the periphery of the inner chamber. The cutting and striking action continues until the material is of the size required and falls out of the bottom of the mill.

FIG. 8-13   Photograph of horizontal-shaft reversible-type hammermill used for the size reduction of solid wastes. (Williams Patent Crusher and Pulverizer Company, Inc.)

Because of the tough and abrasive nature of many of the materials found in solid wastes, frequent rebuilding or replacement of the hammers and breaker plates is routine with high-speed hammer mills. In some installations two hammer mills are operated in series, the first as a coarse shredder and the second to produce particles of the required size.

Hammer Mills (Vertical Shaft) Hammer mills with vertical shafts on which hammers and grinding wheels of various diameters are mounted have also been used. To date (1976) reliability has been a major problem with the vertical-shaft machines.

Hydropulper An alternative method of separating solid waste components involves the use of a hydropulper (see Fig. 8-14). In this system, solid wastes and recycled water are added to the hydropulper. Through the action of the high-speed cutting blades mounted on a rotor in the bottom of the unit, pulpable and friable materials are converted into a slurry with a solids content varying from 2.5 to 3.5 percent. Metal, tin cans, and other nonpulpable or nonfriable materials are rejected from the side of the hydropulper tank (see Fig. 8-14). The rejected material passes down a chute that is connected to a bucket elevator. As the material moves up the elevator, it is given a preliminary washing. The solids slurry passes out through the bottom of the pulper tank and is pumped to the next processing operation. A complete materials recovery system that uses a hydropulper as a first step is discussed in Chap. 9.

Selection of Size-Reduction Equipment

Factors that must be considered in the selection of size-reduction equipment include:

1. Properties of material to be shredded and the characteristics of material after shredding
2. Size requirements for shredded material by component
3. Method of feeding shredder, provision of adequate shredder hood capacity to avoid bridging, and clearance requirements between feed and transfer conveyors and shredder
4. Type of operation (continuous or intermittent)
5. Operational characteristics including energy requirements, routine and specialized maintenance requirements, simplicity of operation, proved performance and reliability, noise output, and air (principally dust) and water pollution control requirements
6. Site considerations including space and height, access, noise, and environmental limitations
7. Material storage after size reduction and in anticipation of next functional operation

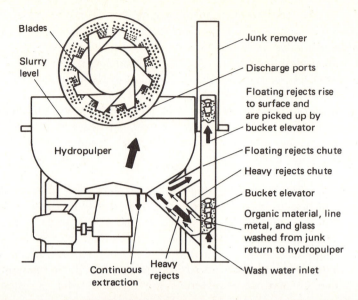

Blades

Slurry level

Hydropulper

Continuous extraction

Heavy rejects

Junk remover

Discharge ports

Floating rejects rise to surface and are picked up by bucket elevator

Floating rejects chute

Heavy rejects chute

Bucket elevator

Organic material, line metal, and glass washed from junk return to hydropulper

Wash water inlet

(*a*) **Section through hydropulper**

(*b*)

FIG. 8-14   Hydropulper used for solid wastes. (*a*) Section through hydropulper. (*b*) Overhead photograph of hydropulper. (Black Clawson Fibreclaim, Inc.)

Typical data on the horsepower requirements for shredding are given in Fig. 8-15. These data were derived from an analysis of information obtained from equipment manufacturers and, to a limited extent, from actual installations [7]. As noted, if preliminary size reduction is used to reduce the size of the wastes before they are processed by hammer mills, an additional 15 hp/ton·h should be added to the estimated horsepower. The use of the data reported in Fig. 8-15 is illustrated in Example 8-2.

EXAMPLE 8-2 *Horsepower Requirements for Size Reduction*

Estimate the horsepower required to reduce municipal solid wastes to a final size of about 3 in for a plant with a capacity of 80 tons/h using the data given in Fig. 8-15.

Solution

1. Using a conservative value of 20 hp·h/ton for the base horsepower, the required horsepower is

    Horsepower = 80 tons/h × 20 hp·h/ton = 1,600 hp

2. Using an estimated product size factor of 1.5 (see Fig. 8-15), the required horsepower is

    Horsepower = 1,600 hp × 1.5 = 2,400 hp (1,789 kW)

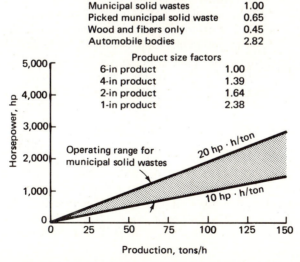

FIG. 8-15 Power requirements for reducing the size of various solid wastes [7].

Comment   An alternative method for estimating the horsepower required for size reductions is given in Problem 8-7. It should be noted, however, that the validity of the expression given as applied to solid wastes remains to be demonstrated.

8-5   COMPONENT SEPARATION

Component separation is a necessary operation in the recovery of resources from solid wastes and where energy and conversion products are to be recovered from processed wastes. The required separation may be accomplished manually or mechanically. When manual separation is used, preprocessing of the wastes is not required; in most mechanical techniques, however, some form of size reduction is required as a first step. Techniques and equipment used for the separation of components from municipal solid wastes are delineated in this section. The techniques considered in this section are listed in Table 8-10 along with information about applications. Of the mechanical techniques reported in Table 8-10, air separation, magnetic separation, and screening are discussed in more detail, because more is known about these operations from their use in other fields. Because few long-term data are currently available on the performance of the equipment used for the separation of specific components from wastes, caution must be exercised in their use.

Handsorting

The manual separation of solid waste components can be accomplished at the source where solid wastes are generated, at a transfer station, at a centralized processing station, or at the disposal site. The number and types of components salvaged or sorted depend on the location and the resale market. Market issues are discussed in Chaps. 9 and 16. Typically, the components include newspaper, aluminum, and glass from residential sources; cardboard and high-quality paper, metals, and wood from commercial and industrial sources; and metals, wood, and bulky items of value from transfer stations and disposal sites.

Air Separation

Air classification has been used for a number of years in industrial operations for the separation of various components from dry mixtures. In solid waste resource and energy recovery systems, air classification is used to separate the organic material—or, as it is often called, the "light fraction"—from the heavier inorganic material, which is called the "heavy fraction." Practically speaking, this involves the separation of paper products, plastic materials, and other light, organic materials from the shredded waste stream. The operation of various air classifiers and some of

TABLE 8-10 SOLID WASTES SEPARATION TECHNIQUES AND APPLICATION INFORMATION

Technique	Material involved	Preprocessing required	Remarks
Source Separation			
Handsorting	Paper, ferrous and nonferrous metals, wood	None	Used to separate corrugated and high-quality paper, metals, and wood at commercial and industrial sites and newspaper at residences; economically feasible if market prices are adequate.
Centralized Separation			
Handpicking and handsorting	Newspaper, corrugated paper	None	May be economical alternative to source separation, depending on labor costs.
Air separation	Combustible materials	Shredding	Used to concentrate metals and glass in a heavy fraction as well as combustible materials in a light fraction.
Inertial separation	Combustible materials	Shredding	Same as air separation.
Screening	Glass	None or shredding, air separation	May be used prior to shredding to remove glass and prior to air separation for similar reasons. May be used to concentrate glass-rich fraction from heavy fraction.
Flotation	Glass	Shredding, air separation	Water pollution control may be expensive.
Optical sorting	Glass	Shredding, air separation, and screening	May be used as an alternative to flotation to separate glass from opaque materials; also used to separate flint from colored cullet.
Electrostatic separation	Glass	Shredding, air separation, magnetic separation, and screening	Experimental.
Magnetic separation	Ferrous metal	Shredding or wet pulping	Proved in numerous full-scale applications.
Heavy media separation	Aluminum, other nonferrous metals	Shredding, air separation	May be used to separate a number of materials by adjusting specific gravity of media; separate units are required for each material to be separated.
Linear induction separation	Aluminum, other nonferrous metals	Shredding, air separation, magnetic separation, and screening	Separate units are required to separate aluminum and other nonferrous metals.

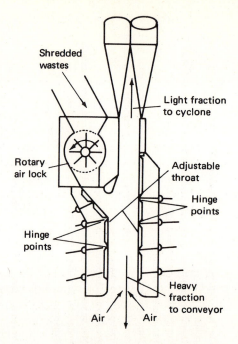

(a) Conventional chute type

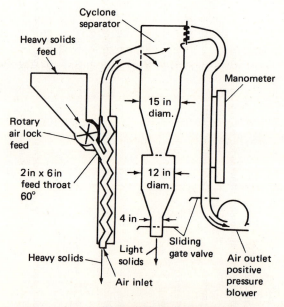

(b) Experimental zigzag type

FIG. 8-16 Typical air classifiers. (a) Chute type (Radar Pneumatics, Inc.) (b) Experimental zigzag type [2]. (c) Open inlet vibrator type. (Triple/S Dynamics System, Inc.)

the factors that must be considered in their selection are presented and discussed in this section.

Air Separation Equipment   In one of the simplest types of air classifiers, processed solid wastes are dropped into a vertical chute (see Fig. 8-16a). Air moving upward from the bottom of the chute is used to transport the lighter materials to the top of the chute. Because the upward airflow is insufficient to transport the heavier materials in the wastes, they drop to the bottom. Control of the percentage split between the light and heavy fractions is accomplished by varying the waste-loading rate, the airflow rate, and the cross-sectional area of the chute. A rotary air-lock feed mechanism is required to introduce the shredded wastes into the classifier.

Another type is known as the "zigzag air classifier." The small experimental unit, shown in Fig. 8-16b, consists of a continuous vertical column with internal zigzag deflectors through which air is drawn at a high rate. Shredded wastes are introduced at the top of the column at a controlled rate, and air is introduced at the bottom of the column. As the wastes drop onto the airstream, the lighter fraction is fluidized and moves upward and out of the column while the heavy fraction falls to the bottom. In theory, each change in direction caused by the zigzags creates turbulence in the airstream which, in turn, causes the wastes to tumble and allows bunched materials to be broken up [2]. Best separation is achieved through proper design of the separation chamber, airflow rate, and influent feed rate. Additional factors and relationships in the design of zigzag air classifiers are discussed in Ref. 2.

Still another type of air classifier is shown schematically in Fig. 8-16c. In this unit the separation of the light fraction is accomplished by the combination of three actions. The first is vibration, which helps to stratify the material fed to the separator into heavy and light components.

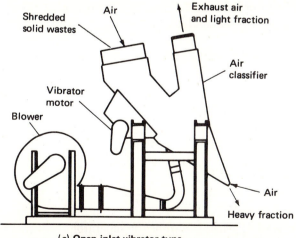

(c) Open inlet vibrator type

Agitation tends to settle the heavier (denser) particles to the bottom as the shredded wastes are conveyed down the length of the separator. The second action on the material is an inertial effect in which the air pulled in through the feed inlet imparts an initial acceleration to the lighter particles while the wastes travel down the separator as they are being agitated. The third action that completes the function of the separator is the injection of fluidizing air in two or more high-velocity, low-mass flow curtains across the bed. This fluidizing air changes the direction of the lighter particles and moves them into a position to be picked up and conveyed out of the unit by the exhaust air. The exhaust air volume is approximately 3 times that of the fluidizing air. For example, in an 80-ton/h municipal solid waste separator, the exhaust airflow would be approximately 60,000 ft³/min. Of this total, approximately 20,000 ft³/min is supplied by fluidizing air blowers built into the separator. The remainder of the air is pulled in through the feed inlet and the discharge chute for the heavy materials [10]. A final stripping of light particles is accomplished at the point where the heavy fraction discharges from the elutriator. It has been reported that the resulting separation is less sensitive to particle size than a conventional vertical air classifier, of either straight or zigzag design [10]. An advantage of the separator shown in Fig. 8-16c is that an air-lock feed mechanism is not required. Wastes are fed by gravity directly into the separator inlet.

The principal components of a complete air classification system are shown in Fig. 8-17. In addition to the air classifier, one or more conveyors are required to transport processed wastes to the loading hopper and into the air classifier. Following the air classifier, a cyclone separator is used to separate the light fraction from the conveying air. Before being discharged to the atmosphere, the conveying air is passed through dust collection facilities. Alternatively, the air from the cyclone separator can be recycled to the classifier with or without dust removal. Air for the operation of the air classifier can be supplied by low-pressure blowers or fans. The heavy fraction that is removed from the air classifier is hauled either to a disposal site or to a subsequent resource recovery system. The light fraction may be stored in bins or transported or conveyed to another shredder for further size reduction before storage or utilization as a fuel or compost material.

Selection of Air Separation Equipment   Factors that must be considered in the selection of air separation equipment include:

1. Characteristics of the material produced by shredding equipment including particle size, gradation, shape, moisture content, tendency to agglomerate, and fiber content
2. Material specifications for light fraction
3. Method of transferring wastes from the shredder to the air separation unit and feeding wastes into the air separator

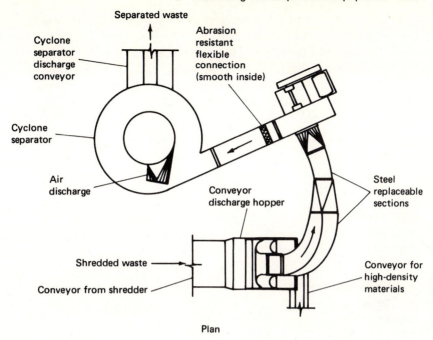

Plan

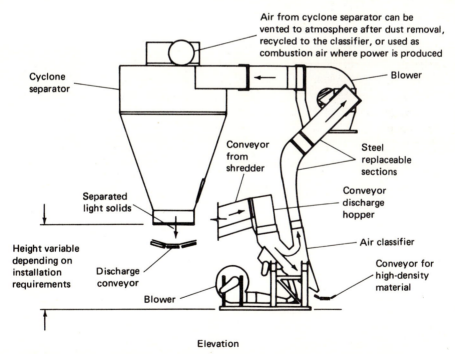

Elevation

FIG. 8-17 Typical air classification system for solid wastes. (Triple/S Dynamics Systems, Inc.)

4. Separator design characteristics including solids-to-air ratio (lb solids/ lb air); fluidizing velocities (ft/min); unit capacity (lb/h); total airflow (ft³/min); and pressure drop (in, of water)

5. Operational characteristics including energy requirements, routine and specialized maintenance requirements, simplicity of operation, proved performance and reliability, noise output, and air and water pollution control requirements

6. Site considerations including space and height, access, noise, and environmental limitations

Data on fluidizing velocities for various solid waste components are presented in Table 8-11. It should be noted that the data reported in Table 8-11 were derived using small pilot-scale equipment and that data derived from full-scale units would be expected to vary with the geometry of the separator, as well as with the loading rate. Based on work in other fields, it would appear that the solids-to-air ratio may prove to be the most

TABLE 8-11 FLUIDIZING VELOCITIES FOR AIR SEPARATION OF VARIOUS SOLID WASTE COMPONENTS*

	Velocity, ft/min	
Component	Zigzag classifier with 2-in throat†	Straight 6-in-diameter pipe
Plastic wrapping (shirt bags)	Less than 400 (electrostatic)	—
Dry, shredded newspaper (25 percent moisture)	400–500	350
Dry, cut newspaper		
1-in rounds	500	350
3-in squares	—	350
Agglomerates of dry, shredded newspaper and cardboard	600	—
Moist, shredded newspaper (35 percent moisture)	750	—
Dry, shredded corrugated cardboard	700–750	450–500
Dry, cut corrugated cardboard		
1-in rounds	980	700
3-in squares	—	1,000
Styrofoam, packing material	750–1,000 (electrostatic)	—
Foam rubber (½-in squares)	2,200	—
Ground glass, metal, and stone fragments (from automobile body trash)	2,500–3,000	—
Solid rubber (½-in squares)	3,500	—

* From Ref. 2.
† See Fig. 8-16b.
Note: ft/min × 0.0051 = m/s

significant design factor. It has been reported that this ratio may vary between 0.2 to 0.8 for light materials and may be as low as 0.02 for shredded paper [2].

Some time ago Dallavalle [5] proposed the following equations for estimating the minimum carrying velocities for pneumatic transport of particulate material in horizontal and vertical ducts. For horizontal ducts:

$$V = 6,000 \frac{S}{S + 1} d^{2/5} \tag{8-3}$$

For vertical ducts:

$$V = 13,300 \frac{S}{S + 1} d^{3/5} \tag{8-4}$$

where V = air velocity, ft/min
S = specific gravity of material being transported
d = diameter of longest particle to be moved, in

Equations 8-3 and 8-4 can be used to estimate the minimum required velocity based on the degree of carryover to be allowed in the light fraction. Typical air velocities necessary to carry various materials are reported in Table 8-12.

Magnetic Separation

The most common method of recovering ferrous scrap from shredded solid wastes involves the use of magnetic recovery systems. Ferrous materials are usually recovered either after shredding and before air classification or

TABLE 8-12 TYPICAL AIR VELOCITY IN DUCTS NECESSARY TO CONVEY VARIOUS MATERIALS*

Material	Air velocity, ft/min
Grain dust	2,000
Wood chips and shavings	3,000
Sawdust	2,000
Jute dust	2,000
Rubber dust	2,000
Lint	1,500
Metal dust (grinding)	2,200
Lead dust	5,000
Brass turnings (fine)	4,000
Fine coal	4,000

* From Ref. 5.
Note: ft/min × 0.3048 = m/min

after shredding and air classification. In some large installations overhead magnetic systems have also been used to recover ferrous materials before shredding (this operation is known as "scalping"). When wastes are mass-fired in municipal incinerators, magnetic separation is used to remove the ferrous materials from the incinerator residue. Magnetic recovery systems have also been used at landfill disposal sites. The specific location(s) where ferrous materials are recovered will depend on the objectives to be achieved, such as the reduction of wear and tear on processing and separation equipment, the degree of product purity to be achieved, and the required recovery efficiency.

Magnetic Separation Equipment   Over the years, various types of equipment have been used for the magnetic separation of ferrous materials. The most common types are the suspended magnet (see Fig. 8-18a), the magnetic pulley (see Fig. 8-18b), and the suspended magnetic drum (see Fig. 8-18c). Two of the magnetic separation systems used most commonly with shredded solid wastes are shown in Fig. 8-19.

In a typical multistage belt system designed to operate at the end of a conveyor (see Fig. 8-19a), three magnets are employed. The first magnet is used to attract the metal. The transfer magnet is used to convey the attracted material around a curve and to agitate it. When the attracted metal reaches the area where there is no magnetism, it falls away freely, and any nonferrous material trapped by the metal against the belt also falls. The metal is then pulled back to the belt by the final magnet and is discharged to another conveyor or into storage containers. To overcome some early problems with belt wear, a specially designed heavy-gauge stainless steel belt has been developed.

Suspended drum separators have been used in a number of large resource recovery installations. Where a single drum is installed at the end of a conveyor, the trajectory of the discharged shredded solid wastes is used to help separate loose, nonmagnetic materials and so improve the recovery of ferrous materials. To achieve the cleanest possible recovered material without secondary shredding or air classification, a two-drum installation such as shown in Fig. 8-19b may be used. The first magnetic drum is used to pick up ferrous material from the shredded waste and toss it forward to an intermediate conveyor. Most of the nonmagnetic material falls to a takeaway conveyor located below the primary separator. Because of the reduced burden on the intermediate conveyor, the second drum separator can be smaller and can be positioned closer to the conveyor. To ensure that no jamming or bridging occurs, the second drum rotates in a direction opposite to the flow of the material.

Selection of Magnetic Separation Equipment   Factors that must be considered in the selection of magnetic separation equipment include:

1. Location(s) where ferrous materials are to be recovered from solid wastes

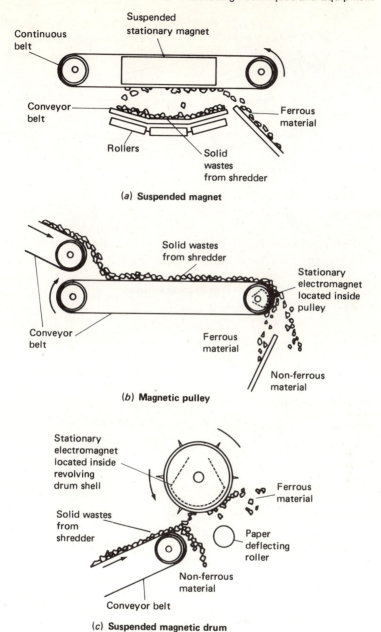

FIG. 8-18 Typical magnet separators. (Eriez Magnetics.)

2. Characteristics of the wastes from which ferrous materials are to be separated, such as the amount of ferrous material present in the waste stream, the degree of compaction, the tendency of the wastes to clump or stick to each other, size (large ferrous items should be reduced in size to about 8 in or smaller), and moisture content

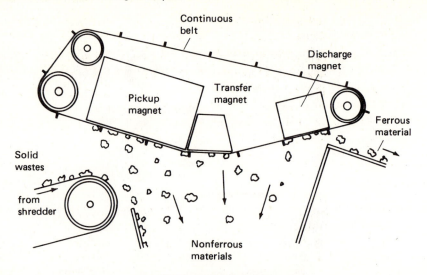

(a) **Belt type magnetic separator**

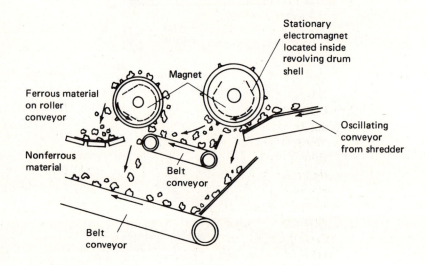

(b) **Two-drum magnetic separator**

FIG. 8-19 Typical magnetic separation systems used with shredded solid wastes. (a) (Dings Company.) (b) (Eriez Magnetics.)

3.   Equipment to be used for feeding wastes to separator and for removing the separated waste streams

4.   Separator-system engineering design characteristics including loading rates for a given size of separator (lb/h), separation efficiency, drum rotational speed (r/min), magnet strength, type of magnet cooling

system (oil or air), conveyor speed, airflow rates if used to improve efficiency, and materials of construction

5. Operational characteristics such as energy requirements, routine and specialized maintenance requirements, simplicity of operation, proved performance and reliability, noise output, and air and water pollution control requirements

6. Site considerations such as space and height, access, noise, and environmental limitations

Screening

Screening involves the separation of a mixture of materials of different sizes into two or more portions by means of one or more screening surfaces, which are used as go or no-go gages [14]. Screening may be accomplished either wet or dry, with the latter being most common in solid waste processing systems. Screening has a number of applications in solid waste resource and energy recovery systems. Screens have been used before and after shredding and after air classification in various applications dealing with both the light and heavy fraction materials. The types of screens that are used, some of the typical applications, equipment selection, and performance evaluation are discussed below.

Screening Equipment   To date, the most common types of screens used for the separation of solid waste components are the vibrating (see Fig. 8-20a) and the rotary drum screens (see Figs. 8-20b and 8-21). Typically they have been used for the removal of glass and related materials from shredded solid wastes. However, their application potential is only now becoming understood more fully. Because the specification of large rotary drum screens appears to be on the increase, two typical applications have been selected for discussion. Full-scale data are not yet available on either of the following installations.

A large rotary screen 10 ft in diameter and 45 ft in length is to be installed in the NCRR Resource Recovery Facility being built near New Orleans which is to be operational in 1976. The screen, similar to the one shown in Fig. 8-21, is equipped with $4\frac{3}{4}$-in round holes. Unshredded solid wastes, as received, are fed to the screen. It is anticipated that 40 percent of the material will pass through the holes. This fine fraction, which will contain a large fraction of the aluminum cans and most of the glass, will be passed around the primary shredder and will go directly to an air classifier. The oversize fraction (approximately 60 percent of the feed) goes to the primary shredder and then to another air classifier where the light fraction is separated. The light fraction from both of these classifiers is to be taken to a landfill. The heavy fraction is taken to a materials recovery facility [11]. The New Orleans resource recovery operation is considered further in Chap. 16.

(a)

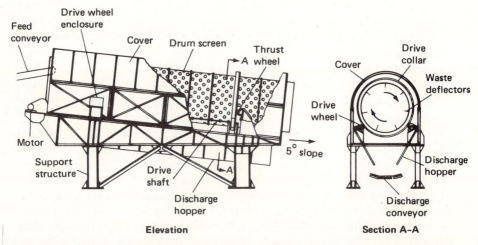

(b) Rotary drum

FIG. 8-20 Typical screens used for the separation of solid wastes. (a) Vibrating screen. (Universal Vibrating Screen Company.) (b) Rotary drum screen. (Triple/S Dynamics Systems, Inc.)

FIG. 8-21 Internal view of large rotary drum screen in operation.
(Triple/S Dynamics Systems, Inc.)

In the Milwaukee resource recovery facility, scheduled to be operational in 1976, a two-drum unit is to be installed. The first 12 ft of the inner drum has 1½-in-diameter openings, and the last 8 ft of the inner drum has 4-in-diameter openings. The outer drum, which is 12 ft long, has ⅜-in-diameter openings. The heavy fraction from an air classifier is to be fed to the unit, and four different size fractions are to be produced. The +4-in material is to be disposed of in a landfill. Aluminum is to be separated from the −4-in and the +1½-in material. The −1½-in and the +⅜-in material goes to a resource recovery facility for the removal of glass. The −⅜-in material is to be disposed of in a landfill [11].

A wire screen developed by a private contractor to separate cardboard from other wastes is shown in Fig. 8-22.

Selection of Screening Equipment   Factors that must be considered in the selection of screening equipment include:

1.   Materials specifications for screened components

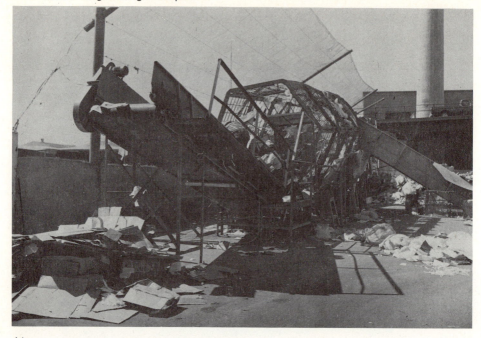

(a)

(b)

FIG. 8-22 Rotary drum wire screen used for the separation of cardboard. (Screen separator was developed by Sacramento Waste Disposal Company.) (a) Overall view of cardboard drum screen separator. Note separated cardboard falling from discharge conveyor in foreground. Feed conveyor is located in right background. (b) End view of cardboard separator in operation showing internal metal rods used to break open plastic bags and cardboard boxes. Separated paper, plastics, and undersized cardboard fall through openings in screen as it slowly rotates.

2. Location where screening is to be applied and characteristics of the waste material to be screened, including particle size, shape, bulk density, and moisture content; particle size distribution; tendency for the waste material to clump or stick together and its rheological properties

3. Screen design characteristics including materials of construction, size of screen openings (usually in inches); configuration of screen openings; total surface screening area (ft²); oscillation rate for vibrating screens (times/min); rotational speed for rotary drum screens (r/min); and loading rates (lb waste/ft²/h), and length (ft)

4. Separation efficiency and overall effectiveness (see following discussion)

5. Operational characteristics including energy requirements, routine and specialized maintenance requirements, simplicity of operation, proved performance and reliability, noise output, and air and water pollution control requirements

6. Site considerations such as space and height, access, noise and environmental limitations

The efficiency of a screen can be evaluated in terms of the percentage recovery of the material in the feed stream by using the following expression [9].

$$\text{Recovery } (\%) = \frac{Uw_u}{Fw_f} (100) \qquad (8\text{-}5)$$

where U = weight of material passing through screen (underflow), lb/h
   F = weight of material fed to screen, lb/h
   w_u = weight fraction of material of desired size in underflow
   w_f = weight fraction of material of desired size in feed

The effectiveness of a screening operation has been defined as [9]:

Effectiveness = recovery × rejection

where   rejection = 1 − recovery of undesired material

$$= 1 - \frac{U(1 - w_u)}{F(1 - w_f)}$$

Using Eq. 8-5 and the above definition for rejection, we can find the screen effectiveness by the following expression:

$$\text{Effectiveness} = \frac{Uw_u}{Fw_u} \left[1 - \frac{U(1 - w_u)}{F(1 - w_f)} \right] \qquad (8\text{-}6)$$

Equations 8-5 and 8-6 can also be used to determine the percentage recovery and effectiveness of any processing operation in which individual components are to be recovered from a waste stream whether liquid or solid. The application of Eqs. 8-5 and 8-6 is illustrated in Example 8-3.

EXAMPLE 8-3 *Determination of Screen Recovery Efficiency and Effectiveness*

Given that 100 tons/h of municipal solid waste with the composition shown in Table 4-2 are applied to a rotary screen for the removal of glass prior to shredding, determine the recovery efficiency and effectiveness of the screen based on the following experimental data.

1. Weight of underflow = 10 tons/h
2. Weight of glass in screen underflow = 7.2 tons/h

Solution

1. Determine the weight fraction of glass in the feed to the screen. From Table 4-2 the percentage of glass is given as 8 percent. Thus, the mass fraction of glass in the feed is

$$w_f = \frac{\text{weight of glass}}{\text{total weight of sample}}$$

$$= \frac{100 \text{ lb} \times 0.08}{100 \text{ lb}}$$

$$= 0.08$$

2. Determine the weight fraction of glass in screen underflow.

$$w_u = \frac{\text{weight of glass}}{\text{total weight of underflow}}$$

$$= \frac{7.2 \text{ tons/h}}{10 \text{ tons/h}}$$

$$= 0.72$$

3. Determine the screen recovery efficiency using Eq. 8-5.

$$\text{Recovery (\%)} = \frac{Uw_u}{Fw_f} (100)$$

$$= \frac{10 \times 0.72}{100 \times 0.08} (100)$$

$$= 90 \text{ percent}$$

4. Determine the effectiveness of the screen using Eq. 8-6.

$$\text{Effectiveness} = \frac{Uw_u}{Fw_f} \left[1 - \frac{U(1 - w_u)}{F(1 - w_f)} \right]$$

$$= \frac{(10 \text{ tons/h})(0.72)}{(100 \text{ tons/h})(0.08)}$$

$$\times \left[1 - \frac{(10 \text{ tons/h})(1 - 0.72)}{(100 \text{ tons/h})(1 - 0.08)} \right]$$

$$= 0.87$$

Other Separation Techniques

Because less is known about the separation techniques to be considered in this section, the following material is meant only to serve as an introduction to the techniques. Specific details must be obtained, as they become available, from the records of full-scale installation, equipment manufacturers, and the literature.

Inertial Separation   Inertial methods rely on ballistic or gravity separation principles to separate shredded solid wastes into light (organic) and heavy (inorganic) particles. The modes of operation of three different types of inertial separators are shown schematically in Fig. 8-23. This type of equipment is used extensively in Europe.

Flotation   In the flotation process, glass-rich feedstock, produced by screening the heavy fraction of the air-classified wastes after ferrous metal

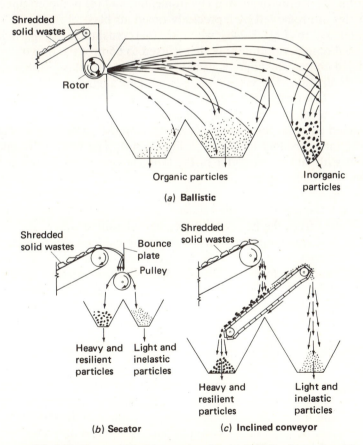

FIG. 8-23 Types of inertial separators [7]: (a) Ballistic; (b) secator; (c) inclined conveyor.

separation, is immersed in water in a suitable tank. Glass chips, rocks, bricks, bones, and dense plastic materials that sink to the bottom are removed with belt scrapers for further processing. Light organic and other materials that float are skimmed from the surface. These materials may be hauled to a landfill for disposal or returned to the head end of the plant and passed through the operation with a new batch of solid wastes. Chemical additives have also been used to improve the capture of light organic and fine inorganic materials.

Optical Sorting   Sorting of glass from opaque particles such as stones, ceramics, bottle caps, and corks can be accomplished optically by idenification of the transparent properties of glass. Optical color sorting can be used to separate flint glass from mixed colored glass. Mixed colored glass can also be sorted into amber and green products. A typical optical sorter is shown in Fig. 8-24. Functionally, four basic operations are involved: (1) particles are fed mechanically, (2) particles are inspected optically, (3) inspection results are evaluated electronically, and (4) predetermined types of particles are removed by a precisely timed air blast.

Referring to Fig. 8-24a, we note that crushed glass particles are hopper-fed onto a vibrating tray that is used to control the rate of feed into an inclined chute. The chute is used to direct the particles to the inspection unit for evaluation. The inspection unit contains a light source and a sensor that is used to examine the free-falling particles. When a rejectable particle is detected, a signal is produced that electronically triggers a timed blast of compressed air from the ejector nozzle causing the particle to be deflected from the main product stream (see Fig. 8-24a). The degree of separation achieved with an optical sorter of the type shown in Fig. 8-24 is usually a function of the feed rate.

Electrostatic Separation   High-voltage electrostatic fields can be used to separate glass from the heavy fraction of air-classified wastes that are free of ferrous and aluminum scrap in the following manner. A vibrating feeder meters feedstock to a negatively charged rotating drum, and a positive electrode near the drum and the feeder induces a charge in the small particles. Nonconductors, such as glass and clay, retain the charge; metals and crystalline materials, such as rock, lose it rapidly. The drum holds the nonconductors, and the remaining material drops off [7].

Heavy Media Separation   Although the removal of aluminum can be accomplished in a number of different ways, heavy media separation is perhaps the process for which the greatest operating experience exists, principally in the automobile recovery industry [7]. In this process a shredded feedstock that is rich in aluminum, such as air-classified solid wastes after ferrous metal and glass have been removed, is dumped into a liquid stream which has a high specific gravity. The specific gravity is maintained at a

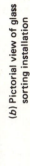

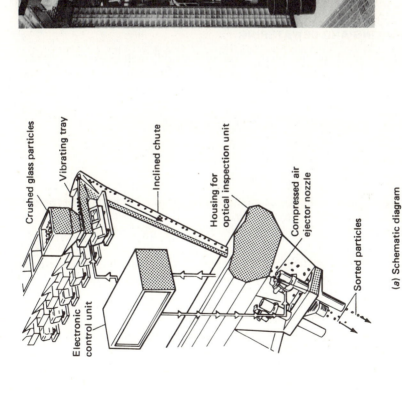

(b) Pictorial view of glass sorting installation

(b) Pictorial view of glass sorting installation. (Sortex Company of North America, Inc.)

Crushed glass particles

Vibrating tray

Inclined chute

Housing for optical inspection unit

Compressed air ejector nozzle

Sorted particles

Electronic control unit

(a) Schematic diagram

FIG. 8-24 Glass sorting unit. (a) Schematic diagram. (b) Pictorial view of glass sorting installation. (Sortex Company of North America, Inc.)

level that will permit the aluminum to float and other materials to remain submerged [7]. At present, the major disadvantage of this process is that the optimum size plant requires about 2,000 to 3,000 tons/day of feedstock.

Linear Induction Separation   A new type of magnet appears to offer great promise for removing aluminum from relatively small quantities (250 tons/day and more) of municipal wastes. There are a few prototype systems operating, and at least one commercial-size system was scheduled to begin operating at a municipal power plant some time in 1976.

Design of these systems is based on fundamental electrical principles. To illustrate: When a moving magnetic field passes through a nonmagnetic metallic conductor, the field induces eddy currents in the metal. This phenomenon is used to drive a rotary induction motor. If a linear induction motor stator—which can be considered a rotary induction motor stator that has been cut and straightened—is placed below a nonmagnetic moving belt, it can create the field necessary to drive nonmagnetic conductors from the belt [7].

8-6   DRYING AND DEWATERING

In many solid waste energy recovery and incineration systems, the shredded light fraction is predried to decrease weight by removing varying amounts of moisture, depending on process requirements. When sludge from waste-water treatment plants is to be incinerated or used as a fuel mixture, some form of dewatering is required.

Drying

Over the years a wide variety of dryer designs have evolved. Before considering any of these designs, however, it may be helpful to review how heat can be applied to the material to be dried. Typically, this is accomplished by one or more of the following methods:

1. Convection, in which the heating medium, usually air or the products of combustion, is in direct contact with the wet material
2. Conduction, in which the heat is transmitted indirectly by contact of the wet material with a heated surface
3. Radiation, in which heat is transmitted directly and solely from the heated body to the wet material by the radiation of heat

Convection Dryers   Of these methods, convection is the one most commonly used for industrial drying. The characteristics of the principal convection dryers are reported in Table 8-13. Because the rotary drum has been used effectively for drying solid wastes, this type of dryer is considered further in the following discussion.

TABLE 8-13   OPERATIONAL CHARACTERISTICS OF CONTINUOUS-CONVECTION DRYERS*

Type of dryer	Method of operation
Rotary tray of hearths	Material to be dried is spread on the top tray of a series of stacked trays and raked to lower trays as it dries.
Endless belt	Material to be dried is spread at the feed end of the dryer on a continuous perforated wire-mesh belt or on conveyor bands which are used to move the material through the dryer. Airflow is usually countercurrent.
Rotary drum	Slow-rotating cylindrical shell, slightly inclined from the horizontal, is provided with a means for continuously feeding material to be dried. The drying medium can be introduced and made to flow either concurrently or countercurrently with the material to be dried.
Fluid bed	Material to be dried is maintained in a fluidized condition. Fluid-bed dryers are usually in the form of vertical cylindrical columns.
Spray	Material to be dried is sprayed into a drying chamber. Movement of feedstock and the drying medium can be cocurrent, countercurrent, or combinations of the two.
Flash	Material to be dried is entrained in the drying medium and is conveyed in the process of drying.

* Adapted in part from Ref. 20.

In its simplest form, a rotary drum dryer is composed of a rotating cylinder, slightly inclined from the horizontal, through which the material to be dried and the drying gas are passed simultaneously (see Fig. 8-25). As the drum rotates, the material to be dried is transported continuously from one end to the other by the lifting action of the internal flights. As the material falls off the flights, it is also broken up so that better drying can be achieved.

The drying of material in a direct rotary dryer is thought to occur in the following stages [20]:

1.   Heating the wet material and its moisture content to the constant-rate drying temperature, which will be approximately that of the wet-bulb temperature of the drying medium

2.   Drying of the material substantially at this temperature

3.   Heating of the material to its discharge temperature and evaporation of the moisture remaining at the end of the stage

Typically, the retention time in the drum varies from 30 to 45 min. An adjustable discharge valve can be used to control the drying time or the retention time of the material or the drying medium. The discharge end of the dryer is fitted with a housing which has an exhaust vent that is used to

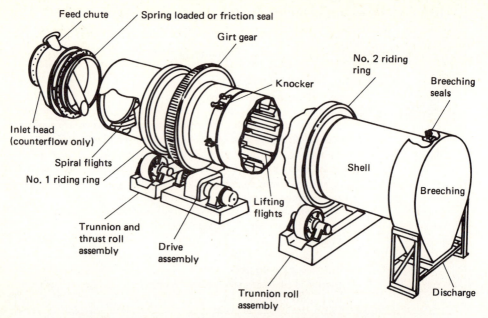

FIG. 8-25   Countercurrent direct-heat rotary drum dryer. (Bartlett-Snow.)

pass the vapor-laden gases into a dust extractor and an air control device before they are discharged to the atmosphere. The dried material drops out of the bottom. Details on other types of drum dryers may be found in Refs. 12 and 14.

Selection of Drying Equipment   Factors that must be considered in the selection of drying equipment include:

1. Properties of the material to be dried as it is fed to and delivered from the dryer
2. Drying characteristics of the material including initial moisture content, type of moisture (whether bound, unbound, or both), maximum material temperature, and anticipated drying time
3. Specification of final product, including moisture content
4. Nature of operation, whether continuous or intermittent
5. Operational characteristics including energy requirements, routine and specialized maintenance requirements, simplicity of operation, proved performance and reliability, noise output, and air and water pollution control requirements
6. Site considerations including space and height, access, noise and environmental limitations

Although the energy requirements for drying wastes vary with local conditions, the required energy input can be estimated by using a value of about 1,850 Btu/lb of water evaporated. The necessary computations to determine the amount of water that must be removed and the required energy input are illustrated in Example 8-4.

EXAMPLE 8-4 *Moisture Content Analysis and Energy Requirements for Drying*

Determine the pounds of water that must be removed per ton of shredded, air-classified solid wastes if the initial moisture content is 25 percent and the final moisture content after drying is to be 10 percent. How much energy will be required to accomplish this?

Solution

1. Determine the pounds of moisture initially present in the shredded solid wastes by using Eq. 4-1.

$$25 = \frac{(a - b)}{2,000 \text{ lb}} (100)$$

$$a - b = W_s = 500 \text{ lb of water initially}$$
$$\text{present in sample}$$

2. Determine the pounds of moisture that must be present in the shredded wastes after drying if the moisture content is to be 10 percent.

$$10 = \frac{W_s}{1,500 + W_s} (100)$$

$$W_s \doteq 167 \text{ lb of water in sample at 10}$$
$$\text{percent moisture content}$$

3. Determine the amount of water to be removed from each ton of solid wastes fed to the dryer.

$$(500 - 167) \text{ lb} = 333 \text{ lb/ton}$$

4. Determine the energy required for drying assuming that 1,850 Btu must be supplied per pound of water removed.

$$\text{Energy required} = 333 \text{ lb/ton } (1,850 \text{ Btu/lb})$$

$$= 616,050 \text{ Btu/ton } (715 \text{ kJ/kg})$$

Dewatering

The problem of sludge disposal from municipal waste-water treatment plants has become critical for many large communities in which the use of drying beds, lagoons, or land spreading is no longer practical or economically feasible. In most cases some form of sludge dewatering has been adopted to reduce the liquid volume. Once dewatered, the sludge can be mixed with other solid wastes. The resulting mixture can be (1) incinerated to reduce the volume, (2) used for the production of recoverable by-products, (3) used for the production of compost, or (4) buried in a landfill. Centrifugation and filtration are the two general methods now used most commonly to dewater treatment plant sludge.

Centrifugation   Bowl, decanting, and horizontal centrifuges have been used to dewater sludge. Although it is possible to produce a reasonably thick sludge (10 to 15 percent) by means of centrifugation, a number of problems have been encountered. The two most critical are (1) the high operation and maintenance costs associated with these units, and (2) the carryover of fines in the filtrate. Details on the application of these units for the dewatering of sludge may be found in Refs. 12 and 13.

Filtration   Both vacuum and pressure filtration have been used to dewater sludge. Vacuum filtration is used more commonly in the United States, whereas pressure filtration is used more commonly in Europe and England. Details on the application of this type of equipment may be found in Refs. 12 and 13.

8-7   DISCUSSION TOPICS AND PROBLEMS

8-1.   Select eight different processing techniques for solid wastes. List their uses, advantages, and disadvantages in outline form.

8-2.   What is the difference between compaction and consolidation? What effect will consolidation have in baled material that has a density of 1,800 lb/yd³?

8-3.   Assuming that the asymptotic value for the curves given in Fig. 8-5 is 1,800 lb/yd³, derived empirical equations to describe the degree of compaction that can be achieved as a function of the applied pressure, starting with an initial solid waste density of 200 and 600 lb/yd³. (Try a rectangular hyperbola.)

8-4.   Using the data given in Fig. 8-5, prepare a plot of the volume reduction in percent versus the applied pressures. How can the use of such a figure be misleading?

8-5.   The maximum amount of solid wastes collected per day for 1 wk is presented below. All the solid wastes are to be burned at a municipal incinerator at a burning rate of 100 tons/day. What is the required capacity of the storage pit that should be designed to accommodate 1.15 times the required capacity?

Day	Solid wastes to be burned, tons/day
Monday	150
Tuesday	130
Wednesday	120
Thursday	120
Friday	100
Saturday	80
Sunday	0

8-6. Estimate the composition of the residue if packaging material wastes with the component distribution reported in Table 4-2 are to be incinerated. What would be the corresponding volume reduction?

8-7. Assume that the energy consumption required for size reduction for solid wastes can be estimated according to the following first-order equation.

$$E = C \ln \frac{l_1}{l_2} \qquad \text{(known as Kick's Law [20])}$$

where E = energy consumption rate, hp·h/ton
C = constant, hp·h/ton
l_1 = initial size
l_2 = final size

If it is found that 10 hp·h/ton is required to reduce the size of the solid wastes from about 6 to 2 in, estimate the energy required to reduce the average size of solid wastes from about 12 to 2 in at a loading rate of about 10 tons/h.

8-8. If an air velocity of 2,000 ft/min is to be used to transport finely ground material with a specific gravity of 0.75 in a horizontal duct, estimate the maximum particle size that can be transported.

8-9. Assuming that an air-to-solids ratio (lb solids/lb air) of 0.6 is required for the separation of the light fraction of shredded solid wastes and that the head loss in a separation column is equal to 4 in of water, estimate the required blower horsepower to separate 50 tons/h of shredded solid wastes. Assume that the weight of air is 0.0750 lb/ft³ and that the following equation can be used to compute the blower horsepower.

$$\text{BHP} = 0.227Q[((14.7 + p)/14.7)^{0.283} - 1]$$

where BHP = blower horsepower, hp
Q = airflow rate, ft³/min
p = pressure drop, lb/in²

8-10. Given that the cost of shredding increases with decreasing particle size and that the cost of air classification increases with increasing particle size, discuss how you would go about determining the tradeoff between cost of the size-reduction and air classification facilities to be used in a solid wastes processing plant. What are the important factors that must be considered?

8-8 REFERENCES

1. American Public Works Association: "Municipal Refuse Disposal," 3d ed., Public Administration Service, Chicago, 1970.

2. Boettcher, R. A.: *Air Classification of Solid Wastes,* U.S. Environmental Protection Agency, Solid Waste Management Program, Publication SW-30c, Washington, D.C., 1972.

3. *Compactor Handbook,* Solid Wastes Management Magazine, New York, 1973.

4. Corey, R. C. (ed.): "Principles and Practices of Incineration," Wiley-Interscience, New York, 1969.

5. Dallavale, J. M.: "The Industrial Environment and Its Control," Pitman, New York, 1958.

6. DeMarco, J., et al.: *Incinerator Guidelines—1969,* Public Health Service, U.S. Department of Health, Education, and Welfare, Washington, D.C., 1969.

7. Drobny, N. L., H. E. Hull, and R. F. Testin: *Recovery and Utilization of Municipal Solid Waste,* U.S. Public Health Service, Publication 1908, Washington, D.C., 1971.

8.　Engdahl, R. B.: *Solid Waste Processing: A State-of-the-Art Report on Unit Operations and Processes,* Bureau of Solid Waste Management, U.S. Department of Health, Education, and Welfare, Publication SW-4c, Washington, D.C., 1969.

9.　Foust, A. S., et al.: "Principles of Unit Operations," Wiley, New York, 1960.

10.　Hill, R. M.: *Effective Separation of Shredded Municipal Solid Wastes by Elutriation,* paper presented at the 78th Annual Meeting of the American Institute of Chemical Engineers, Salt Lake City, Utah, 1974.

11.　Hill, R. M.: Personal communication, 1976.

12.　McCabe, W. L. and J. C. Smith: "Unit Operations of Chemical Engineering," 3d ed., McGraw-Hill, New York, 1976.

13.　Metcalf & Eddy, Inc.: "Wastewater Engineering: Collection, Treatment, Disposal," McGraw-Hill, New York, 1972.

14.　Perry, R. H., C. H. Chilton, and S. D. Kirkpatrick: "Perry's Chemical Engineers' Handbook," 4th ed., McGraw-Hill, New York, 1963.

15.　Reinhardt, J. J. and R. K. Hamm: *Solid Waste Milling and Disposal on Land without Cover,* U.S. Environmental Protection Agency, NTIS Publication PB-234930, Springfield, Va., 1974.

16.　Ross, R. D. (ed.): "Industrial Waste Disposal," Reinhold, New York, 1968.

17.　Schwieger, R. G.: Power from Waste, *Power,* vol. 119, no. 2, 1975.

18.　Stear, J. R.: *Municipal Incineration: A Review of Literature,* U.S. Environmental Protection Agency, Office of Air Programs, Publication AP-79, Washington, D.C., 1971.

19.　Trinks, W. and M. H. Mawhinney: "Industrial Furnaces," 5th ed., vol. 1, Wiley, New York, 1953.

20.　Walker, W. H., et al.: "Principles of Chemical Engineering," 3d ed., McGraw-Hill, New York, 1937.

21.　Williams-Gardner, A.: "Industrial Drying," CRC, International Scientific Series, Cleveland, Ohio, 1971.

9

Recovery of Resources, Conversion Products, and Energy

Solid wastes or selected components of solid wastes, depending on local conditions, may be of value as a source of raw material for industry, fuel for the production of power, and material that can be used for the reclamation of land. The evaluation of these alternatives from a management standpoint is considered in Chap. 16. The primary purpose of this chapter is to delineate the application of the techniques and equipment discussed in Chap. 8 for the recovery of resources, conversion products, and energy. A secondary purpose is to introduce some of the design aspects involved in the implementation of processing systems. A third purpose is to present some of the commercial flowsheets that have been developed for the recovery of materials and energy from solid wastes.

The information is presented in the following sections: (1) materials processing and recovery systems, (2) recovery of chemical conversion products, (3) recovery of biological conversion products, (4) recovery of energy from conversion products, and (5) materials and energy recovery system flowsheets. The sequence for the presentation of this information is indicated in Fig. 9-1, which is a typical flowsheet for the recovery of resources, conversion products, and energy from solid wastes. The term *front-end system* denotes the processes (size reduction, separation, etc.)

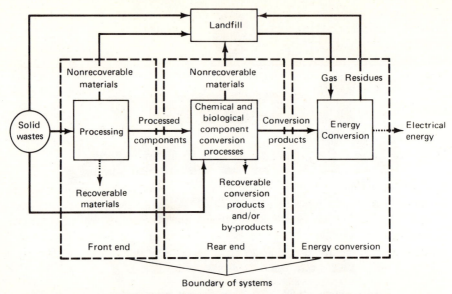

FIG. 9-1   Typical flowsheet for recovery of resources, conversion products, and energy from solid wastes.

used for the recovery of materials and the preparation of individual components for subsequent conversion. The term *rear-end system* denotes the chemical and biological processes (incineration with heat recovery, composting, etc.) and related ancillary facilities used for the conversion of processed solid wastes into various products.

9-1   MATERIALS PROCESSING AND RECOVERY SYSTEMS

Various types of processing techniques and equipment were discussed in Chap. 8. In this section the objective is to show how the individual processes can be combined in alternative flowsheets for the recovery of materials and the preparation of combustible wastes for subsequent processing.

Materials Specifications

Paper, rubber, plastics, textiles, glass, ferrous metals, and organic and inorganic materials are the principal recoverable materials contained in municipal solid wastes. In any given situation, the decision to recover any of or all these materials is usually based on an economic evaluation and on local considerations, which are discussed in Chap. 16. In assessing the economics of materials recovery, the materials specifications will be a critical consideration. The reason is that even though it may be possible to separate the various components, there may be no market for them if they do not meet necessary specifications. Typical materials specifications are presented in Table 9-1.

Raw Materials Typical specifications for eight different materials derived from municipal wastes are reported in Table 9-1. Specific details, such as product purity, density, and shipping conditions, must be worked out with each potential buyer. Whenever possible, it is also beneficial to develop a range of product specifications and product prices. In this way, processing costs to achieve higher product quality can be evaluated with respect to the higher market price obtainable for the higher-quality product.

Fuel Source Energy can be derived from municipal wastes in two forms: by direct use of heat produced by burning and by the conversion of wastes to fuel (oil, gas, pellets) that can be stored and used locally or transported to distant energy markets. Specifications for direct use of wastes for the production of steam are usually not as restrictive as those for the

TABLE 9-1 TYPICAL MATERIALS SPECIFICATIONS THAT AFFECT SELECTION AND DESIGN OF PROCESSING OPERATIONS

Reuse category and materials component	Typical specification items
Raw Material	
Paper and cardboard	Source; grade; no magazines; no adhesives; quantity, storage, and delivery point
Rubber	Recapping standards; specifications for other uses not well defined
Plastics	Type (e.g., ABS, PVC); degree of cleanliness
Textiles	Type of material; degree of cleanliness
Glass	Amount of cullet material; color, no labels or metal; degree of cleanliness; freedom from metallic contamination; quantity, storage, and delivery point
Ferrous metals	Source (domestic, industrial, etc.); density; degree of cleanliness; degree of contamination with tin, aluminum, and lead; quantity, shipment means, and delivery point
Aluminum	Particle size; degree of cleanliness; density; quantity, shipment means, and delivery point
Nonferrous metals	Vary with local needs and markets
Fuel Source	
Combustible organics	Composition, Btu content; moisture content; storage limits; firm quantities; sale and distribution of energy and/or by-products
Wastepaper	Vary with local needs and markets
Land Reclamation	
Organics	Local and state regulations; method of application; control of methane gas migration; leachate control; final land-use designation
Inorganics	Local and state regulations; final land-use designation

production of fuel. However, as firing and storage techniques improve, specifications for direct use may become more stringent.

Land Reclamation   Applying wastes to land is one of the oldest and most used techniques in solid waste management. Land disposal technology has developed to the point that communities can now plan land reclamation projects that use solid wastes without fear of the development of health problems. Some typical land reclamation specifications are listed in Table 9-1. Organic wastes used for landfilling or land reclamation require greater control than inorganic wastes. Land reclamation using either or both types of wastes should not be started until a final land use has been designated.

Processing and Recovery Systems

Once a decision has been made to recover materials and/or energy, process flowsheets must be developed for the removal of the desired components and for processing and combustible materials, subject to predetermined materials specifications.

Process Flowsheets   Two proposed flowsheets for front-end systems for the recovery of specific components and the preparation of combustible materials for use as a fuel source are presented pictorially in Fig. 9-2 and schematically in Fig. 9-3 [3]. In both cases a dry-process flowsheet has been adopted. The principal advantage of dry over wet processing is lower cost. The wet process involves the use of a hydropulper. Another

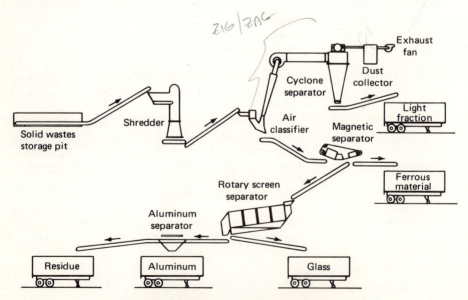

FIG. 9-2   Pictorial flowsheet for materials recovery systems.

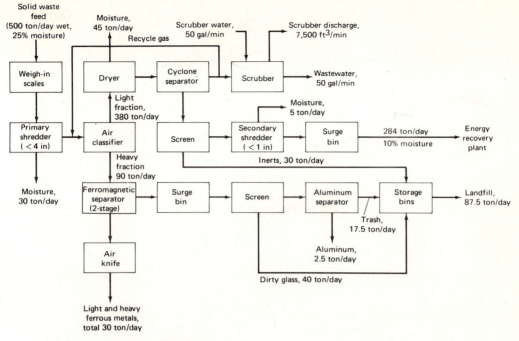

FIG. 9-3 Flowsheet for materials recovery system. (Central Contra Costa Sanitary District and Brown and Caldwell Consulting Engineers [3].)

advantage is that standard equipment used in the mineral processing industries can be adapted for use in this application.

In both flowsheets air classification follows primary shredding, and cyclone separators remove the air from the light fraction. In addition, the flowsheet shown in Fig. 9-3 includes the following: (1) a dryer before the air separator used to meet moisture content specifications for the light fraction, (2) a screening step after the air separator to remove some of the heavier components from the light fraction, and (3) a second step of shredding. The processed light fraction resulting from the flowsheet in Fig. 9-3 would be suitable for direct suspension firing in a steam boiler. Ferrous metal, glass, and aluminum are removed from the heavy fraction in both flowsheets. It is also important to note that there is a residue that must be disposed of in both flowsheets.

Both flowsheets are flexible in terms of adding additional equipment or alternative processing options to meet varying materials specifications. From a review of Figs. 9-2 and 9-3, it is evident that a considerable variety of flowsheets can be prepared. Flowsheets involving manual separation of specific waste components are also commonly used.

System Design and Layout The design and layout of the physical facilities that make up the processing plant flowsheet are an important aspect in the

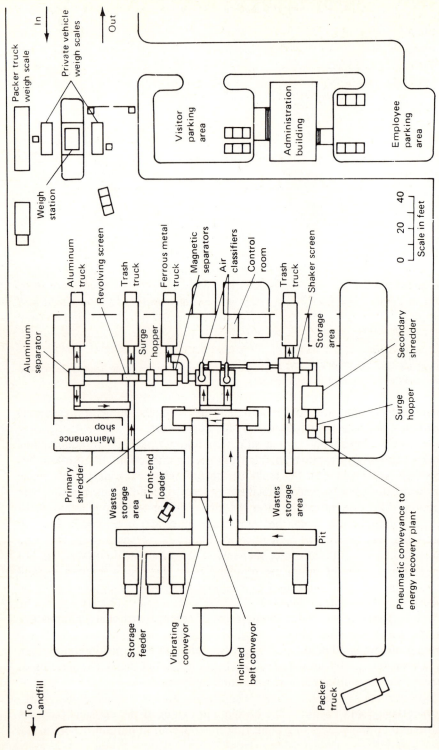

FIG. 9-4 Layout of resource recovery plant. (Central Contra Costa Sanitary District and Brown and Caldwell Consulting Engineers [3].)

implementation and successful operation of such systems. The layout recommended for the system shown in Fig. 9-3 is given in Fig. 9-4. Important factors that must be considered in the design and layout of such systems include (1) process performance efficiency, (2) reliability and flexibility, (3) ease and economy of operation, (4) aesthetics, and (5) environmental controls [3].

Of these factors, the first is the most important in terms of obtaining a plant that will function properly. Although there are a number of ways to evaluate process performance, perhaps the best one relates to the degree of separation achieved for the various components. To achieve optimum performance with respect to component separation, careful planning will ensure that design loading for the various processes is not exceeded. Because solid wastes usually are collected in the morning, adequate storage capacity must be provided at a processing facility so that the feed rate to the process will be uniform and not subject to surges.

The first step in the design of a processing facility will be to decide on the quantity of material to be processed. Where processed wastes are to be used as a fuel, the design quantities will usually depend on the amount of continuous (firm) power that must be developed. Once this has been decided, the individual units are sized according to the loading rates which are determined on the basis of the characteristics of the solid wastes and the separation process to be used.

Materials Balances and Loading Rates   An important aspect in the design of any materials recovery system will involve estimating the quantities of materials that can be recovered and the appropriate design loading rates. Data and information that can be used to estimate the required quantities are presented in Tables 9-2 and 9-3. The components that normally make up the light and heavy fractions after shredding and air classification are identified in Table 9-2. Actual data in this table are from Table 4-2.

It should be noted that the moisture content that may be lost during shredding has not been considered in Table 9-2. The typical moisture content of solid wastes varies from 15 to 40 percent, depending on the geographic location and the season of the year. In the southwestern United States, the average moisture content value is about 25 percent. Under normal circumstances, from 5 to 25 percent of the initial moisture may be lost during shredding. If test data are not available, a value of 15 percent can be used for estimating this loss.

Recoverable quantities of ferrous metals, glass, and aluminum, along with information on the recovery of heavy materials from the light fraction, are reported in Table 9-3. The use of this information in the preparation of a materials balance for a recovery process is illustrated in Example 9-1.

TABLE 9-2   LIGHT AND HEAVY FRACTIONS OF SOLID WASTE COMPONENTS AFTER SHREDDING AND AIR CLASSIFICATION

Component	Percent by weight*	Fraction by weight, percent		Comment
		Light	Heavy	
Food wastes	15	15	—	Components assumed to make up the light fraction after shredding. After air classification the light fraction will contain from 2 to 8 percent of the components from the heavy fraction by weight.
Paper	40	40	—	
Cardboard	4	4	—	
Plastics	3	3	—	
Textiles	2	2	—	
Rubber	0.5	0.5	—	
Leather	0.5	0.5	—	
Garden trimmings	12	12	—	
Wood	2	2	—	
Glass	8	—	8	Components assumed to make up the heavy fraction after shredding. After air classification the heavy fraction will contain from 5 to 20 percent of the light fraction components by weight.
Tin cans	6	—	6	
Nonferrous metals	1	—	1	
Ferrous metals	2	—	2	
Dirt, ashes, brick, etc.	4	—	4	
Total	100	79	21	

* Moisture loss during shredding not considered.

To select component processes properly, the expected loading rates must be known. Loading rates for most processes are expressed in tons per hour. In determining the design loading rates, a careful analysis should be made to determine the actual number of hours per day that the equipment will be operated. The development of loading rates is also illustrated in Example 9-1.

TABLE 9-3   ESTIMATED RECOVERABLE QUANTITIES FOR VARIOUS COMPONENTS IN SOLID WASTES USING MECHANICAL EQUIPMENT

Fraction or component	Recoverable portion of original components, percent		Comments
	Range	Typical	
Light fraction	80–95	90*	Recoverable portion will vary with the composition of the solid wastes and the characteristics of the wastes after shredding.
Heavy fraction	90–98	96†	
Ferrous metal	65–95	85	Varying amounts of light and heavy fraction material will also be removed with these components depending on the specific process and equipment used.
Glass	50–90	80	
Aluminum	55–90	70	

* Varying amounts of the light fraction will be retained with the heavy fraction (see Table 9-2).
† Varying amounts of the heavy fraction will be carried over with the light fraction (see Table 9-2).

EXAMPLE 9-1 *Determination of Quantities and Loading Rates of Material for a Processing System*

Prepare a materials balance for the flowsheet given in Fig. 9-2 and the solid wastes composition given in Table 9-2. Assuming that the processing plant is to be designed to handle 1,000 tons/day, estimate the hourly loading rates for the various separation processes based on a 16-h/day operation.

Solution

1. The following assumptions will be used in the computations:
   a. Initial moisture content = 25 percent.
   b. On the basis of test data, the moisture content lost during shredding = 20 percent of initial value.
   c. Moisture content loss will be from material in light fraction.
   d. Heavy-fraction materials contained in light fraction = 6 percent of heavy fraction (based on weight after shredding).
   e. Light-fraction materials contained in heavy fraction = 15 percent of light fraction (based on weight after shredding).
   f. The initial light fraction = 79 percent (see Table 9-2).
   g. The initial heavy fraction = 21 percent (see Table 9-2).

2. Determine the materials-balance quantities.
   a. Moisture loss during shredding = (1,000 tons/day × 0.25)0.2 = 50 tons/day (45,360 kg/day).
   b. Total weight of light fraction after shredding = (1,000 tons/day × 0.79 − 50 tons/day) = 740 tons/day (671,328 kg/day).
   c. Total weight of light fraction after air classification including carryover corrections = (1 − 0.15)740 + 0.06(1,000 tons/day × 0.21) = 642 tons/day (582,422 kg/day).
   d. Total weight of heavy fraction including light-fraction carryover after air classification = (1 − 0.06)210 tons/day + (0.15)740 tons/day = 308 tons/day (279,418 kg/day).
   e. Weight of ferrous metals removed (assuming that the tin-can category reported in Table 9-2 is included) = 80 tons/day × 0.80 (see Table 9-3) = 64 tons/day (58,061 kg/day). (*Note:* Weight of other material that might be removed along with the ferrous metal is not included.)
   f. Weight of glass removed = 80 tons/day × 0.80 (see Table 9-3) = 64 tons/day (58,061 kg/day). (*Note:* Weight of other material that might be removed along with the glass is not included.)
   g. Weight of aluminum removed = 10 tons/day × 0.70 (see Table 9-3) = 7 tons/day (6,350 kg/day). (*Note:* Weight of other material that might be removed along with aluminum is not included.)
   h. Amount of residue (assuming that the heavy-fraction material contained in the light fraction will not be removed) = (308 − 64 − 64 − 7) tons/day = 173 tons/day (156,946 kg/day).

3.   Determine the loading rates on the individual component processes. The results of the necessary computations are summarized
in Table 9-4.

Equipment Limitations   In general, it has been found that front-end processing operations experience more equipment failures and other operational problems than rear-end and energy conversion systems [23]. The
conveyance of unprocessed solid wastes has proved especially troublesome. Conveyors have been damaged by solid wastes dropped onto them,
especially those containing some of the heavier components often found in
municipal wastes. Problems have also developed at transfer points (for
example, where the wastes are discharged from the conveyor into size-
reduction facilities). Wire and cords in the wastes become snagged on the
equipment, and waste spillage and overflows are common. Binding and
wedging of conveyor systems have also been a problem. Because of the
abrasive nature of many of the components found in solid wastes, the rate
of wear on much of the processing equipment has been greater than
anticipated. This, in turn, has led to longer "down" times.

As a result of these and other equipment limitations, many system
designers now recommend the installation of two or more independent
process trains, especially where power is to be produced on a continuous
basis. Wherever possible, when materials separation systems are being
designed, it is recommended that visits be made to actual operating
installations to obtain first-hand information on performance and maintenance requirements. Because many of the firms in this developing field do
not have a long history, it is recommended that the equipment selected be
such that it can be repaired with standard parts and components which, if
necessary, can be rebuilt or remade locally. The availability of a local
distributor is also important.

TABLE 9-4   SUMMARY OF COMPUTED LOADING
RATES FOR EXAMPLE 9-1

Process	Total amount, tons/day	Loading rate,* tons/h
Shredder	1,000	63
Classifier	950	60
Magnetic separator	308	20
Glass separator	244	16
Aluminum separator	180	12
Residue storage	173	11

* Based on 16-h/day operation. Values have been rounded off.
Note: ton/day × 907.2 = kg/day
        ton/h × 907.2 = kg/h

9-2 RECOVERY OF CHEMICAL CONVERSION PRODUCTS

Chemical conversion products that can be derived from solid wastes include heat, a variety of oils, gases, and various related organic compounds. The principal chemical conversion processes that have been used for the recovery of usable conversion products from solid wastes are reported in Table 9-5. Still others are under development or have been proposed [5]. With the exception of the incineration and pyrolytic processes, few full-scale installations have been in operation that use any of the other processes. Even in the case of pyrolysis, most of the full-scale experience is in the petroleum and wood processing industries. For this reason, most of the information to be presented in this section deals primarily with incineration and pyrolysis. A combined incineration-pyrolysis process is also considered.

In reviewing the data presented on the various processes, it should be noted that the purpose is not to present definitive design information, but rather to introduce and, when possible, to delineate some of the fundamental aspects of the various processes that will be important in evaluating their engineering and economic feasibility.

Incineration with Heat Recovery

Use of the incineration process for volume reduction is discussed in Chap. 8. Heat recovery systems and various combustion computations are considered in this section. Heat contained in the gases produced from the incineration of solid wastes can be recovered by conversion to steam. In addition, the low-level heat remaining in the gases after heat recovery can also be used to preheat the combustion air, boiler makeup water, or solid waste fuel.

Typically, the conversion of the heat contained in the combustion gases to steam is accomplished (1) by installing waste heat systems in which the boiler tubes are located beyond conventionally built refractory-lined combustion chambers, (2) in solid waste incinerators constructed with water-wall combustion chambers (see Fig. 9-5), and (3) in specially designed industrial water-wall boilers (see Fig. 9-6).

With existing incinerators, waste heat boilers can be installed to extract heat from the combustion gases without introducing excess amounts of air or moisture. Typically, incinerator gases will be cooled from a range of 1800 to 2000°F to a range from 600 to 1000°F before being discharged to the atmosphere. Apart from the production of steam, the use of a boiler system is beneficial in reducing the volume of gas to be processed in the air pollution control equipment.

In the water-wall incinerators, the walls of the combustion chamber are lined with boiler tubes that are arranged vertically and welded together in continuous sections (see Figs. 9-5, 9-6, and 9-7). As shown, the tubes are on the inside and are insulated on the outside to reduce radiant-heat losses. Water circulated through the tubes absorbs heat generated in the

TABLE 9-5 CHEMICAL PROCESSES USED FOR THE CONVERSION OF SOLID WASTES

Process	Conversion product	Preprocessing required	Comment
Incineration with heat recovery	Energy in the form of steam	None	Markets for steam must be available; proved in numerous full-scale applications; air-quality regulations may prohibit use.
Supplementary fuel firing	Energy in the form of steam	Shredding, air separation, and magnetic separation	If least capital investment desired, existing boiler must be capable of modification; air-quality regulations may prohibit use.
Fluidized-bed incineration	Energy in the form of steam	Shredding, air separation, and magnetic separation	Fluidized bed incinerator can also be used for industrial sludges.
Pyrolysis	Energy in the form of gas or oil	Shredding, magnetic separation	Technology proved only in pilot applications; even though pollution is minimized, air-quality regulations may prohibit use.
Hydrolysis	Glucose, fufural	Shredding, air separation	Technology on laboratory scale only.
Chemical conversion	Oil, gas, cellulose acetate	Shredding, air separation	Technology on laboratory scale only.

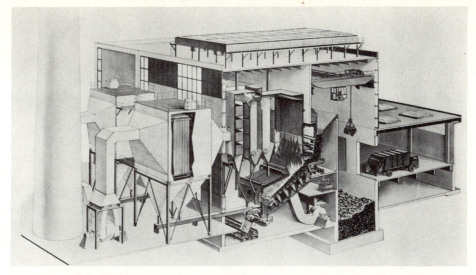

FIG. 9-5 Section through water-wall mass-fired incinerator. (Metcalf & Eddy Engineers, Inc.)

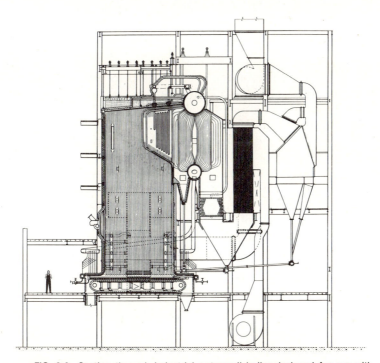

FIG. 9-6 Section through industrial water-wall boiler designed for use with processed solid wastes, natural gas, oil, and coal. (Combustion Engineering, Inc.)

FIG. 9-7   Section through water-wall. (Combustion Engineering, Inc.)

combustion chamber. The heated water is used to produce steam. When water-walls are used in place of refractory materials, they are not only useful for the recovery of steam, but also extremely effective in controlling furnace temperature without introducing excess air; however, they are subject to corrosion by the hydrochloric acid produced from the burning of some plastic compounds.

Prepared solid wastes can also be fired directly in large industrial boilers that are now used for the production of power with pulverized coal or oil (see Fig. 9-6). They also can be fired in conjunction with coal or oil. Although the process is not well established with coal, it appears that about 15 to 20 percent of the heat input can be from prepared solid wastes. With oil as the fuel, about 10 percent of the heat input can be from solid wastes [19].

Firing Methods   Depending on the degree of processing, mass, suspension, spreader-stoker, and double-vortex firing systems are used in solid waste incinerators and boilers [18, 23]. Mass firing is used when unprocessed wastes are to be burned (see Fig. 9-5 and the discussion of incineration in Chap. 8). Typically, reciprocating, traveling, or barrel grates convey the wastes through the boiler furnace (see Table 8-3).

Suspension firing is used with processed wastes (usually primary shredding followed by air classification followed by secondary shredding). In suspension firing, processed solid waste fuel is blown into the furnace of the boiler where it is dried and ignited as it falls. Burn-out grates usually are provided at the bottom of the furnace to handle slower-burning particles. A traveling-grate conveyor removes the ash from the boiler bottom (see Fig. 9-6).

For spreader-stoker firing, processed solid waste fuel is fed onto the traveling grate and incinerated as it travels through the furnace. Typically, a large grate occupying 100 percent of the cross-sectional area of the furnace is necessary because the particle sizes with spreader-stoker firing are usually larger than those for suspension firing. Forced-draft and overfire air, which would be supplied through the grates and the walls, would also be used to distribute the processed solid waste fuel. At the end of the grate, a conveyor is used to remove the ash.

The double-vortex firing system has a double cone-shaped combustion chamber with one end closed and the other end open to exhaust hot combustion gases into the boiler. The burners are housed within a box into which fuel and combustion air enter tangentially. The fuel-air mixture spirals toward the closed end in an outer vortex before traveling toward the open end in an inner vortex. Large fuel particles are recirculated by centrifugal force back into the outer vortex for complete burning. Ash and slag are collected at the bottom of the combustion chamber.

Combustion Computations For proper incinerator operation, sufficient air must be supplied to meet the requirements for (1) primary and secondary combustion and (2) turbulence for mixing the air and solid wastes. In practice, where conventional refractory-lined furnaces are used, it has been found that from 100 to 200 percent excess air must be supplied to meet the combustion and turbulence requirement and to control slagging and the accumulation of other material on the refractory walls. The resulting large gas flow makes the use of such incinerators costly because of the capacity required for the air pollution control equipment. In contrast, when heat recovery systems are used, it has been found that from 50 to 100 percent excess air is adequate. Thus, even though boilers are more expensive, the reduced size and cost of the air pollution control equipment will, in most cases, offset the initial cost.

In Chap. 4 (see Table 4-8), it was noted that the principal elements of solid wastes are carbon, hydrogen, oxygen, nitrogen, and sulfur. Smaller amounts of various other elements will also be found in the ash content. Under ideal conditions, the gaseous products derived from the combustion of municipal solid wastes would include carbon dioxide, water, oxygen, nitrogen, and sulfur dioxide. In actuality, many different reaction sequences are possible, depending on the exact nature of the wastes and the operating characteristics of the incinerator. As a consequence, a variety of sulfur and nitrogenous compounds are also found in the gaseous discharge from an incinerator. However, for purposes of illustration in the following discussion, it will be assumed that the incineration process is ideal.

To determine the amount of air that must be supplied for the complete combustion of solid wastes, it is necessary to compute the oxygen requirements for the oxidation of the carbon, hydrogen, and sulfur contained in the wastes. The basic reactions are as follows.

For carbon

$$C + O_2 \rightarrow CO_2 \tag{9-1}$$
$$(12) \quad (32)$$

For hydrogen

$$2H_2 + O_2 \rightarrow 2 H_2O \tag{9-2}$$
$$(4) \quad (32)$$

For sulfur

$$S + O_2 \rightarrow SO_2 \tag{9-3}$$
$$(32.1) \quad (32)$$

If it is assumed that dry air contains 23.15 percent oxygen by weight, then the amount of air required for the oxidation of 1 lb of carbon would be equal to 11.52 lb [(32/12)(1/0.2315)]. The corresponding amounts for hydrogen and sulfur are 34.56 and 4.31 lb, respectively. In combustion computations, the oxygen requirements for the oxidation of hydrogen usually are based on the net value of hydrogen available. The net value of hydrogen is computed by subtracting one-eighth of the percent of oxygen from the total percent of hydrogen initially present in the sample. This computation is based on the assumption that the oxygen in the sample will combine with hydrogen in the sample to form water.

Heat released from the combustion process is partly stored in the combustion products and partly transferred by convection, conduction, and radiation both to the incinerator walls and to the incoming fuel. If the elemental composition of the solid wastes is known, the energy content can be estimated by using the modified form of the Dulong equation given in Chap. 4 (see Eq. 4-2). Often the energy content of solid wastes is based on an analysis of the heating value of the individual waste components (see Table 4-9).

NRG CONTENT of SW

The necessary combustion computations to determine the amount of air required for complete combustion and to estimate the heat produced from the combustion process that is available for conversion to steam and ultimately to electrical power are illustrated in Example 9-2.

EXAMPLE 9-2 *Materials and Heat Balances in Incineration*

Determine the available heat for steam production from a quantity of solid wastes with the following characteristics, to be incinerated at a rate of 250,000 lb/day.

Component	Percent of total	lb/day
Combustible	60	150,000
Water	20	50,000
Noncombustible	20	50,000

Note: lb/day × 0.4536 = kg/day

Element	Percent
Carbon	28
Hydrogen	4
Oxygen	23
Nitrogen	4
Sulfur	1
Water	20
Inerts	20

Assume that the following conditions are applicable:

1. The as-fired heating value of the solid wastes is 5,000 Btu/lb.
2. Grate residue contains 5 percent carbon.
3. Temperatures:
     Entering air, 80°F
     Residue from grate, 800°F
4. Specific heat of residue = 0.25 Btu/lb·°F.
5. Latent heat of water = 1,040 Btu/lb.
6. Radiation loss = 0.005 Btu/Btu of total furnace input.
7. All oxygen in solid waste is bound as water.
8. Theoretical air requirements based on stoichiometry:
     Carbon: $(C + O_2 \rightarrow CO_2) = 11.52$ lb/lb
     Hydrogen: $(2H_2 + O_2 \rightarrow 2H_2O) = 34.56$ lb/lb
     Sulfur: $(S + O_2 \rightarrow SO_2) = 4.31$ lb/lb
9. The heating value of carbon is 14,000 Btu/lb.
10. Moisture in the combustion air is 1 percent.

Solution

1. Compute the weights of the elements of the solid wastes.

Element		lb/day
Carbon	= 0.28(250,000) =	70,000
Hydrogen	= 0.04(250,000) =	10,000
Oxygen	= 0.23(250,000) =	57,500
Nitrogen	= 0.04(250,000) =	10,000
Sulfur	= 0.01(250,000) =	2,500
Water	= 0.20(250,000) =	50,000
Inerts	= 0.20(250,000) =	50,000

2. Compute the amount of residue.

Inerts = 50,000 lb/day ✓

$$\text{Total residue} = \frac{50,000}{0.95} = 52,600 \text{ lb/day} \checkmark$$

Carbon in residue = 2,600 lb/day $\left[(52,600)(0.05)\right]$ ✓

3. Compute the available hydrogen and bound water.

Available hydrogen = 4% − 23%/8 $(\frac{1}{8} \text{ of } O_2)$
$$= 1.125\% = 2,800 \text{ lb/day}$$
Hydrogen in bound water = (4 − 1.125)%
$$= 2.885\% = 7,200 \text{ lb/day}$$
Bound water = 57,500 + 7,200
$$= 64,700 \text{ lb/day} \checkmark$$

4. Compute the air required.

Element	lb/day
Carbon = (70,000 − 2,600)(11.52)	776,500 ✓
Hydrogen = 2,800(34.56)	96,800 ✓
Sulfur = 2,500(4.31)	10,800 ✓
Total dry theoretical air	884,100 ✓
Total dry air including 100 percent excess	1,768,200 ✓
Moisture 1,768,200(0.01)	17,700 ✓
Total air	1,785,900 (810,084 kg/day) ✓

5. Compute net heat input from solid wastes.

Gross heat input = 250,000 lb/day × 5,000 Btu/lb
$$= 1,250,000,000 \text{ Btu/day} \checkmark$$
Heat input lost in unburned carbon = 2,600 lb/day (14,000 Btu/lb)
$$= 36,000,000 \text{ Btu/day} \checkmark$$
Net heat input = (1,250,000,000 − 36,000,000) Btu/day ✓
$$= 1,214,000,000 \text{ Btu/day (1,280,770,000 kJ/day)}$$

6. Compute latent heat losses.

Inherent moisture = 50,000 lb/day (1,040 Btu/lb)
= 52,000,000 Btu/day ✓
Moisture in bound water = 64,700 lb/day (1,040 Btu/lb)
= 68,000,000 Btu/day ✓
Moisture from oxidation of hydrogen
$$= \frac{9 \text{ lb } H_2O}{\text{lb } H} \, 2,800 \text{ lb/day } (1,040 \text{ Btu/lb})$$
= 27,000,000 Btu/day (28,485,000 kJ/day) ✓

7. Compute reactor losses.

Radiation loss = (0.005 Btu/Btu)(1,250,000,000 Btu/day)
= 6,000,000 Btu/day ✓
Sensible heat in residue
= (52,600 lb/day)[0.25 Btu/lb·°F (800 − 80)°F]
= 9,000,000 Btu/day (9,495,000 kJ/day) ✓

8. Total losses = 162,000,000 Btu/day (170,910,000 kJ/day). ✓

9. Compute the heat available for steam production. Available heat in the hot gases = (1,214,000,000 − 162,000,000) Btu/day = 1,052,-000,000 Btu/day (1,109,860,000 kJ/day). This is the heat above normal air temperature (assumed to be 80°F) in the high-tempera-ture waste gases available at the inlet of a waste heat boiler. The amount of steam produced would depend on the efficiency of the boiler. For example, if the boiler efficiency were 85 percent, the overall efficiency would be about 72 percent. This value is consist-ent with the data obtained in Chicago Northeast Incinerator Plant [22].

Gas and Temperature of Combustion Gases Along with knowing the amount of air required and the quantity of available heat, it is also important to know the composition and temperature of the combustion gases for various amounts of excess air. This is an important factor in the design of heat recovery systems and also for odor control. For example, for some wastes, if the combustion temperature drops below 1400 to 1600°F, the gases emitted from the stack may be odorous because of incomplete combustion. Sample computations to determine the composi-tion and temperature of the gases leaving the combustion chamber for solid wastes with the characteristics considered in Example 9-2 are illustrated in Example 9-3.

EXAMPLE 9-3 *Determination of the Composition of Combustion Gases*

Determine the composition of the combustion gases for the solid wastes in Example 9-2. To simplify the computations, assume that all the

TABLE 9-6   DETERMINATION OF AIR REQUIREMENTS FOR COMPLETE COMBUSTION OF 100 LB OF SOLID WASTES FOR EXAMPLE 9-3

Component	Weight* percent	Atomic weight	Atomic weight units	Moles oxygen required	Combustion reaction and product
Carbon	28	12.0	2.333	2.333	$C + O_2 = CO_2$ (carbon dioxide)
Hydrogen	4	1.0	4.000	1.000	$2H_2 + O_2 = 2H_2O$ (water)
Oxygen	23	16.0	1.438	−0.719	
Nitrogen	4	14.0	0.286	—	
Sulfur	1	32.1	0.031	0.031	$S + O_2 = SO_2$ (sulfur dioxide)
Water	20	18.0	1.111	—	
Inerts	20	—	—	—	
Total	100			2.645	

Moles of air† required per 100 lb of solid wastes = 2.645/0.2069 = 12.78 mol

Pounds of air† required per pound of solid wastes = 12.78(28.7)/100 = 3.67
(mol)(lb/mol)

* See Example 9-2.
† Assumed air composition, in volume fractions: carbon dioxide, 0.0003; nitrogen, 0.7802; oxygen, 0.2069; water, 0.0126. Assuming ideal gases, the volume fractions may be taken as mole fractions and are equal to the percentages of volume divided by 100. The composition as given is for rare gases included with the nitrogen and with moisture content corresponding to 70 percent relative humidity at 60°F. Air of this composition has a weight of 28.7 lb/mol of total gas.

carbon initially present is converted to carbon dioxide. Also estimate the temperature of the combustion gases leaving the combustion chamber.

Solution

1. Determine the moles of oxygen and the pounds of air required per 100 lb of solid wastes. The necessary computations are presented in summary form in Table 9-6.

2. Determine the moles of combustion gases produced from the complete combustion of 100 lb of solid wastes. Also determine the composition of the combustion gases if 50 and 100 percent excess air are used. The necessary computations are illustrated in Tables 9-7 and 9-8.

3. Estimate the temperature of the combustion gases. To do this, data on the enthalpies of various combustion gases are required. The necessary data are presented in Table 9-9. Using the data in Tables 9-8 and 9-9, estimate the heat content in the gas produced from 1 lb of solid wastes by using the following equation [14]:

TABLE 9-7   DETERMINATION OF MOLES OF COMBUSTION PRODUCTS PRODUCED FROM COMPLETE COMBUSTION OF 100 LB OF SOLID WASTES FOR EXAMPLE 9-3

Combustion product	Moles of combustion products			
	From combustion*	From air†	Total	Percent
Carbon dioxide	2.333	0.004‡	2.337	14.8
Water	(2.000 + 1.111§)	0.161	3.272	20.8
Oxygen	—	—	—	—
Nitrogen	0.143	9.97	10.113	64.2
Sulfur dioxide	0.031	—	0.031	0.2
Total			15.753	100.0

Moles of air per mole of gas = 12.78/15.75 = 0.81

* Data derived from Table 9-6.
† Moles of air required per 100 lb of solid wastes = 12.78 (see Table 9-6).
‡ Sample computation 12.78(0.0003) = 0.004 (see Table 9-6, second footnote).
§ Moles of moisture present in original sample.

$$\frac{\text{Btu in product gas}}{\text{lb of solid wastes}}$$

$$= \left[\left(\frac{\text{moles of product gas}}{\text{lb of solid wastes}}\right)\left(\frac{\text{total moles of gas}}{\text{moles of product gas}}\right)\right]$$

$$\times \left[\Sigma(\text{mole fraction of gas component})\right.$$

$$\left.\left(\frac{\text{Btu}}{\text{moles of gas component}}\right)\right]$$

For 1000°F and 50 percent excess, the required computation is

$$\frac{\text{Btu}}{\text{lb}} = \left(\frac{0.1575}{\text{lb}}\right)\left(\frac{140.5}{100}\right)[0.105(10,048) + 0.060(6,974)$$

$$+ 0.682(6,720) + 0.152(26,925)] = 2,245 \ (5,212 \text{ kJ/kg})$$

TABLE 9-8   DETERMINATION OF EFFLUENT GAS COMPOSITION FOR VARIOUS QUANTITIES OF EXCESS AIR FROM COMPLETE COMBUSTION OF 100 LB OF SOLID WASTES FOR EXAMPLE 9-3*

Percent excess air	Moles excess air†	Total moles of gas	Gas composition, percent				
			CO_2	O_2	N_2	H_2O	SO_2
0	0.0	100.0	14.8	—	64.2	20.8	0.2
50	40.5‡	140.5	10.5	6.0§	68.2¶	15.2	0.1
100	81.0	181.0	8.2	9.3	70.4	12.0	0.1

* Refer to Tables 9-6 and 9-7.
† Moles excess air = percent excess air (moles of air/moles of gas).
‡ (50 percent excess air)(0.81 [see Table 9-7]) = 40.5.
§ Percent O_2 = [(40.5 × 0.2069)/140.5]100.
¶ Percent N_2 = {[64.2 + 40.5(0.7802)]/140.5}100.   → WORKS FOR CO_2, H_2O, & SO_2 ALSO

TABLE 9-9   ENTHALPIES FOR VARIOUS COMBUSTION PRODUCTS*
(Btu/lb·mol over standard state†)

Temperature T, °F	CO_2	O_2	N_2	N_2O
1000	10,048	6,974	6,720	26,925
1500	16,214	11,008	10,556	31,743
2000	22,719	15,191	14,520	36,903
2500	29,539	19,517	18,609	42,405

Enthalpy equations

$$CO_3, H = 10,570\left(\frac{T+460}{1000}\right) + 583.3\left(\frac{T+460}{1000}\right)^2 + \frac{667.4}{(T+460)/1000} - 7,085$$

$$O_2, H = 7,160\left(\frac{T+460}{1000}\right) + 278.8\left(\frac{T+460}{1000}\right)^2 + \frac{129.6}{(T+460)/1000} - 4,163$$

$$N_2, H = 6,830\left(\frac{T+460}{1000}\right) + 250.0\left(\frac{T+460}{1000}\right)^2 + \frac{38.9}{(T+460)/1000} - 3,811$$

$$H_2O, H = 7,300\left(\frac{T+460}{1000}\right) + 683.3\left(\frac{T+460}{1000}\right)^2 + 14,810$$

* From Ref. 14.
† Gas, except liquid water, at 1-atm pressure and 77°F.

Tabulated values for temperatures from 1000 to 2500°F and 50 and 100 percent excess air have been computed and are summarized in Table 9-10.

If it is assumed that the energy content of the solid wastes is 5,000 Btu/lb and that 15 percent of the energy is lost, then 4,250 Btu/lb must be contained in the gases. By referring to Table 9-10 one can see that the temperature of the gases with 50 percent

TABLE 9-10   HEAT CONTENT IN COMBUSTION GASES PRODUCED FROM 1 LB OF SOLID WASTES

	Btu	
Temperature, °F	Excess air, 50 percent	Excess air, 100 percent
1000	2,245	2,689
1500	3,184	3,874
2000	4,162	5,108
2500	5,108	6,390

excess air would be about 2000°F and about 1500°F for 100 percent excess air.

Pyrolysis

Of the many alternative chemical conversion processes that have been investigated, excluding incineration, pyrolysis has received the most attention.

Process Description   Because most organic substances are thermally unstable, they can, upon heating in an oxygen-free atmosphere, be split through a combination of thermal cracking and condensation reactions into gaseous, liquid, and solid fractions. *Pyrolysis* is the term used to describe the process. In contrast to the combustion process, which is highly exothermic, the pyrolytic process is highly endothermic. For this reason the term *destructive distillation* is often used as an alternative term for pyrolysis.

At the present time a number of different types of reactors are being evaluated for this application. Depending on the type of reactor used, the physical form of the solid wastes to be pyrolyzed can vary from unshredded raw wastes to the finely ground portion of the wastes remaining after two stages of shredding and air classification (see Fig. 9-3).

Conversion Products   The characteristics of the three major component fractions resulting from the pyrolysis are:

1. A gas stream containing primarily hydrogen, methane, carbon monoxide, carbon dioxide, and various other gases, depending on the organic characteristics of the material being pyrolyzed

2. A fraction that consists of a tar and/or oil stream that is liquid at room temperatures and has been found to contain chemicals such as acetic acid, acetone, and methanol

3. A char consisting of almost pure carbon plus any inert material that may have entered the process

For cellulose ($C_6H_{10}O_5$), the following expression has been suggested as being representative of the pyrolysis reaction [5]:

$$3(C_6H_{10}O_5) \rightarrow 8H_2O + C_6H_8O + 2CO$$
$$+ 2CO_2 + CH_4 + H_2 + 7C \quad (9\text{-}4)$$

In Eq. 9-4 the liquid tar and/or oil compounds normally obtained are represented by the expression C_6H_8O. It has been found that distribution of the product fractions varies dramatically with the temperature at which the pyrolysis is carried out [5]. Representative data on the product as a function of the operating temperature are reported in Table 9-11. Typical

TABLE 9-11   PYROLYTIC PRODUCT YIELD*

Temperature, °F	Wastes,† lb	Gases, lb	Pyroligneous acids and tars,‡ lb	Char, lb	Mass accounted for, lb
900	100	12.33	61.08	24.71	98.12
1200	100	18.64	18.64	59.18	99.62
1500	100	23.69	59.67	17.24	100.59
1700	100	24.36	58.70	17.67	100.73

* From Ref. 5.
† On an as-received basis, except that metals and glass have been removed.
‡ Includes all condensables; figures cited include 70 to 80 percent water.
Note: lb × 0.4536 = kg

TABLE 9-12   GASES EVOLVED BY PYROLYSIS*

Gas	Percent by volume			
	900°F	1200°F	1500°F	1700°F
H_2	5.56	16.58	28.55	32.48
CH_4	12.43	15.91	13.73	10.45
CO	33.50	30.49	34.12	35.25
CO_2	44.77	31.78	20.59	18.31
C_2H_4	0.45	2.18	2.24	2.43
C_2H_6	3.03	3.06	0.77	1.07
Accountability	99.74	100.00	100.00	99.99

* From Ref. 5.
Note: 0.555(°F − 32) = °C

TABLE 9-13   PROXIMATE ANALYSIS OF PYROLYTIC CHAR*

Characteristics	Percent by volume				Pennsylvania anthracite†
	900°F	1200°F	1500°F	1700°F	
Volatile matter	21.81	15.05	8.13	8.30	7.66
Fixed carbon	70.48	70.67	79.05	77.23	82.02
Ash	7.71	14.28	12.82	14.47	10.32
Btu/lb	12,120	12,280	11,540	11,400	13,880

* From Ref. 5.
† Typical values.
Note: 0.555(°F − 32) = °C
       Btu/lb × 2.326 = kJ/kg

characteristics of the gaseous fraction and the char are given in Tables 9-12 and 9-13, respectively. The energy content of pyrolytic oils has been estimated to be about 10,000 Btu/lb. Under conditions of maximum gasification, it has been estimated that the energy content of the resulting gas would be about 700 Btu/ft^3.

NRG CONTENT OF PRODUCTS

In summary, it appears that while the pyrolytic process has great promise, much remains to be known. Background information and data must be gathered on the nature of the problems to be encountered when the process is operated continuously for a sustained period of time. For example, will corrosion or air pollution control problems be surmountable? In view of the number of pilot plant processes currently in operation or about to start operation, answers to some of these questions should become available by 1980.

Incineration-Pyrolysis

A recent development in the chemical conversion of solid wastes is the combination incineration-pyrolysis process developed by Union Carbide [20]. The complete system developed around this process is known as the *Purox system.*

Process Description The process can best be described by referring to Fig. 9-8. Solid wastes are fed through a charging lock located at the top of

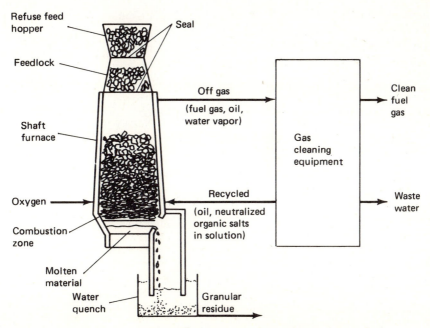

FIG. 9-8 Section through incineration-pyrolysis reactor. (Union Carbide Corporation.)

the reactor. Pure oxygen is injected into the combustion zone at the bottom of the furnace where it reacts with the carbon char residue from the pyrolysis zone. The temperature generated in the hearth is sufficiently high to melt the glass, metal, and other materials into a molten residue. The molten material overflows continuously from the hearth into a water-quench tank where it forms a hard, granular material.

The hot gases formed by the reaction of oxygen and carbon char rise through the descending wastes. In the middle portion of the vertical furnace, organic materials are pyrolyzed under an essentially reducing atmosphere to yield a mixture of gaseous products. As the hot gaseous products continue to flow upward, they dry the entering solid wastes in the upper zone of the furnace.

As the gas mixture leaves the furnace, it contains water vapor, some oil mist formed by the condensation of high-boiling organics, and minor amounts of fly ash. A gas cleaning system is used to remove the oil mist and fly ash solids. The product gas, after cleaning, is passed through a condenser. The resultant dry gas is comparable in combustion characteristics to natural gas [6, 20].

Conversion Products   The gaseous conversion products recovered from this process are reported in Table 9-14. As shown, the gaseous mixture is composed principally of CO, CO_2, and H_2. It is expected that this composition will vary with the characteristics of the solid wastes. In terms of conversion efficiency, it is estimated that typically about 75 percent of the energy contained in the solid waste fuel is recoverable when the incinerator-pyrolysis reactor shown in Fig. 9-8 is used.

Where a market exists, it may be more profitable to upgrade the low-Btu gas obtained from this process to pipeline natural-gas quality (960 to 980 Btu/ft^3) as opposed to using it directly for the production of electrical

TABLE 9-14   TYPICAL COMPOSITION OF GASEOUS PRODUCTS FROM PUROX PROCESS*

Component	Percent by volume†
CH_4	5
CO	40
CO_2	23
H_2	26
Higher organics	1
Nitrogen	1
Heating value, Btu/ft^3	345–370

* From Ref. 6.
† Dry basis.
Note: Btu/ft^3 × 37.259 = kJ/m^3

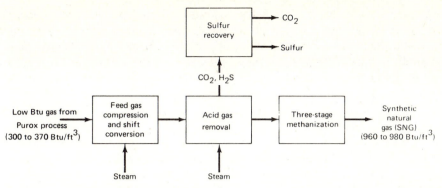

FIG. 9-9   Simplified flowsheet for the conversion of low Btu gas from Purox process to pipeline natural-gas quality. (The Lummus Company.)

energy. Typically this would be accomplished by a methanation process (see Fig. 9-9). As shown in the simplified flowsheet given in Fig. 9-9, the complete process would consist of the following four basic steps or operations: (1) feed gas compression and shift conversion, (2) acid gas removal, (3) methanation, and (4) sulfur recovery.

In the first step, the low-Btu gas is compressed from 2 to 300 lb/in² guage. The shift conversion of carbon monoxide (CO) to carbon dioxide (CO_2) is accomplished in a fixed-bed catalytic reactor.

Shift conversion of CO to CO_2:

$$CO + H_2O \underset{\text{catalyst}}{\rightleftharpoons} CO_2 + H_2$$

This reaction is necessary to achieve carbon monoxide–hydrogen ratios suitable for methanation. In the second step, the bulk of the carbon dioxide (CO_2) and hydrogen sulfide (H_2S) are removed from the cooled shift effluent gas by scrubbing. In the third step, the scrubbed gas is fed to a series of three methanation reactors where the hydrogen gas reacts with carbon oxides to form methane.

Conversion of shift gas to methane:

$$CO + 3H_2 \underset{\text{catalyst}}{\rightleftharpoons} CH_4 + H_2O$$

$$CO_2 + 4H_2 \underset{\text{catalyst}}{\rightleftharpoons} CH_4 + 2H_2O$$

In the fourth step, elemental sulfur is recovered from the condensed CO_2-H_2S gas stream by using the Stretford process.

Other Chemical Conversion Processes

In addition to the various incineration and pyrolytic processes under investigation and/or construction, a variety of other public and proprietary processes are being evaluated. For example, the hydrolytic conversion of cellulose to glucose, followed by the fermentation of glucose to ethyl alcohol, has been demonstrated on a pilot scale [8]. Nothing more definitive can be stated on these processes until sufficient, well-documented data become available.

9-3 RECOVERY OF BIOLOGICAL CONVERSION PRODUCTS

Biological conversion products that can be derived from solid wastes include compost, methane, various proteins and alcohols, and a variety of other intermediate organic compounds. The principal processes that have been used are reported in Table 9-15. Composting and anaerobic digestion, the two most developed processes, are described in detail in this section after the following discussion of process fundamentals.

Some Biological Process Fundamentals

To help the reader understand the biological conversion processes to be discussed later in this section, some fundamentals of microbial systems as they relate to the conversion of solid wastes are presented. The topics include (1) types of microorganisms, (2) dissimilatory and assimilatory processes, (3) aerobic and anaerobic metabolism, (4) nutritional requirements, and (5) environmental requirements.

Types of Microorganisms   The general class of microorganisms that are of interest with respect to the conversion of solid wastes to either cell mass or some by-product of cell metabolism are called *protists*. Microorganisms in this classification may be unicellular or multicellular, but are without cell differentiation. Specifically, the protists of greatest concern in the conversion of solid wastes are bacteria, fungi, yeasts, and actinomycetes. Protozoa and algae are other common protists, but they are not of primary importance.

Typically, bacteria are single cells—cocci, rods, or spirals. Coccal forms vary from 0.5 to 4 μm in diameter; rods are from 0.5 to 20 μm long and 0.5 to 4 μm wide; spirals may be greater than 10 μm long and about 0.5 μm wide [1, 2]. Bacteria are ubiquitous in nature and are found in aerobic (in the presence of oxygen) and anaerobic (in the absence of oxygen) environments. Because of the wide variety of inorganic and organic compounds that can be used by bacteria to sustain growth, bacteria are used extensively in a variety of industrial operations to accumulate intermediate and end products of metabolism. Tests on a number of different bacteria species indicate that they are about 80 percent water and 20 percent dry material, of which 90 percent is organic and 10 percent is inorganic. An approximate empirical formula for the organic fraction is $C_5H_7NO_2$ [13]. On the basis of this formula, about 53 percent by weight of

TABLE 9-15 BIOLOGICAL PROCESSES USED FOR THE CONVERSION OF SOLID WASTES

Process	Conversion product	Preprocessing required	Comment
Composting	Humuslike material	Shredding, air separation	Lack of markets is primary shortcoming; technically proved in full-scale applications
Anaerobic digestion	Methane gas	Shredding, air separation	Technology on laboratory scale only
Biological conversion to protein	Protein, alcohol	Shredding, air separation	Technology on laboratory scale only
Biological fermentation*	Glucose, furfural	Shredding, air separation	Used in conjunction with the hydrolytic process

* For further details see Ref. 8.

the organic fraction is carbon. Compounds that make up the inorganic portion include P_2O_5 (50 percent), CaO (9 percent), Na_2O (11 percent), MgO (8 percent), K_2O (6 percent), and Fe_2O_3 (1 percent). Since all these elements and compounds must be derived from the environment, a shortage of these substances would limit, and in some cases alter, the growth of bacteria [13].

Fungi are considered to be multicellular, nonphotosynthetic, heterotropic protists. Most fungi have the ability to grow under low-moisture conditions which do not favor the growth of bacteria. In addition, fungi can tolerate relatively low pH values. The optimum pH value for most fungal species appears to be about 5.6, but the viable range is from 2 to 9. The metabolism of these organisms is essentially aerobic, and they grow in long filaments composed of nucleated cell units called "hyphae," varying in width from 4 to 20 μm. Because of their ability to degrade a wide variety of organic compounds over a broad range of environmental conditions, fungi have been used extensively in industry for the production of valuable compounds, such as organic acids (e.g., citric, gluconic), various antibiotics (e.g., penicillin, griseofulvin), and enzymes (e.g., cellulase, protease, amylase).

Yeasts are fungi that cannot form in a filament (mycelium) and are therefore unicellular. Some yeasts form elliptical cells 8 to 15 μm by 3 to 5 μm while others are spherical, varying in size from 8 to 12 μm in diameter. In terms of industrial processing operations, yeasts may be classified as "wild" and "cultivated." In general, wild yeasts are of little value, but cultured yeasts are used extensively to ferment sugars to alcohol and carbon dioxide.

Actinomycetes are a group of organisms with intermediate properties between bacteria and fungi. With respect to form, they are similar to fungi, except that the width of the cell is only 0.5 to 1.4 μm. In industry this group of microorganisms is used extensively for the production of antibiotics. Because their growth characteristics are similar, actinomycetes are often grouped with fungi for discussion purposes [2].

Dissimilatory and Assimilatory Processes   To continue to grow and function properly, microorganisms must have a source of energy and carbon for the synthesis of new cellular material. Inorganic elements, such as nitrogen and phosphorus, and other trace elements, such as sulfur, potassium, calcium, and magnesium, are also vital nutrients for cell synthesis. Two of the most common sources of cell carbon for microorganisms are carbon dioxide and organic matter. If an organism derives cell carbon from carbon dioxide, it is called autotrophic; if organic carbon is used, it is called heterotrophic.

Energy is also needed in the synthesis of new cellular material. For autotrophic organisms, the energy can be supplied by the sun, as in photosynthesis, or by an inorganic oxidation-reduction reaction. If the energy is supplied by the sun, the organism is called autotrophic photosynthetic. If the energy is supplied by an inorganic oxidation-reduction reaction, it is called autotrophic chemosynthetic. For heterotrophic organisms, the energy needed for cell synthesis is supplied by the oxidation of organic matter.

In this context, dissimilatory processes may be considered to be those processes associated with the production and/or capture of energy, whereas assimilatory processes are those associated with the production of cell tissue. On the basis of these broad classifications, most industrial fermentations (both aerobic and anaerobic) are dissimilatory in that complex compounds (usually organic) are degraded to simpler compounds or molecules with an attendant release of energy. Assimilatory biological processes currently are used in the formation of complex organic molecules that cannot be synthesized economically by conventional organic chemistry techniques. Most antibiotics fall into this category.

The reason for distinguishing between dissimilatory and assimilatory processes, which always occur simultaneously, is that frequently the optimum conditions for each process may be quite different. Often such considerations affect the process flowsheet and the design of the processing facilities.

Aerobic and Anaerobic Metabolism   Microorganisms that cannot grow or survive in the absence of oxygen are called obligate aerobes. Similarly, obligate anaerobes are those organisms that cannot survive or are inhibited in the presence of oxygen. Organisms capable of growth either in the presence or absence of oxygen are called facultative anaerobes. Many facultative organisms possess both aerobic and anaerobic metabolic systems and can change from one system to another in response to the presence of oxygen. Other facultative organisms have only an anaerobic metabolic system but are insensitive to the presence of oxygen. Aerobic and anaerobic processes are considered further in the discussions on composting and anaerobic digestion later in this chapter.

Nutritional Requirements   To grow and function properly, microorganisms must have all the nutrients necessary to synthesize and maintain their cell tissue. This normally includes a source of carbon, hydrogen, oxygen, nitrogen, inorganic salts, phosphorus, sulfur, and trace amounts of assorted micronutrients [24]. Because these requirements vary with the microorganism in question, a detailed evaluation should be made for each application.

Environmental Requirements   The most important environmental requirements include temperature, moisture content, pH, and the absence of toxicity. The temperature range over which microorganisms have been found to survive varies from -5 to $80°C$. The lower limit is set by the freezing point of water, which is lowered by the contents of the cell [2]. The upper limit is usually established by the characteristics of the constituents that make up the cell tissue. For example, most proteins and nucleic acids are destroyed in the temperature range of 50 to 90°C. For most organisms used for the conversion of solid wastes, the temperature range for optimum growth is much smaller. Microorganisms that grow best in the temperature range of 20 to 40°C are called *mesophiles* and are the largest group found in nature. Those that grow best in the range below 20°C are called *psychrophiles*, and those that grow best above 45°C are called *thermophiles*. These distinctions are not very rigid, and many microorganisms have been identified that can adapt to all three temperature ranges.

Because water is essential for the growth of microorganisms, the moisture content of the wastes to be converted must be known, especially if a dry process, such as composting, is to be used. In many composting operations, it has been necessary to add water to obtain optimum bacterial activity.

The hydrogen ion concentration expressed as pH in and of itself is not a significant factor in the growth of microorganisms within the range from 6 to 9, which represents a thousandfold difference in the hydrogen ion concentration. However, when the pH goes above or below this range, it appears that the undissociated molecules of weak acids or bases can enter the cell more easily than hydrogen and hydroxyl ions, and, by altering the internal pH, damage the cell.

Composting

In Chap. 4 it was noted (see Table 4-2) that the largest fraction of municipal solid wastes is organic in composition. With the exception of plastic, rubber, and leather components, the organic fraction of most municipal solid wastes can be classified as follows [17]:

1. Water-soluble constituents, a group which includes sugars, starches, amino acids, and various organic acids
2. Hemicellulose, a condensation product of five and six carbon sugars
3. Cellulose, a condensation product of the six carbon sugars, glucose
4. Fats, oils, and waxes, which are esters of alcohols and higher fatty acids
5. Lignin, a material the exact chemical nature of which is still not known (present in some paper products such as newsprint and fiberboard)
6. Lignocellulose, a combination of lignin and cellulose
7. Proteins, which are composed of chains of amino acids

If these organic materials are separated from municipal solid wastes and are subjected to bacterial decomposition, the end product remaining after dissimilatory and assimilatory bacterial activity is called *humus*. The entire process involving both the separation and bacterial conversion of the organic solid wastes is known as *composting*.

Decomposition of the organic solid wastes may be accomplished either aerobically or anaerobically, depending on the availability of oxygen. Because anaerobic processes are extremely slow and offensive odors associated with these processes may be difficult to control, most composting operations are aerobic.

In general, the chemical and physical characteristics of the humus vary according to the nature of the starting material, the conditions under which the composting operation was carried out, and the extent of decomposition. Some of the properties of the resulting humus that distinguish it from other natural materials are [17]:

1.   A dark brown to black color
2.   A low carbon-nitrogen ratio
3.   A continually changing nature due to the activities of microorganisms
4.   A high capacity for base exchange and for water absorption

Process Description   Most composting operations consist of three basic steps: (1) preparation of the solid wastes, (2) decomposition of the solid wastes, and (3) product preparation and marketing. Receiving, sorting, separation, size reduction, and moisture and nutrient addition are part of the preparation step. To accomplish the decomposition step, several techniques have been developed. In windrow composting, prepared solid wastes are placed in windrows in an open field. The windrows are turned once or twice per week for a composting period of about 5 wk. The material is usually cured for an additional 2 to 4 wk to ensure stabilization. As an alternative to windrow composting, several mechanical systems have been developed. By controlling the operation carefully in a mechanical system, it is possible to produce a humus within 5 to 7 days. Often the composted material is removed and cured in open windrows for an additional period of about 3 wk. Once the solid wastes have been converted to a humus material, they are ready for the third step of product preparation and marketing. This step may include fine grinding, blending with various additives, granulation, bagging, storage, shipping, and, in some cases, direct marketing. Because a detailed description of the various ways in which these three steps can be accomplished is beyond the scope of this text, Refs. 4, 5, and 7 are recommended.

Process Microbiology   Although they are extremely diverse, the principal microorganisms involved in the aerobic decomposition of solid wastes can be identified as bacteria, fungi, yeasts, and actinomycetes. While members of each of these groups can be found that are capable of decomposing all the raw materials in solid wastes, as a group they prefer different compounds. Typically, bacteria prefer simple water-soluble sugars, while fungi, yeasts, and actinomycetes are particularly effective in the decomposition of celluloses and hemicelluloses.

Aside from metabolic requirements, the predominance of microoganisms varies during the course of the composting process. One of the major factors contributing to this occurrence is the heat released as a result of the dissimilatory and assimilatory activities of the microorganisms in converting solid wastes to a stabilized humus. Initially, the material being composted heats up as a result of the release of energy accompanying the degradation of the readily convertible organic food wastes and simple sugars. When the temperature rises above 45 to 50°C, the mesophilic organisms begin to predominate. These organisms will predominate at about 55°C, which has been observed as the optimum temperature for these organisms. Certain types of bacteria and actinomycetes are common in this temperature range. Under normal conditions, stabilization is more rapid in the thermophilic range than in the mesophilic range.

The amount of oxygen required for the aerobic stabilization of municipal solid wastes can be estimated by using the following equation [17]:

$$C_aH_bO_cN_d + 0.5(ny + 2s + r - c)O_2 \rightarrow$$
$$nC_wH_xO_yN_z + sCO_2 + rH_2O + (d - nx)NH_3 \quad (9\text{-}5)$$

where $r = 0.5[b - nx - 3(d - nx)]$
$s = a - nw$

The terms $C_aH_bO_cN_d$ and $C_wH_xO_yN_z$ represent the empirical mole composition of the organic material initially present and at the conclusion of the process. If complete conversion is accomplished, the corresponding expression is

$$C_aH_bO_cN_d + \left(\frac{4a + b - 2c - 3d}{4}\right)O_2 \rightarrow$$
$$aCO_2 + \left(\frac{b - 3d}{2}\right)H_2O + dNH_3 \quad (9\text{-}6)$$

If the ammonia, NH_3, is to be oxidized to nitrate NO_3^-, the amount of oxygen required to accomplish this can be computed by the following two equations:

$$NH_3 + \tfrac{3}{2}O_2 \rightarrow HNO_2 + H_2O \tag{9-7}$$

$$HNO_2 + \tfrac{1}{2}O_2 \rightarrow HNO_3 \tag{9-8}$$

$$NH_3 + 2O_2 \rightarrow H_2O + HNO_3 \quad \text{(overall)} \tag{9-9}$$

Computation of the amount of oxygen required for the stabilization of prepared solid wastes is illustrated in Example 9-4.

EXAMPLE 9-4 *Oxygen Requirements for Composting*

Determine the amount of oxygen required to compost 1,000 lb of solid wastes. Assume that the initial composition of the material to be composed is given by $[C_6H_7O_2(OH)_3]_5$, that the final composition is estimated to be $[C_6H_7O_2(OH)_3]_2$, and that 400 lb of material remains after the composting process.

Solution

1. Determine the moles of material present initially and at the end of the process.

    Moles present initially:

    $$\frac{1,000 \text{ lb}}{[(30 \times 12) + (50 \times 1) + (25 \times 16)]} = 1.23$$

    Moles present at end:

    $$\frac{400 \text{ lb}}{[(12 \times 12) + (20 \times 1) + (10 \times 16)]} = 1.23$$

2. Determine the moles of material leaving the process per mole of material entering the process.

    $$n = \frac{1.23}{1.23} = 1.0$$

3. Determine the values for a, b, c, d, w, x, y, and z, and then determine the value of r and s in Eq. 9-5.

    For initial compound $(C_{30}H_{50}O_{25})$:

    $$a = 30 \qquad b = 50 \qquad c = 25 \qquad d = 0$$

    For final compound $(C_{12}H_{20}O_{10})$:

    $$w = 12 \qquad x = 20 \qquad y = 10 \qquad z = 0$$

    The value for r is

    $$r = 0.5[b - nx - 3(d - nz)]$$
    $$= 0.5[50 - 1.0(20)] = 15.0$$

    The value for s is

    $$s = a - nw$$
    $$= 30 - 1.0(12) = 18$$

4. Determine the amount of oxygen required.

    $$O_2/\text{lb} = 0.5(ny + 2s + 4 - c)O_2 \quad (\text{mol})(\text{lb/mol})$$
    $$= 0.5[1.0(10) + 2(18) + 15 - 25]1.23(32)$$
    $$= 708 \text{ lb } (321 \text{ kg})$$

5.   Check the computations with the materials balance.

	lb	kg
Process input:		
Organic material	1,000	454
Oxygen	708	321
	1,708	775
Process output:		
Organic material	400	181
Carbon dioxide 1.23(18)44	974	442
Water 1.23(15)(18)	332	151
	1,706	774

Design Considerations   The principal design considerations associated with the biological decomposition of prepared solid wastes are presented in Table 9-16. It can be concluded from this table that the preparation of a composting process is not a simple task, especially if optimum results are to be achieved. For this reason, most of the commercial composting operations that have been developed are highly mechanized and are carried out in specially designed facilities where the design factors reported in Table 9-16 can be controlled effectively. Some of the high-rate commercial composting operations that have been developed are discussed at the end of this chapter.

Although they are not shown in Table 9-16, land area requirements also must be considered. For example, in windrow composting for a plant with a capacity of 50 tons/day, about 2.5 acres of land would be required. Of this total, 1.5 acres would be devoted to buildings, plant equipment, and

TABLE 9-16   IMPORTANT DESIGN CONSIDERATIONS FOR AEROBIC COMPOSTING PROCESSES*

Item	Comment
Particle size	For optimum results solid wastes should be ground finely (1 to 3 in).
Seeding and mixing	Composting time can be reduced by seeding with partially decomposed solid wastes to the extent of about 1 to 5 percent by weight. Sewage sludge can also be added to prepared solid wastes. Where sludge is added the final moisture content is the controlling variable.
Mixing/turning	To prevent drying, caking, and air channeling, material in the process of being composted should be mixed or turned on a regular schedule or as required. Frequency of mixing or turning will depend on the type of composting operation.

TABLE 9-16   (Continued)

Item	Comment
Air requirements	Air with at least 50 percent of the initial oxygen concentration remaining should reach all parts of the composting material for optimum results, especially in mechanical systems.
Total oxygen requirements	The theoretical quantity of oxygen required can be estimated using Eq. 9-5. The actual quantity of air that must be supplied will vary depending on the operation.
Maximum rate of oxygen consumption	The rate can be estimated using the relationship $W_{O_2} = 0.07 \times 10^{0.31T}$, where W_{O_2} is equal to the rate of oxygen consumption in mg O_2/h/g of initial volatile matter and T is equal to the temperature in °C.
Moisture content	Moisture content should be in the range between 50 and 60 percent during the composting process. The optimum value appears to be about 55 percent.
Temperature	The optimum temperature for biological stabilization is between 45 and 55°C. For best results it has been found that the temperature should be maintained between 50 and 55°C for the first few days and between 55 and 60°C for the remainder of the active composting period. If temperatures are allowed to go beyond 66, biological activity can be reduced significantly.
Heat evolution	Heat released during the composting process is equal to the difference in the energy content of the material at the beginning and at the end of the composting process.
Carbon-nitrogen ratio	It has been found that initial carbon-nitrogen ratios (by weight) between 35 and 50 are optimum for aerobic composting. At lower ratios nitrogen is in excess and will be given off as ammonia. Biological activity is also impeded at lower ratios. At higher ratios nitrogen may be a limiting nutrient. After composting the carbon-nitrogen ratio for most municipal solid wastes will usually fall between 10 and 20.
pH	The pH should be prevented from rising above about 8.5 to minimize the loss of nitrogen in the form of ammonia as a gas.
Degree of decomposition	The degree of composition can be estimated by measuring the reduction in the organic matter present using the COD (chemical oxygen demand) test.
Respiratory quotient, RQ	$$RQ = \frac{O_2 \text{ including } CO_2}{O_2 \text{ consumption}}$$ RQ can be used as a measure of the degree of decomposition. When RQ = 1, the total oxygen supply has been used for oxidation of carbon. Where RQ > 1, more CO_2 is being formed than is being supplied, which is indicative of anaerobic decomposition. When RQ < 1, only part of the oxygen is being used to oxidize carbon. Low RQ values are characteristic of aerobic processes.
Control of pathogens	If properly conducted, it is possible to kill all the pathogens, weeds, and seeds during the composting process. To do this, the temperature must be maintained between 60 and 70°C for 24 h.

* Derived in part from Refs. 4, 5, 7, 9, and 17.

roads. For each additional 50 tons, it is estimated that 1.0 acre would be required for the composting operation and that 0.25 acre would be required for buildings and roads [7]. The land requirement for the highly mechanized systems varies with process. An estimate of 1.5 to 2.0 acres for a plant with a capacity of 50 tons/day is not unreasonable; for larger plants, the unit area requirements would be less.

Environmental Concerns   Some important environmental concerns relate to the production of odors, the blowing of loose materials, and the possibility of heavy metal toxicity to the land. Unless proper control is exerted, the production of odors can become a problem, especially in windrow composting. In the highly controlled processes, odor has not been a problem. The blowing of papers and plastic materials is also a problem in windrow composting.

A concern that may affect all composting operations, but especially those where mechanical shredders are used, involves the possibility of heavy metal toxicity. When metals in solid wastes are shredded, metal dust particles are generated by the action of the shredder. In turn, these metal particles may become attached to the materials in the light fraction. Ultimately, after composting, these metals would be applied to the soil. While many of them would have no adverse effects, metals such as cadmium (because of its toxicity) are of real concern. More experimental work needs to be done to quantify the impact of mechanical processing operations on the composition of compost.

Anaerobic Digestion

Conversion of the organic material in solid wastes to methane-containing gases can be accomplished in a number of ways, including hydrogasification, pyrolysis, and anaerobic digestion. Hydrogasification is usually associated with the conversion of petrochemical raw materials. Although the process has been tried with solid wastes, it is not well defined and therefore is not considered in this book. The production of methane from solid wastes by pyrolysis has been considered previously (see Sec. 9-2). The production of methane from solid wastes by anaerobic digestion, or *anaerobic fermentation* as it is often called, is described in the following discussion.

Process Description   In most processes where methane is to be produced from solid wastes by anaerobic digestion, three basic steps are involved. The first step involves preparation of the organic fraction of the solid wastes for anaerobic digestion and usually includes receiving, sorting, separation, and size reduction. The second step involves the addition of moisture and nutrients, blending, pH adjustment to about 6.7, heating of the slurry to between 55 and 60°C, and anaerobic digestion in a reactor with continuous flow, in which the contents are well mixed for a period of time varying from 5 to 10 days. In most operations the required moisture content and nutrients are added to the processed solid wastes in the form of sewage sludge. Depending on the chemical characteristics of the sludge, additional nutrients may also have to be added. Because foaming and the formation of a surface crust have caused problems in the digestion of solid wastes, adequate mixing is of fundamental importance in the design and operation of such systems. The third step involves capture, storage, and, if necessary, separation of the gas components evolved during the digestion process. The disposal of the digested sludge is an additional task that must be accomplished.

Process Microbiology   Carried out in the absence of oxygen, the anaerobic stabilization or conversion of organic compounds is thought to occur in three steps: the first involves the enzyme-mediated transformation (lique-faction) of higher-weight molecular compounds into compounds suitable for use as a source of energy and cell carbon; the second is associated with the bacterial conversion of the compounds resulting from the first step into identifiable lower-molecular-weight intermediate compounds; and the third step involves the bacterial conversion of the intermediate compounds into simpler end products, such as carbon dioxide (CO_2) and methane (CH_4).

   Because the specific organisms involved in the anaerobic fermentation of solid wastes are not well defined, it is common to see the terms *acid formers* and *methane formers* used when referring to the organisms involved in the conversion of the liquefied organic compounds into simpler acid and related intermediates and to carbon dioxide and methane.

   The overall conversion can be represented with the following equation [17]:

$$C_a H_b O_c N_d \rightarrow n C_w H_x O_y N_z + m CH_4 \\ + s CO_2 + r H_2 O + (d - nz) NH_3 \quad (9\text{-}10)$$

where $s = a - nw - m$
$r = c - ny - 2s$

The terms $C_a H_b O_c N_d$ and $C_w H_x O_y N_z$ are used to represent on a molar basis the composition of the material present at the start and the end of the process. If it is assumed that the organic wastes are stabilized completely, the corresponding expression is

$$C_a H_b O_c N_d + \left(\frac{4a - b - 2c + 3d}{4} \right) H_2 O \rightarrow \left(\frac{4a + b - 2c - 3d}{8} \right) CH_4 \\ + \left(\frac{4a - b + 2c + 3d}{8} \right) CO_2 + d NH_3 \quad (9\text{-}11)$$

   In operations where solid wastes have been mixed with sewage sludge, it has been found that the gas collected from the digesters contains between 50 and 60 percent methane. It has also been found that about 10 ft³ of gas is

TABLE 9-17   IMPORTANT DESIGN CONSIDERATIONS FOR ANAEROBIC DIGESTION*

Item	Comment
Size of material shredded	Wastes to be digested should be shredded to a size that will not interfere with the efficient functioning of pumping and mixing operations.
Mixing equipment	To achieve optimum results and to avoid scum buildup, mechanical mixing is recommended.
Percentage of solid wastes mixed with sludge	Although amounts of waste varying from 50 to 90+ percent have been used, 60 percent appears to be a reasonable compromise.
Hydraulic and mean cell residence time, $\theta_h = \theta_c$	Washout time is in the range of 3 to 4 days. Use 7 to 10 days for design or base design on results of pilot plant studies.
Loading rate	0.04 to 0.10 lb/ft³/day. Not well defined at present time. Significantly higher rates have been reported.
Temperature	Between 55 and 60°C.
Destruction of volatile solid wastes	Varies from about 60 to 80 percent; 70 percent can be used for estimating purposes.†
Total solids destroyed	Varies from 40 to 60 percent, depending on amount of inert material present originally.
Gas production	8 to 12 ft³/lb of volatile solids destroyed (CH_4 = 60 percent, CO_2 = 40 percent).

* Adapted in part from Ref. 10.
† Actual removal rates for volatile solids may be less depending on the amount of material diverted to the scum layer.
Note: lb/ft³/day $\times$ 16.019 = kg/m³·day
          ft³/lb $\times$ 0.062 = m³/kg

produced per pound of volatile solids destroyed, or about 7 ft³ of gas produced per pound of material added to the digester [10].

Design Considerations   Although the process of anaerobic digestion of solid wastes is still under development, some of the principal design considerations are reported in Table 9-17. Because of the variability of the results reported in the literature, it is recommended that pilot plant studies be conducted if the digestion process is to be used for the conversion of solid wastes.

Other Biological Processes

Other biological processes that have attracted attention include the conversion of solid wastes to protein and/or glucose and the recovery of gases from existing and newly designed landfill sites. The latter process is considered further in Chap. 10.

In considering both the chemical and biological conversion processes it would be helpful to keep in mind that if all the solid wastes from the 11 largest cities in the United States were converted to methane gas, about 700 billion ft³ would be produced (on the basis of 1971 figures) [16]. This represents about 3 percent of the 1971 United States consumption of 22.8 trillion ft^3 of natural gas.

9-4 RECOVERY OF ENERGY FROM CONVERSION PRODUCTS

Once conversion products have been derived from solid wastes by one or more of the chemical and biological methods listed in Tables 9-5 and 9-14, the next step involves their storage and/or use. If energy is to be produced from these products, then an additional conversion step is required. The purpose of this section is threefold: (1) to present basic flowsheets available for accomplishing this conversion, (2) to present data on the efficiency of the components used in the various conversion process flowsheets, and (3) to illustrate the use of the efficiency data in computing energy output.

Energy Recovery Systems

The principal components involved in the recovery of energy from heat, steam, various gases and oils, and other conversion products are boilers for the production of steam, steam and gas turbines for motive power, and electric generators for the conversion of motive power into electricity. Typical flowsheets for alternative energy recovery systems are shown in Fig. 9-10.

Steam Turbine–Generator Combination   Perhaps the most common flowsheet for the production of electric energy involves the use of a steam turbine–generator combination, shown in Fig. 9-10a. When solid wastes are to be used as the basic fuel source, four operational modes are readily identifiable. In the first, steam is produced from the incineration of processed solid wastes, from solid fuel pellets, or from unprocessed solid wastes. In the second, a boiler is used for the production of steam from the conversion of low-Btu fuel produced from solid wastes. In the third, steam is produced in a boiler fired with low-Btu gas that has been methanated. In the fourth, steam is produced in a boiler fired with oil and related compounds produced from solid wastes. If low-Btu fuel and oils derived from solid wastes are used, it may be necessary to provide a supplementary fuel source.

Gas Turbine–Generator Combination   Two flowsheets using a gas turbine–generator combination are shown in Fig. 9-10b and 9-10c. In Fig. 9-10b, the low-Btu gas is compressed under high pressure so that it can be used more effectively in the gas turbine. In the type of flowsheet shown in Fig. 9-10c, high-pressure and high-temperature exhaust gases are used. The compressor is usually driven by one wheel of the turbine and is used to compress air to maintain some other part of the process, such as a fluidized-bed combustion reactor.

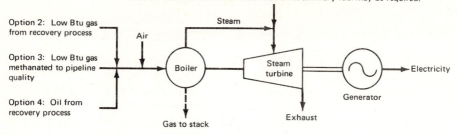

(a) **Options with steam turbine – generator combination**

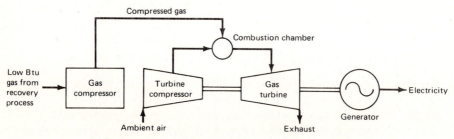

(b) **Options with gas compressor-gas turbine – generator combination**

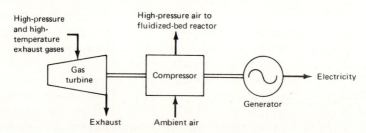

(c) **Option with gas turbine – compressor generator**

FIG. 9-10   Alternative energy recovery systems. (a) Options with steam turbine–generator combination. (b) Options with gas compressor–gas turbine–generator combination. (c) Option with gas turbine–gas compressor generator. (Adapted in part from Ref. 3.)

Process Heat Rate

In the production of power, it is common practice to consider the overall conversion efficiency in terms of the heat rate as expressed in Eq. 9-12 [15]:

$$\text{Heat rate (Btu/kWh)} = \frac{\text{heat supplied in fuel (Btu)}}{\text{energy generated (kWh)}} \tag{9-12}$$

When this equation is used, it is helpful to remember that the theoretical value for the mechanical equivalent of heat is equal to 3,413 Btu/kWh.

TABLE 9-18 TYPICAL HEAT RATES OF REPRESENTATIVE POWER PLANTS*

Type of heat	Plant heat rate, Btu/kWh	Plant thermal efficiency
All stationary steam plants, average	25,000	0.14
Central-station steam plants, average	11,500	0.30
Best large central-station steam plant	8,500	0.40
Small noncondensing industrial steam plant	35,000	0.10
Small condensing industrial steam plant	20,000	0.17
"By-product" steam power plant	4,500–5,000	0.70–0.75
Diesel plant	11,500	0.30
Natural-gas-engine plant	14,000	0.24
Gasoline-engine plant	16,000	0.21
Producer-gas-engine plant	18,000	0.19

* From Ref. 15.
Note: Btu/kWh × 1.055 = kJ/kWh

Thus, if the overall plant efficiency is 20 percent, the heat rate would be equal to 17,065 Btu/kWh [(3,413 Btu/kWh)/0.2]. Typical data for power plants are presented in Table 9-18. For comparative purposes, values for solid waste energy recovery systems range from 15,000 to 30,000 Btu/kWh. Energy efficiency and heat rate computations for a typical energy recovery system using an incinerator and a steam turbine–generator combination are illustrated in Example 9-5, at the end of this section.

Efficiency Factors

To assess the conversion efficiency of the proposed flowsheets given in Fig. 9-10, efficiency data must be known for the individual components. Representative data for boilers, pyrolytic reactors, gas turbines, steam turbine–generators combinations, electric generators, and related plant use and loss factors are given in Table 9-19 and discussed in this section.

Boilers   Important variables that affect the efficiency of boilers used in conjunction with the incineration of solid wastes include the input-heat content of the solid wastes, the moisture content, the exit temperature of the gases, and the configuration of the heat exchange system(s). Although all these variables will tend to be situation-specific, the data presented in Fig. 9-11 and Table 9-19 can be used as a guide in estimating boiler efficiency. As noted in Fig. 9-11, the plotted curves are based on cellulose solid wastes with 50 percent excess air used in the combustion process. The reported boiler efficiencies are assumed to include latent heat losses and radiation, sensible heat, and unburned carbon losses. For boilers

TABLE 9-19   TYPICAL THERMAL EFFICIENCY AND PLANT USE AND LOSS FACTORS FOR
INDIVIDUAL COMPONENTS AND PROCESSES USED FOR THE RECOVERY OF ENERGY FROM
SOLID WASTES

Component	Efficiency* Range	Efficiency* Typical	Comment
Incinerator-boiler	40–68	63	Mass-fired, see Figs. 8-6 and 9-11.
Boiler			
Solid fuel	65–72	70	Mass-fired, see Figs. 9-5 and 9-11.
Solid fuel	60–75	72	Processed solid wastes, see Fig. 9-6.
Low-Btu gas	60–80	75	Burners must be modified.
Oil-fired	65–85	80	Oils produced from solid wastes may have to be blended to reduce corrosiveness.
Pyrolysis reactor			
Conventional	65–75	70	
Purox	70–80	75	
Methanation process	80–90	85	Conversion of low-Btu gas to natural-gas quality.
Turbines			
Combustion gas			
Simple cycle	8–12	10	
Regenerative	20–26	24	Includes necessary appurtenances.
Expansion gas	30–50	40	
Steam turbine–generator system			
Less than 12.5 MW	24–30	29†‡	Includes condenser, heaters, and all other necessary appurtenances, but does not include boiler.
Over 25 MW	28–32	31.6†‡	
Electric generator			
Less than 10 MW	88–92	90	
Over 10 MW	94–98	96	
Plant use and loss factors			
Station service allowance			
Steam turbine–generator plant	4– 8	6	
Purox process	18–24	21	
Methanation process	18–22	20	
Unaccounted heat losses	2– 8	5	

* Theoretical value for mechanical equivalent of heat = 3,413 Btu/kWh.
† Efficiency varies with exhaust pressure. Typical value given is based on an exhaust pressure in the range of 2 to 4 in HgA.
‡ Heat rate = 10,800 Btu/kWh [(3,413 Btu/kWh)/0.316].
Note: Btu/kWh × 1.055 = kJ/kWh

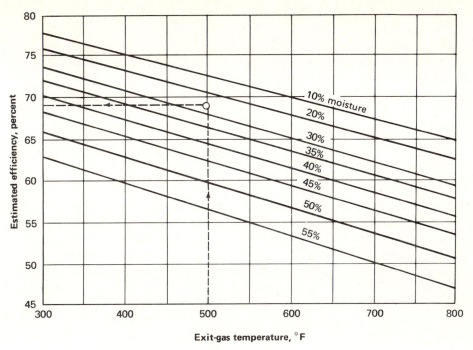

FIG. 9-11 Estimated boiler efficiency for solid waste incineration [11].

burning conventional fuels, efficiencies on the order of 85 percent are common.

Pyrolytic Reactors Typical data for both the conventional reactor and the Purox process incineration-pyrolysis reactor described previously are presented in Table 9-19.

Gas Turbine Data on the thermal efficiency of various gas turbines are given in Table 9-19. The efficiency values include an allowance for the necessary appurtenances.

Steam Turbine–Generator System The data reported in Table 9-19 for the steam turbogenerator are consistent with the best modern practice and reflect all the necessary allowances for condensers, heaters, and other appurtenances. Using the reported typical efficiency factor of 31.6 percent, the corresponding heat factor would be 10,800 Btu/kWh. If a boiler efficiency of 75 percent were achieved, the overall heat rate would be about 14,400 Btu/kWh. This compares well with the value given in Table 9-18 for central-station steam plants. The necessary computations that must be made in evaluating any energy option are illustrated in Example 9-5. Typical efficiency values for electric generators are also included as a separate item.

Other Use and Loss Factors   In any installation where energy is being produced, allowance must be made for the station or process power needs and for unaccounted for process heat losses. Typically, the auxiliary power allowance varies from 4 to 8 percent of the power produced. Process heat losses usually will vary from 2 to 8 percent. Both of these values must be considered in estimating the net heat rate when Eq. 9-12 is used.

EXAMPLE 9-5 *Computation of Energy Output and Efficiency for Energy Recovery System Using a Steam Boiler–Turbine–Generator Plant*

Estimate the amount of energy produced from a solid waste energy conversion system with a capacity of 1,000 tons/day. The system consists of an incinerator–boiler–steam turbine–electric generator combination. Also estimate the heat rate and overall process efficiency, assuming that the station service allowance and the unaccounted heat losses are 6 and 5 percent, respectively, of the total power produced. Assume that the energy value of the solid wastes is 4,500 Btu/lb as incinerated.

Solution

1.   Determine the energy output using data reported in Table 9-19. The required computations are summarized in Table 9-20.

TABLE 9-20 ENERGY OUTPUT AND EFFICIENCY FOR 1,000 TONS/DAY ENERGY RECOVERY PLANT USING A STEAM BOILER–TURBINE–GENERATOR PLANT FOR EXAMPLE 9-5

Item	Value
Energy available in solid wastes, million Btu/h	375
[(1,000 tons/day × 2,000 lb/ton × 4,500 Btu/lb)/(24 h/day × 10^6 Btu/million Btu)]	
Steam energy available, million Btu/h	263
(375 million Btu/h × 0.7)	
Electric power generation, kW	24,352
(263 million Btu/h)/(10,800 Btu/kWh*)	
Station service allowance, kW	−1,461
(24,352(0.06)	
Unaccounted heat losses, kW	−1,218
24,352(0.05)	
Net electric power for export, kW	21,673
Overall efficiency, percent	20
{(21,673 kW)/[(375,000,000 Btu/h)/(3,413 Btu/kWh)]}(100)	

*From Table 9-19 [10,800 Btu/kWh = (3,413 Btu/kWh)/0.316].
Note: Btu/lb × 2.326 = kJ/kg
Btu × 1.055 = kJ
Btu/kWh × 1.055 = kJ/kWh
Btu/h × 1.055 = kJ/h

(handwritten margin notes: NRG (Btu); ELECTRIC POWER; EFFICIENCY)

2. Determine the heat rate for the proposed plant by using Eq. 9-12.

$$\text{Heat rate} = \frac{375,000,000 \text{ Btu/h}}{21,673 \text{ kW}}$$

$$= 17,303 \text{ Btu/kWh } (18,255 \text{ kJ/kWh})$$

3. Determine the overall efficiency.

$$\text{Efficiency} = \{(21,673 \text{ kW})/[(375,000,000 \text{ Btu/h})/$$

$$(3,413 \text{ Btu/kWh})]\}100$$

$$= 20 \text{ percent}$$

Comment   If it is assumed that 10 percent of the power generated is used for the front-end processing system (typical values vary from 8 to 14 percent), then the net power for export would be 19,238 kW and the overall efficiency would be 17.5 percent.

9-5 / MATERIALS AND ENERGY RECOVERY SYSTEMS FLOWSHEETS

Thus far in this chapter, various front-end processing systems, rear-end conversion systems, and energy recovery systems have been discussed. In this section, the discussion is focused on some of the systems that have been proposed or built incorporating different types of front-end, rear-end, and energy conversion systems.

Systems Using Chemical Conversion Processes

Of the many systems using materials recovery and chemical conversion processes, three have been selected for detailed discussion. The first involves the recovery of materials and the production of power from processed solid wastes using a steam boiler and turbine generator. The second involves the recovery of materials and the production of power from processed solid wastes using a pyrolysis reactor. The third involves the recovery of materials and the production of fuel pellets for the generation of power. Systems for the codisposal of treatment plant sludges and solid wastes are also considered. Although the economics will vary with location, it appears that in order for energy conversion to be cost-effective, the capacity of the plant must be in excess of about 1,000 tons/day of solid wastes.

Steam Boiler–Turbine–Generator Plant   A flowsheet proposed for the recovery of raw materials and energy from processed solid wastes is shown in Fig. 9-12. A materials flow schematic is given in Fig. 9-13 [23]. In these figures solid wastes are delivered to the receiving station, which includes weighing and storage facilities. A storage capacity of 2 days is provided. Using an overhead crane, wastes are discharged to the feed conveyor for

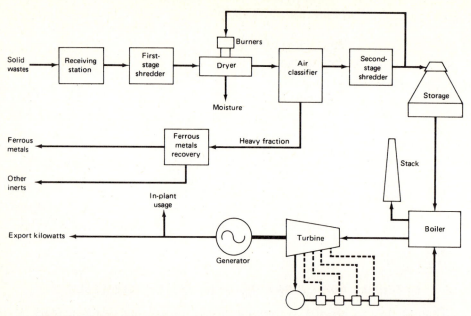

FIG. 9-12 Flowsheet for materials and energy recovery system for solid wastes. (Lunn, Low, Tom, and Hara Inc., and Metcalf & Eddy Engineers, Inc.)

first-stage shredding. After shredding, the wastes are passed through a dryer to remove moisture. An air classifier equipped with a vibrating screen separates the light and heavy fractions. The light materials pass through the induced-draft fan and are blown into a cyclone, which separates the light materials from the air. The exhaust air from the cyclone is cleaned with baghouse filters before being discharged to the atmosphere. The light material that is discharged from the bottom of the cyclone is transferred to a second-stage shredder by a belt conveyor.

The heavy fraction discharged from the classifier is transferred to the magnetic separation system for further processing to separate the ferrous metal from the heavy fraction. Belt conveyors are used to transfer the separated materials to storage containers for hauling from the plant.

After second-stage shredding, a pneumatic conveying system transfers the wastes from the end of the process train into the storage bin and from the storage bin into the boilers. The pneumatic system is a pressure type and includes pressure positive displacement blowers with silencers, rotary feed locks, and piping and cyclone separators complete with dust collectors. Processed solid wastes are stored in a teepee-style bin similar to that used in sugar mills (bridging and compacting problems are minimized with this type of bin). The processed solid wastes are suspension-fired in the steam boiler. Steam is used to produce power with a turbine–generator combination that has a heat rating of 10,695 Btu/kWh [23].

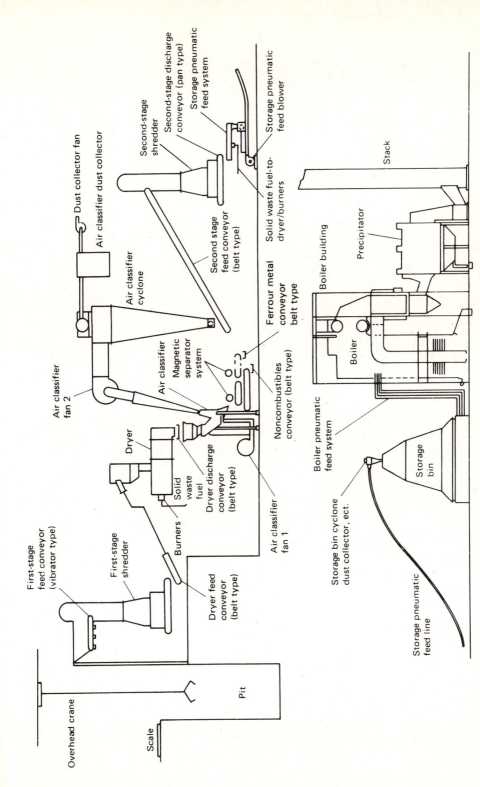

FIG. 9-13   Pictorial representation of materials and energy recovery system shown in Fig. 9-12 (Metcalf & Eddy Engineers, Inc.)

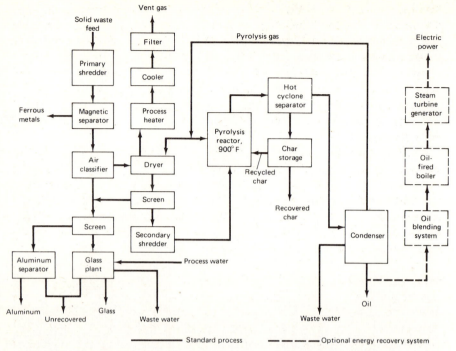

FIG. 9-14 Flowsheet for materials and energy recovery using the Garrett Pyrolysis Process. (Central Contra Costa Sanitary District and Brown and Caldwell Consulting Engineers [3].)

Incineration–Pyrolysis–Gas Turbine–Generator Plant   A variety of systems using the pyrolysis process have been proposed or are under development. Among them are the CPU 400, Garrett, Monsanto, Landgard, Pyrotek, and Wilwerding-Ward systems. A complete flowsheet for a materials and energy recovery system designed around the Garrett pyrolysis system is shown in Fig. 9-14 [3]. In Fig. 9-14, solid wastes are deposited in a receiving pit directly from the collection trucks. From there the waste is passed through a magnetic separator for the removal of ferrous metals. Air classification is then used to remove most of the remaining inorganics, such as glass, metal, dirt, and stones.

From the air classifier the heavy fraction is passed through a screen for the separation of metals and organics from the dirty-glass fraction. The metal fraction is then passed through an aluminum separator. The remaining materials are combined with inorganic rejects from the glass recovery process and stored before disposal. The dirty-glass fraction is applied to a glass reclamation system.

From the air classifier, the light fraction is passed through a rotary drum dryer to reduce the moisture. The dried light fraction is then screened to reduce the inorganic content. The dried, shredded, and essentially organic material is then discharged to a secondary shredder for further pulverization. Pulverized solid wastes from the secondary shredder are introduced into a pyrolytic reactor, designed to flash-pyrolyze the incoming wastes. Oil, off gases, and moisture produced from the pyrolytic process are passed through a hot-cyclone separator for removal of pyrolytic char, and from there through a condenser for separation of the oil and water from the pyrolytic gas. Pyrolytic char, separated from the reactor off gas in the hot cyclone, is quenched and stored to serve as a heat source. It also can be used to further purify treated waste water in an advanced treatment process.

Gas from the condenser is recycled to the process by being applied to the pyrolytic reactor and the light-fraction dryer. Off gases from the dryer are passed through a combustion chamber followed by an air cooler and a bag filter before being discharged to the atmosphere. Waste water condensed from the pyrolytic gas is combined with waste water from the glass separation process and discharged from the plant. The pyrolytic oil can be either sold or used in an oil-fired boiler for the production of power.

Resource Recovery and Fuel Pellet Production   A flowsheet used for the recovery of raw materials and organic fuel in the form of a semisolid pulp is shown in Fig. 9-15. In Fig. 9-15 a hydropulper is used as the first step in the process. Pieces of metal, tin cans, and other nonpulpable materials are ejected from the hydropulper from which ferrous metals are recovered

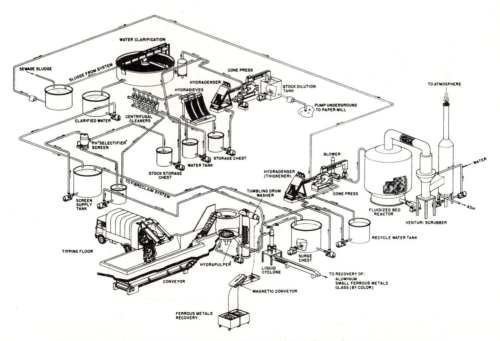

FIG. 9-15   Solid waste recycling system at Franklin, Ohio. (Black Clawson Fibreclaim, Inc.)

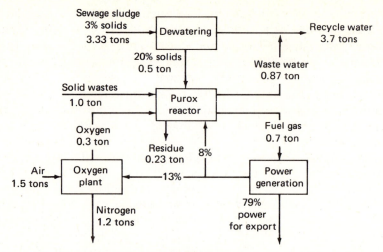

FIG. 9-16 Combined system for disposal of treatment plant sludges with processed solid wastes. (Union Carbide Corporation.)

after washing. The slurry extracted from the bottom of the hydropulper is pumped to a liquid cyclone for removal of heavier materials, of which approximately 80 percent is glass. After the slurry passes through the cyclone, long fibers used for papermaking are separated mechanically from the slurry. The remaining coarse organics such as rubber, textiles, plastics, leather, yard wastes, and small pieces of dirt and glass are then thickened. In the flowsheet shown in Fig. 9-15 this material is mixed with sewage solids from a nearby treatment plant. After an additional thickening step, the combined wastes are burned in a fluidized-bed reactor. When power is to be produced, the thickened organic solids (with or without the removal of fiber and without the addition of sewage solids) could be fired in a water-wall boiler. Alternatively, the resulting fuel could be sold in bulk or in the form of compressed fuel pellets.

A resource recovery and energy production facility using a flowsheet similar to that shown in Fig. 9-15, with the exception that paper fiber will not be recovered, is to be built in Hempstead, New York, and will be completed around 1980. The nonrecoverable inorganic and organic materials remaining in the slurry after passage through the liquid cyclone will be dewatered and converted into a pelletized fuel product. This material will be burned in utility boilers for the production of power. The materials to be recovered include ferrous metals, aluminum, and glass. Paper fiber is not to be recovered because there is a lack of sufficient markets resulting in part from a Federal Drug Administration ruling banning the use of this material in many paper products on the grounds of possible contamination.

Solid Waste–Sludge Processing Systems Because of the problems associated with the disposal of waste water and industrial treatment plant sludges, a number of processes have been proposed for the combined processing of solid wastes and sludges. In most of the processes, thickened or partially thickened sludge is mixed with processed solid waste fuel and burned in a boiler or pyrolyzed. A flowsheet proposed by Union Carbide Corporation using an incinerator-pyrolysis reactor is shown in Fig. 9-16. A similar flowsheet involving the use of multiple-hearth furnaces has been proposed and was under investigation in 1976 [3]. When reliable design information becomes available, it is anticipated that greater use will be made of such combined systems. Processed solid wastes can also be used as a fuel source for drying treatment plant sludges. Thus wastepaper may be more valuable as a fuel than as a raw material for resale.

Systems Using Biological Conversion Processes

Two complete systems using biological conversion processes have been selected for discussion here: the IDC-Naturizer system [5] and a process for the conversion of solid wastes to methane gas.

IDC-Naturizer   In Fig. 9-17, solid wastes are deposited in a receiving area or pit. From there they are conveyed to shredders for size reduction. After

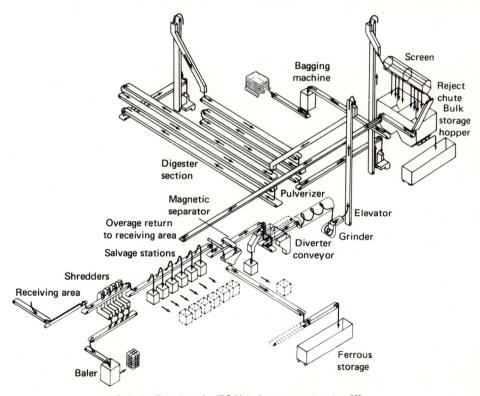

FIG. 9-17   Flowsheet for IDC-Naturizer compost system [5].

size reduction, various components are removed by magnetic separation. Then water is added to the wastes, and they are pulverized and reground before being transported to the digestion system where they are stacked to a depth of about 6 ft on continuous conveyors. These conveyors are about 9 ft wide and about 150 ft long. On the average of once a day, wastes are dumped or transferred to a lower conveyor where fans supply air to the compost. Temperatures within the compost are approximately 140°F, or in the range that is suitable for thermophilic microorganisms. After 2 days of processing, the material is reground and reinserted into the compost conveyor system.

At the end of a detention time of 5 days, the composted material is removed and passed through a screen. The screen separates noncompost materials, such as rags and plastic, from the compost. The separated compost is then reground and conveyed to outdoor curing piles. An additional 10 days is allowed for adequate curing of the composted material, after which it is sold in bulk, or enriched and bagged for retail sales.

Biological Conversion to Methane   A process for the biological conversion of solid wastes to methane gas proposed by Allis-Chalmers, Inc., and Waste Management, Inc., is shown schematically in Fig. 9-18 [16]. The first step involves shredding of the wastes. After ferrous metal separation and air classification, the light fraction is blended with sewage sludge or chemical nutrients, and the pH is adjusted. After being heated to a

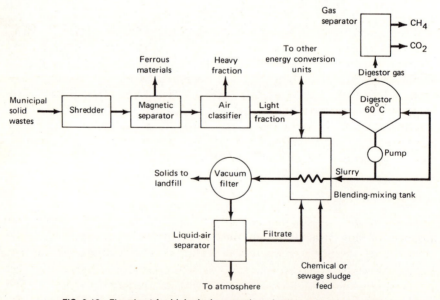

FIG. 9-18 Flowsheet for biological conversion of solid wastes to gas [16].

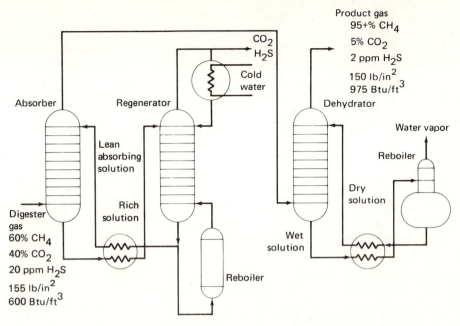

FIG. 9-19   Flowsheet for separation of methane from digester gas [16].

temperature of 130 to 140°F, the resultant slurry is fed to a digester whose contents are completely mixed. The detention time for the process is about 5 days. The gas released during digestion is said to contain about 50 to 60 percent methane by volume and has an energy value of about 600 Btu/ft³ [16]. After digestion the solids are dewatered before disposal.

If digester gas is to be sold to gas utilities, it usually will be necessary to upgrade its heating value from approximately 600 to about 975 Btu/ft³ which corresponds to the quality of pipeline natural gas. Normally, this involves removal of the water vapor and carbon dioxide from the gas. To minimize corrosion, hydrogen sulfide should also be removed. A flowsheet to accomplish this has been proposed by Pacific Gas and Electric Company of San Francisco and the East Bay Municipal Utility District of Oakland, California, as shown in Fig. 9-19 [16]. A gas processing system employing molecular sieves is currently in operation in Los Angeles for processing the gas from a landfill [21].

Review of Process Flowsheets

In reviewing the various process flowsheets presented in this chapter, the absence of any usable design data or information is apparent. This omission was deliberate because little usable or reliable *long-term* information is currently available for these systems, especially those using the chemical conversion processes that have been described. As more pilot-scale and full-scale installations become operative, this situation may change. A number of pilot- and full-scale installations are reviewed in Ref. 21. However, because many of these systems are proprietary, it may still be difficult to obtain actual performance data. If such systems are to be considered as part of a solid waste management plan or study, it is recommended that onsite visits be made to locations where each system under consideration is in operation.

9-6   DISCUSSION TOPICS AND PROBLEMS

9-1. Determine the relative energy consumption caused by the use of returnable and throwaway beverage bottles. The following steps in production and use must be considered: extraction of raw materials, manufacture, bottling, retailing, waste collection and disposal, recycling, and transportation (see Fig. 9-20). Using a 16-fluid-oz bottle for comparison, determine the energy needs per gallon of beverage for returnable and throwaway bottles for two situations: (1) the discarded bottles are disposed of in a landfill, and (2) a portion of the discarded glass is separated and recycled. Summarize the results of the energy computations in a table. The following data may be used.

Weight of 16-oz returnable bottle	1 lb
Weight of 16-oz throwaway bottle	0.656 lb
Life of returnable bottle	8 uses

Transportation energy:
Rail, 640 Btu/ton·mi
Truck, 2,400 Btu/ton·mi

Extraction of raw material	990 Btu/lb
Manufacture of container	7,738 Btu/lb
Manufacture of crown	242 Btu/cap
Bottling	6,100 Btu/gal
Retailer and consumer	neglect
Waste collection	89 Btu/lb
Separation, sorting, return for processing (assuming 30 percent recovery of glass)	1,102 Btu/lb of wastes
Landfill	neglect

Transportation:

From–To	Distance, mi	By rail, percent	By truck, percent	Remainder
Source of extraction– manufacturer	245	79	21	0
Manufacturer–bottler	345	16.3	70.2	Included in manufacturer
Bottler–retailer	231	0	74	Included in bottling

9-2. A city generates 500,000 tons/yr of solid wastes. As the operator of the salvage operation, you are interested in the amount of money that can be made or lost from salvaging various solid waste components. You are limited in the amount of material

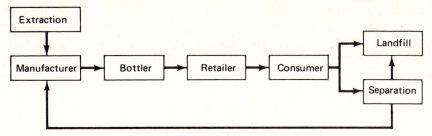

FIG. 9-20 Flowsheet for Prob. 9-1.

that each market can accept in a year. The various market values and limitations are listed below.

Assume labor costs are $100,000/yr and operating costs are $50,000. The building and the separation equipment cost $5,000,000 and is assumed to have an economic life of 10 yr at 10 percent interest (capital recovery factor = 0.16275). Given this information, how much money can be made or lost in 1 yr of operation? (*Note:* The VAM method discussed in Chapter 7 can be used effectively for the solution of this problem.)

1. Price paid at market, $/ton

| | Market | | | |
Item	M_1	M_2	M_3	M_4
Newspaper	5.00	4.50	4.75	5.00
Cardboard	4.00	4.50	4.65	4.85
Glass	18.00	16.00	17.00	15.00
Tin cans	5.50	6.00	6.00	5.75

2. Shipping cost to market, $/ton

| | Market | | | |
Item	M_1	M_2	M_3	M_4
Newspaper	0.50	0.51	0.54	0.48
Cardboard	0.70	0.42	0.54	0.40
Glass	0.25	0.25	0.27	0.16
Tin cans	0.80	0.76	0.72	0.80

3. Market capacity, tons/day

| Market* | | | |
M_1	M_2	M_3	M_4
13	25	12	20

* Total capacity for all components.

4. Composition of solid wastes

Item	Percent by weight
Newspaper	28
Cardboard	10
Glass	16
Tin cans	12

9-3. Using the data in Chap. 4 on moisture content and elemental composition (carbon, hydrogen, oxygen, nitrogen, sulfur, and ash), estimate the theoretical amount of air in pounds that would be required for the complete oxidation of 1 ton of municipal solid wastes with the composition given in Table 4-2, or, if you have sorted some solid wastes as part of your classwork, use your own sample.

9-4. The sludge from a waste-water treatment plant serving 500,000 persons is currently disposed of in a sanitary landfill. Because the capacity of the existing landfill will soon be exhausted, it has been proposed to incinerate the treatment plant sludge using processed solid wastes as a fuel source. Two alternative modes of operation are to be evaluated. In the first, treatment plant sludge with a fuel value of 7,500 Btu/lb (based on dry solids) and a solids content of 5 percent are to be mixed with processed solid wastes and incinerated. In the second, dewatered sludge with a solids content of 20 percent and a fuel value of 6,500 Btu/lb (based on dry solids) is to be mixed with the processed solid wastes before being incinerated. It should be noted that the fuel value of the dewatered solids is lower because of the chemicals added to aid in dewatering.

Assuming that the specific gravity of the combined dry sludge with or without the addition of chemicals is 1.10, the per capita sludge production on a dry basis is 0.35 lb/day, and the moisture content of the processed solid wastes is 20 percent, determine the amount of solid wastes that must be added to the treatment plant sludge to achieve a final moisture content of 60 percent. Would the required quantities of processed solid wastes be available in the wastes from the community? State clearly all your assumptions.

9-5. In Problem 9-4, estimate the expected temperature of combustion gases if the moisture content of the combined mixture of treatment plant sludge and processed solid wastes to be burned is to be 60 percent. Assume that the composition of the dry sludge is given by $C_5H_7NO_2$. Is the resulting temperature sufficient to avoid the production of odors?

9-6. Compute the theoretical amount of oxygen required to completely oxidize biologically 1 lb of solid waste with the following composition: $C_7H_{13}O_2N$. Assume that the nitrogen is converted to ammonia (NH_3) in the first step and that the ammonia is converted ultimately to nitrate (NO_3^-).

9-7. Using the data for municipal solid wastes given in Table 4-2 or your own sample data, estimate the amount of compost that could be produced per ton of solid wastes. Assume that the solid wastes would be sorted prior to composting and that a 40 percent reduction in weight would result from the composting process.

9-8. If the pH of the material being composted in Problem 9-6 were to rise to a value of 10 after the nitrogen had been converted to ammonia, estimate how much of the ammonia might be lost if forced aeration were to continue before the pH was lowered. Assume that the following equation and data apply.

$$NH_3 + H_2O \rightleftharpoons NH_4^+ + OH^-$$

$$K_b = 1.8 \times 10^{-5}(25°C)$$

$$K_w = 10^{-14}$$

9-9. When municipal solid wastes from a community of 100,000 arrive at the composting plant, the moisture content is usually below the desired range of 55 to 70 percent for optimum composting. Rather than adding water to obtain the necessary moisture content, it has been suggested that sludge from the waste-water plant be added to achieve the same result. Determine the required amount of sludge with a solids content of 5 percent that must be added to the solid wastes to achieve the desired moisture content of 55 percent. Assume that the municipal solid waste generation rate is equal to 6.5 lb/capita/day and that the moisture content of the solid waste is 20 percent.

9-10. Estimate the available energy for export from a 1,000-tons/day Purox process plant. Assume that the following data are applicable:

1. Energy content of solid wastes = 4,500 Btu/lb

2. Energy loss in incineration-pyrolysis conversion process = 25 percent

3. Process fuel usage for steam production, building heat, and process maintenance based on the percentage of energy available in conversion gas = 8 percent

4. Gas-turbine thermal efficiency = 24 percent

5. Electrical generator efficiency = 96 percent

6. In-plant electric power usage based on the percentage of total power generated = 21 percent

9-7 REFERENCES

1. Aiba, S., A. E. Humphrey, and N. F. Millis: "Biochemical Engineering," Academic, New York, 1965.

2. Blakebrough, N. (ed.): "Biochemical and Biological Engineering Science," vol. 1, Academic, New York, 1967.

3. Brown and Caldwell Consulting Engineers: *Solid Waste Resource Recovery Study*, Report prepared for Central Contra Costa Sanitary District, San Francisco, 1974.

4. *Composting of Municipal Solid Wastes in the United States*, U.S. Environmental Protection Agency, Waste Management Series, Publication SW-47r, Washington, D.C., 1971.

5. Drobny, N. L., H. E. Hull, and R. F. Testiu: *Recovery and Utilization of Municipal Solid Waste*, U.S. Environmental Protection Agency, Publication SW-10c, Washington, D.C., 1971.

6. Fisher, T. F., M. L. Kasbohm, and J. R. Rivero: *The 'Purox' System*, Presented at the AICHE 80th National Meeting, Boston, 1975.

7. Gotaas, H. B.: *Composting*, World Health Organization, Geneva, Switzerland, 1956.

8. Humphrey, A. E.: Current Developments in Fermentation, *Chemical Engineering*, vol. 81, no. 25, 1974.

9. Jeris, J. S. and R. Regan: Optimum Conditions for Composting, in C. L. Mantell (ed.), "Solid Wastes, Origin, Collection, Processing and Disposal," Wiley-Interscience, New York, 1975.

10. McFarland, J. M., et al.: *Comprehensive Studies of Solid Wastes Management*, Sanitary Engineering Research Laboratory, SERL Report 72-3, University of California, Berkeley, 1972.

11. Meissner, H. G.: Central Incineration of Community Wastes, in R. C. Corey (ed.), "Principles and Practices of Incineration," Wiley-Interscience, New York, 1969.

12. Meller, F. H.: *Conversion of Organic Solid Wastes into Yeast—An Economic Evaluation*, U.S. Department of Health, Education, and Welfare, Public Health Service, Publication 1909, Washington, D.C., 1969.

13. Metcalf & Eddy, Inc.: "Wastewater Engineering: Collection, Treatment, Disposal," McGraw-Hill, New York, 1972.
14. Orning, A. A.: Principles of Combustion, in R. C. Corey (ed.), "Principles and Practices of Incineration," Wiley-Interscience, New York, 1969.
15. Perry, R. H., C. H. Chilton, and S. D. Kirkpatrick: "Chemical Engineers Handbook," 4th ed., McGraw-Hill, New York, 1963.
16. Ricci, L. J.: Garbage Routes of Methane, *Chemical Engineering,* vol. 81, no. 10, 1974.
17. Rich, L. G.: "Unit Processes of Sanitary Engineering," Wiley, New York, 1963.
18. *Seattle's Solid Waste an Untapped Resource,* Departments of Engineering and Lighting, City of Seattle, Washington, 1974.
19. Schwieger, R. G.: Power from Waste, *Power,* vol. 119, no. 2, 1975.
20. *Solid Waste Disposal Resource Recovery,* undated brochure, Environmental Systems Department, Union Carbide, New York.
21. *Status of Technology in the Recovery of Resources from Solid Wastes,* County Sanitation Districts of Los Angeles County, Los Angeles, 1976.
22. Stear, J. R.: *Municipal Incineration: A Review of Literature,* Environmental Protection Agency, Office of Air Programs, Publication AP-79, 1971.
23. Sunn, Low, Tom, & Hara, Inc. and Metcalf & Eddy, Inc.: *Feasibility of Power Generation from Solid Wastes on Oahu,* Honolulu, Hawaii, 1975.
24. Wood, D. K. and G. Tchobanoglous: Trace Elements in Biological Waste Treatment, *Journal Water Pollution Control Federation,* vol. 47, no. 7, 1974.

10

Disposal of Solid Wastes and Residual Matter

Ultimately, something must be done with the solid wastes that are collected and of no further use and with the residual matter after solid wastes have been processed and the recovery of conversion products and/ or energy has been accomplished. There are only two alternatives available for the long-term handling of solid wastes and residual matter: disposal on or in the earth's mantle, and disposal at the bottom of the ocean. Disposal on land is by far the most common method in use today and is therefore the primary subject of this chapter. Although disposal in the atmosphere has been suggested as a third alternative, it is not a viable method because material discharged into the atmosphere is ultimately deposited either on the earth or in the sea by a variety of natural phenomena, the most important of which is rainfall.

Ocean dumping of municipal solid wastes was commonly used at the turn of the century [17] and continued until 1933 when it was prohibited by a United States Supreme Court decision involving New York City. Some industrial solid wastes are still discharged at sea, however, and recently the concept of using the ocean floor as a waste storage location has received some attention. For these reasons, ocean storage is discussed briefly at the end of this chapter.

Based on past experience in cities throughout the United States and elsewhere in the world, land disposal in the form of a sanitary landfill has proved to be the most economical and acceptable method for the disposal of solid wastes. The term *sanitary landfill* means an operation in which the wastes to be disposed of are compacted and covered with a layer of soil at the end of each day's operation (see Fig. 10-1). When the disposal site has reached its ultimate capacity—that is, after all disposal operations have been completed—a final layer of 2 ft or more of cover material is applied. Open dumping, as distinguished from sanitary landfilling, is still used in parts of the country, but is no longer an acceptable means of land disposal from an aesthetic, environmental, or sanitary standpoint. The advantages and disadvantages of sanitary landfills are reported in Table 10-1.

The planning, analysis, and design of modern land disposal systems involve the application of a variety of scientific, engineering, and economic principles. Because of the importance of land disposal, all aspects of the design and operation of sanitary landfills are described in this chapter,

TABLE 10-1   ADVANTAGES AND DISADVANTAGES OF SANITARY LANDFILL*

Advantages	Disadvantages
1. Where land is available, a sanitary landfill is usually the most economical method of solid waste disposal.	1. In highly populated areas, suitable land may not be available within economical hauling distance.
2. The initial investment is low compared with other disposal methods.	2. Proper sanitary landfill standards must be adhered to daily or the operation may result in an open dump.
3. A sanitary landfill is a complete or final disposal method as compared to incineration and composting which require additional treatment or disposal operations for residue, quenching water, unusable materials, etc.	3. Sanitary landfills located in residential areas can provoke extreme public opposition.
	4. A completed landfill will settle and require periodic maintenance.
4. A sanitary landfill can receive all types of solid wastes, eliminating the necessity of separate collections.	5. Special design and construction must be utilized for buildings constructed on completed landfill because of the settlement factor.
5. A sanitary landfill is flexible; increased quantities of solid wastes can be disposed of with little additional personnel and equipment.	6. Methane, an explosive gas, and the other gases produced from the decomposition of the wastes may become a hazard or nuisance and interfere with the use of the completed landfill.
6. Submarginal land may be reclaimed for use as parking lots, playgrounds, golf courses, airports, etc.	

* From Ref. 26.

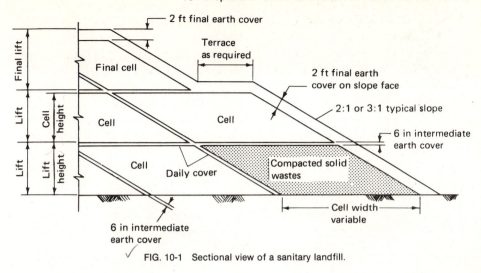

FIG. 10-1 Sectional view of a sanitary landfill.

including (1) factors in landfill site selection, (2) landfilling methods and operations, (3) reactions occurring in completed landfills, (4) gas and leachate movement and control, and (5) design of landfills. Management policies and regulations are discussed in Chap. 17.

10-1 SITE SELECTION

Factors that must be considered in evaluating potential solid waste disposal sites include (1) available land area, (2) impact of processing and resource recovery, (3) haul distance, (4) soil conditions and topography, (5) climatological conditions, (6) surface-water hydrology, (7) geologic and hydrogeologic conditions, (8) local environmental conditions, and (9) potential ultimate uses for the completed site. Because these factors also can be used to screen out unsuitable sites, methods for the preliminary screening and ultimate selection of sites are also presented where appropriate. Final selection of a disposal site usually is based on the results of a preliminary site survey, results of engineering design and cost studies, and an environmental impact assessment. Additional details on site selection from a management standpoint are presented in Chap. 17.

Available Land Area

In selecting potential land disposal sites, it is important to ensure that sufficient land area is available. Although there are no fixed rules concerning the area required, it is desirable to have sufficient area to operate for at least 1 yr at a given site. For shorter periods, the disposal operation becomes considerably more expensive, especially with respect to site

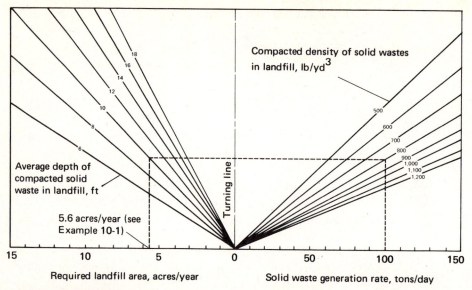

FIG. 10-2   Required landfill area as a function of solid waste generation rate, compacted density, and depth of compacted wastes.

preparation, provision of auxiliary facilities, and completion of the final cover.

   To estimate the amount of land area required for preliminary planning purposes, the curve presented in Fig. 10-2 can be used, as illustrated in Example 10-1.

EXAMPLE 10-1   *Estimation of Required Landfill Area*

Estimate the required landfill area for a community with a population of 31,000. Assume that the following conditions apply:

1.   Solid waste generation = 6.4 lb/capita/day (see Table 4-14)
2.   Compacted density of solid wastes in landfill = 800 lb/yd³
3.   Average depth of compacted solid wastes = 10 ft

Solution

1.   Determine the daily solid wastes generation rate in tons per day.

$$\text{Generation rate} = \frac{(31,000 \text{ people})(6.4 \text{ lb/capita/day})}{2,000 \text{ lb/ton}}$$

$$= 100 \text{ tons/day } (90,720 \text{ kg/day})$$

2. Find the required area by using Fig. 10-2. For the solid waste generation rate computed in step 1, the required landfill area is found to be 5.6 acres/yr.

3. Computationally, the required area is determined as follows:

$$\text{Volume required/day} = \frac{100 \text{ tons/day} \times 2,000 \text{ lb/ton}}{800 \text{ lb/yd}^3}$$

$$= 250 \text{ yd}^3/\text{day} \ (191 \text{ m}^3/\text{day})$$

$$\text{Area required/yr} = \frac{(250 \text{ yd}^3/\text{day})(365 \text{ days/yr})(27 \text{ ft}^3/\text{yd}^3)}{(10 \text{ ft})(43,560 \text{ ft}^2/\text{acre})}$$

$$= 5.66 \text{ acres/yr} \ (2.29 \text{ hectares/yr})$$

Comment The actual site requirements will be greater than the value computed, because additional land is required for site preparation, access roads, utility access, etc. Typically, this allowance varies from 20 to 40 percent. A more rigorous approach to the determination of the required landfill area involves consideration of the compactability of the individual solid waste components (see Sec. 10-5).

Impact of Resource Recovery

In the initial assessment of potential disposal sites, it is important to project the extent of resource recovery processing activities that are likely to occur in the future and determine their impact on the quantity and condition of the residual materials to be disposed of. For example, if 50 percent of the paper were to be recycled, the weight of materials to be disposed of and the landfill area requirements would be reduced. It is also important to know whether the recovery facilities are to be located at the disposal site.

Haul Distance

The haul distance is one of the important variables in the selection of a disposal site. From computations presented in Chaps. 6 and 7, it is clear that the length of the haul can significantly affect the overall design and operation of the waste management system. Although minimum haul distances are desirable, other factors must also be considered. These include collection route location, local traffic patterns, and characteristics of the routes to and from the disposal site (condition of the routes, traffic patterns, and access conditions).

Soil Conditions and Topography

Because it is necessary to provide cover material for each day's landfill and a final layer of cover after the filling is completed, data must be obtained on the amounts and characteristics of the soils in the area. If the soil under the proposed landfill area is to be used for cover material, data will be available from the geologic and hydrogeologic investigation. If cover material is to be obtained from a borrow pit, test borings will be needed to characterize the material adequately. The local topography must be considered because it will affect the type of landfill operation to be used, the equipment requirements, and the extent of work necessary to make the site usable.

Climatologic Conditions

Local weather conditions must also be considered in the evaluation of potential sites. In many locations, access to the site will be affected by winter conditions. Where freezing is severe, landfill cover material must be available in stockpiles when excavation is impractical. Wind and wind patterns must also be considered carefully. To avoid blowing or flying papers, windbreaks must be established. The specific form of windbreak depends on local conditions. Ideally, prevailing winds should blow toward the filling operation.

Surface-Water Hydrology

The local surface-water hydrology of the area is important in establishing the existing natural drainage and runoff characteristics that must be considered. Other conditions of flooding must also be identified.

Geologic and Hydrogeologic Conditions

Geologic and hydrogeologic conditions are perhaps the most important factors in establishing the environmental suitability of the area for a landfill site. Data on these factors are required to assess the pollution potential of the proposed site and to establish what must be done to the site to ensure that the movement of leachate or gases from the landfill will not impair the quality of local groundwater or contaminate other subsurface or bedrock aquifers. In the preliminary assessment of alternative sites, it may be possible to use United States Geological Survey maps and state or local geologic information. Logs of nearby wells can also be used.

Local Environmental Conditions

While it has been possible to build and operate landfill sites in close proximity to both residential and industrial developments, extreme care must be taken in their operation if they are to be environmentally acceptable with respect to noise, odor, dust, and vector control. Flying papers and plastic films must also be controlled.

Ultimate Uses

One of the advantages of a landfill is that, once it is completed, a sizable area of land becomes available for other purposes. Because the ultimate use affects the design and operation of the landfill, this issue must be resolved before the layout and design of the landfill are started. For example, if large, open structures (such as a warehouse) are to be built, footing locations must be established and allowances made for them. If the completed landfill is to be used as a park or golf course, a staged planting program should be initiated and continued as portions of the landfill are completed.

10-2 LANDFILLING METHODS AND OPERATIONS

To use the available area at a landfill site effectively, a plan of operation for the placement of solid wastes must be prepared. Various operational methods have been developed primarily on the basis of field experience. The methods used to fill dry areas are substantially different from those used to fill wet areas.

Conventional Methods for Dry Areas

The principal methods used for landfilling dry areas may be classified as (1) area, (2) trench, and (3) depression. In addition to these methods, which usually are used for unprocessed municipal solid wastes, landfilling using milled (shredded) solid wastes is also discussed.

Area Method The area method is used when the terrain is unsuitable for the excavation of trenches in which to place the solid wastes. Operationally (see Fig. 10-3) the wastes are unloaded and spread in long, narrow strips on the surface of the land in a series of layers that vary in depth from 16 to 30 in. Each layer is compacted as the filling progresses during the course of the day until the thickness of the compacted wastes reaches a height varying from 6 to 10 ft. At that time, and at the end of each day's operation, a 6- to 12-in layer of cover material is placed over the completed fill. The cover material must be hauled in by truck or earth-moving equipment from adjacent land or from borrow-pit areas.

The filling operation usually is started by building an earthen levee against which wastes are placed in thin layers and compacted. The length of the unloading area varies with the site conditions and the size of the operation. The width over which the wastes are compacted varies from 8 to 20 ft, again depending on the terrain. A completed lift, including the cover material, is called a *cell* (see Fig. 10-3). Successive lifts are placed on top of one another until the final grade is reached that was called for in the ultimate development plan. The length of the unloading area used each

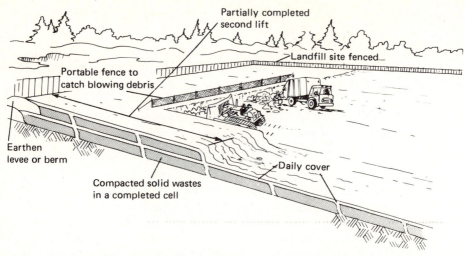

FIG. 10-3   Area method of operation for a sanitary landfill.

day should be such that the final height of the fill is reached at the end of each day's operation.

If a small amount of usable cover material is available at the disposal site, the ramp variation of the area method is often used (see Fig. 10-4). In this method, solid wastes are placed and compacted as described for the area method and are partially or wholly covered with earth scraped from the base of the ramp. Additional soil must be hauled in, as in the area method. Because of increasing costs and the problems associated with

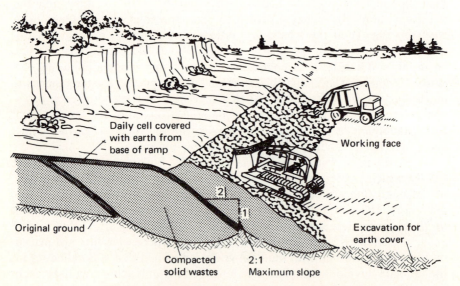

FIG. 10-4   Ramp method of operation for a sanitary landfill.

obtaining usable cover material, the use of the ramp method must be based on a detailed economic feasibility study.

Trench Method   The trench method of landfilling is ideally suited to areas where an adequate depth of cover material is available at the site and where the water table is near the surface. Typically, as shown in Fig. 10-5, solid wastes are placed in trenches varying from 100 to 400 ft in length, 3 to 6 ft in depth, and 15 to 25 ft in width. To start the process, a portion of the trench is dug and the dirt is stockpiled to form an embankment behind the first trench. Wastes are then placed in the trench, spread into thin layers (usually 18 to 24 in), and compacted. The operation continues until the desired height is reached. The length of trench used each day should be such that the final height of fill is reached at the end of each day's operation. The length also should be sufficient to avoid costly delays for collection vehicles waiting to unload. Cover material is obtained by excavating an adjacent trench or continuing the trench that is being filled.

Depression Method   At locations where natural or artificial depressions exist, it is often possible to use them effectively for landfilling operations. Canyons, ravines, dry borrow pits, and quarries have all been used for this purpose. The techniques to place and compact solid wastes in depression landfills vary with the geometry of the site, the characteristics of the cover material, the hydrology and geology of the site, and the access to the site.

If a canyon floor is reasonably flat, the first fill in a canyon site may be carried out using the trench method operation discussed previously. Once

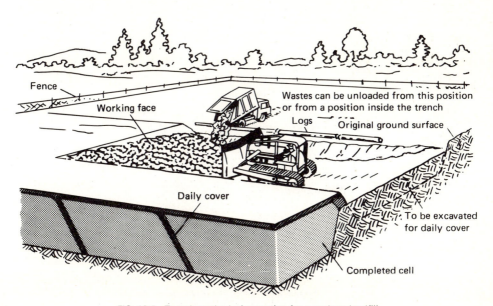

FIG. 10-5 Trench method of operation for a sanitary landfill.

filling in the flat area has been completed, filling starts at the head end of the canyon (see Fig. 10-6) and ends at the mouth. This practice prevents the accumulation of water behind the landfill. Wastes usually are deposited on the canyon floor and from there are pushed up against the canyon face at a slope of about 2 to 1. In this way, a high degree of compaction can be achieved. Compacted densities as high as 1,200 lb/yd³ have been reported. Even higher densities have been recorded in the lower portions of the landfill as the height of the fill increases.

Pit and quarry landfill sites are almost always lower than the surrounding terrain, and so control of surface drainage is often the critical factor in the development of such sites. As with canyon sites, pit and quarry sites are filled in multiple lifts, and the method of operation is essentially the same. A key to the successful use of pits or quarries is the availability of adequate cover material to cover the individual lifts as they are completed and to provide a final cover over the entire landfill when the final height is reached. Because of settlement, it is usually desirable to fill pit and quarry sites to a level slightly above that of the surrounding terrain.

Landfilling with Milled Solid Wastes   An alternative method of landfilling that is being tried in several locations throughout the United States involves milling of the solid wastes before placement in a landfill. The most comprehensive study of this method of operation was conducted at Madison, Wisconsin [21]. From evidence collected to date (1976), it appears that a daily cover of earth is not necessary. In a multilift area-method landfill, the lower layer can be left exposed until the next layer is placed. After the final landfill height is reached, a cover layer of earth should be placed to prepare the site for other uses. Odors and blowing litter have not been a problem. Also, it has been found that rats cannot survive on milled solid wastes containing up to 20 percent food wastes.

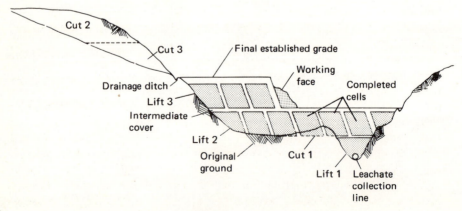

FIG. 10-6   Landfilling in a canyon or ravine [10].

Although flies can grow on milled solid wastes, they have not proved to be a problem. The final density of the milled landfill has been reported to be up to 35 percent greater than that of unprocessed wastes [21].

Although the advantage of this method is significant in areas where available cover material is in short supply, a number of factors must be carefully considered before it can be adopted. First, there is the added cost associated with the milling and related ancillary facilities. Second, even if this method of operation is adopted, some type of landfill will be required for the wastes that cannot be milled effectively. Third, by leaving the landfill uncovered, the movement of leachate may be accelerated and thus may become a limiting factor.

Conventional Methods for Wet Areas

Swamps and marshes, tidal areas, and ponds, pits, or quarries are typical wet areas that have been used as landfill sites. Because of the problems associated with contamination of local groundwaters, the development of odors, and structural stability, the design of landfills in wet areas requires special attention.

In the past, landfilling in wet areas was considered acceptable if reasonably adequate drainage were provided and if nuisance conditions did not develop. The usual practice was to divide the area into cells or lagoons and to schedule the filling operations so that one individual cell or lagoon would be filled each year. Often, solid wastes were placed directly in the water in areas with high groundwater levels. As an alternative, clean fill material was added up to, or slightly above, the water level before waste filling operations were started.

To withstand mud waves and to increase structural stability, dikes used to divide the cells or lagoons were constructed with riprap, trees, tree limbs, lumber, demolition wastes, and related materials in addition to clean fill material. In some cases, to prevent the movement of malodorous leachate and gases from completed cells or lagoons, clay and lightweight interlocking steel or wood-sheet piling have been used.

More recently, because of concern over the possibility of groundwater contamination by both leachate and gases from landfills and the development of odors, the direct filling of wet areas is no longer considered acceptable. If wet areas are to be used as landfill sites, special provisions must be made to contain or eliminate the movement of leachate and gases from completed cells. Usually this is accomplished by first draining the site and then lining the bottom with a clay liner or other appropriate sealants. If a clay liner is used, it is important to continue operation of the drainage facility until the site is filled in order to avoid the creation of uplift pressures that could cause the liner to rupture from heaving. The use of clay liners is considered further later in this chapter (see Sec. 10-4).

Alternative Operational Plans

In addition to the conventional methods of landfilling unprocessed and processed solid wastes, specialized methods are being developed. Alternative plans currently under investigation include (1) the recycling of leachate to accelerate the rate of anaerobic decomposition, and (2) the mixing of sewage sludge and solid wastes to accelerate the anaerobic decomposition of the wastes, with the objective of capturing the conversion gases for use in energy recovery systems.

The impacts of these alternatives on existing methods used for landfilling can be significant. For example, if the gases produced from the anaerobic decomposition of solid wastes are to be captured effectively, the use of deep, individual clay-lined cells, in which the wastes are placed without intermediate layers of cover material, appears to be most effective. This method of operation would, in turn, necessitate the development of new operational methods and landfill designs.

10-3   REACTIONS OCCURRING IN COMPLETED LANDFILLS

To plan and design sanitary landfills effectively, it is important to understand what takes place within a landfill after filling operations have been completed. Solid wastes placed in a sanitary landfill undergo a number of simultaneous biological, physical, and chemical changes. Among the more important of these changes are the following: (1) the biological decay of organic putrescible material, either aerobically or anaerobically, with the evolution of gases and liquids; (2) the chemical oxidation of materials; (3) the escape of gases from the fill and lateral diffusion of gases through the fill; (4) the movement of liquids caused by differential heads; (5) the dissolving and leaching of organic and inorganic materials by water and leachate moving through the fill; (6) the movement of dissolved material by concentration gradient and osmosis; and (7) the uneven settlement caused by consolidation of material into voids [24]. The decomposition and stabilization in a landfill depend on many factors, such as the composition of the wastes, the degree of compaction, the amount of moisture present, the presence of inhibiting materials, the rate of water movement, and temperature.

Because of the number of interrelated influences, it is difficult to define the conditions that will exist in any landfill or portion of a landfill at any stated time. In general, it may be said that the rates of chemical and biological reactions in a sanitary landfill increase with the temperature and the amount of moisture present until an upper limit is reached in each instance [24]. Decomposition, the formation of gases and leachate, and the settlement and structural characteristics of landfills are described further in the following discussion.

Decomposition in Landfills

The organic biodegradable components in solid wastes begin to undergo bacterial decomposition as soon as they are placed in a landfill. Initially, bacterial decomposition occurs under aerobic conditions because a certain amount of air is trapped within the landfill. However, the oxygen in the trapped air is soon exhausted, and the long-term decomposition occurs under anaerobic conditions. The principal source of both the aerobic and the anaerobic organisms responsible for the decomposition is the soil material that is used as a daily and final cover.

The overall rate at which the organic materials decompose depends on their characteristics and, to a large extent, on the moisture content. In general, the organic materials present in solid wastes can be divided into three major classifications: (1) those that contain cellulose or derivatives of cellulose; (2) those that do not contain cellulose or cellulose derivatives; and (3) plastics, rubber, and leather.

Cellulose is a major constituent of organic wastes, such as paper, rags, string, straw, and plant tissues. With the exception of plastics, the principal noncellulose organics are proteins, carbohydrates, and fats. Mineral salts in very limited quantities and moisture are almost always associated with these materials. Plastics that may be found in solid wastes are so many and so varied that no general list is possible in this text.

With the above wastes, the principal end products of anaerobic decomposition are partially stabilized organic materials, intermediate volatile organic acids, and various gases (including carbon dioxide, methane, nitrogen, hydrogen, and hydrogen sulfide). Under normal conditions the rate of decomposition, as measured by gas production, reaches a peak within the first 2 yr and then slowly tapers off, continuing in many cases for periods up to 25 yr or more. If moisture is not added to the wastes in a well-compacted landfill, it is not uncommon to find materials in their original form years after they were buried.

Gases in Landfills

Gases found in landfills include air, ammonia, carbon dioxide, carbon monoxide, hydrogen, hydrogen sulfide, methane, nitrogen, and oxygen. Data on the molecular weight and density of these gases are presented in Table 10-2. Carbon dioxide and methane are the principal gases produced from the anaerobic decomposition of the organic solid waste components. Typical data on the percentage distribution of gases found in a landfill are reported in Table 10-3. The high initial percentage of carbon dioxide is the result of aerobic decomposition. Aerobic decomposition continues to occur until the oxygen in the air initially present in the compacted wastes is depleted. Thereafter, decomposition will proceed anaerobically [3]. As shown, after about 18 months the composition of the gas remains reasona-

TABLE 10-2   MOLECULAR WEIGHT AND DENSITY OF GASES FOUND IN SANITARY LANDFILLS AT STANDARD CONDITIONS (0°C, 1 atm)*

Gas	Formula	Molecular weight	Density g/l	Density lb/ft³
Air			1.2928	0.0808
Ammonia	NH₃	17.03	0.7708	0.0482
Carbon dioxide	CO₂	44.00	1.9768	0.1235
Carbon monoxide	CO	28.00	1.2501	0.0781
Hydrogen	H₂	2.016	0.0898	0.0056
Hydrogen sulfide	H₂S	34.08	1.5392	0.0961
Methane	CH₄	16.03	0.7167	0.0448
Nitrogen	N₂	28.02	1.2507	0.0782
Oxygen	O₂	32.00	1.4289	0.0892

* From Ref. 20.

bly constant. If the landfill is not vented, it would be expected that the percentage of methane would increase over the long term, because carbon dioxide would diffuse into the strata below the landfill.

The volume of the gases released during anaerobic decomposition can be estimated in a number of ways. For example, if all the organic constituents in the wastes (with the exception of plastics, rubber, and

TABLE 10-3   TYPICAL PERCENTAGE DISTRIBUTION OF LANDFILL GASES DURING FIRST 48 MONTHS*

Time interval since start of cell completion, months	Average percent by volume Nitrogen, N₂	Carbon dioxide, CO₂	Methane, CH₄
0–3	5.2	88	5
3–6	3.8	76	21
6–12	0.4	65	29
12–18	1.1	52	40
18–24	0.4	53	47
24–30	0.2	52	48
30–36	1.3	46	51
36–42	0.9	50	47
42–48	0.4	51	48

* From Ref. 18.

leather) were represented with a generalized formula of the form C_aH_b-O_cN_d, then the total volume of gas could be estimated by using Eq. 9-10, with the assumption of complete conversion to carbon dioxide and methane. This method is illustrated in Example 10-2. An alternative method is to assume that (1) the volatile fraction of the total organic portion of the wastes is about 95 percent, (2) 50 percent of the volatile material is carbon, and (3) half the carbon is converted to methane and half to carbon dioxide. In both methods, it must be assumed that some residual amount of the organic material is not decomposed.

✓ **EXAMPLE 10-2** *Estimation of Amount of Gas Produced in a Sanitary Landfill*

Estimate the amount of gas produced in a sanitary landfill per unit weight of solid wastes. Use a weight of 100 lb; assume that the wastes are of the composition shown in Table 4-2 and that the initial moisture content is 25 percent. Also assume that food wastes, paper, cardboard, garden trimmings, and wood are the materials that will decompose.

Solution

1. Determine the amount of organic wastes on a dry basis that will decompose, assuming that the moisture content is associated with the organic components. From Table 4-2, the total weight of organic material in 100 lb of solid wastes is equal to 79 lb.

    $$\text{Organic material (dry basis), lb} = \overset{73}{79} \text{ lb} - (100 \text{ lb})(0.25)$$
    $$= 54 \text{ lb}$$

2. Determine the amount of decomposable organic wastes, assuming that the food wastes, paper, cardboard, 75 percent of the garden trimmings, and 50 percent of the wood are decomposable in a reasonable period of time, say 25 yr. Also assume that, of the decomposable material, 5 percent will remain as an ash (see Table 4-8).

    Decomposable wastes (dry basis), lb

    $$= \frac{[15 + 40 + 4 + (0.75)12 + (0.50)2] \text{ lb } (0.95)}{79 \text{ lb}} \overset{49}{} (54 \text{ lb})$$

    $$= 44.8 \text{ lb}$$

3. Derive an empirical formula for the decomposable organic material. Assume that the organic material can be described with a formula of the form $C_aH_bO_cN_d$; then the coefficients are estimated from the data in Table 4-8. If approximate values are used and the ash

content is neglected, the percent composition and the moles of organic material would be:

Element	Percent	Moles
Carbon	49	4.08 (49/12)
Hydrogen	6	6 (6/1)
Oxygen	44	2.75 (44/16)
Nitrogen	1	0.0714 (1/14)

When the value for nitrogen is set equal to 1, the approximate formula for the solid wastes is $C_{57.1}H_{84}O_{38.5}N$.

4. Using the formula determined in step 3, estimate the amount of methane and carbon dioxide by using Eq. 9-11.

$$C_aH_bO_cN_d + \left(\frac{4a - b - 2c + 3d}{4}\right)H_2O \rightarrow$$

$$\left(\frac{4a + b - 2c - 3d}{8}\right)CH_4$$

$$+ \left(\frac{4a - b + 2c + 3d}{8}\right)CO_2 + dNH_3$$

From step 3 the coefficients are

$$a = 57.1 \quad b = 84 \quad c = 38.5 \quad d = 1$$

The resulting equation is

$$C_{57.1}H_{84}O_{38.5}N + 17.6H_2O \rightarrow 29.05CH_4 + 28.05CO_2 + NH_3$$
$$(1,399.2) \qquad (316.8) \qquad (464.8) \qquad 1,234.2 \qquad (17)$$

5. Determine the weight of methane and carbon dioxide from the equation derived in step 4.

$$\text{Methane} = \frac{464.8}{1,399.2}\ (44.8\ \text{lb}) \qquad (\text{see step 2})$$

$$= 14.9\ \text{lb}\ (6.8\ \text{kg})$$

$$\text{Carbon dioxide} = \frac{1,234.2}{1,399.2}\ (44.8\ \text{lb}) \qquad (\text{see step 2})$$

$$= 39.5\ \text{lb}\ (17.9\ \text{kg})$$

6. Convert the weight of gases, determined in step 5, to volume, assuming that the densities of methane and carbon dioxide are 0.0448 and 0.1235 lb/ft³, respectively (see Table 10-2).

$$\text{Methane} = \frac{14.9 \text{ lb}}{0.0448 \text{ lb/ft}^3}$$

$$= 333 \text{ ft}^3 \ (9.4 \text{ m}^3)$$

$$\text{Carbon dioxide} = \frac{39.5 \text{ lb}}{0.1235 \text{ lb/ft}^3}$$

$$= 320 \text{ ft}^3 \ (9.1 \text{ m}^3)$$

7. Determine the percentage composition of the resulting gas mixture.

$$\text{Methane } (\%) = \left(\frac{333 \text{ ft}^3}{653 \text{ ft}^3}\right) 100$$

$$= 51\%$$

$$\text{Carbon dioxide } (\%) = 49\%$$

8. Determine the total theoretical amount of gas generated per unit weight.

Based on the dry weight of organic material, ft³/lb:

$$\frac{653 \text{ ft}^3}{54 \text{ lb}} = 12.1 \text{ ft}^3/\text{lb} \ (0.75 \text{ m}^3/\text{kg})$$

Based on 100 lb of solid wastes, ft³/lb:

$$\frac{653 \text{ ft}^3}{100 \text{ lb}} = 6.5 \text{ ft}^3/\text{lb} \ (0.41 \text{ m}^3/\text{kg})$$

✓Comment The computed theoretical values for the total volume of gas per pound of organic material and per pound of solid wastes are consistent with the data reported in Chap. 9 and in Ref. 1, but the actual amount that could be recovered would be considerably less. The rate at which the gases are produced varies with local conditions, especially the moisture content. Typically, it is estimated that about 30 to 60 percent of the computed value could, under optimum conditions, be achieved within 2 yr and that perhaps up to 70 percent could be achieved within 5 yr.

Leachate in Landfills

Leachate may be defined as liquid that has percolated through solid waste and has extracted dissolved or suspended materials from it [24]. In most landfills the liquid portion of the leachate is composed of the liquid produced from the decomposition of the wastes and liquid that has entered

TABLE 10-4   DATA ON THE COMPOSITION OF LEACHATE FROM LANDFILLS*

	Value† mg/l	
Constituent	Range‡	Typical *(UNDILUTED LEACHATE)*
BOD₅ (5-day biochemical oxygen demand)	2,000–30,000	10,000
TOC (total organic carbon)	1,500–20,000	6,000 *(NO G.W. MIXING)*
COD (chemical oxygen demand)	3,000–45,000	18,000
Total suspended solids	200– 1,000	500
Organic nitrogen	10– 600	200
Ammonia nitrogen	10– 800	200
Nitrate	5– 40	25
Total phosphorus	1– 70	30
Ortho phosphorus	1– 50	20
Alkalinity as CaCO₃	1,000–10,000	3,000 *–SOLUBILIZATION*
pH	5.3– 8.5	6 *OF VIRGIN SOIL*
Total hardness as CaCO₃	300–10,000	3,500
Calcium	200– 3,000	1,000
Magnesium	50– 1,500	250
Potassium	200– 2,000	300
Sodium	200– 2,000	500
Chloride	100– 3,000	500
Sulfate	100– 1,500	300
Total iron	50– 600	60

* Developed in part from Refs. 1, 3, 4, 11, and 23.
† Except pH.
‡ Representative range of values. Higher maximum values have been reported in the literature for some of the constituents.

the landfill from external sources, such as surface drainage, rainfall, groundwater, and water from underground springs.

When leachate percolates through solid wastes that are undergoing decomposition, both biological materials and chemical constituents are picked up. Representative data on the chemical characteristics of leachate, reported in Table 10-4, indicate that the range of concentration values for the various constituents is rather extreme. For this reason, no average value can be given for leachate. The typical values reported in Table 10-4 are intended to be used only as a guide.

In general, it has been found that the quantity of leachate is a direct function of the amount of external water entering the landfill. In fact, if a landfill is constructed properly, the production of measurable quantities of leachate can be eliminated. When sewage sludge is to be added to the solid wastes to increase the amount of methane produced, leachate control facilities must be provided. In some cases leachate treatment facilities may also be required [11].

Settlement and Structural Characteristics of Landfills

Before a decision is reached on the final use to be made of a completed landfill, the settlement and structural characteristics of landfill must be considered. The settlement depends on the initial compaction, characteristics of wastes, degree of decomposition, and effects of consolidation when water and air are formed out of the compacted material. The height of the completed fill will also influence the initial compaction and the degree of consolidation.

Representative data on the degree of settlement to be expected in a landfill as a function of the initial compaction are shown in Fig. 10-7. It has been found in various studies that about 90 percent of the ultimate settlement occurs within the first 5 yr [8]. The placement of concentrated loads on completed landfills is not recommended. If this must be done, however, it is recommended that load tests be conducted because of the variability of local conditions [9].

10-4 GAS AND LEACHATE MOVEMENT AND CONTROL

Under ideal conditions, the gases generated from a landfill should be either vented to the atmosphere or (in larger landfills) collected for the production of energy. The leachate should be either contained within the landfill or removed for treatment. Unfortunately, these conditions are found only in a few modern landfills, and so the movement of gases and leachate from landfills is an important aspect of solid waste disposal.

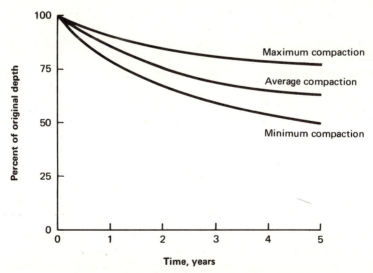

FIG. 10-7 Surface settlement of compacted landfill [8].

Gas Movement

In most cases, over 90 percent of the gas volume produced from the decomposition of solid wastes consists of methane and carbon dioxide (see Table 10-3). When methane is present in the air in concentrations between 5 and 15 percent, it is explosive. However, there is no oxygen in a landfill when methane concentrations in it reach this critical level, and so there is no danger that the fill will explode [1]. Although most of the methane escapes to the atmosphere, both methane and carbon dioxide have been found in concentrations of up to 40 percent at lateral distances of up to 400 ft from the edges of landfills [28]. For unvented landfills, the extent of this lateral movement varies with the characteristics of the cover material and the surrounding soil. If methane is vented into the atmosphere in an uncontrolled manner, it can accumulate (because its specific gravity is less than that of air) below buildings or in other enclosed spaces on, or close to, a sanitary landfill.

With proper venting, methane should not pose a problem. Carbon dioxide, on the other hand, is troublesome because of its density. As shown in Table 10-2, carbon dioxide is about 1.5 times as dense as air and 2.8 times as dense as methane; thus it tends to move toward the bottom of the landfill. As a result, the concentration of carbon dioxide in the lower portions of a landfill may be high for years.

Ultimately, because of its density, carbon dioxide will also move downward through the underlying formation until it reaches the groundwater. Because carbon dioxide is readily soluble in water, it usually lowers the pH, which in turn can increase the hardness and mineral content of the groundwater through solubilization. The reaction of carbon dioxide with water, which results in the formation of carbonic acid, is

$$CO_2 + H_2O \rightarrow H_2CO_3 \qquad (10\text{-}1)$$

If solid calcium carbonate is present in the soil structure, the carbonic acid will react with it to form soluble calcium carbonate, according to the following reaction [16]:

$$CaCO_3 + H_2CO_3 \rightarrow Ca^{2+} + 2HCO_3^- \qquad (10\text{-}2)$$

Similar reactions occur with magnesium carbonates. If a given free carbon dioxide concentration is present, the reaction shown in Eq. 10-2 will proceed until equilibrium is reached, as depicted in Eq. 10-3.

$$\begin{array}{c} H_2O + CO_2 \\ \updownarrow \\ CaCO_3 + H_2CO_3 \rightleftharpoons Ca^{2+} + 2HCO_3^- \end{array} \qquad (10\text{-}3)$$

Thus, any process that increases the free carbon dioxide available to the solution will cause more calcium carbonate to dissolve [16]. The resulting

TABLE 10-5 DATA ON THE ABSORPTION COEFFICIENTS FOR THE GASES FOUND IN SANITARY LANDFILLS*
(Milliliters of gas reduced to 0°C and 760 mm Hg per liter of water when the partial pressure of the gas is 760 mm Hg)

Gas	Formula	Molecular weight	Temperature, °C		
			0	10	20
Air	—	—	29.18	22.84	18.68
Carbon dioxide	CO_2	44.00	1713	1194	878
Carbon monoxide	CO	28.00	35.4	28.2	23.2
Hydrogen	H_2	2.016	21.5	19.6	18.2
Hydrogen sulfide	H_2S	34.08	4670	3399	2582
Methane	CH_4	16.03	55.6	41.8	33.1
Nitrogen	N_2	28.02	23.5	18.6	15.5
Oxygen	O_2	32.00	48.9	38.0	31.0
Vapor pressure of water, mm Hg		—	4.58	9.21	17.5

K_s VALUES

* Adapted from Ref. 6.

increase in hardness is the principal effect of the presence of carbon dioxide in groundwater. The movement of gases is considered further in Refs. 3, 7, 27, and 28.

The solubility in water of the gases as reported in Table 10-2 is shown in Table 10-5. The corresponding concentration of a gas in solution can be computed using Henry's law:

$$c_s = k_s P \qquad (10\text{-}4)$$

where c_s = saturation concentration of the gas in water, ml/l
k_s = coefficient of absorption, ml/l
P = partial pressure of the gas in the gas phase, expressed as a fraction

To apply Henry's law, it is helpful to remember that at standard temperature and pressure (0°C and 760 mm Hg), the molal volume of any gas is 22,412 ml/g·mol, or 359 ft³/lb·mol. The use of the data in Table 10-5 and Eq. 10-4 is illustrated in Example 10-3.

EXAMPLE 10-3 *Saturation Concentration of Carbon Dioxide*

Determine the concentration of carbon dioxide in the upper layers of a groundwater in contact with landfill gases at 760 mm Hg and at 10°C. Assume that the composition of the gas is 50 percent carbon dioxide and 50 percent methane and that the gas is saturated with water vapor.

Solution

1. Determine the partial pressure of the carbon dioxide by correcting for vapor pressure of water.

$$\text{Partial pressure of } CO_2 = 0.50 \, \frac{(760 - 9.21) \text{ mm Hg}}{760 \text{ mm Hg}}$$

$$= 0.49$$

2. Determine the value of c_s in Eq. 10-4 by using the value of k_s given in Table 10-5 and P determined in step 1.

$$c_s = k_s P$$

$$= (1{,}194 \text{ ml/l})(0.49)$$

$$= 585.1 \text{ ml/l}$$

3. Convert the carbon dioxide saturation concentration determined in step 2 to milligrams per liter.

$$\text{Carbon dioxide} = \frac{(585.1 \text{ ml/l})(44 \times 10^3 \text{ mg/g·mol})}{22{,}412 \text{ ml/g·mol}}$$

$$= 1{,}149 \text{ mg/l } (1.15 \text{ kg/m}^3)$$

Control of Gas Movement by Permeable Methods   The lateral movement of gases produced in a landfill can be controlled by installing vents made of materials that are more permeable than the surrounding soil. Typically, gas vents are constructed of gravel, as shown in Fig. 10-8. The spacing of cell vents (see Fig. 10-8a) depends on the width of the waste cells but usually varies from 60 to 200 ft. The thickness of the gravel layer should be such that it will remain continuous even though there may be differential settling; 12 to 18 in is recommended. Barrier vents (see Fig. 10-8b) or well vents (see Fig. 10-8c) also can be used to control the lateral movement of gases. Well vents are often used in conjunction with lateral-surface vents buried below grade in a gravel trench (see Fig. 10-8c).

Where well vents are used, waste gas burners are often installed (see Fig. 10-9), and in these cases it is recommended that the well penetrate into the upper waste cell. The height of the waste burner can vary from 10 to 20 ft above the completed fill. The burner can be ignited either by hand or by a continuous pilot flame. To derive maximum benefit from the installation of a waste gas burner, the pilot flame is recommended [3].

Control of the downward movement of gases can be accomplished by installing perforated pipes in the gravel layer at the bottom of the landfill. If the gases cannot be vented laterally, it may be necessary to install gas wells and to vent the pumped gas to the atmosphere. A gravel layer is often used in conjunction with one or more of the impermeable methods of control.

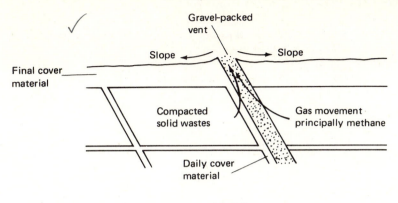

(a) **Cell**

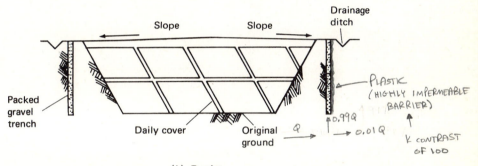

(b) **Barrier**

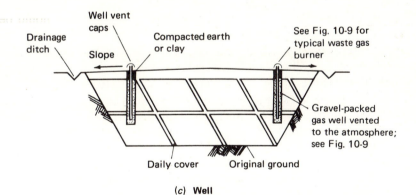

(c) **Well**

FIG. 10-8 Types of vents used to control the lateral movement of gases in sanitary landfills.

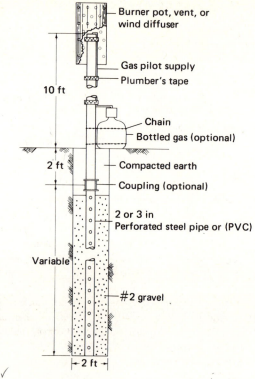

FIG. 10-9   Typical well-type waste gas burner used at a sanitary landfill.

Control of Gas Movement by Impermeable Methods

The movement of landfill gases through adjacent soil formations can be controlled by constructing barriers of materials that are more impermeable than the soil. Some of the landfill sealants that are available for this use are identified in Table 10-6. Of these, the use of compacted clays is most common (see Fig. 10-10). The thickness will vary depending on the type of clay and the degree of control required; thicknesses ranging from 6 to 48 in have been used. If a clay liner is used, it should be constructed as the filling progresses to avoid air drying, which tends to shrink and crack the clay [1]. Another effective method is to install the clay liner first and then to cover it with 1 ft or more of damp, well-compacted soil. The installation of impermeable barriers is of special importance where landfill gas is to be recovered.

Leachate Movement (Seepage)

Under normal conditions, leachate is found in the bottom of landfills. From there its movement is through the underlying strata, although some lateral

TABLE 10-6 LANDFILL SEALANTS FOR THE CONTROL OF GAS AND LEACHATE MOVEMENT*

Classification	Sealant — Representative types	Remarks
Compacted soil		Should contain some clay or fine silt
Compacted clay	Bentonites, illites, kaolinites	Most commonly used sealant for landfills; layer thickness varies from 6 to 48 in; layer must be continuous and not allowed to dry out and crack
Inorganic chemicals	Sodium carbonate, silicate, or pyrophosphate	Use depends on local soil characteristics
Synthetic chemicals	Polymers, rubber latex	Experimental, use not well established
Synthetic membrane liners	Polyvinyl chloride, butyl rubber, hypalon, polyethylene, nylon-reinforced liners	Expensive, maybe justified where gas is to be recovered
Asphalt	Modified asphalt, rubber-impregnated asphalt, asphalt-covered polypropylene fabric, asphalt concrete	Layer must be thick enough to maintain continuity under differential settling conditions
Others	Gunite concrete, soil cement, plastic soil cement	

* Adapted in part from Ref. 2.

movement may also occur, depending on the characteristics of the surrounding material. Because of the importance of vertical seepage in the contamination of groundwater, this subject is considered further in the following discussion.

Darcy's Law   The rate of seepage of leachate from the bottom of a landfill can be estimated by Darcy's law, which can be expressed as

$$Q = -KA \frac{dh}{dL}$$ (10-5)

where Q = leachate discharge per unit time, gpd
K = coefficient of permeability, gpd/ft² (gpd of H_2O passing thru 1 ft² of media with a dh/dl = 1 ft/ft)
A = cross-sectional area through which the leachate flows, ft²
dh/dL = the hydraulic gradient, ft/ft

The minus sign in Darcy's equation arises from the fact that the head loss dh is always negative [5]. The coefficient of permeability is also known as

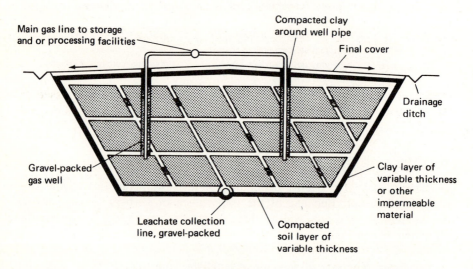

Note: For gas vent options;
see Fig. 10 – 8 a, c

Final cover

Drainage ditch

Compacted soil layer of variable thickness

Gravel lift vents

Clay layer of variable thickness or other impermeable material

Main leachate collection line, use of perforated laterals will depend on topography and aerial extent of landfill; main line and laterals should be gravel-packed.

(a) **Without gas recovery**

Main gas line to storage and or processing facilities

Compacted clay around well pipe

Final cover

Drainage ditch

Gravel-packed gas well

Clay layer of variable thickness or other impermeable material

Leachate collection line, gravel-packed

Compacted soil layer of variable thickness

(b) **With gas recovery**

FIG. 10-10   Use of impermeable liners to control the movement of landfill gases and leachate.

the hydraulic conductivity, the effective permeability, or the seepage coefficient.

The permeability of a soil is influenced by particle size, void ratio, composition degree of saturation, and temperature. From empirical observations, it has been found that the coefficient of permeability can be defined in terms of some characteristic size of the porous medium and the properties of the fluid. The relationship is

$$K = Cd^2 \frac{\gamma}{\mu} \qquad (10\text{-}6)$$

where C = dimensionless constant
 d = diameter of pores
 γ = specific weight of water
 μ = viscosity of water

The term Cd^2 is known as the specific (or intrinsic) permeability k and is thought to be a characteristic of the medium alone. Neglecting the effects of temperature on density, we find that

$$\frac{K_s}{K_t} = \frac{\mu_t}{\mu_{60}} \qquad (10\text{-}7)$$

where K_s = laboratory standard coefficient of permeability defined as the flow of water at 60°F in gallons per day through a medium having a cross-sectional area of 1 ft² under a hydraulic gradient of 1 ft/ft
 K_t = coefficient of permeability at temperature t
 μ_{60} = viscosity at 60°F
 μ_t = viscosity at temperature t

In feet-per-second units, the coefficient of permeability is expressed in gallons per day per square foot, or feet per day. The conversion between

TABLE 10-7 TYPICAL PERMEABILITY COEFFICIENTS FOR VARIOUS SOILS (LAMINAR FLOW)*

| | Coefficient of permeability, K_s | |
Material	ft/day	gal/day/ft²
Uniform coarse sand	1,333	9,970
Uniform medium sand	333	2,490
Clean, well-graded sand and gravel	333	2,490
Uniform fine sand	13.3	100
Well-graded silty sand and gravel	1.3	9.7
Silty sand	0.3	2.2
Uniform silt	0.16	1.2
Sandy clay	0.016	0.12
Silty clay	0.003	0.022
Clay (30 to 50 percent clay sizes)	0.0003	0.0022
Colloidal clay	0.000003	0.000022

* Adapted from Refs. 5 and 24.
Note: ft/day × 0.3048 = m/day
gal/day/ft² × 0.0408 = m³/day·m²

these factors is accomplished by noting that 7.48 gal/day/ft² = 1 ft/day. Typical values for the coefficient of permeability for various soils are given in Table 10-7.

Estimation of Vertical Seepage   Before Darcy's law is applied to the estimation of seepage rates from a landfill, it is helpful to review the physical conditions of the problem by referring to Fig. 10-11. As shown, a landfill cell has been placed in a surface aquifer, composed of material of moderate permeability, which overlies a bedrock aquifer. In this situation, it is possible to have two different piezometric water surfaces if wells are placed in the surface and bedrock aquifers.

In regard to the movement of leachate, two problems are of interest. The first is the rate at which leachate seeps from the bottom of the landfill

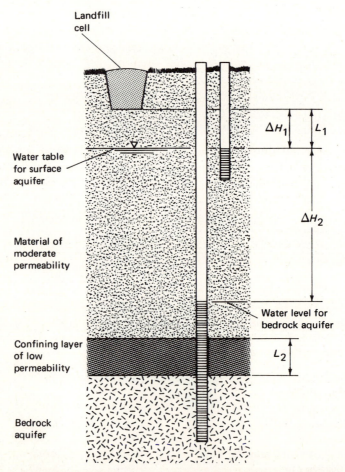

FIG. 10-11   Definition sketch for determination of seepage from landfills and from surface to subsurface aquifers. (Adapted from Ref. 29.)

into the groundwater in the surface aquifer. The second is the rate at which groundwater from the surface aquifer moves into the bedrock aquifer. These two problems will be discussed in the following analysis, but the question of how the mixing of the leachate and groundwater occurs in the surface aquifer will not be considered.

In the first problem, the leachate flow rate from the landfill to the upper groundwater is computed by assuming that the material below the landfill to the top of the water table is saturated and that a small layer of leachate exists at the bottom of the fill. Under these conditions the application of Darcy's equation is as follows:

$$Q(\text{gal/day}) = K(\text{gal/day/ft}^2)A(\text{ft}^2)\frac{h_1(\text{ft})}{L_1(\text{ft})}$$

— SEEPAGE of LEACHATE FROM BOTTOM of LANDFILL TO TOP of GWT

but because $h_1 = L_1$,

$$Q(\text{gal})/\text{day}) = K(\text{gal/day/ft}^2)A(\text{ft}^2)$$

If it is assumed that flow occurs through 1 ft², then

$$Q(\text{gal/day}) = K(\text{gal/day/ft}^2)(\text{ft}^2)$$

Thus, the leachate discharge rate per unit area is equal to the value of K multiplied by square feet.

For example, if the upper strata of material in Fig. 10-11 were sandy clay, the corresponding seepage rate would be equal to 0.12 gal/day/unit area (see Table 10-7). The computed value represents the maximum amount of seepage that would be expected, and this value should be used for design purposes. Under normal conditions, the actual rate would be less than this value because the soil column below the landfill would not be saturated.

In the second problem, the rate of movement of water from the upper aquifer to the lower aquifer would be given by

$$Q(\text{gal/day}) = K(\text{gal/day/ft}^2)A(\text{ft}^2)\frac{h_2(\text{ft})}{L_2(\text{ft})}$$

— MOVEMENT of WATER FROM UPPER AQUIFER TO LOWER AQUIFER

In this case, the thickness of the confining layer is used to determine the hydraulic gradient.

Control of Leachate Movement   As leachate percolates through the underlying strata, many of the chemical and biological constituents originally contained in it will be removed by the filtering and adsorptive action of the material composing the strata. In general, the extent of this action depends on the characteristics of the soil, especially the clay content [4]. Because of the potential risk involved in allowing leachate to percolate to the groundwater, best practice calls for its elimination or containment. When gas is to be recovered, it is especially important to contain the leachate because the initial moisture content must be significantly higher than normal (50 to 60

TABLE 10-8   GENERALIZED RATINGS OF THE SUITABILITY OF VARIOUS TYPES OF SOILS FOR USE AS LANDFILL COVER MATERIAL*

Function	General soil type†					
	Clean gravel	Clayey-silty gravel	Clean sand	Clayey-silty sand	Silt	Clay
Prevents rodents from burrowing or tunneling	G	F–G	G	P	P	P
Keeps flies from emerging	P	F	P	G	G	E‡
Minimizes moisture entering fill	P	F–G	P	G–E	G–E	E‡
Minimizes landfill gas venting through cover	P	F–G	P	G–E	G–E	E‡
Provides pleasing appearance and controls blowing paper	E	E	E	E	E	E
Supports vegetation	P	G	P–F	E	G–E	F–G
Vents decomposition gas (be permeable)§	E	P	G	P	P	P

* From Ref. 1.
† E, excellent; G, good; F, fair; P, poor.
‡ Except when cracks extend through the entire cover.
§ Only if well drained.

percent versus 20 to 25 percent) to achieve the highest production of gas. In some gas recovery systems, this leachate is collected and recycled to the top of the landfill and reinjected through perforated lines located in drainage trenches. Typically, the rate of gas production is greater in leachate recirculation systems. Ultimately, regardless of the system used, it may be necessary to collect and treat the leachate [11].

To date (1976) the use of clay has been the favored method of reducing or eliminating the percolation of leachate (see Table 10-6). Membrane liners have also been used, but they are expensive and require care so that they will not be damaged during the filling operations.

Equally important in controlling the movement of leachate is the elimination of surface water infiltration, which is the major contributor to the total volume of leachate. With the use of an impermeable clay layer, an appropriate surface slope (1 to 2 percent), and adequate drainage, surface infiltration can be controlled effectively. With proper surface control, it may not be necessary to provide an impermeable barrier. Generalized ratings for the suitability of various types of soil for use as a landfill cover are reported in Table 10-8.

10-5 DESIGN OF LANDFILLS

Once a limited number of potential sites have been selected on the basis of a review of the available preliminary information, it will usually be necessary to prepare an engineering design report for each site to assess the costs associated with preparation of the site for filling, placement of solid wastes, and completing the site once filling operations have ceased. The engineering design report in this context is preliminary in nature, as distinguished from a complete evaluation required for the final selection of a site, which includes environmental considerations.

Among the important topics that must be considered in an engineering design report, though not necessarily in the order given, are the following: (1) land requirements, (2) types of wastes that must be handled, (3) evaluation of seepage potential, (4) design of drainage and seepage control facilities, (5) development of a general operation plan, (6) design of solid waste filling plan, and (7) determination of equipment requirements. The more important individual factors that must be considered are reported in Table 10-9. Throughout the development of the engineering design report, careful consideration must be given to the final use or uses to be made of the completed site. Land reserved for administrative offices, buildings, and parking lots should be filled with dirt only and should be sealed against the entry of gases.

The extent to which the engineering computations must be completed for each site will depend on the findings derived in each step. For example, if it is found that the seepage rate will be too great without the use of a clay liner and if no clay or other suitable material is available economically within the area, it may not be necessary to carry out further computations for the site in question. The development of a complete operational plan for a landfill is illustrated at the end of this section in Example 10-7.

Land Requirements

Earlier in the chapter, an approximate method was given for determining the area requirements for landfill (see Example 10-1). In this section consideration is given to the impacts on land requirements of (1) the compactability of the individual solid waste components and (2) the recovery of resources and energy.

Impact of Compactability of Solid Waste Components   The final density of solid wastes placed in a landfill varies with the mode of operation of the landfill, the compactability of the individual solid waste components, and the percentage distribution of the components. Typical compactability data for the components listed in Table 4-2 are reported in Table 10-10. Volume-reduction factors are given for both normally compacted and well-compacted landfills. The use of the data presented in Table 10-10 is illustrated in Example 10-4.

TABLE 10-9 IMPORTANT FACTORS THAT MUST BE CONSIDERED IN THE DESIGN AND OPERATION OF SANITARY LANDFILLS

Factor	Remarks
Design	
Access	Paved all-weather access roads to landfill site; temporary roads to unloading areas.
Cell design and construction	Will vary depending on whether gas is to be recovered; each day's wastes should form one cell; maximum depth of 10 ft; cover at end of day with 6 in of earth; gravel gas vent should be installed every 60 to 200 ft.
Cover material	Maximize use of onsite earth materials; approximately 1 yd^3 of cover material will be required for every 4 to 6 yd^3 of solid wastes; mix with sealants to control surface infiltration.
Drainage	Install drainage ditches to divert surface water runoff; maintain 1 to 2 percent grade on finished fill to prevent ponding.
Equipment requirements	Vary with size of landfill (see Table 10-15).
Fire prevention	Water onsite; if nonpotable, outlets must be marked clearly; proper cell separation prevents continuous burn-through if combustion occurs.
Groundwater protection	Divert any underground springs; if required, install sealants for leachate control; install wells for gas and groundwater monitoring.
Land area	Area should be large enough to hold all community wastes for a minimum of 1 yr but preferably 5 to 10 yr.
Landfilling method	Selection of method will vary with terrain and available cover.
Litter control	Use movable fences at unloading areas; crews should pick up litter at least once per month or as required.
Operation plan	With or without the codisposal of treatment plant sludges and the recovery of gas.
Spread and compaction	Spread and compact waste in layers less than 2 ft thick.
Unloading area	Keep small, generally under 100 ft on a side; operate separate unloading areas for automobiles and commercial trucks.
Operation	
Communications	Telephone for emergencies.
Days and hours of operation	Usual practice is 5 to 6 days/wk and 8 to 10 h/day.
Employee facilities	Restrooms and drinking water should be provided.
Equipment maintenance	A covered shed should be provided for field maintenance of equipment.
Operational records	Tonnage, transactions, and billing if a disposal fee is charged.
Salvage	No scavenging; salvage should occur away from the unloading area; no salvage storage onsite.
Scales	Essential for record keeping if collection trucks deliver wastes; capacity to 100,000 lb.

TABLE 10-10 TYPICAL COMPACTION FACTORS FOR VARIOUS SOLID WASTE COMPONENTS AS DISCARDED

Component	Range	Compaction factors for components in landfills*	
		Normal compaction	Well compacted
Food wastes	0.2–0.5	0.35	0.33
Paper	0.1–0.4	0.2	0.15
Cardboard	0.1–0.4	0.25	0.18
Plastics	0.1–0.2	0.15	0.10
Textiles	0.1–0.4	0.18	0.15
Rubber	0.2–0.4	0.3	0.3
Leather	0.2–0.4	0.3	0.3
Garden trimmings	0.1–0.5	0.25	0.2
Wood	0.2–0.4	0.3	0.3
Glass	0.3–0.9	0.6	0.4
Tin cans	0.1–0.3	0.18	0.15
Nonferrous metals	0.1–0.3	0.18	0.15
Ferrous metals	0.2–0.6	0.35	0.3
Dirt, ashes, brick, etc.	0.6–1.0	0.85	0.75

* Compaction factor = V_f/V_i, where V_f = final volume of solid waste after compaction and V_i = initial volume of solid waste before compaction.

✓ **EXAMPLE 10-4** *Determination of Density of Compacted Solid Wastes*

Determine the density in a well-compacted landfill for solid wastes with the characteristics given in Table 4-2.

Solution

1. Set up a computation table with separate columns for (1) the weight of the individual solid waste components, (2) the volume of the wastes as discarded, (3) the volume-reduction factors for well-compacted solid wastes, and (4) the compacted volume in the landfill. The required table, based on a total weight of 1,000 lb, is given in Table 10-11.

2. Compute the compacted density of the solid wastes.

$$\text{Compacted density} = \frac{1{,}000 \text{ lb} \times 27 \text{ ft}^3/\text{yd}^3}{28.6 \text{ ft}^3}$$

$$= 944 \text{ lb/yd}^3 \ (560 \text{ kg/m}^3)$$

Comment The density value of 944 lb/yd³ would then be used to determine the required landfill area. In some of the literature, a factor is included to account for the volume gain due to decay. Although it is true that there will be a volume gain, seldom, if ever, will this volume

TABLE 10-11 COMPUTATION OF DENSITY OF SOLID WASTES PLACED IN WELL-COMPACTED LANDFILL FOR EXAMPLE 10-4

Component	Weight of solid wastes,* lb	Volume as discarded,† ft³	Compaction factor‡	Volume in landfill, ft³
Food wastes	150	8.3	0.33	2.7
Paper	400	78.4	0.15	11.8
Cardboard	40	12.9	0.18	2.3
Plastics	30	7.5	0.10	0.8
Textiles	20	5.0	0.15	0.8
Rubber	5	0.6	0.3	0.2
Leather	5	0.5	0.3	0.2
Garden trimmings	120	18.5	0.2	3.7
Wood	20	1.3	0.3	0.4
Glass	80	6.6	0.4	2.6
Tin cans	60	10.9	0.15	1.6
Nonferrous metals	10	1.0	0.15	0.2
Ferrous metals	20	1.0	0.3	0.3
Dirt, ashes, brick, etc.	40	1.3	0.75	1.0
Total	1,000			28.6

Compacted density = 944 lb/yd³ (560 kg/m³)

* See Table 4-8.
† See Table 4-6.
‡ See Table 10-10.
Note: lb × 0.4536 = kg
ft³ × 0.02833 = m³
lb/yd³ × 0.5933 = kg/m³

be used for additional filling. It is therefore recommended that this factor not be considered in determining the required volume.

Impact of Resource Recovery The recovery of materials and energy from solid wastes will also reduce the landfill area requirements. The amount of the reduction will depend on the components to be recovered and the amount of residual wastes (see Chap. 9). The necessary computations to determine the impact of resource recovery on landfill area requirements are illustrated in Example 10-5.

EXAMPLE 10-5 *Evaluation of the Impact of Resource Recovery on Landfill Area Requirements*

Determine the impact of a resource recovery program on landfill area requirements in which 50 percent of the paper and 80 percent of the glass and tin cans are recovered. Assume that the wastes have the characteristics reported in Table 4-2.

Solution

1. Prepare a summary table similar to Table 10-11, but one in which the quantities of the components to be recovered are removed, and determine the compacted density of the wastes in the landfill (see Table 10-12).

2. Because the density computed in Table 10-12 is essentially the same as that in Table 10-11, the impact of the materials recovery program can be assessed on the basis of the weight reduction alone.

$$\text{Area with recovery} = \left(\frac{688 \text{ lb}}{1{,}000 \text{ lb}}\right)(\text{area without recovery})$$

$$= (0.69)(\text{area without recovery})$$

Comment In cases where the computed compacted density

TABLE 10-12 COMPUTATION OF DENSITY OF SOLID WASTES IN WELL-COMPACTED LANDFILL AFTER RESOURCE RECOVERY FOR EXAMPLE 10-5

Component	Weight of solid wastes,* lb	Volume as discarded,† ft³	Compaction factor‡	Volume in landfill, ft³
Food wastes	150	8.3	0.33	2.7
Paper	200	39.2	0.15	5.9
Cardboard	40	12.9	0.1	2.3
Plastics	30	7.5	0.1	0.8
Textiles	20	5.0	0.15	0.8
Rubber	5	0.6	0.3	0.2
Leather	5	0.5	0.3	0.2
Garden trimmings	120	18.5	0.2	3.7
Wood	20	1.3	0.3	0.4
Glass	16	1.3	0.4	0.5
Tin cans	12	2.2	0.15	0.3
Nonferrous metals	10	1.0	0.15	0.2
Ferrous metals	20	1.0	0.3	0.3
Dirt, ashes, brick, etc.	40	11.3	0.75	1.0
Total	688			19.3

Compacted density = 962 lb/yd³ (571 kg/m³)

* See Table 4-2.
† See Table 4-6.
‡ See Table 10-10.
Note: lb × 0.4536 = kg
   ft³ × 0.02833 = m³
   lb/yd³ × 0.5933 = kg/m³

changes significantly as a result of a materials recovery program, the required landfill area can also be reduced by the ratio of compacted densities. Large changes in the density value will not be observed with materials recovery where a sizable fraction of the wastes are composed of garden trimmings.

Types of Wastes

Knowledge of the types of wastes to be handled is important in the design and layout of a landfill, especially if hazardous wastes are involved. It is usually best to develop separate disposal sites for hazardous wastes because under most conditions special treatment of the site will be necessary before hazardous wastes can be landfilled. The associated treatment costs are often significant, and it is wasteful to use this landfill capacity for wastes that do not require special precautions.

If significant quantities of demolition wastes are to be handled, it may be possible to use them for embankment stabilization. In some cases, it may not be necessary to cover demolition wastes on a daily basis. (For further discussion see Landfill Operation Plan.)

Evaluation of Seepage Potential

Core samples must be obtained to evaluate the seepage potential of a site that is being considered for a landfill. Sufficient borings should be made so that the stratigraphic formations under the proposed site can be established from the surface to (and including) the upper portions of the bedrock or other confining layers. At the same time, the depth to the surface water table should be determined along with the piezometric water levels in any bedrock or confined aquifers that may be found.

The resulting information is then used (1) to determine the general direction of groundwater movement under the site, (2) to determine whether any unconsolidated or bedrock aquifers are in direct hydraulic connection with the landfill, and (3) to estimate the vertical seepage that might occur under the landfill site.

Drainage and Seepage Control Facilities

In addition to the seepage analysis, it is also necessary to develop an overall drainage plan for the area that shows the location of storm drains, culverts, ditches, and subsurface drains as the filling operation proceeds. In some cases it may also be necessary to install seepage control facilities.

To ensure the rapid removal of rainfall from the completed landfill and to avoid the formation of puddles, the final cover should have a slope of about 1 percent. Where relatively impervious cover material such as clay is used, lesser slope values may be feasible. If it is assumed that (1) the cover material is saturated, (2) a thin layer of water is maintained on the surface,

TABLE 10-13 THEORETICAL VOLUME OF WATER THAT COULD ENTER COMPLETED LANDFILL THROUGH 1 FT² OF VARIOUS COVER MATERIALS IN 1 DAY*

Cover material	Volume of water, gal
Uniform coarse sand	9,970
Uniform medium sand	2,490
Clean, well-graded sand and gravel	2,490
Uniform fine sand	100
Well-graded silty sand and gravel	9.7
Silty sand	2.2
Uniform silt	1.2
Sandy clay	0.12
Silty clay	0.022
Clay (30 to 50 percent clay sizes)	0.0022
Colloidal clay	0.000022

* Adapted from Ref. 24.
Note: gal × 0.003785 = m³

and (3) there is no resistance to flow below the cover layer, then the theoretical amount of water that could enter the landfill per unit area in a 24-h period for various cover materials (listed in Table 10-7) is given in Table 10-13.

Clearly, these data are only theoretical values, but they can be used in assessing the worst possible situation. In actual practice the amount of water entering the landfill will depend on local hydrological conditions, the characteristics of the cover material (see Table 10-8), the final slope of the cover, and whether vegetation has been planted. Use of the rational formula [17] for estimating runoff is usually acceptable for small surface areas such as landfills.

Among the methods to control the seepage into and out of landfills are (1) the use of impermeable cover materials, (2) the interception of high groundwater before it reaches the fill (see Fig. 10-12), (3) equalization of the water levels within and outside the landfill, and (4) the use of an impervious layer of clay material or other sealants (see Table 10-6). The necessary computations for the use of an impermeable clay layer are illustrated in Example 10-6.

EXAMPLE 10-6 *Determination of the Thickness of a Clay Layer Necessary to Limit Seepage of Leachate*

Determine the thickness of a clay layer that must be placed in the bottom of a landfill if the seepage flow rate is to be limited to about 0.05 gal/day/unit area. Assume that the water table is located at the bottom

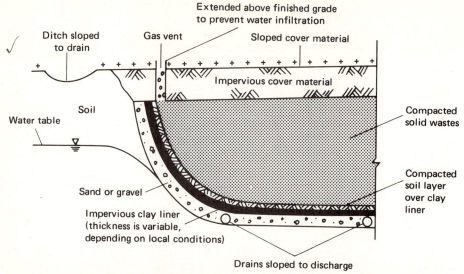

FIG. 10-12   Landfill section showing groundwater control, surface water control, grading, cover material, ditching, and gas vents. (Adapted from Ref. 24.)

of the landfill and that the leachate level in the landfill above the clay layer is to be maintained at 2 ft by pumping. The K value for the clay material to be used is 0.02 gal/day/ft².

Solution

1. Set up Darcy's equation for the specified conditions.

$$Q = KA \frac{dh}{dL}$$

$$0.05 \text{ gal/day} = (0.02 \text{ gal/day/ft}^2)(1 \text{ ft}^2)\left(\frac{2.0 + L_c}{L_c}\right)$$

where L_c = thickness of clay layer

2. Solve for the thickness of the clay layer.

$$2.5L_c = 2.0 + L_c$$

$$L_c = \frac{2.0}{1.5} = 1.33 \text{ ft (0.41 m)}$$

Comment   A similar computation would be made when a clay layer is used to prevent high groundwater from entering a landfill, as in the case of landfills placed in tidal areas.

Landfill Operation Plan

The layout of the site and the development of a workable operating schedule are the main features of a landfill operation plan.

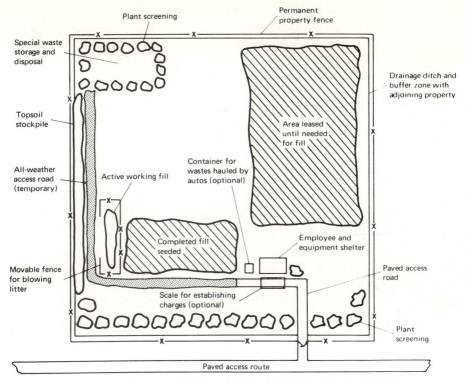

FIG. 10-13 Typical landfill disposal site layout. (Stanley Consultants Inc.)

Site Layout   In planning the layout of a landfill site, the location of the following must be determined: (1) access roads, (2) equipment shelters, (3) scales, if used, (4) storage sites for special wastes, (5) topsoil stockpile sites, (6) the landfill areas, and (7) plantings (also see Table 10-9). A typical layout for a landfill disposal site is shown in Fig. 10-13. Because the layout site is specific for each case, Fig. 10-13 is meant to serve only as a guide.

Operating Schedule   Factors that must be considered in developing operating schedules include (1) arrival sequences for collection vehicles, (2) traffic patterns at the site, (3) the time sequence to be followed in the filling operations, (4) effects of wind and other climatic conditions, and (5) commercial and public access. For example, because of heavy truck traffic early in the morning, it may be necessary to restrict public access to the site until later in the morning. Also, because of adverse winter conditions, the filling sequence should be established so that the landfill operations are not impeded.

Solid Waste Filling Plan

Once the general layout of the landfill site has been established, it will be necessary to select the placement method to be used and to lay out and design the individual solid waste cells. The specific method of filling will depend on the characteristics of the site, such as the amount of available cover material, the topography, and the local hydrology and geology. (Details on the various filling methods were presented earlier in this chapter. See Sec. 10-2.) To assess future development plans, it will be necessary to prepare a detailed plan for the layout of the individual solid waste cells. A typical example of such a plan is shown in Fig. 10-14.

On the basis of the characteristics of the site or the method of operation (e.g., gas recovery), it may be necessary to incorporate special features for the control of the movement of gases and leachate from the landfill. These might include the use of sand drains, plastic films, and/or clay materials.

Equipment Requirements

The type, size, and amount of equipment required will depend on the size of the landfill and the method of operation. Local availability and operator preference are also important factors.

The types of equipment that have been used at sanitary landfills

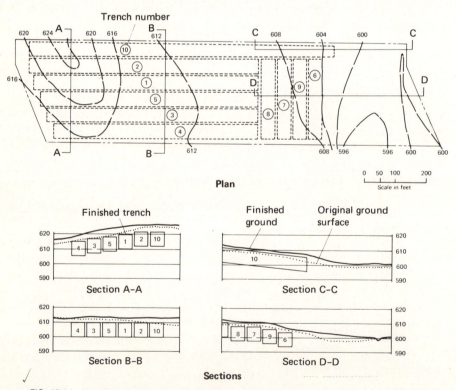

FIG. 10-14   Location plan for fill trenches within a sanitary landfill. (Adapted from Ref. 29.)

TABLE 10-14 PERFORMANCE CHARACTERISTICS OF LANDFILL EQUIPMENT*·†

Equipment	Solid waste		Cover material			
	Spreading	Compacting	Excavating	Spreading	Compacting	Hauling
Crawler tractor	E‡	G	E	E	G	NA
Landfill compactor	E	E	P	G	E	NA
Scraper	NA	NA	G	E	NA	E

* From Ref. 1.
† Basis of evaluation: easily workable soil and cover material haul distance greater than 1,000 ft.
‡ Rating key: E, excellent; G, good; F, fair; P, poor; NA, not applicable.

include both crawler and rubber-tired tractors, scrapers, compactors, draglines, and motorgraders (see Figs. 10-15 and 10-16). Of these, crawler or rubber-tired tractors are most commonly used. Properly equipped tractors can be used to perform all the necessary operations at a sanitary landfill, including spreading, compacting, covering, trenching, and even hauling cover materials [26]. The choice between rubber-tired or crawler tractors should be based on an analysis of local conditions. Some generalized information on the performance of landfill equipment is reported in Table 10-14.

The size and amount of equipment will depend primarily on the size of the landfill operation. Local site conditions will also influence the size of the equipment. Average equipment requirements that may be used as a guide for landfill operations are reported in Table 10-15.

EXAMPLE 10-7 *Design of a Sanitary Landfill Operation*

A city is in the process of closing an existing open dump. A replacement landfill site has been selected and some background data are available. Develop a design and operational plan for the landfill. (Management issues associated with selection of a landfill site are presented in Chap. 17.)

The selected disposal site consists of approximately 160 acres that are presently under private ownership. The site is located about 2 miles south of the city on Empire Road and is presently used for farming and cattle grazing; grains and grasses are the predominant crops. No buildings or other improvements presently exist on the site. The site topography is shown in Fig. 10-17.

Soil tests were made, and it was found that the soil covering the site is of two types—heavy clay topsoil and rocky substrata, mainly sandstones and some shale. The thickness of the topsoil varies from 2 ft on the ridges to 10 to 12 ft in the bottom of the valleys. The rocky

(a)

(b)

FIG. 10-15 Typical equipment used at a medium-sized landfill. (a) Crawler tractor with trash blade. (b) Self-loading scraper.

(a)

(b)

FIG. 10-16 Additional equipment used at landfills: (a) Water truck; (b) steel wheel compactor with trash blade.

TABLE 10-15   AVERAGE EQUIPMENT REQUIREMENTS FOR A SANITARY LANDFILL*

Population	Daily wastes, tons	Equipment			
		Number	Type	Size, lb	Accessory†
0–15,000	0–40	1	Tractor, crawler or rubber-tired	10,000–30,000	Dozer blade Front-end loader (1 to 2 yd) Trash blade
15,000–50,000	40–130	1	Tractor, crawler or rubber-tired	30,000–60,000	Dozer blade Front-end loader (2 to 4 yd) Bullclam Trash blade
		†	Scraper, dragline, water truck		
50,000–100,000	130–260	1–2	Tractor, crawler or rubber-tired	30,000+	Dozer blade Front-end loader (2 to 5 yd) Bullclam Trash blade
		†	Scraper, dragline, water truck		
100,000+	260+	2+	Tractor, crawler or rubber-tired	45,000+	Dozer blade Front-end loader Bullclam Trash blade
		†	Scraper, dragline, steel wheel compactor, road grader, water truck		

* From Ref. 26.
† Optional. Depends on individual need.

material is fractured easily and breaks into a sandy gravel soil upon excavation. Occasional hard areas may be encountered.

The city council agreed that in addition to the use of existing offsite drill logs, one deep test hole should be drilled to determine the level of the groundwater, to obtain a sample of groundwater for analysis, and to establish the nature of the deeper underlying strata. A well was drilled in the North Valley, and groundwater was found about 100 ft below the surface. An artesian effect observed at the drill hole was considered to

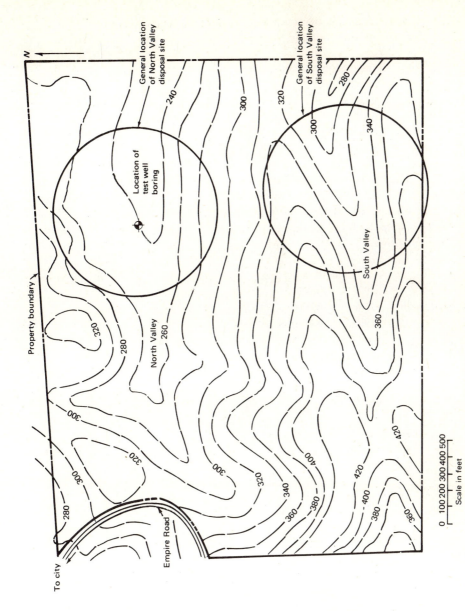

FIG. 10-17 Topographic map of landfill disposal site for Example 10-7.

be an indication that the water originated from a confined aquifer. The results of the laboratory testing of groundwater quality were as follows:

Total dissolved solids, mg/l	7,388
Chlorides, mg/l	1,035
Nitrates, mg/l	40
Hardness, mg/l as $CaCO_3$	788
Alkalinity, mg/l as $CaCO_3$	425
pH	8.0

The boring log for the well is shown in Fig. 10-18.

The principal solid wastes to be placed in the landfill are municipal wastes, but some nontoxic industrial wastes and some agricultural wastes will also be included. The design parameters to be used are as follows:

Area population served
    1970   30,000
    1980   46,000
    1990   66,000

Overall solid waste generation rate, lb/capita/day	6.4
Total lift height, ft	10
Density of solid wastes compacted in landfill, lb/yd³	1,000
Ratio of cover material to solid wastes	1:4

Solution

In analyzing the topographic map of the proposed site, it was decided to establish two fills, one in the North Valley and one in the South Valley, in the general locations shown in Fig. 10-17. The South Valley fill is to be completed first. On the basis of this preliminary assessment, the next step is to consider the important design and operational variables (there are 12).

1. Groundwater protection. The total dissolved-solids value is almost one-quarter the value for seawater, and so the salinity is extremely high. The water hardness is also very high. Water with a hardness of over 300 mg/l expressed as calcium carbonate is considered to be very hard. Because the groundwater is of such poor quality, it is not useful for any purpose other than fire protection and wetting the solid wastes and earth cover.

    If a landfill were placed in either valley without any treatment of the bottom area, the potential for groundwater contamination would be very low. Contamination of the aquifer is not possible unless there is an opening in the confining material and the pressure of the landfill leachate is greater than that of the aquifer.

Upward contamination is possible. To ensure that no interaction occurs, the depth of excavation should be limited to less than 25 ft in the North Valley.

2.   Permanent and temporary roads. The first step in the preparation of the site is the construction of an entry road from Empire Road and an access road to the valley bottom. The road should be of permanent construction because it must be used for the useful life

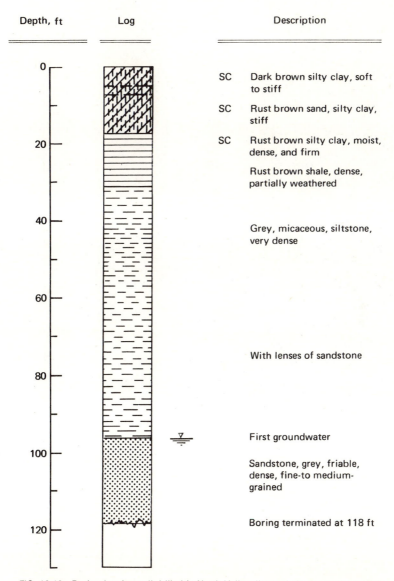

Depth, ft	Log		Description
0		SC	Dark brown silty clay, soft to stiff
		SC	Rust brown sand, silty clay, stiff
20		SC	Rust brown silty clay, moist, dense, and firm
			Rust brown shale, dense, partially weathered
40			Grey, micaceous, siltstone, very dense
60			
			With lenses of sandstone
80			
100			First groundwater
			Sandstone, grey, friable, dense, fine-to medium-grained
120			Boring terminated at 118 ft

FIG. 10-18   Boring log for well drilled in North Valley disposal site for Example 10-7.

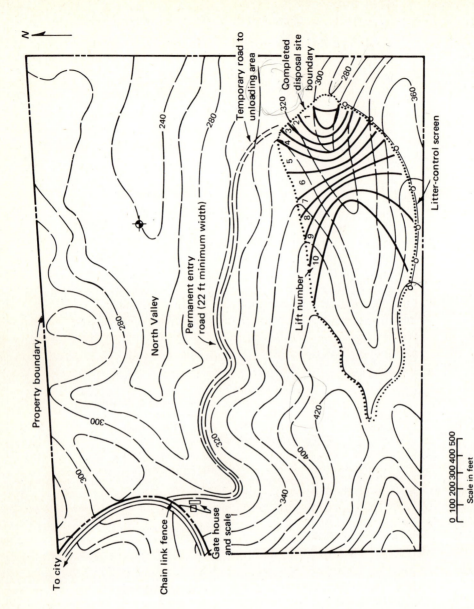

FIG. 10-19 Disposal site layout showing filling sequence for South Valley site for Example 10-7.

Property boundary

To city

Chain link fence

Gate house and scale

North Valley

Permanent entry road (22 ft minimum width)

Temporary road to unloading area

Completed disposal site boundary

Lift number

Litter-control screen

240

280

280

300

320

340

420

400

360

300

280

320

0 100 200 300 400 500

Scale in feet

N

of the site. It should be at least 22 ft wide with the necessary side ditches for drainage. It should be constructed of crushed rock and oiled to make the surface permanent. Topsoil should be removed to the stockpile site as the road is built. When the filling operation is completed and the disposal site is closed, the access road should be removed and the area covered with clean fill.

The temporary haul roads to the operating area of each lift can be constructed of a solid waste and soil mixture. The haul roads need not be permanent because they will be covered by successive lifts of solid wastes.

3.   Filling plan and operation. The landfilling method selected is a combination area-depression fill. The first lift in the South Valley should be started at an elevation of 300 ft, as shown in Fig. 10-19, and should be filled with solid wastes to a depth of about 9.5 ft. A 6-in layer of cover material should be placed over the compacted fill at the end of each day to provide a surface for vehicular traffic and to eliminate rodent access to the compacted wastes. To start the landfill operation, the topsoil should be stripped away in the lower portions of the South Valley and removed to the stockpile at the eastern end of the landfill site. The stockpile serves as a dam

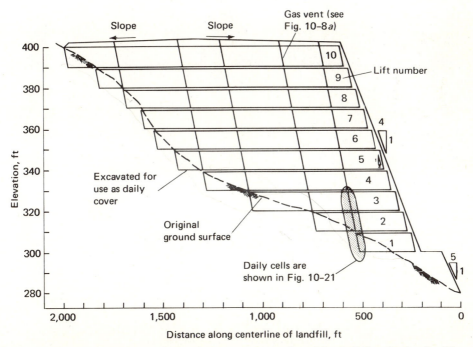

FIG. 10-20   Typical section through approximate centerline of South Valley site showing lift details for example 10-7. (Note different scales used for horizontal and vertical distances.)

to retain runoff as well as a site for topsoil storage. The west end of the lift should be excavated as filling proceeds, and each lift should be extended as shown in Figs. 10-19 and 10-20.

Vents for the discharge of gases from the decomposing solid wastes must be incorporated into the construction of the landfill. The vents should be constructed of a 12- to 18-in layer of granular material so that the gases can travel to the surface easily (see Figs. 10-20 and 10-21). The rocky nature of the cover material makes it suitable for use as a venting medium. The ground surface around the vents should be shaped so that rainwater drains away from openings to prevent excessive amounts of water from entering the landfill.

The width of the unloading area should be limited to a distance of 150 ft to prevent indiscriminate dumping and to maintain better operational control. Traffic at the unloading area will consist of the collection vehicles with experienced drivers, as well as vehicles from commercial and industrial establishments and private vehicles driven by residents who may not be familiar with a landfill operation. Areas for regular haulers and other vehicles should be kept separate so that traffic problems will not develop. The areas can be adjusted on weekends when regular haulers are not working and private vehicle traffic is at a maximum.

4. Determination of site capacity. Once a filling plan is selected, it is possible to calculate the site capacity. The sequence for filling the South Valley is shown in Figs. 10-19 and 10-20. The necessary data needed to determine the capacity of the South Valley disposal site are summarized in Tables 10-16 and 10-17.

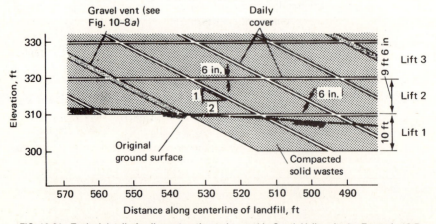

FIG. 10-21   Typical detail of cell construction to be used in South Valley site for Example 10-7.

TABLE 10-16 ESTIMATION OF WASTE QUANTITIES FOR EXAMPLE 10-7

Year	Population (000)	Daily volume,* yd³	Yearly volume, yd³	Cumulative total, yd³
1975	38	243.2	88,800	88,800
1976	39.6	253.4	92,500	181,300
1977	42.1	269.4	98,300	279,600
1978	42.8	273.9	100,000	379,600
1979	44.4	284.2	103,700	483,300
1980	46	294.4	107,500	590,800
1981	48	307.2	112,000	702,800
1982	50	320.0	116,800	819,600
1983	52	332.8	121,500	941,100
1984	54	345.6	126,100	1,067,200
1985	56	358.4	130,800	1,198,000
1986	58	371.2	135,500	1,333,500
1987	60	384.0	140,200	1,473,700
1988	62	396.8	144,800	1,618,500
1989	64	409.6	149,500	1,768,000
1990	66	422.4	154,200	1,922,200

* Based on end-of-year population and a compacted in-place solid waste density of 1,000 lb/yd³. For example: (38,000 persons × 6.4 lb/person/day)/(1,000 lb/yd³).
Note: yd³ × 0.7646 = m³
  lb/person/day × 0.4536 = kg/person·day

The expected daily, yearly, and cumulative yearly total waste quantities are given in Table 10-16. As noted, the daily and yearly waste quantities were computed on the basis of the projected end-of-year population. This procedure is recommended even though it is on the conservative side. The volume was computed by using an assumed value of 1,000 lb/yd³ for the in-place compacted density of the solid wastes. The computed values can be scaled for any other assumed density values.

The volumetric capacity of the South Valley landfill site in cubic yards is computed in Table 10-17. The area at each contour interval is obtained from Fig. 10-19 by using a planimeter. Alternatively, the contour area can be determined by placing a see-through grid over Fig. 10-19 and counting the squares. To determine the total volume of the landfill, the two end areas are averaged, and then the corresponding value is multiplied by 10 ft (the lift height) and divided by 27 to convert to cubic yards. The volume of solid wastes is determined by multiplying the total

TABLE 10-17 ESTIMATION OF CAPACITY OF SOUTH VALLEY LANDFILL SITE FOR EXAMPLE 10-7

Lift number	Elevation	Area, ft² At contour interval*	Area, ft² Average between contours	Capacity between contours,† yd³
	300	23,000		
1			35,500	13,200
	310	48,000		
2			64,000	23,700
	320	80,000		
3			115,000	42,600
	330	150,000		
4			197,500	73,200
	340	245,000		
5			272,500	101,000
	350	300,000		
6			310,000	114,800
	360	320,000		
7			342,500	126,900
	370	365,000		
8			402,500	149,100
	380	440,000		
9			450,000	166,700
	390	460,000		
10			447,500	165,700
	400	435,000		
Total capacity, yd³				976,900 (747,328 m³)
Solid waste capacity,‡ yd³				781,520 (597,863 m³)

* From Fig. 10-19.
† Volume = (average area, ft²) × (10 ft)/(27 ft³/yd³).
‡ Solid waste capacity = total capacity × 0.8.
Note: ft × 0.3048 = m
         yd³ × 0.7646 = m³
         ft² × 0.0929 = m²
         ft³ × 0.02833 = m³

volume by 0.8, on the assumption that 1 yd³ of cover material will be required for every 4 yd³ of solid wastes.

When the quantities given in Tables 10-16 and 10-17 are compared, the useful life of the South Valley disposal site is found to be about 7 yr (1975 to 1981). At that time it would be necessary to develop the North Valley landfill site.

5. Cover material. The cover material for each day's accumulation of solid wastes would be taken from the sides of the ridges and from the areas where the subsequent lifts will be placed. Some materials should be available from excavation of the lift in operation. The amount of daily cover will vary from about 60 yd^3 at the start to 105 yd^3 at the end of the filling operation. The total cover material requirement for the 7 yr of operation is estimated to be about 195,380 yd^3 (see Table 10-17).

6. Erosion and drainage control. The excavation of material from the sides of the ridges and the placing of cover material on finished slopes will create surfaces that will be susceptible to erosion during the rainy winter months. Planting of grain or grass crops on higher slopes will help reduce the amount of runoff reaching the lower borrow areas, thus stabilizing erosion. Periodic filling and regrading of finished slopes will help prevent exposure of underlying solid waste fill.

    A drainage ditch should be constructed around the operation area to carry away runoff water. The ditches should be located at the 300-ft elevation at the North Valley site and near the 400-ft elevation for the South Valley site. The ditch should slope to the east and terminate at the toe of the landfill area. The natural slope of the offsite lands will carry the water away from the site. The high point of a lift should be at the center so that water runs off toward the sides of the landfill and into the drainage ditch.

7. Wet-weather operation. Operations during the rainy period of the year, November to April, can continue if the haul roads are passable. The addition of a quick-draining gravel to the road surface should help keep it operational. Culverts under the road should transmit the water across the roadway to prevent flooding of the road. Spreading and compacting equipment can operate in wet-weather conditions. If the haul roads become impassable, it may be necessary to develop a wet-weather unloading site near the paved entrance road.

8. Water supply and fire prevention. A supply of water is necessary for landfill operations. Wetting the solid wastes decreases the blowing and spreading of debris by the wind. Applying water to the face of the fill before covering greatly decreases the possibility of fire in the fill. The water supply should also be used for dust control in the operation area and on the roads.

    Potable water should be provided at the gatehouse for operator consumption. Because of the poor quality of the onsite water, potable water must be obtained from a bottled-water supplier. A water tank truck should be used to supply water for dust control, wetting, fire protection, and miscellaneous use. The truck should be filled with water from a city fire hydrant and should fill up as required during the day. The use of a truck would be less costly than extending the city water system and putting in a booster pump station.

9. Operation control. An entrance scale and gatehouse are recommended. The gatehouse would be used by personnel who weigh trucks. If the weight of the solid wastes delivered is known, then the in-place density of the wastes can be determined and the performance of the operation can be monitored. The weight records would also be used as a basis for charging participating agencies and private haulers for their contributions.

The gatehouse can be a relatively simple building since only a few people will be using it and no complicated system will be required to operate the site. A 10 × 20 ft trailer office or prefabricated building should be adequate.

The recommended time schedule for operation of the site is 8 A.M. to 5 P.M., 7 days/wk, as the site should be open for the convenience of the public.

10. Equipment requirements. The equipment requirements for a sanitary landfill operation depend on the quantity and type of solid wastes to be handled, the type of cover material, and the distance that the cover material must be transported. At this site, cover material must be excavated and hauled to the fill location, the solid wastes must be spread and compacted, and cover material must be placed and compacted over the solid wastes. The borrow area and the fill location should also be cleaned up occasionally. The surface of the lift should be graded periodically to eliminate ruts caused by equipment, to remove holes caused by differential settlement as the solid wastes decompose, and to maintain proper drainage. Information on landfill equipment and needs is provided in Table 10-15.

The amount of solid wastes delivered to the landfill will vary from 122 to 154 tons/day over the 7-yr life of the site. The following equipment, which should be able to handle 130 to 250 tons in 8 h (see Table 10-15), will be required:

a. One crawler dozer with landfill blade, 150 to 180 hp
b. One landfill compactor
c. One scraper, 15 yd³
d. One tank truck for water hauling and distribution, 1,200-gal capacity

The crawler dozer can be used full-time for spreading and covering wastes and for general maintenance of the site. The landfill compactor can be used to compact the wastes and to serve as backup of the crawler dozer. The scraper can be used to excavate and transport cover material.

11. Personnel. Personnel requirements at the landfill disposal site are as follows:

    a. One scale attendant and fee collector who will be located at the gatehouse to control access to the site, weigh incoming vehicles, collect fees, and keep records.

    b. Two equipment operators. The tank truck will be used intermittently as will the landfill compactor and scraper. One person can operate these pieces of equipment. The other person can handle the crawler dozer on a full-time basis. The two can alternate jobs on the site to provide variety in work tasks. The operators can aid in the unloading of solid wastes by directing vehicles to the appropriate unloading areas.

12. Land use after disposal. The total area of the landfill site presently is used for grazing and crop farming. The final landfill lift will be prepared so that agricultural operations can be resumed. This will be accomplished through the movement of stockpiled topsoil to the finished landfill area. The finished topsoil dressing should be 2 ft thick. Once completed, periodic filling and regrading will be required for the next 2 to 5 yr to maintain proper surface drainage conditions.

10-6 OCEAN DISPOSAL OF SOLID WASTES

Although ocean dumping of municipal solid wastes was abandoned in the United States in 1933, the concept has persisted throughout the years and is still frequently discussed today. Within the past few years, the idea that the ocean is a gigantic sink, in which an infinite amount of pollution of all types can be dumped, has been discarded. On the other hand, it is argued that many of the wastes now placed in landfills or on land could be used as fertilizers to increase the productivity of the ocean. It is also argued that the placement of wastes in ocean-bottom trenches where tectonic folding is occurring is an effective method of waste disposal. Currently (1976) a number of industrial solid wastes and certain other wastes are disposed of at sea (see Chap. 11).

Industrial Solid Wastes [25]

The usual method used for the disposal of industrial wastes at sea consists of transporting the wastes in bulk or containers aboard towed or self-propelled barges to the point of discharge, usually on the high seas. Bulk tank barges range in capacity from 1,000 to 5,000 tons. They must be of double-skinned bottom construction and certified by the U.S. Coast

Guard. Discharge rates for conventional industrial wastes vary between 4 to 20 tons/min. The discharge hose is trailed at a depth of 1 to 2.5 fathoms while under way at a speed of 3 to 6 kn.

Wastes in containers are either weighted and sunk or ruptured and allowed to sink. In some cases, chemical wastes are carried to sea as deck cargo on merchant ships. Once the ships are on the high seas, the containers are dumped overboard.

Municipal Solid Wastes

With the exception of some isolated cases and excluding sewage sludge, municipal solid wastes from the United States are not now discharged into the ocean environment [25]. One of the principal reasons is that many solid waste components, including paper, wood, plastics, and rubber, will float to the surface. The presence of large quantities of floating solid wastes is unacceptable from an aesthetic, marine craft, and environmental standpoint. Even if wastes are baled before ocean disposal, it is almost certain that, over a period of time, the bales will disintegrate and floatables will rise to the surface. For these reasons, the disposal of municipal solid wastes in the ocean is not a viable alternative at this time.

10-7 DISCUSSION TOPICS AND PROBLEMS

10-1. A medium-sized community has three major solid waste generation areas (two residential areas and the downtown business district). This city is very fortunate because it has available for its use four different disposal sites. The closest site to the city has a daily capacity D_1 and a useful life of 5 yr. The next two disposal sites have a daily capacity of D_2 and D_3, respectively. They are both the same distance from the city, and both have a useful life of 20 yr. The fourth disposal site is the farthest from the city. Its daily capacity, D_4, is sufficient to handle all the solid wastes, and its useful life is 100 yr. Solid wastes are now delivered to all the sites and covered or burned intermittently.

A recent proposal to the city council by a private collector calls for abandoning the use of the three smaller disposal sites and hauling all the solid wastes to the farthest disposal site. If the only criterion is to be cost, outline in detail how you would arrive at the best method of operation (i.e., which site or sites should be used and for how long) to yield the lowest annual cost for the next 20 yr. Assume that the following conditions prevail and that all the required data are available.

1. The three smaller disposal sites are to be operated as modified sanitary landfills (covered every other day). The larger site will be operated as a sanitary landfill (solid wastes will be covered every night) if it is used alone or as a modified landfill if it is used jointly.

2. Because of the distances involved, a transfer station is not economically feasible.

10-2. 1. Using the volume-reduction data reported in Table 10-10, estimate the in-place density of solid wastes with the following composition:

Component	Weight, percent
Food wastes	12
Cardboard	5
Paper	50
Tin cans	10
Glass	7
Garden trimmings	16
	—
	100

2. If 80 percent of the paper were removed, what would be the resulting in-place density?

3. By what factor would the useful life of the landfill site be increased if 80 percent of both the paper and cardboard were removed?

10-3. Prepare a lift diagram and determine the capacity of the North Valley landfill disposal site shown in Fig. 10-19.

10-4. Given the site plan shown in Fig. 10-22 for a parcel of land near the Fallen Oak River, prepare a sanitary landfill operation plan for the following conditions:

1. Number of collection services = 2,800 (average over 20 yr)   *19.6 TON/d*

2. Amount of solid wastes generated per service = 14.0 lb/day

3. Compacted density of solid wastes in landfill = 800 lb/yd³

4. Maximum allowable finish grade elevation above surrounding ground = 5 ft

Include the following in your plan analysis:

1. Required site preparation work, if any

2. Placement operation plan (e.g., the proposed method to be followed in filling the site)

3. Estimated useful life of site

4. Equipment and storage facility requirements

5. Workforce and requirements

6. Operational plan

10-5. Assuming that the curves shown in Fig. 10-7 can be approximated by a first-order equation, estimate the surface settlement after 10 yr in a well-compacted sanitary landfill (use maximum compaction curve). What will the maximum surface settlement be after 50 yr?

10-6. A sanitary landfill 50 ft deep has been completed for several years in an alluvial gravel. The normal groundwater level is 150 ft below the surface, or 100 ft below the bottom of the fill. A special sampling well at the edge of the landfill shows that the atmosphere in the interstices of the soil 20 ft above the water table contains 48 percent CO_2, 28 percent CH_4, 20 percent N_2, 2 percent O_2, 1 percent H_2S, and 1 percent other gases, analyzed and calculated on a dry basis at 0°C and 760 mm pressure. On the basis of a long period of contact (i.e., equilibrium) at 10°C, compute the concentration in mg/l of each of these five gases to be expected in the upper layers

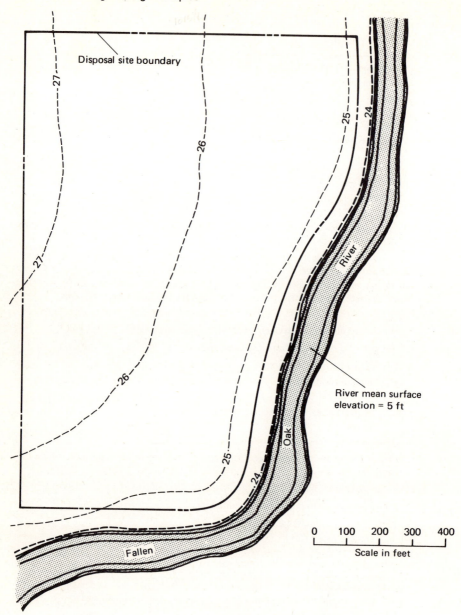

FIG. 10-22   Solid waste disposal site for Prob. 10-4.

of the groundwater under a total pressure of 1 atm at 10°C. Assume saturation with respect to vapor pressure. (Problem, courtesy of Dr. Paul H. King.)

10-7.   If municipal solid wastes with the composition given in Table 4-2 are to be mixed with waste-water treatment plant sludge containing 5 percent solids to achieve a final moisture content of 55 percent, estimate the ultimate amount of leachate that would be produced per cubic yard of compacted solid waste if no surface infiltration were allowed to enter the completed landfill. Assume that the following data and information are applicable.

1. Initial moisture content of municipal solid wastes = 21 percent.

2. In-place density of compacted mixture of solid wastes and sludge = 1,200 lb/yd³.

3. Chemical formula for decomposable portion of the combined wastes = $C_{60}H_{85}O_{40}N$.

4. Decomposable wastes will be converted totally according to Eq. 9-5.

5. Final moisture content of wastes remaining in landfill = 35 percent.

6. Neglect surface evaporation.

10-8. In Problem 10-7, if final in-place density after all the decomposable wastes have been converted and the leachate has been removed is 1,400 lb/yd³, estimate the total percentage volume reduction. State clearly all the assumptions used in solving this problem.

10-9. Determine the effect of a 10° rise in temperature on the rate of percolation of leachate from a sanitary landfill.

10-10. On your first day at work for a solid waste consulting organization, your superior asks you to prepare a proposal (in outline form) to evaluate the feasibility of ocean dumping of baled solid wastes. The only information available is that the Press-It-Tight Baling Co. claims that it can produce bales with an average density of 70 lb/ft³ and that if these bales of solid wastes are dumped in the ocean, they will sink to the bottom because of their greater density and remain there, causing no problems. Structure your proposal by asking yourself what kinds of information, data, and criteria would be required to protect the environment and to formulate public policy concerning ocean dumping.

10-8 REFERENCES

1. Brunner, D. R. and D. J. Keller: *Sanitary Landfill Design and Operation,* U.S. Environmental Protection Agency, Publication SW-65ts, Washington, D.C., 1972.

2. Clark, D. A. and J. E. Moyer: *An Evaluation of Tailing Ponds Sealants,* U.S. Environmental Protection Agency, Publication 660/2-74-065, Corvallis, Oreg., 1974.

3. County of Los Angeles, Department of County Engineer, Los Angeles, and Engineering-Science, Inc.: *Development of Construction and Use Criteria for Sanitary Landfills, An Interim Report,* U.S. Department of Health, Education, and Welfare, Public Health Service, Bureau of Solid Waste Management, Cincinnati, 1969.

4. Cummins, R. L.: *Effects of Land Disposal of Solid Wastes on Water Quality,* U.S. Department of Health, Education, and Welfare, Public Health Service Publication SW-2ts, Cincinnati, 1968.

5. Davis, S. N. and R. J. M. DeWiest: "Hydrogeology," Wiley, New York, 1966.

6. Dean, J. A. (ed.): "Lange's Handbook of Chemistry," 11th ed., McGraw-Hill, New York, 1973.

7. Dickason, O. E.: *A Study of Gases in the Zone of Aeration,* Unpublished Ph.D. thesis, Stanford University, Stanford, Calif., 1970.

8. Eliassen, R.: Decomposition of Landfills, *American Journal of Public Health,* vol. 32, no. 3, 1942.

9. Eliassen, R.: Load Bearing Characteristics of Landfills, *Engineering News Record,* vol. 129, no. 11, 1942.

10. Eliassen, R.: Refuse Collection and Disposal, in R. W. Abbett (ed.), "American Civil Engineering Practice," Wiley, New York, 1956.

11. Foree, E. G. and E. N. Cook: *Aerobic Biological Stabilization of Sanitary Landfill Leachate,* Department of Civil Engineering, University of Kentucky, Publication UKY TR58-72-CE21, Lexington, 1972.

12. Glasstone, S.: "Textbook of Physical Chemistry," 2d ed., Van Nostrand, Princeton, N.J., 1946.

13. *Guidelines for Local Government on Solid Waste Management,* U.S. Environmental Protection Agency, Publication SW-17c, Washington, D.C., 1971.

14. Hughes, G. M., R. A. Landon, and R. N. Fairolden: *Hydrogeology of Solid Waste Disposal Sites in Northeastern Illinois,* U.S. Environmental Protection Agency, Publication SW-12d, Washington, D.C., 1971.

15. Jumikis, A. R.: "Introduction to Soil Mechanics," Van Nostrand, Princeton, N.J., 1967.

16. Krauskopf, K. B.: "Introduction to Geochemistry," McGraw-Hill, New York, 1967.

17. Linsley, R. K., Jr., M. A. Kohler, and J. L. H. Paulhus: "Hydrology for Engineers," McGraw-Hill, New York, 1958.

18. Merz, R. C. and R. Stone: *Special Studies of a Sanitary Landfill,* U.S. Department of Health, Education, and Welfare, Washington, D.C., 1970.

19. Parsons, H. de B.: "The Disposal of Municipal Refuse," 1st ed., Wiley, New York, 1906.

20. Perry, R. H., C. H. Chilton, and S. D. Kirkpatrick: "Chemical Engineers Handbook," 4th ed., McGraw-Hill, New York, 1963.

21. Reinhardt, J. J. and R. K. Ham: *Solid Waste Milling and Disposal on Land without Cover,* U.S. Environmental Protection Agency, NTIS Publication PB-234930, Springfield, Va., 1974.

22. *Refuse Collection and Sanitary Landfill Operational Methods,* Texas State Department of Health, Division of Sanitary Engineering, Austin, Tex., 1954.

23. *Report on the Investigation of Leaching of a Sanitary Landfill,* California State Water Pollution Control Board, Publication 10, Sacramento, Calif., 1954.

24. Salvato, J. A., W. G. Wilkie, and B. E. Mead: Sanitary Landfill—Leaching Prevention and Control, *Journal Water Pollution Control Federation,* vol. 43, no. 10, 1971.

25. Smith, D. D. and R. P. Brown: *Ocean Disposal of Barge-Delivered Liquid and Solid Wastes from U.S. Coastal Cities,* U.S. Environmental Protection Agency, Solid Waste Management Series, Publication SW-19c, Washington, D.C., 1971.

26. Sorg, T. J. and H. L. Hickman: *Sanitary Landfill Facts,* 2d ed., U.S. Public Health Service, Publication 1792, Washington, D.C., 1970.

27. State Water Resources Control Board: *In-Sites Investigation of Movements of Gases Produced from Decomposing Refuse,* The Resources Agency, Publication 31, State of California, Sacramento, 1965.

28. State Water Resources Control Board: *In-Sites Investigation of Movements of Gases Produced from Decomposing Refuse,* Final Report, The Resources Agency, Publication 35, State of California, Sacramento, 1967.

29. Wall, T. E. and J. C. Young: *Design Guide for Sanitary Landfills in Iowa,* Presented at the Eleventh Annual Water Resources Design Conference, Iowa State University, Ames, Iowa, 1973.

11

Hazardous Wastes

Hazardous wastes have been defined by the EPA as wastes or combinations of wastes that pose a substantial present or potential hazard to human health or living organisms because (1) such wastes are nondegradable or persistent in nature, (2) they can be biologically magnified, (3) they can be lethal, or (4) they may otherwise cause or tend to cause detrimental cumulative effects [12]. Discussion of hazardous wastes has been reserved for a separate chapter because the technology and management of these wastes are both highly specialized and controlled. Yet so much more needs to be known about this subject that the intention here is only to introduce the reader to hazardous waste management; specific design data and information are not presented.

In this chapter, the identification and classification of hazardous wastes and the important local, state, federal, and international regulations related to their control are described. Then the role of each of the functional elements (considered previously in Chaps. 4 through 10) in the management of hazardous wastes is discussed.

11-1 IDENTIFICATION OF HAZARDOUS WASTES

To assess whether a given substance or material is hazardous, a preliminary decision model for screening and selecting hazardous compounds and ranking hazardous wastes has been developed [12]. The screening model is presented in the form of a flowchart in Fig. 11-1 (p. 378). Terms and abbreviations used in Fig. 11-1 are defined in Table 11-1. The hazardous waste criteria used in the screening model relate to only the intrinsic hazard of the waste on uncontrolled release to the environment, regardless of quantity or pathways to humans or other critical organisms. For this

TABLE 11-1 DEFINITION OF TERMS USED IN HAZARDOUS WASTE SCREENING MODEL SHOWN IN FIG. 11-1*

Term	Abbreviation	Definition
Maximum permissible concentration	MPC	Levels of radioisotopes in waste streams which, if continuously maintained, would result in maximum permissible doses to occupationally exposed workers and which may be regarded as indices of the radiotoxicity of the different radionuclides.
Bioconcentration (bioaccumulation, biomagnification)		The process by which living organisms concentrate an element or compound to levels in excess of those in the surrounding environment.
National Fire Protection Association category 4 flammable materials	NFPA	Materials including very flammable gases, very volatile flammable liquids, and materials that in the form of dusts or mists readily form explosive mixtures when dispersed in air.
NFPA category 4 reactive materials		Materials which in themselves are readily capable of detonation or of explosive decomposition or reaction at normal temperatures and pressures.
Lethal dose 50	LD_{50}	A calculated dose of a chemical substance which is expected to kill 50 percent of a population of experimental animals exposed through a route other than respiration. Dose concentration is expressed in milligrams per kilogram of body weight.

reason, criteria such as toxicity, phytotoxicity, genetic activity, and bioconcentration are used [12]. It must be remembered that as our knowledge increases, other factors may be added to the model, and the critical limits may be revised.

In the development of a system for ranking potential hazard, the threat to public health and to the environment for a given hazardous waste is strongly dependent on the quantity of the waste involved. The extent to which present treatment technology and regulatory activities mitigate against threat, and the pathways to humans or other critical organisms, is considered further in Ref. 12.

11-2 CLASSIFICATION OF HAZARDOUS WASTES

From a practical standpoint, there are far too many compounds, products, and product combinations that fit within the broad definition of hazardous wastes just stated to list individually in this text. For this reason, groups of wastes are considered in five general categories: (1) radioactive substances, (2) chemicals, (3) biological wastes, (4) flammable wastes, and (5) explo-

TABLE 11-1   (Continued)

Term	Abbreviation	Definition
Lethal concentration 50	LC_{50}	A calculated concentration which, when administered by the respiratory route, is expected to kill 50 percent of a population of experimental animals during an exposure of 4 h. Ambient concentration is expressed in milligrams per liter.
Grade 8 dermal irritation		An indication of necrosis resulting from skin irritation caused by application of a 1 percent chemical solution.
Median threshold limit	96-h TLm	That concentration of a material at which it is lethal to 50 percent of the test population over a 96-h exposure period. Ambient concentration is expressed in milligrams per liter.
Phytotoxicity		Ability to cause poisonous or toxic reactions in plants.
Median inhibitory limit	ILm	That concentration at which a 50 percent reduction in the biomass, cell count, or photosynthetic activity of the test culture occurs compared to a control culture over a 14-day period. Ambient concentration is expressed in milligrams per liter.
Genetic changes		Molecular alterations of the deoxyribonucleic or ribonucleic acids of mitotic or meiotic cells resulting from chemicals or electromagnetic or particulate radiation.

* Adapted from Ref. 12.

sives. The characteristics of each category and a list of some typical examples are presented in the following discussion.

Radioactive Substances

Substances that emit ionizing radiation are defined as being radioactive. Such substances are hazardous because prolonged exposure to radiation often results in damage to living organisms. Radioactive substances are of special concern because they persist over long periods of time. The time period in which radiation continues to occur is commonly measured and expressed in half-lives. The *half-life* of a radioactive substance is defined as the time required for the radioactivity of a given amount of the substance to decay to half its initial value. For example, uranium compounds have half-lives that range from 72 yr for U_{232} to 23,420,000 yr for U_{236} [12].

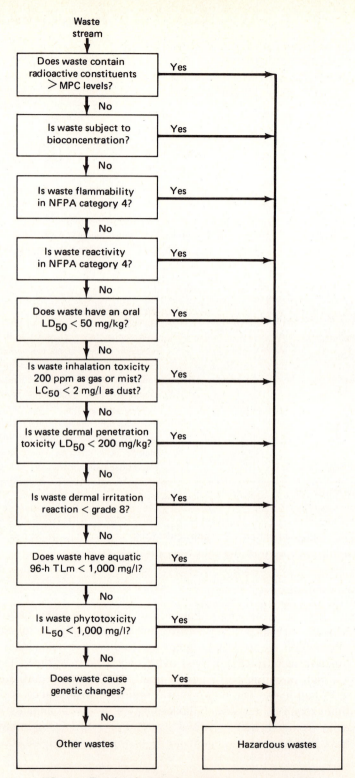

FIG. 11-1 Flowchart for hazardous waste screening model [12].

The management of radioactive wastes is highly controlled by federal and state regulatory agencies that are not associated with management agencies and have little in common with them. Disposal sites that are used for the long-term storage of radioactive wastes are not used for the disposal of any other solid wastes. Because radioactive waste management is a highly specialized activity and generally will not be the responsibility of the municipal solid waste system manager, it is not covered in this text. For more information on this subject, Refs. 11 and 13 are recommended.

Chemicals

Most hazardous chemical wastes can be classified into four groups: (1) synthetic organics; (2) inorganic metals, salts, acids, and bases; (3) flammables; and (4) explosives. Flammables and explosive chemicals are discussed separately in this section because they present an especially difficult storage, collection, and disposal hazard.

Typical chemical hazardous wastes are reported in Table 11-2. Most of the chemicals listed are hazardous because they are highly toxic to most life forms. The most common approach used in characterizing wastes containing the compounds listed in Table 11-2 is based on the assumption that the pure compound is most hazardous. When hazardous compounds are present in a waste stream at levels equal to or greater than the threshold levels established for them, the entire waste stream is identified as hazardous. A more complete listing of toxic substances is given in Ref. 2.

Biological Wastes

The principal sources of hazardous biological wastes are hospitals and biological research facilities. The ability to infect other living organisms and the ability to produce toxins are the most significant characteristics of a hazardous biological waste. Included in this group of solid wastes are malignant tissues taken during surgical procedures and contaminated materials, such as hypodermic needles, bandages, and outdated drugs. Hazardous biological wastes are also generated as a by-product of industrial biological conversion processes.

Flammable Wastes

Most flammable hazardous wastes are also identified as hazardous chemical wastes. This dual grouping is necessary because of the high potential hazard in storing, collecting, and disposing of flammable wastes. These wastes may be in liquid, gaseous, or solid form, but most often they are liquids. Typical examples include organic solvents, oils, plasticizers, and organic sludges. Many of the compounds listed as toxic chemicals in Table 11-2 are also flammable.

TABLE 11-2 A SAMPLE LIST OF NONRADIOACTIVE HAZARDOUS COMPOUNDS (from Ref. 12)

Miscellaneous inorganics:
Ammonium chromate
Ammonium dichromate
Antimony pentafluoride
Antimony trifluoride
Arsenic trichloride
Arsenic trioxide
Cadmium (alloys)
Cadmium chloride
Cadmium cyanide
Cadmium nitrate
Cadmium oxide
Cadmium phosphate
Cadmium potassium
    cyanide
Cadmium (powdered)
Cadmium sulfate
Calcium arsenate
Calcium arsenite
Calcium cyanides
Chromic acid
Copper arsenate
Copper cyanides
Cyanide (ion)
Decaborane
Diborane
Hexaborane
Hydrazine
Hydrazine azide
Lead arsenate
Lead arsenite
Lead azide
Lead cyanide
Magnesium arsenite
Manganese arsenate
Mercuric chloride
Mercuric cyanide
Mercuric diammonium
    chloride
Mercuric nitrate
Mercuric sulfate
Mercury
Nickel carbonyl
Nickel cyanide
Pentaborane-9
Pentaborane-11
Perchloric acid (to 72%)
Phosgene (carbonyl
    chloride)

Potassium arsenite
Potassium chromate
Potassium cyanide
Potassium dichromate
Selenium
Silver azide
Silver cyanide
Sodium arsenate
Sodium arsenite
Sodium bichromate
Sodium chromate
Sodium cyanide
Sodium
    monofluoroacetate
Tetraborane
Thallium compounds
Zinc arsenate
Zinc arsenite
Zinc cyanide
Halogens and interhalogens:
Bromine pentafluoride
Chlorine
Chlorine pentafluoride
Chlorine trifluoride
Fluorine
Perchloryl fluoride
Miscellaneous organics:
Acrolein
Alkyl leads
Carcinogens (in general)
Chloropicrin
Copper acetylide
Copper chlorotetrazole
Cyanuric triazide
Diazodinitrophenol (DDNP)
Dimethyl sulfate
Dinitrobenzene
Dinitro cresols
Dinitrophenol
Dinitrotoluene
Dipentaerythritol hexanitrate
    (DPEHN)
GB (propoxy (2)-
    methylphosphoryl fluoride)
Gelatinized nitrocellulose
    (PNC)
Glycol dinitrate
Gold fulminate
Lead 2,4-dinitroresorcinate

Lead styphnate
Lewisite (2-chloroethenyl
    dichloroarsine)
Mannitol hexanitrate
Nitroaniline
Nitrocellulose
Nitrogen mustards (2,2',2"
    trichlorotriethylamine)
Nitroglycerin
Organic mercury compounds
Pentachlorophenol
Picric acid
Potassium
    dinitrobenzfuroxan (KDNBF)
Silver acetylide
Silver tetrazene
Tear gas (CN)
    (Chloroacetophenone)
Tear gas (CS) (2-
    chlorobenzylidene
    malononitrile)
Tetrazene
VX (ethoxy-methyl
    phosphoryl N,N dipropoxy-
    (2-2), thiocholine)
Organic halogen compounds:
Aldin
Chlorinated aromatics
Chlordane
Copper acetoarsenite
2,4-D (2,4-
    dichlorophenoxyacetic
    acid)
DDD
DDT
Demeton
Dieldrin
Endrin
Ethylene bromide
Fluorides (organic)
Guthion
Heptachlor
Lindane
Methyl bromide
Methyl chloride
Methyl parathion
Parathion
Polychlorinated biphenyls

Explosives

Explosive hazardous wastes are mainly ordnance materials and the wastes resulting from ordnance manufacturing. Also included in the group are some industrial gases. As previously noted, explosives like flammable wastes have a high potential for hazard in storage, collection, and disposal, and therefore they should be considered separately in addition to being listed as hazardous chemicals. These wastes may exist in solid, liquid, or gaseous form.

11-3 REGULATIONS

Regulations developed on the international, national, state, and local levels have had, and will continue to have, a great impact on hazardous waste management. While most of these regulations deal with generation and final disposal, they affect all aspects of solid waste management. General information on regulations for the management of hazardous wastes is presented in Table 11-3.

International Agreements

Industrial developments in many nations have led to the widespread production and use of synthetic chemicals. Wastes from these production facilities are often discharged in the ocean or burned, resulting in combustion gas emissions to the atmosphere. In both cases, the final points of disposal are often outside national boundaries.

Presently, there are only a few international agreements affecting the management of hazardous wastes. The best known is the agreement between the United States and the Soviet Union which bans the atmospheric detonation of nuclear devices. There are continuing efforts to develop international criteria and controls on chemical wastes discharged into the oceans. Therefore, additional demands for land disposal sites can be expected in the future.

Federal Regulations

The greatest number of regulations concerning hazardous materials are formulated at the federal level. Many of these regulations are designed to control the packaging, storage, and movement of hazardous materials. Because most federal regulations cover hazardous goods and not hazardous wastes, they are of only casual interest to the solid waste system manager. A partial list of regulations passed by the United States Congress is presented in Ref. 12.

The most comprehensive regulations for controlling hazardous wastes at the federal level are related to water discharges and air emissions. The amounts of waste discharges allowable to either water or air are usually set after the ambient level of a particular compound is known. In other cases,

TABLE 11-3 REGULATIONS FOR THE MANAGEMENT OF HAZARDOUS WASTES

Type	Legal form	Typical contents	Impacts on solid waste management
International	Treaties and agreements	Deals mostly with the oceans and the atmosphere.	Negligible to nonexistent
Federal	Directly legislated public laws and administrative procedures developed by implementation agencies	Primary emphasis is on the safe packaging, storage, and movement of hazardous compounds, not hazardous wastes; the ability of hazardous compounds to become hazardous wastes is of concern: secondary emphasis is on the protection of waterways and the atmosphere.	Moderate to extremely high economic impact associated with the land disposal of concentrated sludges
State	Directly legislated state laws and administrative procedures developed by implementation agencies	Primary emphasis is on the protection of waterways and the atmosphere; some facility review to determine types of hazardous wastes discharged; normally includes some designation of wastes that are acceptable at designated disposal sites.	Moderate to extremely high economic impact on transport and disposal; some operational impact caused by location of disposal sites
Local	Ordinances and administrative procedures developed by local agencies	Primary emphasis is on the protection of community water and waste-water treatment facilities; large economic penalties are often specified to prevent violations.	Moderate economic impact on disposal site operation if the site can be used for hazardous wastes; moderate to high impact on administrative agencies caused by adverse community reaction to hazardous solid wastes

complete restrictions are set, and all discharges are prohibited. An example is the Federal Water Pollution Control Act (Public Law 92-500). The prohibition of ocean disposal suggested in that law could cause the county of Los Angeles to divert nearly 800,000 tons/yr of sewage sludge (at 75 percent equivalent moisture content) from the ocean to local landfills.

It should be noted that the emphasis in federal regulations is directed toward the elimination of the disposal of hazardous wastes in either the water or air environment. Consequently, the management of hazardous wastes that may occur as solids, liquids, or gases has become a solid waste management problem by default. It is for this reason that all three forms are commonly grouped and evaluated together with solid wastes in typical solid waste management systems.

State Regulations

State regulations regarding the control of hazardous wastes follow federal regulations closely. This similarity is expected, for most federal funding for waste-water treatment facilities is available only if federal discharge requirements are met. Because most hazardous wastes are in liquid form, they must be first removed from waste waters and then concentrated, stored, collected, and transported to landfills for treatment and/or disposal.

Perhaps the most comprehensive hazardous waste management program established by a state is that in California. The State Department of Health has established a comprehensive list of hazardous wastes, and the State Water Resources Control Board has defined the class of landfill at which these wastes can be discharged (see Chap. 17). Further regulation of hazardous wastes is achieved by means of the discharge requirements set for individual disposal sites by the regional water resources control boards.

Local Regulations

The regulations of local governments are necessarily narrow in scope. When hazardous wastes are identified, they are restricted by ordinance from local sewers and waste-water treatment facilities. This restriction leads to the removal and concentration of liquid hazardous wastes for delivery to an acceptable solid waste disposal site.

11-4 GENERATION

Hazardous wastes are generated in limited amounts throughout a community. An aerosol can discarded by a community resident is a potential hazard. However, the degree of risk represented by aerosol cans is low. Therefore, these cans are allowed to be collected, stored, transported, and disposed of in the same manner as other nonhazardous residential wastes. In terms of generation, the concern is with the identification of the amounts and types of hazardous wastes developed at each source, with emphasis on those sources where significant waste quantities are generated.

Sources of Hazardous Wastes

Unfortunately, very little information is available on the quantities of hazardous wastes generated within a community and in various industries. The production records of industry are proprietary and not generally accessible to waste system managers. Without production information, it is impossible to develop unit waste generation data. Hazardous waste generation outside industry is irregular, rendering waste generation parameters meaningless. The only practical means to overcome these limitations is to conduct detailed inventory and measurement studies at each potential source in a community.

As a first step in developing a community inventory, potential sources of hazardous wastes must be identified (see Table 11-4). The information in Table 11-4 is intended as only a guide (not a complete list) in identifying sources where hazardous wastes might be generated. The total annual quantity of hazardous wastes at any given source in a community must be established through data inventory completed during onsite visits.

Hazardous Waste Spillage

The spillage of containerized hazardous wastes must also be considered an important aspect of generation. The quantities of hazardous wastes that are

TABLE 11-4 SOME COMMON HAZARDOUS WASTE SOURCES FOUND IN TYPICAL COMMUNITIES IN THE UNITED STATES

Waste category	Sources
Radioactive substances	Biomedical research facilities, college and university laboratories, dentists' offices, hospitals, nuclear power plants
Toxic chemicals	Agricultural chemical companies, battery shops, car washes, chemical and paint storage warehouses, city and county equipment corporation yards, city police stations, college and university laboratories, construction companies, county sheriff stations, crop-dusting firms, dry cleaners, electric utilities, electronic and radio repair shops, fire departments, hospitals and clinics, industrial cooling towers, industrial plants too numerous to list, newspapers (photographic solutions), nuclear power plants, pest control agencies, photographic processing facilities or shops, plating shops, service stations, tanker-truck cleaning stations
Biological wastes	Biomedical research facilities, drug companies, hospitals, medical clinics
Flammable wastes	Dry cleaners, petroleum reclaiming plants, petroleum refining and processing facilities, service stations, tanker-truck cleaning stations
Explosives	Construction companies, dry cleaners, munitions production facilities

involved in spillages usually are not known. However, the wastes from a spill requiring collection and disposal are often significantly greater than the amount of spilled wastes, especially where an absorbing material, such as straw, is used to soak up liquid hazardous wastes or if the soil into which a hazardous liquid waste has percolated must be excavated. Both the straw and the liquid and the soil and the liquid are then hazardous wastes.

The effects of spillages are often spectacular and visible to the community. Because the occurrences of spillages cannot be predicted, the potential for human and environmental impact from them is greater than from routinely generated hazardous wastes. The effects of spillage are shown pictorially in Fig. 11-2. Additional information on the spillage and control of hazardous materials is found in Refs. 3 and 7.

11-5 ONSITE STORAGE

Onsite storage practices are a function of the types and amounts of hazardous wastes generated and the time period over which waste genera-

FIG. 11-2 Spillage of hazardous liquid. Depending on the characteristics of the spilled liquid, it may be necessary not only to remove the ponded liquid but also to excavate the soil into which the liquid has infiltrated. In this situation the total material that is now classified as a hazardous waste is significantly greater in volume than the original volume of the spilled liquid. (State of California, Department of Transportation, District 4.)

tion occurs. Usually, when large quantities are generated, special facilities are used that have sufficient capacity to hold wastes accumulated over a period of several days. When only small amounts of hazardous wastes are generated on an intermittent basis, they may be containerized, and limited quantities may be stored for periods covering months or years.

Containers and facilities used in hazardous waste storage and handling are selected on the basis of the characteristics of the wastes. For example, corrosive acid or caustic solutions are stored in fiber glass or glass-lined containers to prevent deterioration of metals in the container. Great care must also be exercised to avoid storing incompatible wastes in the same containers or locations. Codisposal of incompatible wastes can lead to the development of hazardous situations through heat generation, fires, explosions, or release of toxic substances. General information on the storage containers used for hazardous wastes and the conditions of their use is

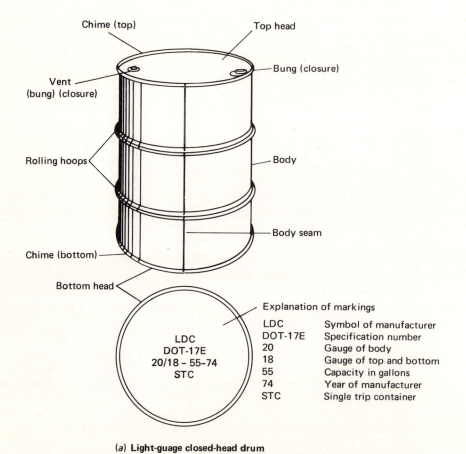

Chime (top)

Top head

Vent
(bung) (closure)

Bung (closure)

Rolling hoops

Body

Chime (bottom)

Body seam

Bottom head

LDC
DOT-17E
20/18 – 55–74
STC

Explanation of markings

LDC	Symbol of manufacturer
DOT-17E	Specification number
20	Gauge of body
18	Gauge of top and bottom
55	Capacity in gallons
74	Year of manufacturer
STC	Single trip container

(a) **Light-guage closed-head drum**

FIG. 11-3 Typical steel drum containers used for the storage of hazardous wastes.

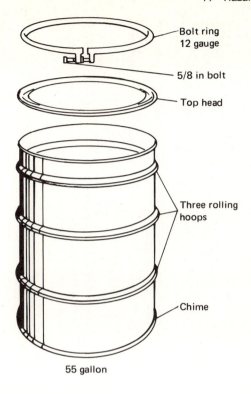

Bolt ring
12 gauge

5/8 in bolt

Top head

Three rolling
hoops

Chime

55 gallon

(b) **Light-guage open-head drum**

presented in Table 11-5. Typical drum containers used for the storage of hazardous wastes are shown in Fig. 11-3.

11-6   COLLECTION

The collection of hazardous wastes for delivery to a treatment or disposal facility normally is done by the waste generator or a specialized hauler. Typically, the loading of collection vehicles is completed in one of two ways: (1) wastes stored in large-capacity tanks are either drained or pumped into collection vehicles, and (2) wastes stored in sealed drums or other sealed containers are loaded by hand or by mechanical equipment onto flatbed trucks. All storage containers collected with the wastes are transported unopened to the treatment or disposal facility. At no time in the collection cycle should the collector come directly in contact with the wastes. To avoid accidents and the possible loss of life, two collectors should always be assigned when hazardous wastes are to be collected.

The equipment used for collection varies with the waste characteristics. Typical collection equipment is listed in Table 11-6. For short haul

TABLE 11-5 CONTAINERS USED FOR THE STORAGE OF HAZARDOUS WASTES

| Waste category | Container | | Auxiliary equipment and conditions of use |
	Type	Capacity, gal	
Radioactive substances	Lead encased in concrete	Varies with waste	Isolated storage buildings; high-capacity hoists and lighting equipment;
	Lined metal drums	55	special container markings
Toxic chemicals	Metal drums	55	Washing facilities for empty containers; special blending precautions to
	Lined metal drums	55	prevent hazardous reactions
	Storage tanks	Up to 5,000	
Biological wastes	Sealed plastic bags	32	Heat sterilization prior to bagging; special heavy-duty bags with hazard
	Lined metal drums		warning printed on sides
Flammable wastes	Metal drums	55	Fume ventilation; temperature control
	Storage tanks	Up to 5,000	
Explosives	Shock-absorbing containers	Varies	Temperature control; special container markings

Note: gal × 0.003785 = m³

TABLE 11-6 EQUIPMENT USED FOR THE COLLECTION OF HAZARDOUS WASTES

Category of waste	Collection equipment and accessories
Radioactive substances	Various types of trucks and railroad equipment, depending on characteristics of wastes; special markings to show safety hazard; heavy-duty loading equipment to handle concrete-encased lead containers
Toxic chemicals	Flatbed trucks for wastes stored in drums; tractor-trailer tank truck combinations for large volumes of wastes; railroad tank cars; special interior linings, such as glass, fiber glass, or rubber; stainless steel trailer tanks
Biological wastes	Standard packer collection trucks with some special precautions to prevent contact between wastes and the collector; flatbed trucks for wastes stored in drums
Flammable wastes	Same as for toxic chemicals, with special colorings or safety warning printed on vehicles
Explosives	Same as for toxic chemicals; some restrictions on transportation routes, especially through residential areas, when transporting wastes to treatment or disposal sites

distances, drum storage and collection with a flatbed truck is often the preferred method. As hauling distances increase, larger tank trucks, trailers, and railroad tank cars are used.

11-7 TRANSFER AND TRANSPORT

The economic benefits derived by transferring smaller loads into larger transport vehicles, as discussed for municipal solid wastes in Chap. 7, are also applicable to hazardous wastes. The facilities of a hazardous waste transfer station are quite different, however, from those of a municipal solid waste transfer station. Typically, hazardous wastes are not compacted (mechanical volume reduction), discharged at differential levels, or delivered by numerous community residents. Instead, liquid hazardous wastes are generally pumped from collection vehicles, and sludges or solids are reloaded without removal from the collection containers, for transport to processing and disposal facilities.

It is unusual to find a hazardous waste transfer facility at which wastes are simply transferred to larger transport vehicles. Some processing and storage facilities are often part of the materials handling sequence at a transfer station. For example, neutralization of corrosive wastes might result in the use of lower-cost holding tanks on transport vehicles. As in the case of storage, great care must be exercised to avoid the danger of mixing incompatible wastes.

11-8 PROCESSING

Processing of hazardous wastes is undertaken for two purposes: (1) to recover useful materials, and (2) to prepare the wastes for disposal.

TABLE 11-7   HAZARDOUS WASTE TREATMENT OPERATIONS AND PROCESSES*

Operation/process	Functions performed†	Types of wastes‡	Forms of waste§
Physical Treatment			
Aeration	Se	1, 2, 3, 4	L
Ammonia stripping	VR, Se	1, 2, 3, 4	L
Carbon sorption	VR, Se	1, 3, 4, 5	L, G
Contrifugation	VR, Se	1, 2, 3, 4, 5	L
Dialysis	VR, Se	1, 2, 3, 4	L
Distillation	VR, Se	1, 2, 3, 4, 5	L
Electrodialysis	VR, Se	1, 2, 3, 4, 6	L
Encapsulation	St	1, 2, 3, 4, 6	L, S
Evaporation	VR, Se	1, 2, 5	L
Filtration	VR, Se	1, 2, 3, 4, 5	L, G
Flocculation/settling	VR, Se	1, 2, 3, 4, 5	L
Flotation	Se	1, 2, 3, 4	L
Reverse osmosis	VR, Se	1, 2, 4, 6	L
Sedimentation	VR, Se	1, 2, 3, 4, 5	L
Thickening	Se	1, 2, 3, 4	L
Vapor scrubbing	VR, Se	1, 2, 3, 4	L
Chemical Treatment			
Calcination	VR	1, 2, 5	L
Ion exchange	VR, Se, De	1, 2, 3, 4, 5	L
Neutralization	De	1, 2, 3, 4	L
Oxidation	De	1, 2, 3, 4	L
Precipitation	VR, Se	1, 2, 3, 4, 5	L
Reduction	De	1, 2	L
Solvent extraction	Se	1, 2, 3, 4, 5	L
Sorption	De	1, 2, 3, 4	L
Thermal Treatment			
Incineration	VR, De	3, 5, 6, 7, 8	S, L, G
Pyrolysis	VR, De	3, 4, 6	S, L, G
Biological Treatment			
Activated sludges	De	3	L
Aerated lagoons	De	3	L
Anaerobic digestion	De	3	L
Anaerobic filters	De	3	L
Trickling filters	De	3	L
Waste stabilization ponds	De	3	L

* Adapted from Ref. 12.
† Functions: VR, volume reduction; Se, separation; De, detoxification; and St, storage.
‡ Waste types: 1, inorganic chemical without heavy metals; 2, inorganic chemical with heavy metals; 3, organic chemical without heavy metals; 4, organic chemical with heavy metals; 5, radiological; 6, biological; 7, flammable; and 8, explosive.
§ Waste forms: S, solid; L, liquid; and G, gas.

Processing can be accomplished either onsite or offsite. Variables affecting the selection of the processing site include the characteristics of the wastes; the quantity of the wastes; the technical, economic, and environmental aspects of available onsite treatment processes; and the availability of the nearest offsite treatment facility (haul distances, fees, and exclusions).

The treatment of hazardous wastes can be accomplished by physical, chemical, thermal, and biological means. The various individual processes in each category are reported in Table 11-7. Clearly, the number of possible treatment process combinations is staggering. In practice, the physical, chemical, and thermal treatment operations and processes are the ones most commonly used. Biological treatment processes are used less often because of their sensitivity. Selection of specific treatment method(s) for use in a given situation is a complex matter in which the assistance of a chemist is essential. Depending on the nature of the wastes, the services of other specialists—such as biologists and chemical, combustion, and sanitary engineers—may also be required. Details on the unit operations and processes reported in Table 11-7 are found in Refs. 1, 8, and 10.

A materials flow diagram for a typical processing, recovery, and disposal facility is shown in Fig. 11-4 [9]. The operational sequence is as follows. Hazardous wastes unloaded from collection vehicles are placed in separate storage containers or tanks, or some other holding facility. (In most receiving facilities, separate storage tanks are used for specific types of wastes to avoid mixing wastes that may produce undesirable reactions.) Burnable wastes with no potential value in other treatment processes are routed directly to the incinerator. Other nonburnable wastes are routed to the treatment facility.

Depending on the types of wastes being treated, one or more of the following processes may be used: neutralization (acid base), heavy-metal precipitation, and vapor stripping with steam. Sludges withdrawn from the treatment process may be treated further in biodegradation ponds, spread on land, and disked into the soil, or disposed of directly in a landfill. Effluent from the treatment process is discharged to holding ponds. Vapors are incinerated following scrubbing to remove inorganic gases. Effluent from the holding ponds is discharged, after chlorination, to solar evaporation ponds. Sludge that may accumulate in the holding and solar evaporation ponds is removed periodically and disposed of in the landfill. Ashes from the incinerator are also disposed of in the landfill. Skimmed oil from the treatment process and from the holding and solar ponds is recovered for sale.

In reviewing the operational sequence of the facility shown in Fig. 11-3, the key item is knowledge of the characteristics of the wastes to be treated. Without this information, effective treatment is impossible. For this reason, the characteristics of the wastes must be known before they are accepted and hauled to a treatment or disposal site. In California,

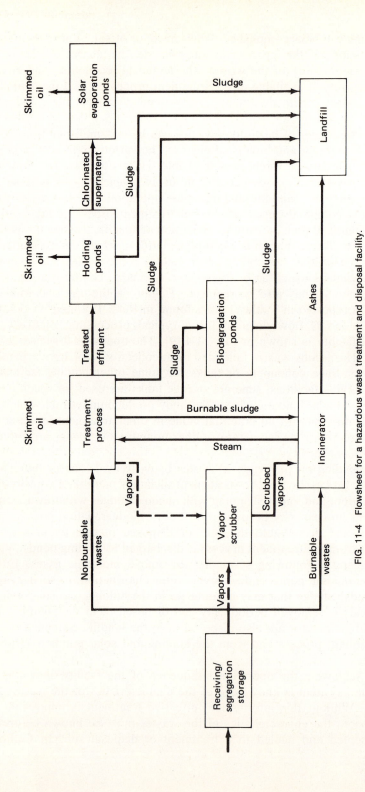

FIG. 11-4 Flowsheet for a hazardous waste treatment and disposal facility.

proper identification of the constituents of the waste is the responsibility of the waste generator.

11-9 DISPOSAL

Regardless of their form (solid, liquid, or gas), most hazardous wastes are disposed of either near the surface or by deep burial (see Table 11-8). An exception is ocean disposal [14], but this practice has come under greater restriction in recent years. Although controlled landfilling methods have proved adequate for municipal wastes and limited amounts of hazardous wastes, they are not as suitable for the disposal of large quantities of hazardous wastes. Some of the reasons are as follows: (1) the possible percolation of toxic liquid wastes to the groundwater; (2) the dissolution of solids followed by leaching and percolation to the groundwater; (3) the dissolution of solid hazardous wastes by acid leachates from solid wastes, followed by leaching and percolation to the groundwater; (4) the potential for undesirable reactions in the landfill that may lead to the development of explosive or toxic gases; (5) the volatilization of hazardous wastes leading to the release of toxic or explosive vapors to the atmosphere; and (6) corrosion of containers holding hazardous wastes [7]. Consequently, great care must be taken both in the selection of a hazardous waste disposal site and in its design. In general, disposal sites for hazardous wastes should be separate from those for municipal solid wastes. In situations in which separate sites are not possible, great care must be taken to ensure that separate disposal operations are maintained.

The operation of a landfill for hazardous wastes is considerably different from that for municipal wastes. When containerized hazardous wastes are to be disposed of, precautions must be taken to avoid (1) the rupturing of containers during the unloading operation and (2) the place-

TABLE 11-8 HAZARDOUS WASTES DISPOSAL AND STORAGE METHODS*

Operation/process	Functions performed†	Types of wastes‡	Forms of waste§
Deep-well injection	Di	1, 2, 3, 4, 5, 6, 7	L
Detonation	Di	6, 8	S, L, G
Engineered storage	St	1, 2, 3, 4, 5, 6, 7, 8	S, L, G
Land burial	Di	1, 2, 3, 4, 5, 6, 7, 8	S, L
Ocean dumping	Di	1, 2, 3, 4, 7, 8	S, L, G

* Adapted from Ref. 12.
† Functions: Di, disposal; St, storage.
‡ Waste types: 1, inorganic chemical without heavy metals; 2, inorganic chemical with heavy metals; 3, organic chemical without heavy metals; 4, organic chemical with heavy metals; 5, radiological; 6, biological; 7, flammable; and 8, explosive.
§ Waste forms: S, solid; L, liquid; and G, gas.

ment of incompatible wastes in the same location. To avoid rupturing them, the containers are unloaded and placed in position individually. The covering of the containers with earth should be monitored and controlled carefully to ensure that an earth layer exists between each container and that the equipment placing the earth does not crush or deform the container. To avoid the codisposal of incompatible wastes, separate storage areas within the total landfill site should be designated for various classes of compatible wastes.

In addition to the general engineering aspects of sanitary landfill designs considered in Chap. 10, provision should be made to prevent any leachate from escaping from landfills used for hazardous wastes. Typically, this requires a clay liner. In some cases it may be advisable to use both a clay liner and an impermeable membrane liner (see Table 10-6). A layer of limestone should be placed in the bottom of the landfill to neutralize the pH of the leachate. A well-sloped impervious cover liner should also be provided over the completed landfill. A final earth cover 2 ft thick or more should then be placed over the liner. The completed site should be monitored continuously both visually and with wells. Any depressions that develop should be filled, or the landfill cover should be regraded, to avoid unnecessary surface water infiltration. For additional details on the operation and design of landfills for the disposal of hazardous wastes, local and state regulations should be consulted. EPA and other engineering literature on this subject should also be reviewed [3–7].

11-10   PLANNING

Planning for hazardous waste management involves documentation of the types, quantities, and sources of wastes and the selection of a disposal site. The lack of options in disposal complicates the problem of finding an acceptable site. The fact that there is, at present, a high degree of uncertainty surrounding the long-term effects of burying hazardous wastes in the land further complicates the problem. Both the public and regulatory agency personnel are suspicious of most proposed sites. Consequently, only a limited number of acceptable sites are available in areas as large as a state. In a 1975 nationwide inventory, the management practices of 64 hazardous waste management facilities known to the EPA were summarized [4]. On the average, this means that there are only 1.28 facilities for each of the 50 states.

If there is sufficient demand for a hazardous waste disposal site in or near a community, an appropriate course of action is to (1) identify a number of technically feasible sites, (2) document fully the demand for a local site, and (3) submit the site proposals, one at a time, to hearings by regulatory agencies and the public for acceptance or rejection.

11-11 DISCUSSION TOPICS AND PROBLEMS

11-1. If possible, identify 10 sources of hazardous wastes in your community. On the basis of your knowledge, into which categories should the wastes be placed?

11-2. Identify and discuss how hazardous wastes are handled in your community. Do the procedures used seem adequate? If not, what recommendations would you make to improve them?

11-3. How are hazardous wastes handled at your school? What about the wastes from chemical analysis laboratory classes? Are they discharged to the waste-water collection system, or are they collected separately? Analyze the various tests that are performed during the quarter or semester and whether toxic chemicals are used.

11-4. What type of labeling does the local control agency require for containers used for the storage of hazardous wastes? Does it seem adequate? If not, why not?

11-5. Identify the types of vehicles used for the collection of hazardous wastes in your community. What safety provisions are followed? In your opinion, are they adequate? Discuss.

11-6. Because it would be practically impossible to prevent all hazardous materials from entering a municipal incinerator, what precautions must be taken to minimize the danger?

11-7. Develop a treatment flowsheet for a waste stream containing cadmium, arsenic, and mercury. How would you dispose of the sludge?

11-8. Develop a treatment flowsheet for a waste stream containing hexavalent chromium. How would you dispose of the sludge?

11-9. Based on a review of the literature, develop an appropriate set of design criteria for a landfill to be used for disposal of sludge from a hazardous waste treatment facility.

11-10. Review your state's regulations dealing with the management of hazardous wastes. Are adequate provisions included for each of the functional elements of generation, storage, collection, transport, processing, and disposal as they relate to hazardous wastes? If not, which areas should be strengthened?

11-12 REFERENCES

1. Aware, Inc.: "Process Design Techniques for Industrial Waste Treatment," Enviro Press, Nashville, Tenn., 1974.

2. Christensen, H. E., et al. (eds.): *The Toxic Substances List*, 1972 ed., U.S. Department of Health, Education, and Welfare, Rockville, Md., 1972.

3. Dawson, G. W., A. J. Shuckrow, and W. W. Swift: *Control of Spillage of Hazardous Polluting Substances,* U.S. Department of the Interior, U.S. Government Printing Office, Washington, D.C., 1970.

4. Farb, D. and S. D. Ward: *Information about Hazardous Waste Management Facilities*, U.S. Environmental Protection Agency, Publication SW-145, Washington, D.C., 1975.

5. Fields, T. and A. W. Lindsey: *Landfill Disposal of Hazardous Wastes: A Review of Literature and Known Approaches,* U.S. Environmental Protection Agency, Publication SW-165, Washington, D.C., 1975.

6. *Hazard Waste Management: Laws, Regulations, and Guidelines for the Handling of Hazardous Wastes*, California Department of Public Health, Sacramento, 1975.

7. Lindsey, A. W.: Ultimate Disposal of Spilled Hazardous Materials, *Chemical Engineering*, vol. 82, no. 23, 1975.

8. Metcalf & Eddy, Inc.: "Wastewater Management: Collection, Treatment, Disposal," McGraw-Hill, New York, 1972.

9. Metcalf & Eddy, Inc.: *Contra Costa County Solid Waste Management Report,* Prepared for Public Works Department, Contra Costa County, Calif., 1975.

10. Nemerow, N. L.: "Liquid Waste of Industry: Theories, Practices, and Treatment," Addison-Wesley, Reading, Mass., 1971.
11. Pittman, F. K.: Management of Radioactive Waste, *Water, Air, and Soil Pollution,* vol. 4, no. 3, 1975.
12. *Report to Congress: Disposal of Hazardous Wastes,* U.S. Environmental Protection Agency, Publication SW-115, Washington, D.C., 1974.
13. Schneider, K. J.: High Level Wastes, in L. A. Sagan (ed.), *Human and Ecological Effects of Nuclear Power Plants,* Charles C Thomas, Springfield, Ill., 1974.
14. Smith, D. D. and R. P. Brown: *Ocean Disposal of Barge-Delivered Liquid and Solid Wastes from U.S. Coastal Cities,* U.S. Environmental Protection Agency, Publication SW-19c, Washington, D.C., 1971.

PART III

MANAGEMENT ISSUES

The management framework in which the engineering principles presented in Part II must be applied is the subject of Part III. Attention is focused on the planning aspects of solid waste management rather than on the day-to-day management operations mentioned briefly in Part I. It is thus possible to highlight some of the more important management problems that must be resolved to improve the operation of existing systems, to implement proposed systems, and to develop local, regional, and state plans.

The fundamental role of planning in solid waste management is described in Chap. 12. A detailed five-step planning procedure is also delineated. Planning for the functional elements of onsite storage, collection, transfer and transport, processing and recovery, and disposal is then described separately in the following five chapters. In each of these chapters the topics considered are the same, and they correspond roughly to the first three steps in the planning process: (1) discussion of management issues, as related to problem identification; (2) delineation of the needs and appropriate methods for inventory and data accumulation; and (3) illustration of the evaluation process by means of case studies. The final two steps in the planning process involve the selection of programs and the development of implementation plans and schedules. Those are the most important planning steps for decision-makers, and they are discussed in the last chapter.

12

Planning in Solid Waste Management

Planning in the field of solid waste management may be defined as the process by which community needs regarding waste management are measured and evaluated and workable alternatives are developed for presentation to decision-makers. Planning is accomplished by applying the engineering principles (presented in Part II) to the needs, capabilities, and goals of the community. This planning is both exciting and challenging because most of the technical, environmental, economic, social, and political factors, and the interrelationships that are involved, are now only partially understood. Moreover, all these factors and interrelationships are so dynamic in nature that the accumulation of data concerning them is both difficult and time-consuming, if even possible; consequently, few guidelines for planning can be given. Nevertheless, in response to the need to develop workable solutions to solid waste management problems, some general planning considerations have been identified, and methods of approach have evolved.

The purpose of this chapter is (1) to explore some of the important considerations in the planning process, (2) to describe what constitutes waste management programs and plans, (3) to define a general methodology for planning and the preparation of planning reports, and (4) to examine the nature of the decision-making process in the field of solid waste management. The basic ideas and approaches that are briefly outlined in this chapter are then expanded upon in Chaps. 13 through 18.

Planning with respect to five of the six functional elements of a solid waste management system, as described earlier in this text, is considered in detail in Chaps. 13 through 17. A separate chapter is not devoted to the functional element of waste generation because at present most collection agencies, system managers, and engineers have little or no direct control over the quantities and composition of wastes that are generated (both of these factors are affected more significantly by legislation).

12-1   IMPORTANT CONSIDERATIONS IN THE PLANNING PROCESS

Solid waste management encompasses a wide range of individual activities. To deal effectively with these activities from a technical planning and management standpoint, it has been found helpful, as noted in Chap. 2, to group the activities into six functional elements: waste generation, onsite storage, collection, transfer and transport, processing and recovery, and disposal. In this way the public, politicians, decision-makers, and planners are able to recognize and understand more easily the important relationships that must be evaluated in the planning process. The combination of functional elements is known as a solid waste management system. Thus, one of the goals of management is to provide the best possible system subject to the constraints imposed by its users and by those who are affected by it or who control its use.

In general terms, the planning process involves the collection, evaluation, and presentation of data relevant to some problem. In the field of solid waste management, the problem usually requires some type of action by a decision-maker who probably is an elected official. Therefore, to understand the nature of the planning process in this application, it is important to consider (1) the framework in which planning activities are usually conducted, (2) the effect of planning time periods, (3) the jurisdictional levels at which planning studies are conducted, (4) the impact of alternative concepts and technologies on the planning process, and (5) the definitions of programs and plans. Additional details on the planning process may be found in Refs. 1, 4, 9, and 11.

Framework for Planning Activities

Planning activities in the field of solid waste management are generally undertaken in response to the recognition of some community need. The community problem-solving cycle and the interrelationships of planning that are involved are shown in Fig. 12-1.

The planning activity commences once a community need has been articulated and the problem has been recognized. This problem recognition is important because if meaningful planning is to result, it must be related to some community need. Otherwise the planning process becomes self-serving and is of little value. It is the responsibility of the planner, however, to call to the attention of the decision-maker all the potential problem areas that may be identified during the planning process.

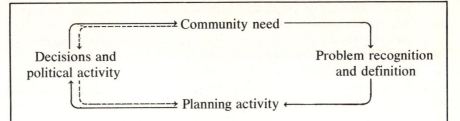

Note: Dashed line indicates feedback.

Community need A community need is identified by the public, usually in response to issues related to costs, service being provided, resources, and the environment. The extent of the need is often determined by the social standards of residents, institutions, and businesses.

Problem recognition and definition Responsible decision-makers perceive and interpret community needs. They also are responsible for problem definition and specification.

Planning activity The planning activity is undertaken by agency staff or consultants as directed by decision-makers. Alternative programs are developed to solve specific problems.

Decisions and political activity This is the action step in the problem-solving cycle. Decision-makers review alternatives, select equipment, provide manpower, and make financial, operation, and political decisions.

FIG. 12-1 Community problem-solving cycle.

At the same time as the problem-solving cycle moves forward, there is feedback from the community to the decision-maker and then to the planner. In Fig. 12-1, feedback terminates at the planning activity because it is assumed that the true problem is being dealt with. In cases where incorrect definitions are selected, the feedback loop would extend to problem recognition, and a redefinition would be necessary. The presence of feedback in the problem-solving cycle is essential to the development of responsive management plans.

Planning Time Periods

Planning in solid waste management can be either short term or long term. A precise time division is not fixed, although 5 to 7 yr into the future presently is accepted as the upper limit for short-term planning; long-term planning extends for all time periods past that.

Short-term planning is generally carried out to fulfill immediate needs. The EPA "Mission 5,000" project is an example [8]. Conceived in 1970, the intent was to close some 5,000 open burning dumps throughout the United States by 1975. The framers of this project planned it for only the short term, although it did have long-term implications in that the replacement disposal facilities were designed to meet the long-term need of each community.

One difficulty that arises in selecting an appropriate time period for planning is that while short-term planning may be limited to 5 to 7 yr, the payback period for equipment and facilities may be considerably greater, often running to 20 yr or more for facilities such as incinerators used for waste volume reduction. When energy recovery options are being evaluated, long-term payback periods usually will be required. Yet, at present, little or no long-term information is available on the useful life of such facilities. This limitation can have serious implications on the economic feasibility of energy options. In situations in which the uncertainty is high and long-term payback periods are required, the best approach is to develop multiple analyses based on estimates of both average and least optimistic conditions. This will give the decision-maker a better understanding of the risks involved. Also a planner will seldom be questioned if the process performs better than anticipated.

Planning Levels

Planning activity in the field of solid waste management can be associated with three jurisdictional planning levels: (1) local, (2) subregional or regional, and (3) state or federal. The characteristics of these three levels are identified in Table 12-1. The distinctions are sharpest between the first

TABLE 12-1 PLANNING LEVELS AND THEIR CHARACTERISTICS

Planning level	Agencies involved	Characteristics
Local	City County Special districts	Normally a single agency conducts the planning study to solve a localized problem.
Subregional or regional	Cities Counties	Several communities join together to form an efficient planning and operational activity.
State and federal	Cities Counties States	Several cities, counties, and states can form a planning unit. Typically, the concern will be with issues such as materials use from origin in the natural environment, through processing for reuse, and the disposal of unusable residues. Planning efforts often result in demonstration projects.

two planning levels. Local agencies usually find responsibilities for management fragmented among many organizations, some of which do not even recognize that their activities are part of the solid waste system. Economic resources usually are dispersed widely at the local level and are unavailable for effective planning and plan implementation.

By contrast, at the subregional and regional planning level, the political and social awareness of waste management has caused several agencies to join forces to achieve a common goal. Joint objectives are worked out along with estimates of the economic resources necessary to develop comprehensive plans. As natural resources become scarce or depleted, there is a very strong tendency to combine the second and third planning levels. At present, the political, social, and economic variables for the state or federal level are not yet understood well enough to permit clear descriptions of materials markets, equipment, and a time sequence for implementation. Comprehensive planning is considered further in Refs. 5, 6, and 7. Solid waste management from a federal perspective is considered in Ref. 3.

Planning Concepts and Technologies

Engineers and planners in the field of solid waste management today are confronted with breakthroughs, both in public awareness and in technological advancements, that make comprehensive planning for the future an especially difficult task. Many of the recent technological advancements are as yet too new to have been proved in full-scale installations. Thus, decision-makers are often faced with a choice of whether to use long-established and well-proved equipment and techniques that may or may not be the optimum ones for present and future conditions, or to use new and unproved equipment and techniques that may or may not work as expected. A helpful perspective in performing planning studies is to consider the various solid waste management concepts and technologies in three categories: (1) developing, (2) transitional, and (3) long term.

In the broadest sense, all planning is based on predictions of future conditions. Developing concepts and technologies in solid waste management are based on new ideas derived from the public's awareness of resources, economics, and the quality of the environment. Typical developing concepts include the reuse of food containers, limiting of product packaging containers, control and standardization of materials used in packaging, and development of low-energy-demand products. Much of the technology used for the recovery of materials, conversion products, and energy from solid wastes, as discussed in Chap. 9, must still be categorized as developing. When developing concepts or technologies can be defined specifically in terms of function, operation, and equipment, they become transitional.

Solutions to immediately perceived needs (the needs may be only temporary) usually are accomplished with proved but relatively static transitional concepts and technologies. Typical equipment and facilities that are a part of the transitional technology now used include plastic or metal waste containers, packer collection trucks, stationary compactors, transfer stations, shredders, and incinerators. In most cases, transitional technologies are subject to replacement as developing concepts and technologies are proved adequately.

In environmental planning, long-term concepts and technologies are those that will remain viable for a long period of time (decades and generations) and include preservation of public health, onsite storage, surface transportation, and landfill disposal of wastes.

Because of the fundamental significance of the various concepts and technologies, it is important to consider their application in planning. For example, which ones are used in short-term planning? Typically, developing concepts and technologies are rarely involved directly because, in most cases, they cannot survive the hard economic reality associated with making short-term changes. Thus, the Mission 5000 project was not presented on the basis of emerging resource shortages or energy needs. Its primary emphasis was on community pride, aesthetics, and public health [8].

Solutions of short-term projects usually involve the use of transitional concepts and technologies. The immediate needs of a community can be met through the application of available and proved waste management facilities and techniques. Because immediate needs will spawn expedient short-term solutions, better solutions will be derived when the selection of equipment and facilities is influenced by the consideration of long-term planning concepts and technologies.

In long-term planning, short-term needs must still be satisfied, but the acceptability and success of long-term plans usually are based on the adoption of both developing and long-term concepts. Thus, one key to the development of successful long-term plans is the proper identification of the function of transitional concepts and technologies, for they are often the means by which developing concepts can be incorporated into long-term plans, even though the optimum technology is not available. For example, much of the equipment now used for resource recovery was not developed for that purpose; yet it can be used to achieve our objectives until more optimum equipment becomes available.

12-2 PROGRAMS AND PLANS

In general terms, programs and plans represent blueprints for achieving solid waste management objectives. The fundamental difference between programs and plans is in the scope of activities involved.

Programs

As used in this text, the term *program* encompasses all the activities associated with the solution of a problem within a functional element of a solid waste management system. Thus, typical program areas of concern within a functional element may involve operation budgets, financing, rate structures, staffing requirements, contracts, equipment procurement and replacement, and maintenance.

Programs dealing with specific problems related to a functional element may or may not involve policy issues and objectives. If they do, they must be presented to the appropriate decision-maker, such as a city council or county board. For example, if a program is designed to assess the feasibility of a specific truck replacement, the final decision would be made at the operating level. On the other hand, if a program deals with the frequency with which solid wastes are to be collected from residential areas, the final decision would be made at the decision-maker's level because it involves a policy issue. The distinction between programs and plans is considered further in the following discussion.

Plans

Solid waste management *plans* are developed to define and establish objectives and policies. Typically, a plan will encompass one or more functional elements and may be composed of one or more program areas. For example, a local plan for collection may involve only the program area associated with the setting of rates. Most plans, however, are made up of many, many programs, and each one may be considered individually during the development of a final plan. For example, suppose a management plan has been developed that encompasses all the functional elements except waste generation. In turn, the plan is based on a number of individual programs. If objections are raised to the plan, it is now possible to isolate them with respect to the individual programs. By focusing on an individual program, it usually will be possible to reach a workable compromise without having the entire plan rejected. In fact, this is the decision-making process.

In many planning situations it is beneficial to identify more than one program or set of programs that can be used to solve a given problem—in other words, to develop alternatives. Alternatives are used as a means of demonstrating the impact of the various programs on the solid waste management system. They serve only as an aid to the decision-maker in understanding the impact of management choices described in a waste management plan. Often, a preliminary plan is developed in which two or more alternatives involving several programs are presented to decision-makers, and the final plan adopted is in fact composed of programs taken from one or more of the original alternatives.

Types of Plans

In most cases, the types of plans that are developed in the field of solid waste management correspond in effort and scope to the planning levels identified previously (e.g., local, subregional or regional, state and federal).

TABLE 12-2   TYPICAL PROGRAM AREAS ASSOCIATED WITH PLAN PREPARATION

Type of plan	Functional element	Typical program areas
Local	Onsite storage	Container size, location, and frequency of collection
	Collection	Collection vehicles, crew size, and routes
	Transfer	Transfer station layout, transport vehicle type, and highway routes
	Processing and reuse	Restricted market research for recovered resources and minimal homeowner programs
	Disposal	Soil characteristics, equipment requirements, excavation and cover cycles, etc.
Regional or subregional	Onsite storage	Standards and criteria for container usage in multiple communities
	Collection	Operating criteria and cost levels for multiple types of workforce and equipment systems
	Transfer	Station and truck sizes, potential benefit of large-scale processing combined with transfer
	Processing and reuse	Processing equipment and location, research of available markets for recovered resources
	Disposal	Disposal-site details, financing responsibility among communities, general standards for operating equipment and crews
State or federal	Generation	Legislative pressure on manufacturing industries that use excess packaging, taxes on throw-away beverage containers
	Onsite storage	Same as regional or subregional
	Collection	Same as regional or subregional
	Transfer	Assess resource recovery potential at the transfer facilities, other details the same as regional or subregional
	Processing and reuse	Research, pilot plant facilities, transportation incentives, tax penalties, etc., general definition of scarce materials and resource priorities, purchasing specifications that encourage resource recovery
	Disposal	Same as regional or subregional

Typical program areas for the various functional elements are reported in Table 12-2. Local management plans usually involve one or more of the functional elements of a solid waste management system and at least one program area. Subregional or regional plans also involve multiple functional elements and programs, but here the effort is concentrated on using them more efficiently in coordination with other communities. State or federal plans usually involve total resource management, including the functional element of waste generation. In fact, state and federal planning offers the only opportunity in which to consider waste generation programs.

Any individual problem requiring solution through action can be interpreted in relation to these planning levels, hopefully resulting in a better understanding of the planning effort needed by individuals and communities. The steps involved in the preparation of programs and plans, as discussed in the following section, are the same for all planning levels.

12-3   PLANNING STUDY METHODOLOGY

Developing or understanding solid waste management programs and plans of varying complexities is now an integral part of the job of many persons, including the owners or managers of solid waste collection agencies (such as those in numerous county, state, and public works and environmental health agencies), engineering consultants, land-use or community planners, industrial plant managers, and members of environmental or resource groups. Because of the importance and far-reaching impacts of solid waste management programs and plans, it is essential to have a clear understanding of the steps involved in their preparation. The purpose in this section is (1) to identify a general methodology for the development of management programs and plans, and (2) to illustrate how the various tasks that must be completed are organized and monitored.

Planning Steps

In most cases, the planner and the decision-maker do not have an opportunity to study the entire solid waste management system and develop a total knowledge of the community under all conditions. Time and economic constraints resulting from social and political needs often lead to decisions based on little or no information. In order for planners and decision-makers to be able to respond to these situations and to ensure that the best use is made of time and available funds in the resolution of solid waste management problems, the following step-by-step planning procedure is recommended. An expanded discussion of planning methodology may be found in Ref. 9.

Step 1: Problem Definition and Specification   The first and most critical step in any planning study is to obtain a clear problem statement and corresponding specifications from the people responsible for making decisions about solid waste management. Problem statements and specifications usually are derived from the concerns of the public.

Difficulties often arise because solid waste systems are not well understood at most levels of decision making. Consequently, the planner may have to redefine a problem that was originally specified by a decision-maker. Management issues and concerns from which problem definitions are derived for the various functional elements are discussed in Chaps. 13 through 17.

Step 2: Inventory and Data Accumulation   In this step an inventory is made of all pertinent factors about the community, and data are collected as needed to meet the problem specifications. The main purpose of the inventory is to define the existing solid waste system(s) as completely as needed and as accurately as possible and to collect certain other basic information (such as population data)—a task that requires a considerable amount of judgment. It is an important step in planning because all subsequent recommendations for action will be based on the findings of this step. Therefore, it is essential that at any level of planning all the functional elements that make up the solid waste management system be considered.

Inventory methods for all the functional elements, with the exception of waste generation, are presented in Chaps. 13 through 17. Inventory methods for waste generation may be found in Chap. 4.

Step 3: Evaluation and Alternative Development   This step involves the detailed evaluation and analysis of the data accumulated in step 2. It is during this step that the programs of the plan begin to be formed. In some cases, it may be necessary to collect additional data and information. However, before the programs are formed, it is important to review the original problem statement and specifications. Often it will be found that some revisions have to be made in light of the data gathered during the inventory.

Since a problem can have more than one solution, it is beneficial for decision-making purposes to develop alternatives comprised of one or more programs. When practical, these alternatives should be documented for presentation in the plan. A simple plan may deal with only one or two programs. A more complex plan includes more of the functional elements, and its alternatives may include numerous programs. In either case, both administrative and engineering activities must be evaluated. Examples are presented in Chap. 18.

In developing alternatives, it is especially important that all functional elements be coordinated to ensure system continuity—from onsite storage through processing and final disposal. By evaluating the coordinated programs, the planner is able to recommend viable alternatives.

Step 4: Program and Plan Selection   In this step, a limited number of alternatives are selected by the planner for inclusion in the plan. The alternatives are reviewed by the planner, the decision-maker, and members of the public. The logic of the individual programs that make up the alternatives is reviewed, and programs are changed as necessary to include review comments. The administrative control of all programs is identified and evaluated during this step. This is important because solid waste management will not function properly without responsive control. Hence, the planner must develop a thorough knowledge of the social and political structure of the community.

The final action in this step is the selection of a preferred set of programs to form the plan. The programs can be selected from a single alternative, or they can be selected from various alternatives. The final selection will be made by decision-makers.

Step 5: Development of Implementation Schedule(s)   When planning failures have occurred, the lack of a well-defined implementation schedule acceptable to administrative and management organizations has been the principal contributing factor. The degree of documentation in any implementation schedule depends on the type of programs developed in the plan. If possible, the degree of documentation that will be required for implementation should be set by the planner and decision-maker during the problem-specification stage (step 1) of the plan development. The following general observations can be used as a guide to the required degree of documentation: (1) Local plans developed for a single functional element usually require only a simple implementation schedule, possibly without a step-by-step sequence for program changes; (2) local and regional plans developed for more than one functional element require the development of comprehensive implementation schedules that must be read, interpreted, and used by numerous operational and administrative agencies; and (3) state and federal plans contain limited but very specific implementation schedules because these plans normally deal with multiple political and jurisdictional boundaries. Because these plans are developed in response to specific legislation, they also contain specific completion dates. Details and examples are presented throughout the following chapters, with the most comprehensive to be found in Chap. 18.

With the completion of step 5 and proper documentation, the planner will have completed the most demanding work. The planner continues to be involved in the planning process during plan implementation and when the plan requires updating. The principal work for implementation now shifts to the decision-maker.

Organization of Work Effort

While knowledge of the planning steps and their sequence is useful, organization of the required work effort is of critical importance if the planning project is to meet the stated project goals and objectives and is to be completed on time and within the budget. The organization of the planning work in most solid waste management studies is the responsibility of the plan director, who may be in a public agency or in a private consulting firm. That person will have to be concerned about the time available for completing the work and the amount of money committed to pay for it. Typically, local plans are simple and are developed for only one functional element; staffing and work organization are usually straightforward. Regional and state and federal plans, however, are complex and usually include most of the functional elements. A broad staff capability is needed to assess the diverse activities that must be considered, and the work organization must be controlled closely to meet cost and time limits.

The control of planning work effort becomes a management task. An activity chart, such as the one shown in Fig. 12-2, can be a valuable aid to the plan director. The major work tasks and their interrelationships, and the time limits in which those work tasks must be completed, are displayed on the chart. The plan director uses the chart to stay up to date on work progress as well as to assign the planning team their task assignments. The chart shown in Fig. 12-2 is a typical one for comprehensive plans involving multiple functional elements.

The suggested procedure for completing the planning work is to assign work tasks according to functional elements. The procedure is as follows (see Fig. 12-2):

1. Assign separate inventory tasks for generation, onsite storage, collection, transfer and transport, and disposal through month 3.

2. Decrease team members after month 3 by eliminating field survey crews (weighing and monitoring).

3. Continue work on task assignments by functional element through the program development phase at month 6.

4. Receive written preliminary program descriptions from each team for each functional element at month 6.

5. Reduce team members after consolidation of programs into plan alternatives at month 9.

6. Submit documented plan alternatives to decision-makers and to the public for review and comment at month 9.

7. Receive comments from the public and decision-makers, and modify programs accordingly at month 11.

8. Select the preferred programs for the plan, and develop the implementation schedule by month 13.

9. Reduce the planning staff to a minimum, but retain staff capability to assist in implementation and plan updating.

The plan director will have the opportunity to shift team members as special work problems arise, particularly when unproved technologies are being assessed, and additional time and workforce beyond that budgeted may be required. However, the work tasks and time schedules of the activity chart should be followed as closely as possible so that original work organization goals are preserved.

12-4 THE DECISION PROCESS

The purpose of planning has been established as the accumulation, evaluation, and presentation of data relevant to a problem that requires

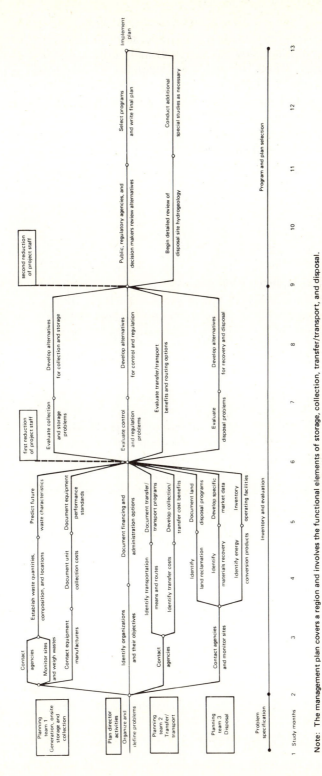

Note: The management plan covers a region and involves the functional elements of storage, collection, transfer/transport, and disposal.

FIG. 12-2 Activity chart for the preparation of a solid waste management plan.

some type of action by a decision-maker. As such, the planning process is an important part of the decision-making process; thus, the relationship of the planner and decision-maker is normally quite close during plan development. Decision-making needs and events as related to the selection and implementation of solid waste management programs are discussed in this section. The decision process is considered and discussed further in Ref. 2.

Requirements for Decision Making

It is clear that one of the fundamental and perhaps most important requirements for decision making is sound planning. Another is an understanding of the goals of the community. Consider the community in which you live. What do you perceive are the solid waste problems that need corrective action? If these problems are put in the framework of the planning methodology described earlier, a decision regarding problem solution should be possible when the results of the planning study are available. Individuals with responsibility for selecting and implementing solid waste management systems must perform in precisely the same manner. In addition, the decision-maker must use the results of planning to follow through with capital expenditures, workforce allocations, and system implementation.

The availability of newly developing concepts and technologies and the recent broadening of social awareness concerning solid waste management and the value of resources make the decision-making process in this field very uncertain. Without effective decision-making guides that result from good planning, many decision-makers respond to this dynamic condition by putting off any implementation action through an endless cycle of additional studies. Although this type of activity is sometimes politically expedient, it is rarely responsive to community needs. A more practical approach is to develop a dynamic solid waste plan and an appropriate updating technique that will allow solid waste systems to be modified as social values, concepts, and technologies change [1, 4].

Important Decision Events

As previously described, planning is an activity that leads to the development of management alternatives. Decision making is an activity that results in actions to implement equipment and workforce systems. Although there is an inherent danger of oversimplifying the decision-making process, the following four decision events are considered essential in completing solid waste management actions:

1. Adoption of a solid waste management plan, including specific programs

2. Adoption of an appropriate implementation schedule

3. Selection of an agency or agencies to administer the plan and operate the system

4. Selection of staff and funding sources and means

It should be noted that not all the decision events are required to initiate an action program. For example, administration, operation, and staffing are normally the responsibility of local management. Thus, decision events 1 and 2 are most important at the local planning level. In the implementation of subregional or regional plans, however, it is often necessary to create new staffs and funding sources. Thus, at these levels, all four events are important.

12-5 DISCUSSION TOPICS

12-1. A small Midwestern subregional government has received a proposal from a private collector to take over and operate the existing solid waste collection systems now operated by public agencies. The proposal contains a firm rate for the collection of solid wastes that is 20 percent lower than the existing public agency rate. The proposal also contains the provision that the private collector would become the sole owner of all wastes collected. Within the subregional governmental boundaries are four cities of equal size with a total population of 150,000 and an unincorporated urban area with a population of 50,000.

Develop a planning format in outline form to evaluate this proposal. Indicate in terms of percentages how the responsible government agency should allocate its planning budget and workforce resources. Briefly discuss what you consider to be the key issues to be considered in the development of the problem specifications.

12-2. Planners in a rural county with a population of 17,000 have identified three possible alternative action programs for waste management. The county commissioners have reviewed the alternatives and have selected the one incorporating (1) a hauled container (drop box) collection system (see Chap. 6), (2) several convenience transfer stations, (3) hand picking of salvageable materials at the disposal site, and (4) a single sanitary landfill. The system cost will be offset through a combination of service rates and a general property tax subsidy. No countywide waste collection system existed prior to the commissioner's decision to implement this program. What are the implications of this decision? What is the potential for community acceptance?

12-3. What is the planning level of Discussion Topic 12-1? Discussion Topic 12-2? Compare the planning levels and list the key program areas that might be found in each plan.

12-4. What local governmental agency has responsibility for planning and establishing comprehensive community goals and objectives for solid waste management? Identify the public and private agencies that are responsible for the development of programs and plans not involving community policy issues.

12-5. A state regulatory agency has requested that regional planning agencies develop solid waste management plans that will cover a 20-yr period. Programs for short-, medium-, and long-term plan time periods have been delineated. Assume that you are the project director for a local community planning effort, and list the programs you would put into (1) the short-term section, (2) the medium-term section, and (3) the long-term section.

12-6. Using Discussion Topic 12-5, develop a step-by-step work planning schedule to resolve the problems of your community and to meet regulatory agency mandates. Assume that your planning effort covers a community with multiple cities and a population of 275,000.

12-7.   Review a typical solid waste management report selected by your instructor with respect to plan organization and planning methodologies presented in this chapter. Prepare a brief summary of your findings.

12-8.   In Discussion Topic 12-7, how were developing, transitional, and long-term concepts and technologies dealt with in the plan?

12-6   REFERENCES

1.   Brandt, H. T.: Preparation of a Comprehensive Solid Waste Management Plan, *Public Works,* vol. 106, no. 5, 1975.

2.   Colonna, R. A. and C. McClaren: *Decision-Makers Guide in Solid Waste Management,* U.S. Environmental Protection Agency, Publication SW-127, Washington, D.C., 1974.

3.   Eliassen, R.: *Solid Waste Management: A Comprehensive Assessment of Solid Waste Problems, Practices, and Needs,* Office of Science and Technology, Executive Office of the President, Washington, D.C., 1969.

4.   Bluckman, L. A. (ed): *Planning for Solid Waste Management, Symposium of State and Interstate Solid Waste Planning Agencies,* U.S. Environmental Protection Agency, Publication SW-2p, Washington, D.C., 1971.

5.   Golueke, C. G. and P. H. McGauhey: *Comprehensive Studies of Solid Wastes Management,* First Annual Report, Sanitary Engineering Research Laboratory, SERL Report 67-7, University of California, Berkeley, 1967.

6.   Golueke, C. G. and P. H. McGauhey: *Comprehensive Studies of Solid Wastes Management,* Second Annual Report, Sanitary Engineering Research Laboratory, SERL Report 69-1, University of California, Berkeley, 1969.

7.   McFarland, J. M., et al.: *Comprehensive Studies of Solid Wastes Management,* Final Report, Sanitary Engineering Research Laboratory, SERL Report 72-3, University of California, Berkeley, 1972.

8.   *Mission 5000: A Citizens Solid Waste Management Project,* U.S. Environmental Protection Agency, Publication SW-115ts, Washington, D.C., 1972.

9.   Theisen, H., P. L. Maxfield, and G. E. Lynch: Solid Waste Management Planning: A Methodology, *Journal of Environmental Health,* vol. 38, no. 3, 1975.

10.   Toftner, R. O.: *Developing a State Solid Waste Management Plan,* U.S. Environmental Protection Agency, Publication SW-42ts, Washington, D.C., 1970.

11.   Toftner, R. O: *Developing a Local and Regional Solid Waste Management Plan,* U.S. Environmental Protection Agency Publication SW-010ts.1, Washington, D.C., 1973.

13

Choices in Onsite Handling, Storage, and Processing

Traditionally, the onsite handling, storage, and processing of solid wastes has been considered a personal matter, and the choices involved have been left largely to individual preferences. As a result, the role of management in this area is not well defined. Nevertheless, because of the need to improve overall efficiency of solid waste management systems, especially in urban areas, planning for this functional element is becoming increasingly important. Thus, the purpose of this chapter is to examine onsite handling, storage, and processing from a management standpoint.

In keeping with the method of approach described in Chap. 12, the three main sections in this chapter (as well as in Chaps. 14 through 17) are related to steps 1, 2, and 3 in the planning process. The discussions in these sections deal with the topics of problem definition (step 1), inventory and data accumulation (step 2), and evaluation and alternative development (step 3). In regard to step 1, it is noted that there is no way of knowing what types of problems will be encountered in the field; thus, for the purposes of planning, the approach taken here is to focus on the general areas of management issues and concerns from which most problem specifications will be derived. Therefore, this "step 1 section" is Management Issues and Concerns (Sec. 13-1). Similarly, in regard to step 3, it is noted that only an introduction to this complex step can be presented in this book, and here a case study format is used. In this chapter, for example, the "step 3 section" is entitled Case Studies in Onsite Handling, Storage, and Processing (Sec. 13-3). As mentioned in Chap. 12, the activities associated with plan selection (step 4) and implementation (step 5) are the same for each functional element, and these topics are discussed separately in Chap. 18.

13-1   MANAGEMENT ISSUES AND CONCERNS

Each resident in a community can and should contribute to the successful operation of the solid waste management system. The improper storage of solid wastes in homes can lead to the breeding of flies and other insects, provide food and harborage for rats, cause odors, and result in home accidents. All these conditions have a health significance [5]. By maintaining sanitary onsite conditions, the resident becomes an important participant in the community's solid waste management system. Furthermore, the future roles of the resident may be even more significant because the best place to sort wastes for recovery, from an economic standpoint, is at the source of generation.

During the past 20 yr onsite storage conditions have changed considerably [2, 3]. Because the quantity of food wastes collected has been reduced by changes in the composition of consumer goods, the impact of changes in the food packing industry, and the use of home grinders, the need for uniform containers for the protection of human health is no longer as important as it once was. Changes in the types of collection services provided (such as from backyard to curb service) have also lessened the need for container standardization. Because of these and other related factors, the decision-maker is faced with an extremely difficult task in attempting to deal with onsite storage. Often problem specifications are misstated, and planning efforts are wasted.

Onsite storage problems typically are caused by one or more of the following factors: (1) the existence of numerous diverse waste sources, (2) public health, odor, aesthetic, and rodent conditions, (3) sizes and conditions of wastes in relation to storage container size and location, and (4) lack of proper administrative control through ordinances and standards. The continuity relationships between onsite storage and the other functional elements of the system can also lead to problems.

Diverse Waste Sources

Managing onsite storage for waste generation sources as diverse as a street on which candy-bar wrappers are discarded and a ranch feedlot on which solid manures accumulate from 2,000 beef involves widespread concerns. The greatest management concern is the ease with which people can move, accumulate, and discard heterogeneous wastes. Solid wastes generally are not stored in vessels or moved in pipe networks. Rather, every pocket and every automobile serves as a storage container whose contents might be emptied anywhere along a traveled route or in the home. As a result, public relations appeals and various types of incentive programs are often developed in attempts to discourage uncontrolled storage (litter). A different type of management incentive program is one that involves a city-supplied home storage container, as discussed in Refs. 8 and 10. City-supplied containers are an incentive for the use of uniform equipment at every waste source (which will affect operation costs). Some common onsite storage conditions are shown pictorially in Fig. 13-1.

Public Health and Aesthetics

Public health concerns are related to the disease and injury potential of inadequately stored solid wastes. Aesthetic concerns are related to human sensual reactions to wastes that are inadequately stored or to wastes that are objectionable even though they are adequately stored (objectionable usually because of odor and container size, location, color, etc.). Management should strive to eliminate both health and aesthetic concerns. If that is not possible, measures to mitigate their impacts should be investigated.

The resolution of this management issue and its associated concerns is usually accompanied by local community action, but studies of multi-community action are under way. An example is the recent study entitled *California Litter, A Comprehensive Analysis and Plan for Abatement*, which was prepared for the California State Assembly Committee on Resources and Land Use [9]. Although litter control is emphasized in this study, recommendations are made for the modification of onsite storage facilities on a statewide level.

Sizes and Conditions of Wastes

Wastes are discarded in various sizes and in varying conditions (decomposing food wastes, partially ripped bags, etc.). Management concerns are for employee safety and system efficiency. The greatest potential for employee injury is to collectors providing collection service to single-family detached residential dwellings. Irregular-shaped and oversize wastes that protrude from containers are potential causes of injury to eyes and hands (see Fig. 13-1). Food wastes can also be a problem because they are very dense. A 32-gal container filled to the top with food wastes is heavy, and the lifting of excessive weights can injure the employee or damage the container. This subject is considered further in Ref. 2.

Ordinances and Standards

The issues and concerns that have been described are usually resolved through administrative and operational actions. The basis for administrative actions is ordinances, regulations, and standards. Management con-

(a)

FIG. 13-1   Wastes protruding from overloaded containers are a potential hazard to collectors.

(b)

(c)

cerns are related to the development of comprehensive and complete administrative regulations. A typical example of regulations that supplement a community ordinance regarding the preparation and onsite storage of residential solid wastes is presented in Fig. 13-2. Communities with a different type of collection service would have different regulations, but the subjects considered in the regulation should be similar to those in most communities. A digest of selected community ordinances may be found in Ref. 6.

System Continuity

System continuity may be defined as the interrelationship of the functional element in question to the other functional elements in the system. Onsite handling, storage, and processing is the second functional element in the solid waste management system (see Fig. 2-3). As such, it has a significant impact on the first element, waste generation, and on every element that follows it, but its greatest impact is on the third element, waste collection.

Generation The most important factor in the relationship between waste generation and onsite storage is the amount of waste generated. What sizes of containers should be used? Does the use of multiple types and sizes of containers cause inefficiencies in the collection operation? At what rate of

REGULATIONS FOR SUPPLEMENTAL REFUSE COLLECTION SERVICE
FOR SINGLE FAMILY, DUPLEXES, TRIPLEXES AND FOURPLEXES ONLY

We hope you will take a minute to familiarize yourself with these easy-to-remember regulations.

We want to do the best job possible for you and would appreciate your co-operation.

1. WHERE DO I PLACE MY REFUSE?

Two standard garbage cans will be picked up from a side or backyard location a maximum distance of 125 feet from an accessible roadway. Additional refuse PROPERLY CONTAINERIZED will be picked up at curbside. Do not place refuse in such a manner as to obstruct any drainage ditch, culvert, or other water way, sidewalk, or regularly traveled footpath.

2. WHEN DO I PUT REFUSE OUT FOR COLLECTION?

All refuse and trash must be available *BY 5:30 a.m.* on scheduled collection day — *HOLIDAYS INCLUDED!* Cans must be readily accessible for servicing and, if practical, should be placed where visible from the street. (Maximum of two (2) containers by the house — additional at curb.) Sorry, but we cannot service cans located in the garage.

3. WHAT KIND OF CONTAINER MAY I USE?

All putrescible garbage must be in standard garbage containers.

A. Containers: Standard Garbage Containers not to exceed 32 gallons with a 60 pound total weight limit per container.

Heavy Duty Plastic Bags and Cardboard Boxes (at curbside only) — must be of a size and shape to permit easy handling by one man. Total weight of each not to exceed 40 pounds.

B. Disposable containers must be secured to prevent spillage or littering and, during wet or rainy weather, must be protected against moisture so they will not come apart when lifted.

C. *What Kinds of Containers are NOT ACCEPTABLE?*

WASTE BASKETS, BARRELS, OIL DRUMS AND FIBERBOARD DRUMS *ARE NOT* ACCEPTABLE — PLEASE DO NOT USE THIS TYPE OF CONTAINER.

D. *WHAT ABOUT TREE LIMBS OR BRUSH TRIMMINGS?*

Loose branches, cuttings, trimmings, etc. that are not containerized must be secured into bundles that will remain intact without separation while being removed by one man. Bundles may not exceed 3 feet in length, 2 feet in diameter, or 40 pounds in weight. The maximum diameter of any limb shall not exceed 4 inches. Larger limbs may cause damage to the packing mechanism of the collection truck.

4. WHAT WILL BE COLLECTED?

A. Normal household trash and garbage.

B. Animal droppings, ashes, sawdust and similar materials ONLY WHEN SECURELY WRAPPED.

C. Leaves, grass and yard refuse.

5. WHAT WILL NOT BE COLLECTED?

A. Dirt, sod, rocks, concrete, large metal objects, furniture, etc.

B. Construction and remodeling debris.

C. Flammables and liquids.

D. Live ammunition.

E. Paints, oils, acids.

6. WHAT ELSE SHOULD I KNOW?

A. On occasion, we may not be able to provide service due to various reasons such as: Gate locked or difficult to open, animals loose in yard, can too heavy or overfull, water or dirt in can. If so, a tag will be left explaining the reason for non-service.

B. *MISSED SERVICE:*

Trucks break down . . . new employees may be unfamiliar with your area, etc. If we miss your regular collection day, call by noon of the following day.

C. Where do I call for more information?

Please call between 8:00 a.m. and 4:30 p.m., Monday through Friday for:

Refuse Collection Service and General Information 363-6531

Billing Information 363-6537

NOTE: If you are calling from a 725 or 726 exchange, Dial Operator and ask for Enterprise 1-3623. Your call will be placed toll free.

D. Sacramento County has two disposal sites at which you may dump refuse not collected under normal service. See reverse side for details.

E. SPECIAL NOTICE:

Your refuse collector's job is difficult and dangerous. Please help him by not placing loose glass, television tubes, live ammunition, or open cans of paint or oil out for collection. Ashes set out for collection must be cold and securely wrapped to prevent blowing into a collector's eyes, thereby causing injury.

F. KEEP CANS AWAY FROM CARS, BOATS, TRAILERS — AVOID RISK OF DAMAGE!

G. FOR YOUR CHILDREN'S SAFETY DO NOT ALLOW THEM NEAR THE COLLECTION TRUCKS!

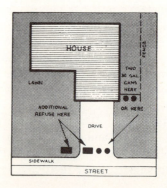

FIG. 13-2 Typical regulations for controlling onsite storage of residential wastes.

SACRAMENTO COUNTY WASTE DISPOSAL FACILITIES

Sacramento County operates two facilities at which you may dispose of practically any form of refuse not collectable under normal residential service: (1) North Area Transfer Station at 4640 Roseville Road, and (2) County Disposal Site at Grant Line Road and Kiefer Blvd. (NOTE: For Health and sanitation reasons, you may not dispose of garbage in lieu of regular collection service.) Disposal fees are charged.

DISPOSAL SITE NO. 1   **12701 Kiefer Blvd.**

Open 7 days a week, except January 1st
May 1 — August 31 . . . . 7 AM — 7 PM
Sept 1 — Apr 30 . . . . 7 AM — 5 PM

TRANSFER STATION   **4640 Roseville Road**

Open 7 days a week
Monday—Friday . . . . 6 AM — 6 PM
Sat—Sun—Holiday . . . . 8 AM — 6 PM

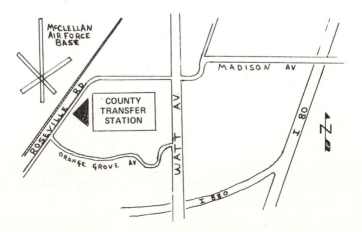

waste generation does it become more economical to change containers so that more efficient collection trucks can be used?

Collection   Possibly the most critical relationship between onsite storage and collection is the compatability of storage containers with the collection methods and vehicles. The capacity of the receiving hopper on the collection vehicle determines the sizes of wastes that can be stored in containers. The lifting capability of crew members limits the amount of wastes that can be placed in one container. This subject is considered further in Chap. 14.

Transfer and Transport   The principal impact of onsite storage on transfer operations is from container size, which must be matched with the unloading clearances of the transfer station. Coordinated design can mitigate this potential problem. The storage and collection of demolition wastes can also present a special problem when disposal areas are remote from sources of generation. Direct haul is an expensive option, and transfer facilities do not normally accept extremely dense wastes, such as concrete masonry, broken pavement, and dirt. The solution often is to coordinate waste handling at the source so that the light organic fractions are stored separately from the dense inorganic fractions. In many cases, clean, dense wastes can be disposed of in nearby ravines, washouts, or depressions without harm to the environment.

Processing and Recovery   There is a significant potential for conflict between the onsite handling, storage, and processing of solid wastes and their possible recovery and reuse. The primary reason is related to the basic purpose of each function. The function of onsite handling, storage, and processing is concern with wastes that are to be thrown away. For this reason, the wastes are made as inconspicuous as possible at the least cost. For example, compaction is often used to inprove the efficiency of storage. Unfortunately, compaction leads to contamination, which is undesirable from the standpoint of materials separation (the first step in recovery and reuse).

13-2   INVENTORY AND DATA ACCUMULATION

Coming to grips with the question of inventory and data accumulation concerning onsite handling, storage, and processing conditions is both difficult and time-consuming, especially when one considers that a city with a population of 100,000 may have as many as 25,000 individual sources of solid wastes (residences, schools, parks, industry, stores, etc.). It is therefore important to establish specific inventory requirements before any data collection program is initiated.

Inventory Requirements

To limit the scope of inventory of a community's onsite handling, storage, and processing of wastes, the first step is to develop information groups. Four groupings have proved useful: (1) geographic area to be covered, (2) quantities and types of wastes generated, (3) identification of equipment, and (4) service conditions and regulations.

The second step is to consider the following questions: What data concerning each of the above groupings are actually needed? How will the data be used? If the data are not available, how will the outcome of the planning effort associated with this functional element be affected? The third step is to estimate how much information and data can be collected for specific amounts of money, such as $10,000, $5,000 and $1,000, and to determine whether the value of this information is worth the cost of obtaining it.

Geographic Area

The geographic area to be covered in an inventory is significant in setting data requirements. The larger the geographic area, the greater the diversity of waste sources. After the geographic area to be studied is selected, it is important to identify land uses within that area. Many communities have zoning regulations and possibly a general plan that can be used to identify land uses. As shown in Fig. 13-3, typical land-use categories of interest in solid waste management planning are as follows: low-density residential, high-density residential, commercial, industrial, open areas, treatment plant sites, and agricultural. These categories correspond to those given in Table 4-1 which deals with waste generation. It is important to know the distribution patterns of these land uses because they will serve as the basis of establishing data requirements for the quantities and types of wastes as well as for potential waste processing equipment.

Quantities and Types of Wastes

The selection of land-use categories as the basic investigative units for onsite storage of wastes is an attempt to make inventory easier and more efficient. In most planning problems, it can be assumed that typical quantities and types of solid wastes (see Table 4-1) are generated in residential, commercial, and open recreational areas. However, for individual sources that generate significantly large quantities of wastes, this assumption is not valid. As a general rule, any individual source that contributes more than 5 percent to the total community wastes should be inventoried separately. In applying this rule, it should be remembered that there are many sources of wastes within any one land-use category, and the 5 percent value is related to only a single source within that category.

There also are exceptions to the 5 percent rule. The exceptions

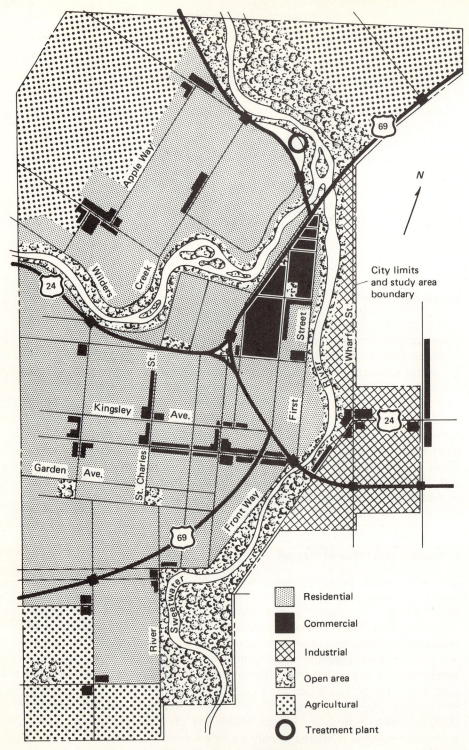

FIG. 13-3 Community land-use plan.

include any source, regardless of the amount of wastes involved, in which toxic or hazardous wastes are generated or any source that generates an unusual type of waste. The greatest potential sources of unusual types of wastes are in industry and agriculture. Typically, most light industries do not have hazardous waste disposal problems. The storage of pesticide and insecticide containers and their residues should be assessed carefully.

Identification of Equipment

Waste storage containers, source processing equipment, and other related facilities are the waste handling components that must be identified. Both existing equipment and available equipment not now in use in the system should be considered. The key for effective management is to establish an inventory technique that is consistent with economic constraints and that will yield the needed information. Some useful techniques are provided in Table 13-1.

TABLE 13-1 TECHNIQUES FOR IDENTIFYING EQUIPMENT USED AT VARIOUS WASTE GENERATION SOURCES

Source	Identification technique
Residential	Drive through typical areas (low and high density) and observe the existing storage equipment. Interview personnel of the waste collection agency. Review building codes and ordinances. (Visiting each source would provide the most comprehensive results but would probably be too costly and impractical.)
Commercial	In addition to the techniques listed for residential sources, contact the local chamber of commerce or better business bureau to obtain a list of businesses and public institutions. Restaurants, hotels, clubs, and hospitals are examples of sources that require special storage because of the rapid decomposition of food wastes.
Industrial	When possible, visit each industrial source. Monitor large containers at disposal sites. Send out questionnaires. Each industry generally uses distinctly different raw materials, and an inventory of each one often affords an opportunity to define much more than source storage conditions. Data on resource recovery for reusable materials as well as the potential for power, heat, or steam reuse may also be obtained at the same time.
Open areas	Visit each major open area because storage control is probably exercised by a nontypical waste handling agency, such as a park department or a park district. Identify any adaptations that make containers vandalproof.
Agricultural	Contact local officials in agriculture and forestry for information on crop acreages, animal populations, and forest product harvesting. Conduct field investigations. Observe processing and disposal operations.

Service Conditions

It is important to identify existing service conditions that must be met. The various types of services provided are identified in Chap. 5. Local ordinances and regulations must also be reviewed, and persons or agencies responsible for operating the waste handling and storage equipment must be identified (see Table 5-1). Special service conditions may be involved where large parks or recreational areas are owned and operated by a community [7].

13-3   CASE STUDIES IN ONSITE HANDLING, STORAGE, AND PROCESSING

Two case studies are presented to illustrate in detail some of the techniques that can be used in assessing onsite storage problems. The first case deals with the problem of reducing costs of solid waste handling and storage in the setting of a growing community. The second case deals with onsite handling and storage problems in the core area of an old city. It is noted that the resolutions address only those management issues and concerns listed for each case.

CASE STUDY 13-1   *Onsite Handling and Storage of Wastes in a Growing Community*

The public works director of a Western community is dissatisfied with the high cost of collection service in the city and has asked that a study be conducted to determine whether the combined waste storage and collection systems could be modified to reduce labor requirements to collect wastes. The director believes that a reduction in labor requirements will reduce system costs.

Management Issues

The management issues relate to modifying an existing collection service:

1. How strong is community identification with existing services now provided, and what community objections would have to be overcome to change the type of storage containers now used and the service received?

2. What types of storage containers and collection vehicles are available that can be used to reduce labor-intensive operations?

Information and Data

The following data are available:

1. City population: 100,000

2. Container size (existing): 30-gal metal or plastic

3. Collection location: curbside or alley

4. Collection trucks: 20-yd³ capacity, rear-loading packers, and three-person crews

5. Persons responsible for onsite waste handling: homeowners or occupants

6. Waste collection agency: Department of Public Works

Resolution

The adequacy or inadequacy of the existing onsite handling and storage system was not an issue that would provide justification for change. Still, important aspects of the existing system were identified and recorded, primarily to form a baseline of technical and other characteristics against which the results of change could be measured.

The principal means of reducing labor costs is through some type of automated or mechanical collection equipment. It is extremely difficult to use such equipment in conjunction with waste storage containers of irregular sizes. Therefore, it was necessary to question community residents to establish their reactions to any changes in storage containers (either size or location). Curiously, residents were most opposed to a change that would result in two homes using the same storage container. The principal reason given was that residents did not want their neighbors to be able to identify their wastes. The anonymity of discarded materials remains a strong social force! Another reason given was that residents did not want anyone else's wastes to be placed on their property. The evaluation and resolution of the issues in this case were as follows.

1. Because an expanding community does not have the social and environmental constraints that exist in old, crowded metropolitan centers, improved waste management systems can be planned as community expansion occurs. In this case, changes were made that involved approximately half of the residents of the community. Resident-owned containers and curbside or alley setout collection services were changed to city-owned containers and two different setout conditions based on neighborhood street arrangements. Homes with curbside (street) setout were provided with 80-gal containers mounted on wheels, which are rolled to curbside on collection day. Homes on an alley were provided with 300-gal containers that were conveniently located in permanent areas on the public rights-of-way. Each container provides storage space for four homeowners. All containers are provided by the city.

2. All new housing subdivisions in the community must be planned to accommodate either of these storage systems. In the older half of the community, the use of standard 30-gal containers, plastic bags, and other miscellaneous storage containers was continued. Even though the use of common containers for multiple homes is a revival of an old transitional concept, the lasting concept (source storage) was continued unchanged. The greatest technology change occurred through the interfacing of common container storage for multiple homes with mechanized collection vehicles.

3. The collection vehicles for the portion of the community that would have city-supplied containers were designed so that the driver could serve as the collector without leaving his truck cab. This was a radical change in technology from any equipment available on the market, and development of the vehicles required special cooperation with a manufacturer. By combining the storage container and collection truck in this manner so as to permit one-person crews, the city accomplished the following:

   a. Costs of collection in the affected areas were reduced by approximately 50 percent.
   b. The level of service was generally improved. Most homeowners preferred the large city-furnished containers after they were installed.
   c. Employee working conditions were improved.
   d. Safety problems were reduced.
   e. Sanitation was upgraded.

Comment The solution developed in this case represents a significant change. However, some of the traditional arguments against the use of large storage containers apply to this case. Specifically, the level of service has not improved for customers, especially for the resident on an alley who has given up a more convenient container location and is now carrying solid wastes for some of the distance the collector formerly covered. Safety problems might have been reduced for the collector (no more heavy lifting), but the storage container is now large enough for a person to crawl inside. Another potential safety problem has been created.

CASE STUDY 13-2   *Waste Handling and Storage in the Core Areas of a Large City*

Building owners and apartment managers in the core area of an old city are receiving constant complaints from tenants that the building is nothing more than a solid waste dump. In addition, the operator of the collection system is trying to reduce labor costs and is asking that all wastes be placed in a central location for convenient truck access. The housing authority officials have contracted for a study of the problem, and they have requested that solutions be developed.

Management Issues

The issues derive directly from the complaints of the tenants and the operator of the collection system:

1. How can the existing community onsite storage facilities, which are objectionable aesthetically and can provide potential harborage for rodents, be improved or modified?

2. What administrative controls are needed to reverse the trend of deteriorating storage conditions?

Information and Data

The following information and data were available before the inventory was undertaken:

1. City population: 2,500,000

2. Storage containers: 30-gal containers, cardboard boxes, and plastic bags

3. Container location: outside apartment doorways

4. Building size: three-story maximum

5. Persons responsible for onsite waste handling: tenants

6. Waste collection agency: private industry

The following conditions were observed and recorded as part of the inventory assessment:

1. Solid wastes were being stored in suitable containers, and tenant complaints of container inadequacy were not excessive. (This established the acceptabilty of the existing storage containers and encouraged the investigators to look for other conditions contributing to system deficiencies.)

2. The locations of containers for collection were causing indirect property damage. In moving from a street stop, between parked cars, and then up narrow stairs or through narrow halls, the collectors were scratching cars, breaking light bulbs, and generally disturbing private property. (The investigators discovered an important waste handling condition: the interface between storage and collection was a key deficiency in this solid waste management system.)

3. In the social environment of the core area of the city, the need to deal with food wastes and solid wastes was accepted, but the benefit of the proper management of solid waste equipment and

systems was not recognized. Unemployed persons searched containers for anything of value, and children used the containers as toys.

4. Living conditions are extremely crowded. All land areas are taken up by dwelling units, and all the space within the dwelling units is occupied.

Resolution

Conditions that may result from proposed changes cannot be observed unless some type of pilot operation is undertaken, but predictions of probable results can be made. Some of the evaluations made and alternatives considered were as follows:

1. Changes in onsite storage equipment. Onsite storage equipment changes would not improve the system unless large ground-level bins could be installed, but crowded streets and narrow lot frontages eliminated this possibility. Additional drawbacks to changing to bin storage included the lack of capital (both the community's and the collector's) to purchase bins, the inability of the administrative agency (housing authority) to enlist public support for change, and the prolongation of the social environment related to disregard for equipment and systems (scavenging and vandalism would shift from doorway containers to bins).

2. Change in container location. Residents could be asked to move containers from doorways to a curbside location on collection day. The greatest potential problem with this change is the social environment. Vandalism could increase significantly because of the increased ease of reaching the containers. Such a system would provide no solution for residents who are physically unable to carry heavy containers down two or three flights of stairs.

3. Use of onsite processing equipment. Onsite processing to reduce the volume and weight of wastes would be a change with good potential for increased benefits. Possible processing units would include incineration in the building, wet pulping and pumping, and building storage bins with attached compaction devices. However, the housing area is old, and the administrative agency (housing authority) would not undertake large capital improvements exclusively for improving the waste management system. Other technical problems would include the air discharge problems associated with incinerators and building space limitations for storage bins. This alternative would likely be defeated on the basis of economics alone.

Comment In this case there were no acceptable solutions. In fact, the older core areas of most major urban centers have an acute waste storage problem. Planning has lagged so far behind technological development that only very costly and outdated transitional concepts are used. Not unexpectedly, the reason for this lag is the reliance on the collectors to pick up greater quantities of wastes while competing for street space with other transportation systems. Costs of collection have increased dramatically as a result of increased labor requirements and time lost in street delays.

Although there was no satisfactory solution to this case, it must be understood that the problem and the community cannot be ignored. Social programs related to improving community pride are an essential first step. The existing containers can then be used for an additional time, but eventually a reconstruction effort must be made in which waste storage is provided in a secure and sanitary manner.

13-4   DISCUSSION TOPICS

13-1.   Group the various solid waste generation sources in your community into the appropriate land-use categories, and identify the approximate number of sources in each category.

13-2.   Assume that you are a city engineer with administrative responsibility for solid waste collection. At the last city council meeting, the mayor reported that a manufacturer's representative proposed a solid waste management system that would save the city about $100,000/yr. The proposal involved the purchase, by the city, of storage containers that would be used by all city residents. The city council has given you one month to evaluate the proposal and to make a firm recommendation. What issues are involved? How would you organize an inventory to complete the evaluation in the short time span? What are your recommendations? Clearly state all your assumptions.

13-3.   Resources Unlimited, a nonprofit environmental organization, has presented your private solid waste collection company with a proposal stipulating that the residential customers you serve would separate and store paper, ferrous metals, aluminum, and glass for recovery. Assume that you are one of five owners of the company and you are the only one who has a technical background in data accumulation, evaluation, and report writing. The public agency that administers your franchise has endorsed the proposal and has asked your company to evaluate its feasibility. What are the management issues concerning onsite storage? What are the management issues for your solid waste collection company?

13-4.   A city located in central California wants to evaluate the impact of banning all residential burning of yard and tree wastes. Assume that your agency is responsible for completing a plan to evaluate the impact of a backyard burning ban and for recommending a new policy for solid waste storage and handling at residential sources. Which city agencies would you contact in organizing this planning effort? Which state or regional agencies might you contact? Outline a work program to guide you in completing the plan.

13-5.   Referring to Discussion Topic 13-4, identify and list ordinance and regulation changes that might be needed to implement an expanded residential storage and collection system for yard wastes. Estimate the percentage increase in solid wastes requiring collection that you expect from the implementation of a backyard burning ban (see Chap. 5).

13-6.   Agencies with administrative and operational responsibilities for solid waste storage and collection are being criticized for charging a single rate for collection service without regard for the number of storage containers used by the generator of the wastes. List the positive and negative aspects of such a rate schedule (consider the entire solid waste management system in your analysis).

13-7.   As the manager of a solid waste collection agency, you are developing a plan to evaluate the benefits of placing large-capacity storage containers in the residential areas of your city. Each container would hold the wastes of six residential dwelling units. In the present system, 32-gal containers are used at each residence. Does this plan deal in policy issues? Document the reasons for your answer.

13-8.   What inventory method would you recommend to determine the number of 32-gal garbage cans that would be abandoned if the container system of Discussion Topic 13-7 were implemented? Why would you conduct such an inventory? Assume that 30,000 residential dwelling units are involved.

13-5   REFERENCES

1.   American Public Works Association: "Municipal Refuse Disposal," 3d ed., Public Administration Service, Chicago, 1970.

2.   American Public Works Association, Institute for Solid Wastes: "Solid Waste Collection Practice," 4th ed., American Public Works Association, Chicago, 1975.

3.   Greenleaf/Telesca, Planners, Engineers, and Architects: *Solid Wastes Management in Residential Complexes,* U.S. Environmental Protection Agency, Publication SW-35c, Washington, D.C., 1971.

4.   *Guidelines for Local Governments on Solid Waste Management,* U.S. Environmental Protection Agency, Publication SW-17c, Washington, D.C., 1971.

5.   Hanks, T. G.: *Solid Waste/Disease Relationships,* U.S. Department of Health, Education, and Welfare, Solid Wastes Program, Publication SW-1c, Cincinnati, 1967.

6.   Powell, M. D., B. P. Fiedelman, and M. J. Rose: *Digest of Selected Local Solid Waste Management Ordinances,* U.S. Environmental Protection Agency, Publication SW-38c, Washington, D.C., 1972.

7.   Spooner, C. S.: *Solid Waste Management in Recreational Forest Areas,* U.S. Environmental Protection Agency, Publication SW-16ts, Washington, D.C., 1971.

8.   Stragier, M. G.: Mechanized Residential Refuse Collection, in *Solid Waste Demonstration Projects,* U.S. Environmental Protection Agency, Publication SW-4p, Washington, D.C., 1972.

9.   Syrek, D. B.: *California Litter, A Comprehensive Analysis and Plan for Abatement,* Institute for Applied Research, Carmichael, Calif., 1975.

10.   Twenty-three Municipalities Adopt Cart Systems, *Refuse Report,* International City Management Association, vol. 1, no. 5, 1975.

14

Collection Alternatives

The identification and evaluation of alternatives for waste collection usually are the most important part of solid waste management planning, because an estimated 60 to 80 percent of all solid waste system expenditures are for collection. Within the functional element of collection, more than in any other functional element, it is important to understand the nature of the operations and operational variables, especially those that are responsive to planning and management. For this reason, a significant amount of operational detail is presented in this chapter.

14-1 MANAGEMENT ISSUES AND CONCERNS

Management issues and concerns with regard to collection relate primarily to the multiple and diverse equipment and workforce programs that can be developed. The major concerns are related to labor efficiency and customer service levels (frequency of collection and location of containers). Collection problems typically are caused by one or more of the following factors: (1) the existence of strong community identification with existing collection service levels (e.g., inherited policies), (2) the involvement of both public and private agencies in collection operations, (3) the lack of uniform and complete management records that can be used to evaluate labor and equipment efficiency (productivity) and to develop strong performance standards, (4) financing difficulties, (5) labor constraints, (6) collection of special wastes, and (7) the technological constraints of surface transportation. The close relationship between collection and other functional elements of onsite handling, storage, and processing, transfer and transport, and processing and recovery leads to additional management concerns.

Community Service Levels

Collection is the functional element in which the concentration of wastes for efficient handling and protection of public health begins. Because collection is an extremely visible community activity in which crews are placed in daily contact with residents, the social structure and standards of the community bear directly on operating conditions. Community pride and aesthetics also influence collection systems. Because collection operations are widespread within communities, there is a high probability that numerous problems will require some type of planned solution. For these reasons, planners need to monitor community reaction to programs and plans for collection. This communication usually is accomplished through public membership on planning committees, public meetings, and media presentations.

Solid waste collection operations affect everyone in a community. Residents can easily relate to them and communicate their views to politicians. The development of a solid waste collection ordinance is the most direct action the political system takes regarding collection operations. Additional action involves the setting of policy regarding collection rates or taxes to pay for service.

All these factors affect community collection service. The greatest demands on management occur when a change in service level is proposed. Questions and resistance can be expected from every part of the community. As discussed previously, this resistance can be overcome most effectively through good public communications.

Public and Private Agencies

Collection service can be provided by either public or private agencies. In the United States, private agencies provide the greater part of the commercial and industrial waste collection service, and public agencies provide the greater part of the residential service. Additional information on private agency collection is available in Ref. 3. The primary management concern is to provide the most efficient collection service at the desired community service level. Typical contractual conditions for private agency collection are listed in Table 14-1.

One of the significant developments in solid waste management during the past few years has been the establishment and growth of national, private waste collection agencies. The full impact of these recently formed companies (the earliest ones appeared around 1970) cannot be evaluated yet. Although their success or failure cannot be assured at this time, their existence cannot be denied, and in many cases they represent a viable alternative as an operating agency with a large funding capability. This capability can be important in resolving problems related to system improvement.

Management concerns with respect to agencies include the regulation

TABLE 14-1   TYPICAL CONTRACTUAL CONDITIONS FOR COLLECTION SYSTEMS OPERATED BY PRIVATE AGENCIES

Type of contract	Typical conditions
Franchise	Used where quality of service takes precedence over cost of service. Competition is eliminated. The franchise area is set within specific geographic boundaries. The franchise administrator must monitor rates and service complaints. Elimination of overlapping competitive routes allows higher collection efficiency. Contract stability allows long-term investment in equipment.
Limited permits	Used where limited competition is preferred but the costs of franchise administration are undesirable. Rates are self-regulating. There are no permit boundaries. The number of permits is set by criteria such as potential revenue and population in the permitted area. Difficulties in administration arise when the logic of a set number of permits is questioned.
Unlimited permits	Used where completely open competition is desired. Rates are self-regulating. Efficiency drops because of overlapping routes. Competition can become very intense with a resulting high number of company failures. Administrative difficulties increase as business failures occur.

and control of collection. A digest of typical ordinances that are used to control collection is presented in Ref. 1. Operational regulations are closely related to onsite storage (see Fig. 13-2). Regulations for public agencies are established by political bodies. Regulations for private agencies are self-imposed to some extent, but complete self-regulation by private agencies is rare and impractical. In most cases, public agencies resolve public concerns through the regulation of private agencies. The political systems under which the collection agencies have developed will strongly influence future public and private interactions. In all collection systems, a local public health agency usually regulates matters concerning health and safety.

Management Records

The lack of uniform and complete management records for collection is a concern because documented comparisons need to be made to resolve problems, and frequently data are not available against which change can be measured. The present technology of collection requires a high labor input for each unit of output (labor-intensive activity). Any management actions to reduce collection costs must include programs for reductions in crew labor.

The resolution of problems may require records of any of the following: hours worked, tonnage collected, number of service stops, and miles of route. Additional information dealing with collection records and data can be found in Refs. 1 and 2. As a general rule, the data contained in detailed records are used directly only in local plans. Detailed records might be used in regional and state plans only to ensure that records of all communities are comparable.

Financing

A change in the collection system normally requires a capital investment and an increase in annual costs. Capital normally is obtained through bonds (for public agencies) or lending institutions (for private agencies). Under these conditions, the bonding capacity of a public agency and the borrowing capacity of a private agency are limiting criteria that represent a major management concern. An alternative to capital outlay is the leasing of collection equipment. Increased annual costs associated with leasing are offset by increased rates or taxes. If the economic benefit of a change in the collection system accrues to only a select and discrete number of customers, it may be advisable to revise only their specific collection rates. The financial option selected will depend to a large extent on the political situation.

Funding concerns have caused many of the problems related to inadequate development of alternative transitional concepts and equipment for collection. Several authors define and discuss the role of funding capability in collection operations [4, 12]. Historically, the rate charged for collection service has been low. The low rates can be traced to the use of low-capital-investment equipment and low wages for crews. Communities encouraged these low-cost collection systems because residents believed that throwing something away should not be costly; or, stated in another way, people generally seek to minimize investment in an operation that will return nothing. As a result, it has been difficult to obtain capital for the purchase of collection equipment. The large national waste collection agencies may be able to resolve this problem.

Labor Constraints

Labor unions are an integral part of many collection operations because most collection crew members are affiliated with a union. Union contracts often contain clauses defining route work conditions that cannot be changed without contract renegotiations. Labor contracts therefore are a significant management concern. As previously mentioned, collection route work conditions are a concern because collection is a labor-intensive activity, and any improvements that will significantly reduce collection cost must reduce the use of labor on collection routes.

One of the most difficult problems regarding collection routes is maintenance of a balanced work load for the individual crews. The primary cause of unbalanced work loads is the irregular week-to-week pattern of waste generation. Homeowners will follow a regular pattern in setting out food wastes but will follow no pattern in setting out accumulations of magazines, plant trimmings, and other nonputrescible wastes. Another cause of unbalanced work loads is seasonal change. Spring cleanup in the community, as well as the autumn leaf season, will result in heavy waste loads on residential routes. Winter operations in all communities where snow and ice are common will be less productive than operations in other seasons because crews must slow down to avoid injury.

Management techniques used to resolve unbalanced work-load conditions include (1) assigning an average quantity of waste as a standard work task, recognizing that the crews will finish early on some days and late on other days, (2) reassigning crews that finish light routes to assist crews on heavy routes, (3) paying overtime for crews to pick up additional wastes on heavy routes, and (4) hiring extra crews during peak seasonal waste loads. To form a flexible system, it may be necessary to use various combinations of these techniques.

In planning any combination of crew assignments it is important to evaluate both labor and equipment requirements. The main equipment factors are (1) capacity of vehicles, (2) loading position (top, rear, or side loading), (3) unloading method (hydraulic, tilt-body, or by hand), (4) maneuverability (turning radius, street grades, traction in snow or ice), (5) reliability (maintenance, parts availability), (6) safety and comfort of crew, (7) adaptability to changing route conditions, and (8) vehicle appearance. Additional technical criteria are discussed in Ref. 2.

Special Wastes

The collection of special wastes causes special management concerns. As defined earlier in this text, *special wastes* are those with special or unusual characteristics (see Fig. 14-1). Here management concerns are centered on the special administrative and operational handling procedures that are necessary to dispose of these wastes. The handling of abandoned vehicles is a typical special waste handling problem. The disposal-reclaiming cycle for abandoned vehicles and seven special waste problems that must be resolved by the waste system manager are identified in Fig. 14-2.

The adaptability of collection trucks for multiple collection needs, including special waste handling, should be considered when trucks are being evaluated. Many small communities must purchase multiple-purpose collection equipment for the sake of economy; in some cases, dump trucks used for special waste collection are also used for snow removal. Under such joint-use conditions, equipment costs can be shared with another agency that is not involved in waste collection.

Technology of Surface Transportation

The development of cost-effective surface transportation technology for solid waste collection is at an early stage and is limited by restricted funding. As a result, few choices of a technological nature are available to

(a)

(b)

FIG. 14-1   Examples of special wastes found in most communities: (a) Abandoned bulky wastes; (b) unusable tires.

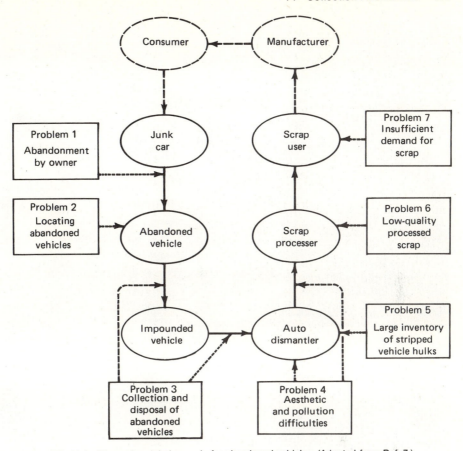

FIG. 14-2 Disposal-reclaiming cycle for abandoned vehicles. (Adapted from Ref. 7.)

operating agencies attempting to establish more efficient and reliable collection systems.

Many transportation technology concerns in collection arise from restricted street or road access conditions for collection vehicles (see Fig. 14-3). A special technological constraint exists in the collection of wastes from park and recreation areas. Large trucks are uneconomical because of the long distance between campgrounds and cabins. Special management techniques must be considered. A comprehensive discussion of some special problems encountered in recreational areas is presented in Ref. 6.

System Continuity

Collection is an intermediate functional element in most solid waste management systems. As such, its relationship to other elements must be clearly understood.

(a)

(b)

FIG. 14-3 Limited road and street access on collection routes: (a) In San Francisco; (b) in Sacramento County.

Onsite Storage   The interrelationship of collection operations and onsite storage is a sensitive area for action because of the great numbers of people directly involved. Any change in collection procedures may alter the service level to the customers (homeowners, businesses, etc.). In changing the service level, the collection agency must be concerned with the following interactions with onsite storage: (1) compatibility of containers and collection equipment, (2) waste container movement by generators versus waste container movement by collection crews, and (3) aesthetics and other social considerations.

Transfer and Transport   The purpose of transfer and transport is to improve the efficiency of waste handling (see Chap. 15). Individual waste generators, as well as drivers of collection agency vehicles, can bring waste loads to a transfer station at a convenient urban location where the wastes are transferred to larger transport vehicles. Management efficiency results from the elimination of multiple vehicle trips to a remote landfill.

The significant management and planning considerations with respect to system continuity include (1) the compatibility of collection vehicles with transfer facilities and (2) the economics of direct haul by collection vehicles to the processing or disposal site versus use of the transfer station. Here compatibility means that collection vehicles must be able to use the transfer facilities without loss of efficiency. Direct haul by collection vehicles to the disposal area is preferred when haul distances are short.

Processing and Recovery   The principal relationship between waste collection and waste processing and recovery is centered on the collection vehicle. As with transfer and transport, the processing and recovery facilities should be constructed so that collection vehicle movements can be completed with minimum delay. The type of collection vehicle used has a unique relationship to the recovery of waste materials. Recovery programs that provide immediate quantities of recoverable materials involve separation of wastes at the source of generation. In order for the wastes to be kept separate as they are collected, collection vehicles must be modified to carry separated wastes. When modified vehicles are available, additional processing and recovery equipment may not be needed for waste separation.

14-2 INVENTORY AND DATA ACCUMULATION

Many of the inventory techniques for onsite handling, storage, and processing presented in Chap. 13 also apply to collection. In most cases, the types of data required are different, and the breadth of inventory is reduced. The principal reasons for this are that (1) owners of collection systems are not nearly so proprietary about data as are waste generators, (2) the number of inventory contacts is reduced dramatically, and (3) collection operations generally are better understood and better documented than onsite handling storage and processing. In fact, collection data are being accumulated throughout the United States. It is therefore feasible to gather and evaluate a limited amount of collection data, to compare them with published standards and data, and to use the resulting data in the development of collection system alternatives for most communities in the United States. For planning purposes, the data may be divided into two parts: (1) collection system operations and (2) collection system constraints.

Collection System Operations

Collection system inventories typically include identification of (1) the agencies responsible for collection, (2) collection system records, (3) crew size and working conditions, (4) collection equipment, (5) special collection operations, and (6) the handling of special wastes.

Agencies Responsible for Collection   As discussed earlier in this chapter, collection agencies may be either public or private. In an inventory for a plan that involves collection, it is important to establish the relationship between the agencies responsible for collection. The number of agencies involved in collection usually varies with the size of the community or area under study. Consider the results of a management plan completed in 1974 for a four-county area surrounding Portland, Oregon [5]. Approximately 100 public and 300 private agencies were identified as having a responsibility in the collection system!

  The best place to obtain data is directly from each agency participating in the system. The most rewarding first contact point is the local public health agency. Health agencies for many years have been instrumental in developing regulations for collection operations and will continue to provide strong regulatory guidance wherever putrescible wastes are stored, collected, and moved to disposal [2]. Another useful starting point for inventory is the billing log from the disposal site. Any collector operating within reasonable driving distance (within 20 mi) of the disposal site might use it. Public agencies that may be involved in the collection of solid wastes in urban communities are listed in Table 14-2.

Collection System Records   The operating divisions of most collection agencies have the largest budgets and are responsible for the most activity in the collection system. Although reliable documented information is not available to verify relative expenditures, it can be assumed for planning purposes that more than 90 percent of collection expenditures are made by the operating divisions and less than 10 percent are made by administrative divisions. For this reason, planning efforts to improve system economics should concentrate on the operating divisions.

  Record keeping and data accumulation with respect to collection are nonuniform and vary with each collection agency, although efforts are

TABLE 14-2 TYPICAL PUBLIC AGENCIES AND THEIR FUNCTION IN THE COLLECTION SYSTEM

Agency	Function
City public works department	Collects residential solid wastes and sweeps streets
City contract administration	Administers contracts with private collectors for commercial and industrial waste collection
County health agency	Regulates all collector containers, trucks, and site storage; enacts ordinances
Special sanitation district	Enacts ordinances and administers contracts
City police department	Tags and removes abandoned automobiles and property
County and state highway divisions	Clean roadways and collect refuse from highway and park litter containers
State environmental agency	Serves as clearinghouse for financial grants; coordinates state and federal programs; sets standards for operations

under way to implement the use of standard methods [1]. Traditionally, each agency division also keeps private financial records and labor statistics. This makes the inventory of collection operations time-consuming and complex and causes difficulty when attempts are made to develop comparative statistics. Categories for equipment, labor, and other expenditures should be established prior to inventory so that meaningful questions regarding financial and labor reports can be prepared. A useful list of expenditure categories, and the type of agency (operating or administrative) that is the source of expenditure, is presented in Table 14-3. As a general rule, the collection system inventory for a local plan must be more comprehensive than for a regional or state plan.

In many planning situations it is beneficial to compare the costs of collection operations as reported by other communities. Such comparisons are often used to evaluate the efficiency of public agency operations as measured against private industry operations. The expenditure categories identified in Table 14-3 should be used as the basis for data accumulation in these comparisons. If a mailed questionnaire is used to accumulate cost data, it should list the expenditure categories and request that the respondents mark those expenditures included in their completed cost questionnaires. If the data are obtained by telephone interview, then verification of the expenditure categories should be completed on an item-by-item basis. In completing this type of cost comparison, it is also important to identify the type of service provided by each collection agency (backyard, curbside, alley, one-can, two-can, unlimited, etc.) and to determine whether the operational costs reported totally support the collection services being compared (e.g., residential collection operations might be subsidized by profits from commercial collection operations).

Crew Size and Working Conditions Two of the most critical planning

TABLE 14-3   USEFUL EXPENDITURE CATEGORIES FOR COLLECTION
SYSTEM INVENTORY

Expenditure category	Division
Direct labor salaries and benefits	Operating
Equipment maintenance services	Operating
Equipment maintenance supplies	Operating
Disposal–transfer station costs	Operating/administrative
Fuel and lubricants	Operating
Miscellaneous operation expenses (tools, training, etc.)	Operating
Equipment rentals	Operating
Rent and leases—land and buildings	Operating/administrative
Laundry services	Operating
Billing service	Operating/administrative
Postage	Operating/administrative
Data processing services	Operating
Bad debts	Operating
Depreciation or replacement fund (equipment)	Operating
Indirect administrative services (labor, planning, etc.)	Operating/administrative

factors for operating divisions of collection agencies, whether public or private, are crew size and crew working conditions. In assessing crew size and working conditions it is often common to use production standards or productivity measures, such as collector-minutes per ton of solid waste collected, as a general guide to overall effectiveness of collection operations [10]. When such measures are used, great care must be exercised to ensure that valid comparisons are made between similar operations.

In management decisions involving change, personal satisfaction and crew spirit should be considered along with production standards. Personal satisfaction can be enhanced through the assignment of work tasks based on discrete daily routes. This allows crews to work at their own speed on assigned routes with the day's work considered done when the route is completed. Crew spirit, while impractical to measure, must be recognized as a factor in assessing collection operations

Collection Equipment   An inventory of the types of equipment now used in the collection operation as well as of the types of equipment available and/ or used in other communities or operations is necessary in evaluating equipment and system costs and in making comparisons between collection operations. Often equipment limitations will affect employee morale. Among the factors that should be inventoried are (1) size and capacity, (2) compaction efficiency, (3) loading height, (4) unloading devices, (5) turning radius, (6) watertightness of the collection bodies, (7) safety and comfort of collection crews, (8) adaptability of equipment to other services, (9) general appearance, and (10) age and purchase price. The specific items to be inventoried will depend on the use to be made of the data. Data on the equipment used by private agencies are best obtained by interviews.

Special Collectors and Informal Haulers   As previously mentioned, the collection function is identified easily in the traditional solid waste system because it is a routine community activity that people can observe daily. In addition to the regular collectors, however, there are two classes of special collectors: (1) recycling groups, scavengers, cardboard recoverers, scrap collectors, and other assorted entrepreneurs; and (2) informal haulers, usually employees of commercial and industrial companies who haul their companies' wastes.

Many recycling collectors use the same waste setout system established for scheduled residential collection. Consider the implications when two separate agencies use the same routes, storage containers, and setout locations. Who has responsibility for litter, vandalism, and missed pickups? With respect to resale of recovered materials, most recyclers act as brokers, selling materials to other processors who prepare products for resale, and it is difficult to inventory and monitor these collectors. A business license is normally the only requirement they must meet to begin operations, and equipment is rarely inspected. The existence of the recycling collector is related directly to the market demand and the price paid for recoverable wastes.

As noted, informal haulers are made up of a large group of individuals and agencies that move wastes independently of formally scheduled collections. The group is characterized by heterogeneous equipment (pickup trucks, trailers, produce vans, etc.) and inexperienced crews. The most feasible technique for inventory is to observe transfer stations and disposal sites. Billing records from these operations are also useful in that they show the names of repetitive informal haulers.

Special Wastes   Solid wastes that cannot be handled by normal collection equipment and preset collection routes require special handling. Industrial sludges, large dead animals, litter, and bulky items (refrigerators, furniture, etc.) are examples of special wastes. Inventory of special waste collection can be completed in the same way as the inventory of regular waste collection. Agency interviews and field investigations can be conducted to identify most operations.

Collection System Constraints

Important planning constraints for collection operations can be derived from a review of existing operating conditions. Areas of investigation and inventory that should be considered include (1) jurisdictional and budgetary, (2) social, and (3) economic constraints.

Jurisdictional and Budgetary Constraints   In performing their management tasks, public agencies responsible for the administration of collection operations often face two distinct and restrictive conditions: legislated jurisdictional functions and limited budgetary capacity. Legislated jurisdictional functions are the rules by which a public agency can operate legally. The spectrum of functions ranges from the very broad powers of an incorporated city in regulating all utility services to the very limited powers of a special utility district created to regulate a single utility, such as one that provides water service or waste-water collection and treatment. Many existing agencies have the legal authority to expand and undertake the administration of solid waste collection systems. The value of such an expansion is realized in lower system costs as well as in coordinated utility planning. The jurisdictional functions of administrative agencies must be determined and considered in the evaluation of operating conditions. Typically, the agencies listed in Table 14-2 also are involved in administration. A selected digest of ordinances used by administrative agencies is found in Ref. 11.

The budgetary capacity of an administrative agency is the amount of money it regularly sets aside for the enforcement of its jurisdictional function. In planning for change in a collection system, the variables associated with a budget must be considered. Because many administrative agencies serve multiple utility functions, competition for available funds is often intense. An example of an administrative agency faced with budgetary constraints is a health agency assigned the responsibility to inspect and certify solid waste collection vehicles. The agency also regulates hospitals, restaurants, and other community health programs. The solid waste system must compete directly with these other programs for available funds.

Social Constraints   Collection crews are expected in a given community on certain days and usually at a regular hour. The emptying of containers and the operation of a truck-packer mechanism produce loud street noises. Through its visual impact and noise level, the collection operation becomes a part of the community's social structure. The study, evaluation, and planning of collection system modifications must include an inventory of the social relationships of community residents and collections crews. Operating conditions are affected measurably by two factors: storage container locations and community pride.

As mentioned earlier, the locations normally used for placement of solid waste containers on collection days are alleys, backyards or side-yards, and street curbs (see Chap. 5). Whichever location presently is being used in a community has become a part of the social structure of that community. The trend in setout locations is toward curb collection. The basis for this trend is economics. The benefits that can be derived by switching to curb collection can be seen in the results obtained from such a program in Cleveland, Ohio [9]. Prior to curbside collection, the city's collection and disposal division had 1,475 employees, and backyard service was provided. After the operations were modified to emphasize curbside collection, the number of employees was reduced by 50 percent, to approximately 740. Because the economic benefits were so great, the community accepted the change. However, not all communities will accept the economic benefit as sufficient justification for changing an existing socially acceptable system. The planning process must include a measurement of community standards as well as economics regarding the issue of setout locations.

Economic Constraints   Economic constraints are important in all aspects of solid waste management. Costs are a critical factor for the decision-maker at every step in the system. However, there are two economic constraints which affect the administrative and operating divisions of collection agencies that must be recognized separately in the decision process: (1) the rate paid for collection service, and (2) the funding capabilities of the solid waste collection agencies.

Any change in operating conditions that results in a significant increase in collection charges will be resisted. The cause of that resistance in many cases will be the fact that there is no economic return for an investment in the waste collection system. Sources of resistance (commonly the taxpayer leagues and apartment owners' associations) should be identified during the inventory.

The funding capability of a collection agency is its capability to raise capital for investment in equipment and to maintain adequate funds for daily operations. The inventory of funding capability should concentrate on bonding capacity for a public agency and debt-to-asset ratio for a private agency.

The length of time for which a permit or franchise is awarded must also be considered when economic constraints are reviewed. Funding capability is determined in part by the length of time over which capital can be repaid. As collection system costs increase and funding needs increase accordingly, it is often beneficial to set permit or franchise time limits to match the depreciated capital time period for the equipment. This period normally varies from 5 to 7 yr.

14-3 CASE STUDIES IN COLLECTION

To illustrate some management issues related to collection and their resolution, two case studies are presented. The first deals with the problem of combining two separate solid waste collection operations in a medium-sized city, and the second deals with improving solid waste collection operations at a major airport.

CASE STUDY 14-1   *Combining Separate Solid Waste Collection Operations*

The city council of a Midwestern city (population 300,000) has authorized a study of its solid waste management operations (1) to determine the feasibility of consolidating two separate operating divisions with responsibilities for the collection of solid wastes into a single division and (2) to develop a plan for implementation. The existing system includes a waste removal division that collects household solid wastes and a street cleaning division that collects yard and garden debris and cleans the city streets. Each division maintains separate crews, equipment, and management. An engineering consulting firm has been retained to conduct the study.

Management Issues

The principal management issues relate to the transformation of a labor-intensive dual collection system into a more mechanized combined system that will still provide an adequate level of service to city residents. The following specific issues were identified by the project director for resolution during the study:

$ VS. SERVICE

1.   What is the relationship between the cost savings (economic benefit) and level of service, especially for a modified system in which the existing operations are combined?

HABIT + CONVENIENCE

2.   What are the implications for collection crews and city customers if a modified system is implemented?

Information and Data

To inventory the two existing collection operations and the levels of service provided by each, the study area was divided into 46 "waste removal areas." Six of these areas, which were selected because they were representative of the various types of neighborhoods found throughout the city, were used to develop and verify collection data for both divisions. Detailed observations, measurements, and time-and-motion studies were conducted to determine waste quantities and trends, collection efficiency and production, volume-to-weight ratios, and costs versus revenue.

It was found that the routes of the waste removal division crews and the street cleaning division crews were covering the same geographic area. Crews from the waste removal division were collecting household wastes from containers in side-yard and rear-yard locations. Crews from the street cleaning division were collecting yard and garden debris stored loosely at curbside locations.

The waste removal division serviced 68,000 accounts. A typical collection crew consisted of three people using a two-axle rear-loaded packer truck. All crews were dispatched from the sanitary landfill site. Single- or multiple-container service was available, and residents could request collection twice per week. Nearly 90 percent of the residents requested that one or two containers be collected each week. A summary of the relevant data for the waste removal division was as follows:

Collection truck capacity, yd³	20
Average weight per account per week, lb	43.0
Average time per service, person-min	2.3
Average crew workday, h	4.6
Average number of accounts serviced per week	1,750
Average number of collectors	150
Average monthly collection cost per account, $	2.93
Average monthly revenue per account, $	2.70
Average haul distance to landfill, mi	3

The street cleaning division operated with three different crew configurations: (1) a one-person crew with a hand-loaded dump truck, (2) a three-person crew operating a front-scoop tractor and two dump trucks in shuttle, or (3) a one-person crew using a side-loader packer truck. Crews were dispatched from three locations on a fleet basis rather than on specific routes. Where streets had curbs and gutters, residents were permitted to place yard and garden debris loose in the street gutter. Where streets did not have curbs and gutters, residents were required to place the debris in containers.

The average weekly quantity of collected yard and garden debris ranged from approximately 25 lb per account during February to over 65 lb per account during November. Other data for the street cleaning division were as follows:

One-person hand-loaded dump truck or side loader	
Average time per service, person-min	2.2
Average number of accounts served per week	815
Three-person mechanically loaded dump truck	
Average time per service, person-min	2.8
Average number of accounts served per week	1,930
Average number of collectors	80
Average monthly collection cost for all types of crews, $	1.75
Average monthly revenue per account, $	1.52

Resolution

Working with the operating divisions allowed the consulting firm to develop a comprehensive plan. The resolution of the two cited management issues is described in the discussion that follows.

1. Evaluation of economic benefit of consolidation and development of alternatives. After the inventory and evaluation of the existing collection operations were done, it was apparent that any economic benefit from consolidation of the separate divisions was dependent, at least in part, on requiring that all yard and garden debris be placed in containers. Even without consolidated operations, it was evident that any improvement in the efficiency of separate collection operations was dependent on this requirement and on the relocation of household containers to side-yard or curbside locations. The basic collection programs were grouped into the following alternatives:

   a. Alternative A. Require that all yard and garden debris be bundled or placed in containers and consolidate the debris collection with household waste collection.

   b. Alternative B. Continue separate collection of household wastes and yard and garden debris, but require that yard and garden debris be bundled or placed in containers.

   c. Alternative C. Maintain the separate existing systems with minor equipment and service modifications.

   Alternative A. Several collection operation programs were evaluated for alternative A: (1) the use of two-person crews to collect both household wastes and yard and garden debris, (2) the use of curbside locations for wastes from household containers, and (3) the use of containers for yard and garden debris and their placement at curbside locations. Both disposable (plastic bags or boxes) and nondisposable containers for yard and garden debris were studied because of the significant collection efficiency advantages that would result. Collection crews could be reduced from three to two persons, and all the collection trucks would be the rear-loaded type. An exception to the requirement for placing yard and garden debris in containers would be made during seasonal periods of heavy leaf drop when residents would be allowed to rake leaves into gutters. The leaves would be collected by vacuum leaf collection units mounted to the front of street cleaning division dump trucks.

   Alternative B. In alternative B the two divisions would be combined. Two-person crews using rear-loaded trucks would collect household wastes from side-yard or curbside locations. Yard and garden debris would be collected by a one-person crew using a right-hand-drive, side-loading packer. The yard and garden debris would be placed in containers and set at the curbside. Significant collection efficiency advantages would be realized through the use of disposable containers.

   Alternative C. Without the use of containers for yard and garden debris, significant changes to the existing separate operations were prohibited. Minor modifications that could be implemented included the elimination of the twice-per-week collection of household wastes, regulation of container size, and equipment modifications to achieve higher payloads.

2. Presentation of recommended plan: The implications for collection crews and the public. The consulting firm recommended that alternative A be implemented. The most significant factors affecting implementation included reduction in personnel, capital outlay for new equipment, collection service changeover, and changes in rates for services. The recommended programs included the transition from 5-ton- to 10-ton-capacity, rear-loaded packers for consolidated waste collection. It should be noted that all four factors affecting implementation involved policy issues requiring political hearing and action. Consequently, the implementation of the programs of this alternative became extremely complex.

A 5-yr phased implementation program was developed to allow normal personnel attrition to keep pace with gradually reduced workforce requirements. In this manner, no laying off of personnel would be necessary. The number of street cleaning division personnel would be reduced from 80 to 40 under the recommended programs. Duties retained by this division would include leaf collection, street sweeping and flushing, and general shopping mall and street island cleanup. The number of personnel in the waste removal division would be decreased from 150 to 124 when the recommended programs were fully implemented. Two-person crews would replace the existing three-person crews. Waste removal division employees would collect household wastes at side-yard locations and yard and garden debris in containers at curbside locations. Use of disposable containers for yard and garden debris was recommended but was left optional.

With a 5-yr implementation program, new and old collection systems would be used simultaneously during the changeover period. This would reduce initial capital outlay but would result in higher annual costs and reduced efficiency when compared to a complete one-time changeover. The extended implementation program was recommended as a means of meeting political and social constraints. Average monthly cost for the combined two-container service provided once per week under alternative A was estimated to be $2.78 per residential account. This compared favorably to the average existing combined collection costs of $4.68 per residential account.

Despite the economic and technical advantages of the consulting firm's recommended collection system changes, the programs were rejected by the city council. The implementation plan became a "political football." City residents and the families of collection crews angrily complained at council hearings about the reductions in the level of service and the reduction in jobs required to improve collection efficiency. The requirement for placing yard and garden debris in containers became the central issue. As discussed previously, without this requirement, major modifications of the existing separate collection operations could not be implemented. Residents believed that with the mandatory containers, yards and gardens would not be kept up, trees would not be pruned, and in general the natural beauty of the city would deteriorate.

In this situation, the residents, at least the vocal ones, expressed a ''willingness to pay'' to maintain the existing separate systems and considered the level of service more important than cost reduction. Council members eventually rejected the consulting firm's recommendation and selected the minor modifications available within the existing dual collection operations (alternative C).

Comment   Rejection of collection system modifications was unfortunate but not wholly unexpected because of the number of complex issues involved. At least one issue, the level of service preferred by city residents, could have been assessed more clearly if a public-attitude survey had been conducted during the inventory phase of plan development. In addition, the plan might have been more successful if the issues had been clearly identified and presented separately to the political body. Elected officials act most positively when the issues are simple and easy to understand and when the solutions are positive and relatively straightforward.

CASE STUDY 14-2   *Evaluation of Solid Waste Collection Operations at an Airport*
(Adapted from Ref. 8)

Airport traffic and aircraft security become increasingly important as passenger loadings rise and capital investment in a single aircraft reaches a level exceeding $20 million. Major airports across the nation are developing into commercial complexes, and in many cases the airport locations are quite remote from urban areas. These factors have major impacts on the management of solid wastes at airports.

In this case, the public agency (called the Airport Authority) responsible for operations of a major international airport is concerned about litter on airfield pavements and the security of aircraft positioned alongside passenger loading ramps near solid waste storage containers. The Airport Authority, representing all airlines and associated activities that lease building space as airport tenants, has authorized a consulting firm to conduct a study (1) to identify the need for changes in existing collection equipment and (2) to develop a plan for a future collection system. Present collection services are being provided by a private collector.

Management Issues

1. What surface transportation equipment is available that can be used to eliminate the presence of large waste collection trucks near

aircraft and to reduce the number of trips collection trucks now make in the terminal area?

2. Should the Airport Authority operate the new collection equipment, or should the services of the private collector be continued?

Information and Data

During the inventory conducted for this study, some solid waste containers at the airport were observed along passenger loading piers within the influence zone of air blasts from parking aircraft. Paper and debris were seen in runway and taxiway areas. The airport management suspected that much of the debris problem was originating from waste collection operations that included the use of a front-loading compactor truck. Rear-loaded containers were being used in the terminal building. Still another type of container, a roll-on debris box, was being used at aircraft service centers and at a large aircraft maintenance base.

Administration and regulation of the solid waste collection system was practically nonexistent. There was no solid waste management ordinance, and there were no written regulations. Each airport tenant was contracting for the removal of wastes directly with the single private collector serving the airport complex. Collection rates were being negotiated separately with each tenant. The collector was removing all wastes from the airport and delivering them to a remote landfill. The following data were developed during the study:

Total solid waste quantity, tons/wk	300
Passenger loadings per year	18,000,000
Air cargo, tons/yr	436,000
Number of airport employees	10,000
Number of storage containers	
1 yd³	60
6 yd³	38
14 yd³	9
32 yd³ (equipped with self-contained compaction mechanism)	1
40 yd³ (equipped with self-contained compaction mechanism)	2
Collection frequency	3 to 4 times per day for some back-loaded containers; once per week for others
Number of collection trucks	
Back-loaded	1
Front-loaded	1
Tilt-frame	1 (part-time, also served nearby off-airport customers)

Resolution

In dealing with this local planning problem, it was determined that either the Airport Authority or the private collector could operate any new collection equipment, and that the Airport Authority would adopt ordinances and regulations regarding solid waste collection. The resolution of the problem involved three steps.

The first step was to determine whether the airport tenants could be grouped into sources that generated a common type of waste. During the inventory it was found that similar wastes were generated at four sources: passenger terminals, air freight areas, aircraft service centers, and aircraft maintenance base. It was then possible to continue the evaluation by investigating storage and collection equipment in large and more efficient sizes.

The second step was to identify potential collection and handling equipment. The equipment considered suitable for use in the potential system included the following "shopping list" of system components:

Item	Capacity
Rolling containers	2 yd^3
Stationary compactor with debris box	40 yd^3 (box)
Debris boxes	20 yd^3
Wheel-mounted containers	4 yd^3
Compactor on wheel-mounted containers	5 yd^3 (containers)
Front-loaded truck	30 yd^3
Towing tractor	2 containers
Transfer station and transport vehicles	50 tons/day
Wood shredder	7.5 tons/h
Wet pulping system	3,200 lb/h

The third step was to develop alternative systems by combining equipment components and administrative functions so that the Airport Authority would have more than one choice for system improvement. The following guidelines for the development of alternative systems were derived from the initial problem specification and statement of objectives for the study.

1. Flexibility. Because the airport is constantly changing and adapting to new tenants, flight equipment, and passenger service (especially around the terminal and passenger loading piers), the solid waste system must be flexible.

2. Noninterference with aircraft operations. The loading piers, aprons, taxiways, and runways are reserved for aircraft movement. The transfer of wastes from a storage container to a collection truck should be eliminated in those areas.

3. Onsite waste handling by tenants. Most major airlines have ground support vehicles that are used to handle the wastes entering the solid waste systems at least once. Maximum use of these in-house

vehicles to deliver wastes to central collection containers was considered desirable because it would eliminate outside collection vehicle traffic from congested areas.

As a result of the evaluation of all the foregoing information, two alternatives were developed for presentation to the Airport Authority:

Alternative A. The basis of this alternative was that 38 front-loading containers and 60 back-loading containers would be abandoned, and the wastes formerly stored in those containers would be consolidated into seven compactor debris boxes and five ordinary debris boxes. The private collector would continue to collect wastes, but the Airport Authority would expand its administrative functions. The expanded functions would include specifications for types and locations of containers as well as regulations of collection rates. The estimated annual cost was $5.20/ton. The capital required to purchase the replacement containers would be provided by private sources. Such funding normally allows rapid system change since bonding and public approval are not necessary.

Alternative B. The basis of this alternative was that all existing storage containers and collection vehicles would be abandoned. The wastes formerly collected in those containers would be stored in 5 wheeled containers of the compactor type and 48 wheeled containers of the uncompacted type. These containers are built to be towed at a maximum speed of 15 mi/h. Additional system equipment would include four electric-powered towing tractors and one transfer station with a capacity of 50 tons/day. The Airport Authority would have complete system control (both operational and administrative). Airport employees would operate the towing tractors, the transfer station, and the transport vehicles. The estimated annual cost was $5.60/ton.

This alternative involved a severe change from existing operations. Such change normally causes management and political upheaval. The Airport Authority would need to expand its staff to accomplish waste collection, the private collector would have to be terminated, and the airport tenants would have to accept new collection contracts. Although alternative B was the most responsive solution for compatibility with aircraft-related operations, it was the least compatible with the structure of the existing system.

In this case, the development of the two alternatives completed the planning study. A specific recommendation for implementation was not made because the Airport Authority and others involved wanted a longer time to consider the alternatives.

Comment This case is typical of the existing planning conditions in many communities in the United States where the collection of wastes is accomplished through a contract between individual waste generators and a private collector. In such situations, community controls related to equipment, rates, and service level are almost nonexistent, and competition among collectors is usually intense. Planning under these uncontrolled conditions is often terminated when the alternatives are developed. The alternatives are then used as a basis for suggesting system modifications. The actual changes that will be made are decided upon through negotiations between the parties directly involved.

14-4 DISCUSSION TOPICS

14-1.   You have just been elected to the city council of your community, and one of your campaign promises was to improve solid waste collection. Assume that the remaining council members agree with your position. On the basis of your knowledge of the community, prepare a problem specification to be used by the city public works department in dealing with the problem. ("Identify and solve problem" is an unacceptable problem specification.)

14-2.   A city contracts for the collection of its solid wastes with a private company, and the city administrator has reported that the collection cost is $12.30/ton. Because the city provides all other utilities and handles the customer billings, it also handles the billing for solid waste collection service. In addition, the city provides the company with rent-free land on which to locate parking and truck maintenance facilities. The city-operated landfill receives all the solid wastes hauled by the company without charge because city residents are already paying for the landfill in the general property tax. The city also operates a streetside rubbish pickup service as a part of its operations funded by the general tax. Is $12.30/ton the full system cost? If not, identify the expenditure categories and costs that the city administrator might have overlooked.

14-3.   Contact the agencies responsible for solid waste collection in your community. Identify and list the various types of collection service available to community residents. Have there been any changes in the level of residential collection service in the past 2 yr?

14-4.   A public agency has been providing solid waste collection service to a residential area with a population of 80,000. As an engineer employed by the public agency, you have been assigned the task of evaluating a proposal from a large national private waste collection company to take over residential collection service. The proposal was solicited by the community because the public agency collection trucks were worn out and the community could not raise capital to finance the acquisition of replacement units. What factors would you look for in the proposal? What form of contract is preferred for the given conditions?

14-5.   Refer to Discussion Topic 14-4. What changes would you recommend in existing administrative procedures and controls for solid waste collection? Discuss the economic impacts for the community. Assuming that collection rates are increased by this change, how would you prepare the community residents to accept higher rates?

14-6.   As collection costs increase, it becomes more important to measure the efficiency of a given set of equipment, crews, and service levels. List the expenditure categories that could be used to make a rapid estimate of costs and efficiency. Where are comparative cost and operating data available against which any community's operating costs and productivity might be measured?

14-7.   Labor is the highest-cost item in solid waste collection operations. Any changes to an existing collection system that are expected to result in significant cost savings must, therefore, reduce the number of employees collecting wastes. What are the management issues involved in reducing the number of employees? What are the differences in the issues involved for a public agency versus a private company?

14-8. A city is evaluating the benefits to be achieved by changing the level of waste collection service. The change would result in (1) reduced service to the customer, (2) an increase in the number of houses served daily by each crew, and (3) a cost savings that would put off a 10 percent collection rate increase for 1 yr. The existing rate for collection service (two containers per house) is $4.20/month. List the important variables for (1), (2), and (3) that must be evaluated when this change is planned. Discuss the probability for successful implementation of the change.

14-9. As a member of a regional solid waste management planning team, you are supervising a task group responsible for developing the collection programs. The regional area covered by the plan has a mixed collection system involving both regular collection agencies and informal haulers. Your planning director has given you a $20,000 budget and 5 months in which to develop collection programs that can be combined with other functional element programs to form plan alternatives. What special inventory concerns have an impact on your task group? Develop a detailed work plan for your task group that shows your budget allocation and a time sequence to complete your task.

14-10. Many solid waste collection agencies, both public and private, are asked to implement operations for the recovery of newspaper, cans (ferrous and aluminum), and glass. The point of separation of these wastes is the source of generation (home or business). The collection agencies complete the recovery step by keeping the wastes separated during collection operations and selling the recovered materials. Discuss the impact of this type of resource recovery program on waste collection operations from the standpoint of the management issues presented in this chapter.

14-11. Refer to Discussion Topic 14-10. Assume that you have implemented a community-wide newspaper separation and recovery program. Suddenly, the price paid for waste newsprint triples from $10/ton to $30/ton. Church groups, boy scouts, and other organizations recognize the possibility of raising funds by conducting paper drives. Would you resist this type of competitive activity? What are the implications for your newspaper recovery program?

14-5 REFERENCES

1. *A Collection Management Information System for Solid Waste Management (COLMIS)*, U.S. Environmental Protection Agency, Publication SW-57c, Washington, D.C., 1974.
2. American Public Works Association, Institute for Solid Wastes: "Solid Waste Collection Practice," 4th ed., American Public Works Association, Chicago, 1975.
3. Applied Management Sciences, Inc.: *The Private Sector in Solid Waste Management*, U.S. Environmental Protection Agency, Publication SW-51d.1, Washington, D.C., 1973.
4. Colonna, R. A. and C. McLaren: *Decision-makers Guide in Solid Waste Management*, U.S. Governmental Printing Office, Washington, D.C., 1974.
5. COR-MET: *Solid Waste Management Action Plan*, vol. 1, Metropolitan Service District, Portland, Oreg., 1974.
6. Little, H. R.: *Design Criteria for Solid Waste Management in Recreational Areas*, U.S. Environmental Protection Agency, Publication SW-91ts, Washington, D.C., 1972.
7. Management Technology Inc.: *Automobile Scrapping Processes and Needs for Maryland*, U.S. Department of Health, Education, and Welfare, Publication 2027, Washington, D.C., 1970.
8. Metcalf & Eddy, Inc.: *Analysis of Airport Solid Wastes and Collection Systems*, U.S. Environmental Protection Agency, Publication SW-48d, Washington, D.C., 1973.
9. New Pickup Practice Saves $6.3 Million in Two Years, *Solid Wastes Management*, vol. 16, no. 1, 1973.

10. *Opportunities for Improving Productivity in Solid Waste Collection,* National Commission on Productivity, Washington, D.C., 1974.

11. Powell, M. D., B. P. Fiedelman, and M. J. Roe: *Digest of Selected Local Solid Waste Management Ordinances,* U.S. Environmental Protection Agency, Publication SW-38c, Washington, D.C., 1972.

12. Resource Planning Associates: *Financing Methods for Solid Waste Facilities,* U.S. Environmental Protection Agency, NTIS Publication PB-234 612, Springfield, Va., 1974.

15

Transfer and Transport Options

The role of transfer and transport in a solid waste management system is to increase efficiency, particularly when haul distances to disposal sites or processing facilities are long. Although a transfer facility can include waste processing and recovery operations, it is primarily used to receive wastes from small vehicles and transfer them to larger vehicles. Planning an efficient solid waste management system involving long-distance transport requires the selection of cost-effective means of surface conveyance. Highway, railroad, and water transport are all potential means for moving solid wastes, as indicated in the case histories presented at the end of this chapter.

15-1 MANAGEMENT ISSUES AND CONCERNS

Transfer and transport equipment and facilities are expensive to construct and operate. They may be designed to handle large or small quantities of wastes. The most efficient operations are those used to handle the large quantities normally associated with subregional or regional areas. Small, rural transfer facilities, which are used to replace open dumps, are also costly to operate, but they are usually less expensive than small sanitary landfills. Complete technical descriptions of transfer and transport facilities may be found in Chap. 7.

The principal management issues that must be considered in any transfer and transport option are (1) identification of potential users of transfer stations, (2) economic feasibility, (3) technical feasibility, and (4) political and social constraints. Additional management issues arise in connection with maintaining system continuity.

Transfer Station Users

Transfer facilities can be designed to be used (1) by the general public (homeowners and others who might deliver waste in automobiles and pickup trucks), (2) by collection agencies (which may be public or private), or (3) by a combination of both. Although the larger facilities often are designed to serve both groups, their economic feasibility depends on collection-agency use because the agencies deliver the largest quantities of wastes.

When a regional agency is responsible for planning transfer facilities, a major issue that must be resolved is whether, and under what conditions, the separate agencies responsible for collection will use the transfer facilities. The problem is much less complex when the principal agency responsible for collection builds and operates the transfer station. In this situation, the agency must assess what additional use, if any, will be made by other agencies (both public and private).

Economic Feasibility

The economic feasibility of transfer operations is dependent on the ability of the responsible agency to find capital for facility construction and then to maintain operating revenues of a sufficient level to meet annual operating costs. The issue of capital funds for construction is not easily resolved because the benefits of transfer are not widely understood by capital-lending institutions. Contracts for facility use, which are often required as evidence of the ability of the agency responsible for the construction to repay capital, are not normally provided by transfer station users

The combination of institutional uncertainty and lack of operational contracts decreases the potential for private financing of transfer facilities. Consequently, economic feasibility is increased when funding is provided through public agencies. Raising capital through tax-exempt bonds has at least two benefits: (1) the cost of capital (interest) is lower, and (2) community investment in a transfer facility demonstrates to potential users that the community considers the facility a beneficial part of the waste management system. Such investment should therefore encourage collection agencies to use the transfer station.

The critical factor in maintaining a profitable transfer facility is a continuous waste loading rate, usually expressed in tons per day. Because the use of a transfer facility is optional for the operating agencies, the rate charged for unloading significantly affects the use of the station. For example, a station designed to process 1,000 tons/day must set an unloading rate that will provide revenues to offset the cost of handling that amount. If waste loadings drop below 1,000 tons/day, economic viability is threatened. One solution is to raise the unloading rate, but that might cause loadings to drop and thus further weaken the economic viability of the operation. Planners and decision-makers must evaluate total revenue for each increment of rate increase.

The traditional microeconomics price-and-demand model can be used to illustrate the cost-use relationship for a transfer facility [7]. The relationship is shown in Fig. 15-1. The demand curves (*A* and *B*) and the operating-cost curve intersect at point *a*, with the design capacity of 1,000 tons/day at a rate of $5/ton. If it is determined from actual operation that the quantity delivered is 900 tons/day, the total revenue would be $4,500 (900 tons/day multiplied by $5/ton). However, at an operating capacity of 900 tons/day, operating costs are $4,950/day, resulting in a net loss of $450/day (the new operating cost is shown as point *c*).

Can the transfer operator increase rates to make up for losses? The answer depends on the demand curve of the agencies that use the facilities. If there are numerous substitutes available (other competing transfer stations or landfills), the demand curve will be similar to curve *B*. Increasing the unloading rate would result in a larger revenue loss because

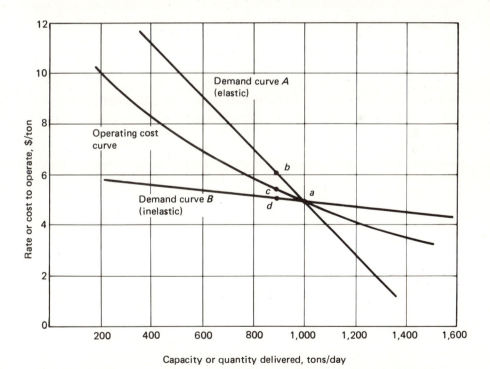

Capacity or quantity delivered, tons/day

Note: The curves shown have been made up for illustration purposes. The demand curve for a specific community must be derived from operating and economic data.

FIG. 15-1 Price and demand curves for a transfer station operation.

more agencies would switch to other facilities. The correct action would be to lower rates and attract additional agency use. However, if demand curve *A* is representative of the waste collection agency demand for transfer, then the unloading rate can be increased to $5.50 at point *c,* or even to $6.00 at point *b.* The point *c* rate would normally be charged by a public agency operating the facility; the point *b* rate might be charged by a private agency operating the facility.

In most transfer and transport planning situations, data for the empirical demand relationships are rarely available because of the lack of systematic and uniform record keeping by the agencies involved. When the agency responsible for the transfer operations is also the principal collection agency, it is extremely important that separate cost and management records be maintained for each operation. This is especially vital where additional transfer facilities are to be built in the future or an existing facility is to be enlarged.

Technical Feasibility

Management concerns with respect to technical feasibility are related to (1) the question of whether, and to what extent, the use of transfer facilities will improve the efficiency of the solid waste management system, (2) the operational characteristics of the transfer operation that may affect its implementation, and (3) the adaptability of transfer facilities for other uses, such as the processing and recovery of resources.

Improved Efficiency   In most operations, improved system efficiency ultimately is measured in terms of cost savings. Thus, while a transfer operation may be technically feasible from an engineering standpoint, its implementation will seldom be undertaken unless cost savings can be demonstrated. In some cases, transfer facilities have been provided as a means of reducing the time spent by the collectors on the job even though they are paid for a full day's work.

Operational Characteristics   The operational characteristics of a transfer facility will often have a significant impact on whether such facilities are used. For example, noise, traffic congestion, and aesthetic conditions may be unacceptable to local residents or commercial establishments. In most cases, such problems can be overcome or eliminated through proper design and site location.

Adaptability to Processing and Recovery   An idea that often receives widespread public support is hand separation and recovery of discarded appliances (salvage). In a technological society, this idea derives from a human desire to tinker with mechanical devices to extract any remaining usefulness. A transfer facility is a logical location for salvage although hand separation is inefficient for processing large quantities of wastes. The assumed inefficiencies have prevented the widespread development of salvage as a part of waste management operations [1]. In 1973, however, the cost of energy increased dramatically, and this development, combined with a significant downward turn in the economy, caused a reassessment of these inefficiencies. The economic cycle is not yet complete (as of 1976), but high unemployment and federal employment payments for public works projects have resulted in renewed interest in hand-separation activities.

Although current emphasis is on the use of mechanical separation of wastes, there are concerns that available processing and separation equipment will not be compatible with future recovery equipment and facilities. One way of overcoming these concerns is to build a basic transfer facility and to provide for future expansion to receive processing and separation equipment. The location of both operations on the same site is preferred because the transfer facility usually is a focal point for collection routes.

Political and Social Aspects

As noted previously, the economic viability of a transfer station can be maintained only through a continuous loading rate at reasonable charges. For this reason, politics can affect the economic viability of transfer stations, especially when the stations must serve large regions with multiple political jurisdictions. Social acceptability can be achieved only by overcoming the traditional image of a transfer facility as "just another dump."

Political Suitability   Regional considerations lead inescapably to issues associated with the interactions of numerous small political entities. Political forces in a community often will affect the use of transfer facilities. Factors such as the number of jobs, capital expenditures, transfer site location in relation to political jurisdictions, and facility sponsorship will enter into the decisions. These factors seem to apply independently, whether the agency that is planning or managing the transfer facility is public or private.

Political action has accomplished waste system improvements where it was apparent that the costs of improving the system to meet minimum public health or performance levels were prohibitive. Consider a small incorporated city that has no nearby land area suitable for landfill. Transfer to a regional landfill is a suitable answer. However, the city residents cannot afford to pay the capital costs necessary to construct and operate the transfer station. Under these conditions, a broader political body, such as a county board of supervisors or state legislature, can provide economic relief through general taxes, special land-use fees, or subsidies collected from waste management operations in other jurisdictions. Thus, the planner must be aware of the important political interactions in evaluating the feasibility of transfer.

Social Acceptability   The public recognizes transfer stations in the tradi-tional image of a dump. All social concerns related to dumps (smoke, rats, etc.) will be attached to the transfer facility. Only time and a history of properly operated facilities can be used to overcome this image.

It must also be recognized that sound management practices alone will not prevent a transfer facility from incurring periodic aesthetic problems, such as those shown in Fig. 15-2. In this facility, the well-managed conditions when the weather was fair were completely disrupted when high winds were blowing. The resolution of aesthetic problems at this facility included strategic fencing and immediate litter pickup when the winds subsided. The key is to find a site that will meet public or community standards as set by zoning laws. This is normally accomplished as part of the preparation of an environmental impact report.

To achieve social acceptance, it is often helpful to provide a transfer facility that can be used by the general public. This requires a station operation that permits unloading from private automobiles and pickup trucks 7 day/wk. There should be a large number of spaces for unloading along the storage pit, or a large unloading area should be provided if a storage pit is not used. Also, traffic should be routed so that large commercial collection trucks are not competing for unloading spaces with private automobiles. (Collection trucks can be intimidating in the confined spaces of a closed transfer building.)

System Continuity

Planning for transfer and transport must be coordinated with the other functional elements in the solid waste management system—onsite hand-ling, storage, and processing; collection; processing and recovery; and disposal

Onsite Handling, Storage, and Processing   The containers and other equip-ment used for storing wastes are often hauled to a transfer facility. Therefore, building clearances must be adequate to handle them, and adequate spaces must be allowed for unloading.

Collection   The interaction between collection and transfer is most impor-tant. In any waste management plan involving changes in either collection or transfer, the two should be considered as one continuous operation. This interaction, or interface, extends to operating hours and crew assign-ments. Transport equipment movement must be closely coordinated with collection routes. Collection truck deliveries peak for a 3- or 4-h period about mid-morning. The combined transfer station pit capacity (if a pit is used) and transport vehicle capacity must be adequate for these peak periods. In view of the need for this close management coordination, a waste collection agency is in many cases the most suitable agency to construct and operate a transfer facility.

(a)

(b)

FIG. 15-2 Aesthetics of a storage discharge transfer station operation: (a) Under fair-weather conditions; (b) under high-wind conditions.

Processing and Recovery   The increased public demand for more efficient use of resources is causing a significant change in how solid wastes are processed. Increasingly transfer facilities are being viewed as control points in the waste management system where wastes from widely dispersed sources of generation are accumulated, repackaged, and then sent on. Thus processing is becoming an extension of the repackaging step. It can be shown through conceptual materials-flow diagrams that all wastes can be processed and repackaged at a transfer facility so that no materials are "wasted." Unfortunately not all the necessary techniques of converting conceptual diagrams into economical operations are widely known or documented.

Disposal   If transport trucks deliver wastes to landfill disposal sites, it is important to have adequate roads and coordinated unloading equipment. Disposal site roads must be capable of supporting large transport trailers without causing damage to tires and vehicle undercarriages. Special operating equipment is often used at disposal sites to unload the transport trailers. Examples are crawler tractors pulling cables and nets from parked trailers (see Fig. 17-4) and hydraulically operated tipping ramps that lift, tip, and empty transfer trailers (see Fig. 7-13).

15-2   INVENTORY AND DATA ACCUMULATION

To develop a sound management program for transfer and transport, inventory information must be available on the following items: (1) sources, types, and quantities of wastes; (2) collection system operations including agency policies, collection equipment, and collection routes; (3) disposal sites; and (4) transfer and transport operations, including potential transfer station locations, transport means, routes, and equipment, and transfer station facilities. Because the inventory requirements will vary, depending on the level of the planning effort, this subject is considered first.

With respect to planning procedures, it is noted that transfer and transport are very closely interrelated to all the other functional elements, and the accumulation of inventory data (step 2 in the planning procedure) is often undertaken in conjunction with (not separately from) the evaluation and development of alternatives (step 3), because each step affects the other. In many cases, the feasibility of transfer and transport can be demonstrated on the basis of data available in the literature. In this situation inventory data are collected for the purpose of transfer station location, design, and implementation.

Requirements for Different Planning Levels

Planning at the local level is normally site-specific. That is, the need for transfer facilities will have been established, and the purpose of planning is to identify the site at which facilities are to be constructed. Inventory efforts at this planning level should be concentrated on the types of wastes, collection system characteristics, transportation routes, the transfer facility construction details, and the numbers and types of transport vehicles required. Environmental impact assessments will normally be prepared once the engineering evaluation of alternatives has been completed.

Planning at the regional level requires data of a broader nature. The study emphasis is shifted from specific details to the identification of regional problems or constraints that may necessitate the use of transfer facilities. Thus, data should be collected that can be used to establish the technical feasibility and cost-effectiveness of facility use by multiple agencies. The inventory need only be specific enough to identify total costs, average daily waste tonnages, and unit costs, such as dollars per ton. By developing unit values, a comparative base is established that can be used to demonstrate economic benefits. Planning for a regional transfer facility is illustrated in Case Study 15-2.

State and federal planning for transfer facilities is not well understood. Generally it is assumed that processing and recovery facilities will be a part of resource planning, but the proper methodology for arriving at the most feasible locations for these operations is not yet known.

Sources, Types, and Quantities of Wastes

Knowledge of the sources, types, and quantities of wastes collected is useful in establishing the size and type of both the transfer station and transport vehicles. Wastes that are not adaptable to unloading, processing, compacting, reloading, and transporting must be identified. Wastes that should be excluded from transfer are those with a high density, such as demolition materials. Eliminating these wastes will usually permit the construction of smaller, less expensive transfer facilities.

Information on the sources, types, and quantities of wastes can be obtained from (1) disposal site records, (2) records kept by the agencies responsible for the collection of wastes, (3) projections using generalized data (see Chap. 4), or (4) combinations of the first three methods. Often, it will be necessary to interpret the records kept by the agencies to extract the necessary information, especially with respect to specific details on sources and types of wastes.

Collection System Operation

Aspects of collection that are most closely involved with transfer and transport are (1) the policies of the agencies responsible for collection, (2) collection equipment, and (3) collection routes.

Agencies Responsible for Collection   As mentioned earlier, the agencies responsible for waste collection have the greatest influence on whether transfer facilities are to be used. For this reason, these agencies and their policies must be identified clearly. The purpose of the collection agency inventory is to establish the need for transfer facilities and the conditions for their use. To accomplish this, it will be necessary not only to obtain factual records but also to conduct detailed discussions with the people involved in the management of the collection and disposal operations, to elicit their views regarding the use of transfer operations and, if possible, to identify any specific concerns that they might have.

Collection Equipment   An inventory of types of collection equipment is useful in locating transfer facilities within the community. Information on the collection container size, vehicle size, compaction efficiency, and the age of equipment has the most important influence on selecting the type and location of transfer station. Data on the equipment used by agencies should be obtained by interviews. If the transfer facilities are to serve individuals hauling wastes in trailers and pickup trucks, the inventory must include a field check of disposal sites and monitoring of illegal dumps to establish the type of equipment individuals are using and to identify the routes used in delivering waste loads.

Collection Routes   The route a collection vehicle travels is made up of two parts: (1) the collection route itself, which includes all travel from generation source to generation source until the vehicle is loaded, and (2) the route traveled from that point where the vehicle was loaded to a processing or disposal site. Both parts are important and they should be identified in the inventory.

The best source of data for collection routes is the agency responsible for collection. Route books in which daily collections are noted according to streets and house addresses are used by many agencies, especially for residential waste collection. However, industries and some businesses often prefer an on-call service, which results in irregular pickup days and nonuniform routes. This irregular collection routing makes the inventory of industrial and commercial routes difficult and often inaccurate. Approximation of these routes, along with the identification of the centers of the industrial areas served, usually is sufficient.

Disposal Sites

Disposal site options must be established before a detailed inventory and evaluation are made of transfer and transport options. Normally, this will involve the identification of the operational characteristics and useful life of (1) all existing disposal sites and (2) all proposal sites. Alternative transportation means and routes should then be identified for each of the disposal sites.

A planning inventory for a transfer facility designed to be used only by the general public must include a spot check of existing disposal sites to monitor traffic and to identify the need and probable level of facility use. Inventory requirements for larger facilities designed to be used by collection agencies and the general public must include data on waste quantities, collection vehicles, and routes. This information can usually be obtained from surveys conducted at existing disposal sites.

Transfer and Transport Operations

Once information is available on the sources, types, and quantities of wastes, collection system operations, and disposal sites, the remaining inventory that must be completed deals with transfer and transport operations. The principal issues that must be considered include (1) transfer station location, (2) transport means, routes, and equipment, and (3) transfer facilities.

Transfer Station Location Generally, to achieve the greatest economic benefits, transfer facilities should be located close to the collection routes, and transport routes should be easily accessible. A logical and technically sound selection for a site location is at, or near, an existing or former waste system facility, such as an unusable incinerator or a closed landfill. Waste collection agencies will have already established routes by which to deliver materials to the site. In such cases, the greatest difficulties may be encountered in the conversion from one type of facility to another and in the accessibility of transport routes from the converted facility to the disposal site.

One of the best ways to identify potential transfer station locations is to drive through the community, following, wherever possible, the transportation routes now used by the drivers of the collection vehicles and to note all potentially suitable locations, such as empty lots and industrial buildings. Once a suitable number of sites have been identified, the economic feasibility with respect to haul costs can be assessed using VAM (Vogel's approximate method), as outlined in Chap. 7. The next step is to apply all the other criteria associated with a cost-effective analysis. Additional details concerning the design and location of transfer stations may be found in Chap. 7.

Transport Means, Routes, and Equipment There are three principal means of surface transport: highways, railroads, and water (see Chap. 7 and Refs. 3 and 5). The means used depends on economics and availability. Generally, highways are most likely to be available. Railroads are less likely to be available than highways but can be used to transport significantly greater loads of wastes per car. Waterways are the least likely to be available, but they can be used to transport the greatest loads (per barge or other vessel) of all. For purposes of planning transfer and transport facilities, the availability and the potential application of each of these means of transport must be identified.

Inventory data for highway transport must include speed limits, weight per axle and total load weight limits, and vehicle length limits, since the design of transport vehicles (see Chap. 7) and the transport routes that are selected normally will be governed by these limits. In addition, the expected traffic conditions on highway transportation routes must be delineated. For example, although the posted speed limit for large arterial highways may be 55 mi/h, the true speed of daily commuter traffic between 7 and 9 A.M. and 4 and 6 P.M. is typically closer to 30 mi/h. For this reason, it will be necessary to drive any proposed routes for several days during these hours so that average round-trip speeds can be estimated. Information on transportation route speed and the time used for a round trip from transfer station to disposal site is vital in determining the number of highway vehicles that must be purchased to move wastes.

The inventory of railroad facilities for the transfer of wastes must include finding a transfer station location and specifying the equipment for loading, identifying the types of vehicles to be used, describing the route over which wastes must move, and identifying unloading facilities at the disposal site. Any product that can be backhauled on the return trip to improve waste system economies must also be identified.

The inventory for a waterways transport system must include highway transportation routes over which wastes could be carried to the dock loading area, identification of available vessels and routes to move the wastes from loading docks to the disposal site, the degree of containerization necessary to place the wastes on the vessel, unloading requirements, and identification of any products that can be backhauled, such as sand and gravel.

Transfer Facilities In most situations where transfer facilities are being evaluated, there are no existing facilities. Where they do exist, a site visit should be made and the operator should be interviewed. The investigation of installations in other communities is also important. Factors that should be considered include similarities regarding climate, geography, transfer station function, reliability, and types of wastes transferred. Additional guidance can be derived from transfer station standards set by regulatory agencies, such as those of the Solid Waste Management Board of California [8]. The California standards are particularly helpful because they contain performance standards; nationwide data on transfer station performance are not yet available (as of 1976). The sections on health and safety, site controls, and equipment are most useful for inventory purposes.

15-3 CASE STUDIES IN TRANSFER AND TRANSPORT

Two case studies involving transfer and transport are presented to illustrate some of the management issues involved and the procedures used to develop solutions. The first case deals with the evaluation of highway, railroad, and water transport alternatives, and the second case deals with planning a countywide transfer system.

CASE STUDY 15-1 *Evaluation of Highway, Railroad, and Water Transport Alternatives for the Landfill Disposal of Solid Wastes at Remote Sites*

In the late 1960s San Francisco began to evaluate alternative disposal sites to replace the existing sanitary landfill that was scheduled to be closed. The existing landfill, located in a city just outside San Francisco city limits, was used to dispose of all municipal solid wastes. It was owned and operated by the two private waste collection agencies that provided collection service in the city. Replacement disposal sites could not be located in the city because there were no suitable land areas. Incineration had been evaluated in previous plans but was rejected because gas emissions could not meet local air pollution control regulations. Thus, the emphasis of the disposal site evaluation was to find locations that could be used as an immediate disposal solution (short-term planning).

Management Issues

In this case the city had to investigate the very broad issues of waste ownership and the impact of the use of disposal sites located in other political jurisdictions. The management issues involved in transfer and transport options would evolve after a disposal site had been selected. The principal issues were:

1. Which disposal site and transfer facility options would result in an economical and reliable system and would provide the greatest accessibility for waste collection agencies and other potential transfer station users?

2. What would be the long-term implications of the recommended short-term plan?

Information and Data

Traditionally, the city used its solid wastes as fill material for land reclamation along San Francisco Bay. As the land within the city was developed, these reclamation projects were forced outward beyond the city limits. Disposal sites were developed on privately owned land and were operated by private contractors. In the late 1960s the cities adjacent to San Francisco passed zoning laws that prevented the development of landfill disposal sites by private landowners. The owners and operators of the existing site were given a fixed time period in which to terminate all disposal operations.

San Francisco is an urban center with densely developed land areas. It is located on the tip of a peninsula with water boundaries on three sides and other city boundaries along its fourth side. Highway, railroad, and waterway transport systems terminate in the city. Waste collection is conducted under franchise by two private agencies which provide service to residential, commercial, and industrial customers. The predominant collection vehicle used is the 20-yd³ rear-loaded compactor.

In developing its short-term disposal system plan, the city worked very closely with the collection agencies on the disposal site inventory that led to the evaluation of transfer and disposal facilities. During inventory, the Western Pacific Railroad presented data and a proposal to resolve the disposal problem. In addition, several other proposals, from interested agencies, involving remote disposal sites were received. These disposal sites varied in purpose and size as follows: desert reclamation (the Western Pacific proposal), 350 mi²; canyon fill, 1,000 acres; soil stabilization, 10 mi²; park development, 540 acres. The transport distance to these sites ranged from 40 to 350 mi. After inventory, the disposal sites that were identified as having the best potential for immediate implementation were the desert reclamation project and the park development project. These sites were used in completing the transfer and transport evaluation and problem resolution.

The following additional data were developed:

Population	750,000
Municipal wastes generated, tons/day	2,000
Collection truck capacity, yd³	20
Average haul distance to existing landfill, mi	5
Hours of collection	
Residential	7 A.M.–6 P.M.
Commercial	11 P.M.–6 A.M.
Disposal rate at existing landfill, $/ton	2.40

Resolution

In resolving this case, each of the proposed disposal sites was evaluated with respect to transfer and transport. The sequence of resolution was as follows:

1. The evaluation of the economics and reliability of the proposed systems was made jointly by private industry and the city.

   Desert reclamation proposal. The Western Pacific Railroad performed the economic evaluation for its desert reclamation proposal after the city had set the following minimum reliability

requirements:
a. Unloading facilities at the transfer station must be compatible with the existing collection trucks.
b. No wastes were to be stored overnight at the transfer station.
c. Dust and odors must be controlled at the transfer station.
d. The transfer station must be located in reasonable proximity to existing city collection routes.

The most economical system included waterway and railroad transportation networks. System reliability was to be ensured through close coordination between the private collection agencies and Western Pacific in the equipment layout and facility design. All reliability requirements were to be met as follows:

a. A transfer station would be constructed on Western Pacific's property near its barge dock in the city. This location would provide a shorter haul for collection agencies and easy switching to Western Pacific's freight car barge.
b. Wastes would be accepted 24 h/day to meet the needs of collection agencies serving both residential and commercial customers.
c. An automatic scale would be used to weigh all loaded trucks, and the weights would be furnished for billing purposes.
d. To control dust, the station would be equipped with a water spray to be directed on the wastes as they are removed from collection vehicles.
e. Wastes would be discharged from collection vehicles directly into containers on flatcars and would be compacted by backhoe tampers. This dumping operation would be designed to accommodate without delay at least 100 collection trucks per hour.
f. The wastes would be sealed in watertight containers for shipment.
g. Flatcars with loaded containers would be moved by barge from the city to a railroad terminal located approximately 2 mi across the bay. Trains would then have a 15-h trip to the desert disposal site (a transport distance of about 350 mi).
h. At the disposal site a modified rubber-tired loader would lift the containers from the flatcars, empty them by tipping, and replace them on the flatcar.

The waterway and railroad transport proposal was an adaptation of the unit-train concept used by railroads for moving raw materials, usually from mines. In this case, three unit trains would be used as follows: one would be loading in the city, one would be on its way to the disposal site, and one would be emptying its wastes at the disposal site.

The cost of this transfer and disposal option was estimated to be $4.50/ton of wastes delivered to the transfer station (existing disposal costs were $2.40/ton). The city was prepared to pay higher costs for disposal at remote sites; therefore, negotiations were started between the city and Western Pacific. The ownership of wastes would be transferred to Western Pacific at its transfer station, and Western Pacific would be responsible for maintaining disposal operations.

Park development proposal. The second proposal involved a different transport option in that the disposal site was to be a land reclamation project located about 40 mi from the city, and access to the site would be by highway. The disposal site was identified during the inventory as a result of newspaper publicity on the need for importation of earth fill for the site, which was to be used as a regional park. Initial contact was established between the two cities, and then San Francisco conducted its own transfer facility and disposal site evaluation. The proposal for the use of solid wastes for park development was completed by the city, its two private collection agencies, and the private landfill operator.

The unique manner in which this proposal was developed eliminated the need for the city to document reliability standards, such as were used for the desert reclamation proposal. Standards were communicated directly to the potential operators and were included in the plan as it was prepared. The transfer station layout that was proposed is shown in Fig. 7-8.

The key operational features of the proposal were as follows:

a. The transfer station would be constructed on the old disposal site. The use of this location would allow collection agencies to maintain routes existing prior to site closure.

b. Wastes would be unloaded in a storage pit, broken and homogenized with a crawler tractor, and then pushed into trailers for transport.

c. Water spray would be used to control dust.

d. A backhoe would be used to tamp wastes in the trailers to achieve maximum allowable highway loads.

e. The transport vehicles would travel on an interstate highway to the disposal site.

f. Unloading at the disposal site would be by tipping, using hydraulically operated ramps (see Fig. 7-13).

g. There would be strict monitoring of leachate and gas movement.

The cost of this transfer and disposal option was estimated to be $6.50/ton of wastes delivered to the transfer station.

2. Both proposals offered the collection agencies good accessibility to the transfer station.

3. The long-term implications of the two proposals were different, however. The desert reclamation proposal offered a disposal capacity in excess of 600 yr, while the park development proposal offered disposal capacity of only 5 to 8 yr. This difference in capacity was

viewed as follows: (1) while desirable, the long-term disposal capacity of the desert reclamation site might tend to limit the consideration given to resource recovery; and (2) the limited capacity of park development would force the city to evaluate new options continuously, including resource recovery.

The city selected the park development proposal, but only after the costs for the desert reclamation proposal had escalated and it had come under attack by environmental groups and the cities through which the train would pass while moving to the disposal site. The Western Pacific financial proposal was very adversely affected by a severe shift upward in the interest rates charged by lending institutions that occurred during the evaluation of the proposals. As a result, the original transfer and disposal cost increased to $7/ton to $8/ton, making the two proposals about equal in terms of economics.

Comment It is worthwhile to note the sequence followed during the development of this disposal plan. A disposal site inventory and evaluation was the first step undertaken even before an inventory of transport routes and transfer station locations could be started. An inventory of the types of wastes and potential transfer station users was not necessary because of the direct involvement of the collection agencies in the evaluation of proposals. This sequence is typical when transfer facilities are being planned in a community with a limited number of collection agencies and a well-developed solid waste management system.

CASE STUDY 15-2 *Countywide Transfer Feasibility Study (Adapted from Ref. 4)*

A county located in a Western state is faced with a comprehensive solid waste management problem involving collection, transfer and transport, and disposal. The technical problems are compounded by significant differences in collection and disposal services provided by the two political jurisdictions, a city of 300,000 people and a countywide unincorporated area of 400,000 people. The increasing population in the unincorporated areas is creating increasing quantities of wastes, and the county's elected officials are concerned about the continued use of numerous small landfills. The county public works staff has been directed to evaluate existing landfills and to develop recommendations, including transfer and transport, to relieve any immediate disposal problems.

Management Issues

The basic question in this case is the economic feasibility of transfer and transport as compared with direct hauling to existing or new disposal sites by collection vehicles. The following major issues are derived from this basic question:

1. Is there a need for additional transfer stations or new disposal sites?

2. Who would use any new or expanded facilities?

3. What transfer station facilities and equipment are suitable for use in this situation?

4. What financing methods would result in reasonable waste disposal charges at either transfer stations or disposal sites?

Information and Data

Jurisdictional boundaries and the major existing facilities and those that were proposed to resolve this case are shown in Fig. 15-3. The public

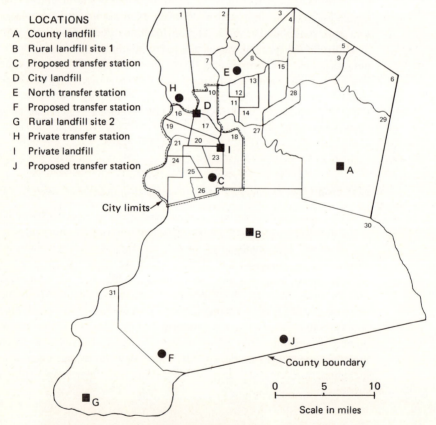

LOCATIONS

A   County landfill
B   Rural landfill site 1
C   Proposed transfer station
D   City landfill
E   North transfer station
F   Proposed transfer station
G   Rural landfill site 2
H   Private transfer station
I   Private landfill
J   Proposed transfer station

Note: Numbered areas indicate population centers
of approximately 25,000 persons each.

FIG. 15-3   Location map for Case Study 15-2.

agency responsible for countywide solid waste management was the board of supervisors. The city within the county had the authority to select and implement its own waste management system. Regulation by the county health agency was strong throughout the county, but administration was weak in the rural areas that constituted approximately 50 percent of the land area.

Four of the five existing disposal sites were under the administrative control of the board of supervisors. All sites except one (the city landfill) were available for public use. Through administrative action, the city opened its site only to city collection vehicles. The north transfer station was controlled by the board of supervisors.

The following determinations were made as a part of the inventory step: (1) the board of supervisors should administer the construction and operation of any new transfer facilities; (2) county, city, or private agencies all have the capability to operate any new transfer facilities; (3) the number of landfill sites should be kept to an absolute minimum; (4) the collection agencies have the choice of delivering wastes to any transfer station or disposal site; (5) the total charge to any county resident for waste transfer and/or disposal should not exceed $10/ton.

The inventory resulted in the identification of three important factors concerning the existing waste management system. First, the county collection service was unlimited, but city collection service was limited to two containers per week. Second, the city had not yet developed plans showing the final fill elevations for its disposal site; therefore, the capacity of the city landfill was unknown. Third, the rural landfill, site 2, was operating in violation of state regulations for sanitary landfills. In fact, all land within a 15-mi radius of that site had a shallow water table and therefore could not be used for sanitary landfill of residential wastes unless impermeable linings were installed to control leachate movement.

The following additional data were developed:

Total waste quantity, excluding agricultural wastes, tons/day	2,000
Total population	700,000
Number of waste collection agencies (one municipal, one county, and five private)	7
Public automobile and pickup truck transactions at the landfills, number/yr	
County landfill	120,000
City landfill	none
Rural landfill site 1	44,000
Rural landfill site 2	1,000
Private landfill	none
Capacity of collection vehicles, yd^3	
Residential collection	13–25
Commercial collection	20–35

Capacity available of existing disposal
sites, yd^3

County landfill (650 acres)	30,000,000
City landfill (80 acres)	unknown
Rural landfill site 1 (15 acres)	50,000
Rural landfill site 2 (10 acres)	unsuitable for landfill
Private landfill (30 acres)	60,000

Capacity of transfer facilities, tons/day

North transfer station	400
Private transfer station	100

Costs of transfer and disposal, $/ton

North transfer station	5.00 (includes disposal site charge)
Private transfer station	6.25
County landfill	1.50
City landfill	1.20
Rural landfill site 1	4.80
Rural landfill site 2	5.00
Private landfill	2.50

Resolution

The manner in which each issue was resolved is described in the following step-by-step procedure.

1. There was a need for new disposal sites or transfer stations because rural landfill site 1 would be filled within 2 yr and rural landfill site 2 would be closed within 1 yr (its operation was violating state regulations). It was concluded from a field investigation that no suitable substitute rural landfill site was available within 20 mi of the majority of the rural population. Thus, transfer was the apparent solution to the problem of a lack of nearby landfills for residents in the rural portions of the county (population centers 30 and 31 in Fig. 15-3). A new disposal site was also needed by the commercial collectors because of the limited capacity remaining in the private landfill (location I in Fig. 15-3). Finally, it was concluded from the large number of transactions at existing landfills (165,000) that the public wanted to haul to disposal facilities even though a well-organized collection service was available.

2. The greatest immediate need for new or expanded transfer and disposal facilities was in population centers 30 and 31 (see Fig. 15-3). The largest number of facility users—private haulers serving from residential areas—was from these areas. Additional private users were from population centers 20 through 26. Private collec-

tion agencies would deliver some commercial wastes from these same centers within approximately 1 yr when the private landfill (location I) was filled to capacity

A survey was conducted of the community collection agencies to estimate the extent of potential transfer station use and to develop standards for layout and equipment (see Chap. 7 for layout and equipment description). Because of the many transactions occurring at landfills, it was concluded that a significant number of haulers would use any new transfer station. However, transfer station use by collection trucks in the 25- to 35-yd^3 size was considered unlikely. Collection agencies had adapted to the long haul distance (15 mi one way) to the county landfill by operating large-capacity trucks in which it is economical to haul up to about 15 mi one way (see Chap. 7 for economic comparison procedures). Therefore, wastes hauled by large-capacity collection trucks would be excluded from transfer stations but would be delivered to any new landfills.

3. Alternative waste management systems were developed by combining equipment and financing options. Through this combined analysis, it was possible to present a plan to decision-makers from which they could determine the technical, social, and financial impact of transfer stations or disposal sites on the community.

It was found during the technical evaluation that only well-designed and controlled landfills could be placed in the land areas of population centers 30 and 31 because of high groundwater conditions. The high cost of engineering and installing leachate controls would result in sanitary landfill costs of $10/ton to $12/ton of wastes in that area of the county. All other land areas, except those included in population centers 6 and 29, were eliminated after investigation for use as disposal sites because of strong citizen objections voiced at environmental impact hearings.

The facilities and equipment considered suitable for use in alternative transfer station systems included:

a. Transfer station of 250-tons/day capacity

b. Transfer station of 50-tons/day capacity

c. Transfer station of 10-tons/day capacity

d. Optional equipment at all transfer stations included:
   Compaction devices on trailers
   Compaction device in station
   Push tractors and a dumping pit
   Direct dumping into the transport vehicle
   Hand separation for resources recovery

e. Optional equipment at two stations (50- and 10-ton capacity) included transport trailers and drop boxes.

A prime consideration in a system where equipment and facilities are added to similar existing facilities is to maintain system efficiency. Thus, it was important to consider the value of having future transfer equipment match present equipment.

The impetus for transfer station construction was public convenience. In developing alternatives it was important to establish a method of financing transfer operations in lightly populated areas where station costs could not be offset by direct charges to station users. In this case the alternatives were developed so that different site locations, different station sizes, and multiple financing methods could be evaluated.

Feasible transfer station site locations were identified in population centers 26 and 30. The financing methods considered included general taxes, land-use taxes, and gate fees at each transfer station and disposal site.

In completing the evaluation, 12 alternatives were identified. Expanding the number of sanitary landfills was not included among the alternatives because the high cost of landfills ($10/ton to $12/ton) exceeded the board of supervisors' imposed maximum cost of $10/ton. One transfer station alternative was recommended, and an in-depth financial evaluation was presented in the planning report.

4. The recommended alternative included a facilities program and a financing program.

   a. *Facilities program.* Two new transfer stations were recommended, one with a capacity of 250 tons/day and one with a capacity of 10 tons/day. Transfer trailers and personnel were to be shared. The larger station would have a storage pit (for public wastes) and a direct-discharge ramp (for compactor trucks and debris boxes). The smaller station would have only a direct discharge ramp.

   b. *Financing program.* In projecting transfer station use and costs, probable waste loading rates were estimated from discussions with personnel in waste collection agencies operating in the area as well as from the transaction records at the landfills. Both high and low probability loading rates were determined. These data allowed the decision-makers to perform their own risk analysis before choosing among the alternatives. The unit cost was estimated to be $8.25/ton. Since a transfer station was already operating in the north urban area and was charging an unloading rate of $5.00/ton, it was concluded that a charge of $8.25 would not be used in the new stations, and therefore a subsidy from another source would be needed. The potential subsidy sources were the general fund (taxes), the unincorporated area fund (taxes), solid waste collection revenue (rate payers), and solid waste disposal revenue (rate payers). The preferred tax subsidy was the general fund because about 75 percent of the station users would come from the incorporated city. The preferred rate-payer subsidy was disposal revenue because the greatest number of persons who would benefit from the transfer station were presently driving to the disposal sites.

The accepted alternative included constructing two stations, primarily to serve as convenience stations for the public. Hand separation of recoverable wastes was included. The rate for dumping was set at $5.50/ton, and any needed subsidy was to be derived from a surcharge on disposal fees. Rural landfill sites 1 and 2 were closed.

Comment The accepted alternative was not demonstrated to be cost-effective. However, the cost of energy used by thousands of vehicles being driven to a remote disposal site was not considered. Individual automobile use is difficult to evaluate and document accurately because many people do not identify their trip to the disposal site as added cost. Whether the planning agency should consider such costs is open to question. However, a potential energy saving was achieved when the transfer station alternative was selected and a subsidy applied. In taking this action the regional political entity (county) established a waste management policy that served overall community needs and protected public health without total reliance on economic justification for each separate facility.

15-4 DISCUSSION TOPICS

15-1. An incorporated city with a population of 2,000 has no nearby land area suitable for a sanitary landfill. Transfer to a regional landfill is a feasible solution. However, the city residents cannot pay the rate of $10/month charged by the regional agency for transfer and disposal at the regional landfill. What are the community conflicts here? If the regional agency chooses to subsidize the city transfer, what are the most common sources for subsidy? Assume that the regional agency already operates several large urban transfer stations.

15-2. Contact solid waste management agencies in your community and determine if any are presently using or planning transfer stations. If they are, what is the purpose of the transfer facility?

15-3. Assume that you are planning a regional solid waste management system. List the major concerns you would address in establishing whether transfer stations should be used. Briefly discuss each concern.

15-4. As the manager of a private solid waste collection company, you are considering the construction of a transfer station to improve the efficiency of your collections operations. Your collection costs have increased significantly because the nearby disposal site you owned has been filled and is now closed. Because of community resistance to new landfills, a public agency has acquired land and opened a landfill about 25 mi from your collection routes. Your company provides commercial collection service; the public agency that owns the landfill provides residential collection service. You believe that a transfer station would be profitable if the wastes hauled by the public agency could be routed through the station. What are the implications of public agency participation? Assume that your company would finance and own the station. Is this a high-risk venture? Explain your position.

15-5. Refer to Discussion Topic 15-4. Assume that the public agency is planning the conversion of its solid wastes to energy. Does this change the economic feasibility of your transfer station? If energy conversion is implemented, what processing operations might be installed at your transfer station?

15-6. List the important functional element interactions that must be identified in a planning inventory for transfer and transport. Why are these interactions important?

15-7. Highways, railroads, and waterways are the three principal means for transporting solid wastes to remote disposal or recovery facilities. Study the solid waste management system in your community, and rank the transport means from most feasible to least feasible. Discuss your reasons for the ranking, and list the regulations and controls that would be important to consider before selecting equipment, workforce, and routes for a transfer and transport operation.

15-5   REFERENCES

1. American Public Works Association: "Municipal Refuse Disposal," 3d ed., Public Administration Service, Chicago, 1970.
2. American Public Works Association, Institute for Solid Wastes: "Solid Waste Collection Practice," 4th ed., American Public Works Association, Chicago, 1975.
3. American Public Works Association: *Rail Transport of Solid Wastes,* U.S. Environmental Protection Agency, NTIS Publication PB-222-709, Springfield, Va., 1973.
4. Division of Solid Waste Management: *Feasibility Study, County-Wide Transfer Station,* Department of Public Works, Sacramento County, Sacramento, Calif., 1974.
5. Hegdahl, T. A.: *Solid Waste Transfer Stations: A State-of-the-Art Report on Systems Incorporating Highway Transportation,* U.S. Environmental Protection Agency, NTIS Publication PB-213-511, Springfield, Va., 1972.
6. Largest Transfer Point Now in Full Operation, *Solid Wastes Management,* vol. 14, no. 1, 1971.
7. Mansfield, E.: "Microeconomics Theory and Applications," Norton, New York, 1970.
8. *Minimum Standards for Solid Waste Handling and Disposal,* California Administrative Code, Title 14, Division 7, chapter 3, Sacramento, 1974.

16

Choices in Processing and in Materials and Energy Recovery

Processing equipment and operations described in Chap. 8 are used for two purposes in solid waste management systems: to improve the efficiency of the system and to recover materials, conversion products, and energy from solid wastes. While the use of processing equipment to improve system efficiency is reasonably well documented, its use in the recovery of materials and energy is not well documented. The planning and management aspects associated with both these applications are discussed in this chapter.

Because of the uncertainties associated with materials and energy recovery, presentation in this chapter is slightly different from that in the preceding management chapters: (1) management issues and concerns are extended beyond traditional areas to include materials specifications and markets and market sensitivity; (2) inventory and data accumulation are also expanded beyond the traditional areas associated with waste systems; and (3) in-depth evaluation details are not presented in the case studies because many of the necessary relationships regarding recovery are still unknown. System continuity is discussed only briefly because most of the relevant details have been considered in Chaps. 12 through 15.

16-1   MANAGEMENT ISSUES AND CONCERNS

Processing as part of a solid waste management system should be chosen only if it provides some measurable benefit (usually economic, although environmental and social benefits are also important). Usually the greatest potential benefit is from the sale of recovered materials.

The principal issues and concerns with respect to processing and recovery are as follows: (1) to determine the appropriate time and place to use processing, (2) to establish priorities, (3) to recognize that most of the available recovery technology has not been proved under full-scale operation, (4) to identify markets for the sale of recovered materials, (5) to determine materials specifications and the extent of the effort that should be made by the solid waste management agency to attempt to meet the specifications, and (6) to assess the impact of market stability on the waste management system.

When and Where to Use Processing

Processing should be used where it will benefit the solid waste management system. There are two general areas: (1) within the other functional elements to improve their operations and (2) in a completely separate facility to assist in materials and energy recovery. The principal applications within functional elements are summarized in Table 16-1. With the exception of the uses listed in Table 16-1, the benefits to be derived from the application of further processing options in the various functional elements are limited. Processing adds another set of costs without providing offsetting cost reductions. Some exceptions do exist, but typically only in communities with restricted disposal options. For this reason, the discussions of issues and concerns in the remaining sections focus on the economically attractive resource recovery options.

Priorities

Community, national, and international resource needs have caused uncertainties and a sense of urgency about changing waste management systems. Reuse opportunities are being suggested by homeowners, charitable groups, and environmental action groups. In the absence of action by solid waste agencies, many groups have started their own recycling projects. This intensive activity has resulted in the identification of numerous waste processing and recovery opportunities. Yet in most instances these groups have spent little time considering the energy requirements and costs associated with the implementation of the various recycling programs suggested. Often, recycling programs may cost more and utilize more energy compared with other disposal means. For this reason it is important to establish community priorities with respect to processing and materials recovery.

TABLE 16-1 APPLICATION OF UNIT OPERATIONS AND PROCESSES TO IMPROVE EFFICIENCY OF FUNCTIONAL ELEMENTS

Unit operation or process	Functional element	Remarks
Separation	Onsite storage	Homeowner hand separation
	Transfer/transport	Hand sorting; magnetic belt or drum; air classifier
	Disposal	Hand salvage; portable magnetic separation machine
Mechanical size reduction	Onsite storage	Paper shredder; wet pulper; wood chipper
	Transfer/transport	Rasper; shredder; hammer mill
	Disposal	Solid waste shredder
Mechanical volume reduction	Onsite storage	Compactors; hydropulper with dewatering by mechanical press or vacuum filter
	Collection	Truck-mounted packers
	Transfer/transport	Hydraulic rams (fixed and portable); high-compression hydraulic baling
	Disposal	Crawler tractors; wheeled compactors
Chemical and biological volume reduction	Onsite storage	Building incinerator; open burning; composting
	Transfer/transport	Incinerator; composting
	Disposal	Anaerobic decomposition in landfill; composting

Typically priority questions that must be answered are the following: What materials should be recovered? When should a program be undertaken and how large should it be? Should community economic resources be devoted to the program? In establishing priorities it is especially important that they be based on the social and economic circumstances of the community as viewed jointly by the planner, decision-maker, politician, and the people of the community.

Unproved Technology

Technology concerns with respect to materials and energy recovery are centered on the required equipment. As noted in Chap. 9, it has been found that the front-end systems are less reliable than most rear-end systems. Often, duplicate front-end systems may be required because of "down" time for repairs or maintenance. Uncertainty also exists concerning the disposition of the by-products of conversion (gases and residues) [10, 11, 12]. Not meeting stack discharge requirements after energy conversion has caused management problems in some of the large power and heating facilities.

Technological problems will be resolved by more testing and modifications of processing and recovery equipment. Extended time periods and large capital investments will be required. Meanwhile, the solid wastes system manager must be concerned about these problems because they affect the need for multiple backup systems, including disposal options.

Identification of Markets

Traditionally, identification of materials recovery opportunities has been left to private industry; planning and coordinating activity by public agencies has been missing. These past practices are undergoing a slow but significant change, caused partially by resource needs and partially by governmental regulation. A good perspective from which to consider the concerns for resource recovery is the market that will use recovered materials. There are three broad categories of markets: (1) raw materials for industry, (2) raw materials for the production of energy or fuel, and (3) land reclamation. It is possible to group all recovery (with appropriate processing) facilities so that wastes will be used by one of these markets. Some of the markets and common recovery uses are listed in Table 16-2. Market stability is considered later in this section.

Raw Materials for Industry   The specifications for waste materials vary with the market, but usually they are quite restrictive (see Table 9-1). Unfortunately, these specifications are not generally known by planners and

TABLE 16-2   MARKETS AND TYPICAL USES FOR MATERIALS RECOVERED FROM SOLID WASTES

Market category	Materials	Typical uses
Raw materials for industry	Newspaper	Newsprint stock
	Corrugated cardboard	Fiberboard and roofing material
	Ferrous metal	Reinforcing bars
	Rubber tires	Sandals, paving materials, and oil
	Oil	Rerefined oil and road surfacing
	Broken glass	New glass and paving materials
	Beverage bottles	Refilled beverage containers
	Textiles	Wiping rags
	Aluminum cans	New aluminum products
	Organic wastes	Chemicals
Raw materials for energy or fuel production	Organic wastes	Steam production
	Organic wastes	Fuel for burning waste sludges
	Organic wastes	Methanol, methane
	Organic wastes	Fuel oil
Land reclamation	Organic wastes	Compost
	All solid wastes	Fill material for mined lands
	Demolition wastes	Fill materials for eroded lands
	All solid wastes	Fill material for creation of recreational lands

decision-makers in the field of waste management because their objective has been the disposal of wastes and not the marketing of materials. However, if materials are to be recovered, system managers must develop an understanding of materials specifications.

The specific details of product purity, density, and shipped conditions must be worked out with each potential buyer. An example of private industry actions in the primary metals industry is shown in Fig. 16-1. The relationship of the waste system manager to the metals cycle is also identified. As shown, obsolete scrap and its management are only a small increment of the cycle. In this situation the market buyer is the scrap processor who also sets materials specifications.

It is beneficial to develop a range of materials specifications and recovered waste prices jointly with the buyer. The waste system manager then has the data necessary to evaluate the added processing costs to achieve higher materials quality versus the higher market price for the quality product. By dealing in a price-specification range, the market buyer and the waste system manager will be negotiating to establish the most

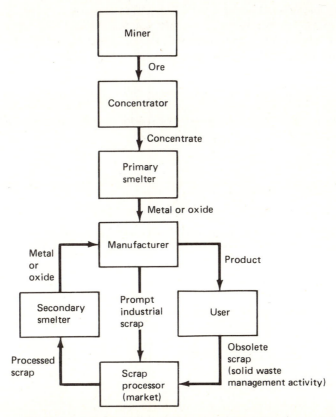

FIG. 16-1 Primary metals-processing cycle and its relationship to solid waste management.

beneficial location for the processing system. Additional relationships are identified in Case Study 16-3 at the end of this chapter.

Raw Materials for Energy or Fuel Production   Energy is derived from solid wastes in two forms: (1) directly, by burning as a fuel to produce steam, and (2) indirectly, through the conversion of wastes to fuel (oil, gas) or fuel pellets that can be stored for subsequent use. The specifications for direct use are more widely known than those for indirect use involving fuel products. This situation has developed because direct-heat processes (incineration) have been used on heterogeneous wastes for many years, but the creation of fuel from wastes is a recent development. When solid wastes are to be used for energy or fuel production, the waste system manager must arrange for the sale of recovered inorganic wastes separately from an energy or fuel recovery system. Related management issues are considered further in Case Study 16-4.

Land Reclamation   Applying waste materials to land is the oldest and most used technology in solid waste management. In recent years the emphasis of management has changed from using the land for disposal to using the wastes to reclaim otherwise useless lands. Through the conversion of dumps to sanitary landfills, the safe and efficient use of wastes has been demonstrated. This upgrading has made it possible for communities to plan land reclamation projects that use solid wastes without concern for the development of health problems. An example of such a project is shown in Fig. 16-2. Because the concerns related to land reclamation are similar to those for sanitary landfill disposal, the reader is referred to Chap. 17 and to Ref. 2.

Materials Specifications

The organizations that contract for purchase of recovered materials and energy will, in most cases, use the converted product in manufacturing a new product for resale (see Fig. 16-1). Stringent materials specifications are usually essential in maintaining the required quality control in manufacturing. Thus, major questions arise at the interface of the waste management system and the product manufacturing system. Should the manager of a solid waste system purchase the necessary equipment and pay the required salaries to improve the quality of wastes and thereby meet manufacturing specifications? Or should the manufacturer purchase raw wastes and pay the costs of improving quality? The answers must come from communication and negotiation between the waste system manager and the potential purchaser. An illustration of the evaluation procedures to use in such negotiations is presented in Case Study 16-2 at the end of this chapter. A more complex evaluation and negotiation procedure was used to resolve the issues in a New Orleans case, as discussed in Ref. 8.

FIG. 16-2
Land reclamation
in a residential
community. (*a*)
Ravine before fill-
ing. (*b*) Landfill-
ing operation in
progress. (*c*)
Completed land
reclamation pro-
ject. (County San-
itation Districts of
Los Angeles
County.)

(*a*)

(*b*)

(*c*)

(a)

(b)

FIG. 16-3   Impact of market stability on the sale of recovered materials. (a) Stockpiled baled cardboard. (b) Stockpiled ferrous metal.

Market Stability

The resolution of all types of market problems involves long-term contracts for recovered materials and energy. The impact of unstable markets is shown pictorially in Fig. 16-3. The large stockpiles in both cases were caused by termination of existing market contracts. In most cases, the solid waste system manager cannot create a market. Therefore, market outlets must be found either directly with materials and energy users or indirectly through brokers or wholesalers. Of specific concern under these conditions is the manner in which the materials markets fluctuate and the impact of these fluctuations on the solid waste management system. Market demands, legal aspects, resource shortages, and political aspects must all be considered in assessing market stability.

Market Demands   Potential markets for recovered materials are most readily available in the salvage industry and the utility fuel industry. Another growing market that represents a third potential demand is the sale of recovered materials directly to industry for use as raw materials.

The salvage market is highly variable. Costs can increase or decrease dramatically over a short time period. For example, consider the market prices for ferrous scrap from 1972 through 1975, as shown in Fig. 16-4 [1]. In late 1974, the price for high quality scrap was about $100/ton; by March 1975, it had fallen to about $55/ton. Utility fuels do not follow such a rapid change, but they are subject to competitive pricing from other fuels. Demand for virgin raw materials is increasing, while known reserves are decreasing. This action could force prices for virgin materials up and increase the potential for competitive uses from waste materials.

Legal Aspects   The traditional views of raw materials and wastes have resulted in industry practices and the passage of regulatory laws that lower

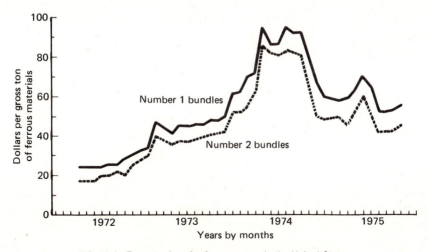

FIG. 16-4 Recent prices for ferrous scrap in the United States.

the potential of recycling wastes. Examples include (1) tariffs that prohibit the importation of products containing high amounts of recovered wastes, (2) differential freight rates that are higher for hauling recovered or salvaged products, (3) taxes that favor the mining of virgin materials, and (4) monetary incentives (depletion allowances, etc.) that favor new virgin materials. As resources become more scarce, it is anticipated that both industry practices and regulatory laws will be modified.

Resource Shortage   The demand for the raw materials available in wastes will increase as natural supplies decrease. As a result, the potential for recycling is affected in the following ways: (1) the publicity of shortages will cause changes in public attitudes, and (2) the extractive or mining industries will expand their processing operations to include waste recycling. Shortages can be of local origin, nationwide, or worldwide. A resource that normally is available close to a community may become depleted. Continued use of that resource from more distant sources will result in higher transportation costs to that community. The potential for solid wastes recycling may increase as a result. Additional discussions of recovery potential can be found in Refs. 4, 9, 10, and 11.

Political Aspects   The political nature of solid waste management has been identified and discussed in numerous sections of this text. Its importance to recycling is manifested in the vested interests of organizations operating salvage systems, waste collection and disposal systems, and raw-material mining enterprises. Both public and private organizations are involved. An example of a public agency with a vested interest is a community that has a large capital investment in an incinerator. In most cases, the potential of recycling is reduced to only heat-related functions because public officials are reluctant to abandon a tax-funded incinerator. Vested interests of private organizations are focused more on profitability. An example is a privately owned and operated waste collection system. The potential of recovery (homeowner separation of paper, glass, etc.) that would aid in recycling is lowered because the operator is reluctant to change profitable collection operations. The political influence of both public and private agencies can thus have a strong impact on markets and market prices.

System Continuity

System continuity with respect to the functional elements of onsite handling, storage and processing, collection, and transfer and transport has been considered in the previous three chapters. The most feasible processes for waste management systems are those that reduce the volume and weight of wastes (see Table 16-1). Communities where land is either scarce or of high value will derive the greatest benefit from volume reduction.

Because most of the technology used for materials and energy recovery is now in the elementary stages of development, system continuity is not well defined and therefore is not considered further in this discussion. Waste management system continuity and resource recovery systems are considered further in Ref. 3.

16-2 INVENTORY AND DATA ACCUMULATION

Management must develop the data and information needed to help the decision-maker and the public understand the role of processing and recovery in solid waste management. The planning is complex because most communities do not have a shortage of resources (hence local markets) that would lead to local issues and concerns. The concern is typically at the state, national, or international level; yet the planner must develop inventories and data that will result in programs for solutions at local or regional levels. It is extremely important that this relationship be understood so that management does not attempt to force projects on a protesting community. Exceptions to the state, national, and international market situation normally are found in energy conversion markets and land reclamation markets.

The manager or planner investigating resource recovery for a community must first assess recovery potential with respect to materials and markets. If no markets exist for recovered wastes, then that phase of the inventory can stop. The next step is to determine what potential exists for marketing materials subject to the constraints of the community. Four key areas should be investigated for both processing and recovery: (1) local community needs; (2) state, national, and international needs; (3) economics; and (4) community priorities.

Local Community Needs

The benefits of processing in the solid waste management system are identified most directly and accurately through inventory of the functional element affected. The best technique is that used for the planning of each element. Refer to Chaps. 13, 14, 15, and 17 for detailed inventory procedures.

The most logical place to begin the inventory necessary to establish the need for processing and recovery systems is within the community. The resource needs of a community are often vague and hard to define, making that inventory difficult. Good sources of data are those industries that presently use wastes as a component in manufacturing. The chamber of commerce will normally supply business statistics on local industry. Combining that information with published data from associations whose members are users of waste materials should result in a list of community contacts who may be able to define local market potentials for raw materials. Additional markets outside the community can be identified by reviewing available research literature and industry publications [1, 7].

In inventorying energy users who might become a part of the waste management system, it is best to concentrate on the larger operations because of the economy of scale. Urban commercial centers with combined heating and cooling piping networks and electric utilities are possible candidates found in most communities. If any exist, the best opportunities for the conversion of wastes to energy normally are found within the community. The economics of conversion in remote plants are affected adversely by materials preparation and transportation costs.

Land reclamation is closely tied to the surrounding community. A prime candidate for reclamation is a local quarry or gravel pit (see Fig. 16-2). A good source of information is the local mineral aggregate industry. The conditioning of depleted soils is another type of land reclamation.

The best methods of inventory involve interviews with local public agencies, such as an agricultural department or a park district, and driving through or flying over the area in question.

State, National, and International Needs

Currently, there are only a few solid waste management systems with state, national, or international political boundaries. Therefore, in this section discussions of inventory for processing are restricted to its use with recovery. The progression of inventory for nonlocal markets should be from state to national and then to international levels. The strongest markets at these levels will be for raw materials. Some current examples of nonlocal markets are (1) the shipment of recovered ferrous cans from northern California to copper mines in Nevada for use in precipitating copper, (2) the shipment of recovered ferrous metal from Portland, Oregon, to Japan, and (3) the shipment of recovered newspapers from Salt Lake City to Los Angeles.

The critical participants in all nonlocal recovery markets are private industries and the supportive network of distribution companies (paper brokers, scrap dealers, etc.). Generally, a waste system manager should deal directly with the buying industry. However, brokers should be used when they have an established processing system that meets market specifications and when they have transport services already operating.

Economics

Economic value is an important criterion for evaluating processing and recovery opportunities. Social value and political advantage are less important but must also be considered. The economic value of operating a waste processing facility is derived by developing cost comparisons with existing system operations and then projecting future processing benefits. Existing costs can be obtained from waste management agencies in the community. Future processing benefits can be developed by setting processing costs and estimating market values for processed wastes (this procedure is illustrated in Case Study 16-2). An inventory and analysis procedure for home source separation is discussed in Ref. 5.

The sources of market price data are the salvage or secondary materials industries. If the community does not have these industries, useful pricing data can be obtained from trade publications such as *Official Board Markets* [7] and *American Metal Market* [1]. Although local and national markets should be surveyed first, foreign buyers often will pay the highest prices, especially if their communities are located near sea shipping ports.

As a general rule, the more the wastes are processed, the higher the price paid for the recovered materials. In fact, the specifications for the recovered materials, which are governed by the subsequent reuse, should be established during the inventory.

16-3 CASE STUDIES IN PROCESSING AND RECOVERY

Because there are numerous choices in processing and in materials and energy recovery available to a community, the following cases are presented to illustrate typical procedures for the selection of processing equipment and the evaluation of alternatives in materials recovery and energy recovery.

CASE STUDY 16-1 *Selection of a Mechanical Volume-Reduction Process*

The disposal site for a medium-sized community in the Midwest is located in flat terrain. Earth cover materials must be purchased and imported from a distant quarry. Land areas suitable for sanitary landfill are scarce. The city's engineering staff has been directed to investigate and evaluate volume-reduction processes that might be used to extend the life of the existing landfill. Incineration is not feasible because of air pollution restrictions.

Management Issues
Volume reduction can be undertaken at numerous points within the solid waste management system, and in this case all the possibilities had to be evaluated. The main issues were:

1. What are the possible functional elements within which volume-reduction processes might be applied?

2. What are the impacts of each choice?

3. What is the cost per ton for landfill using the selected process?

Information and Data
The community had just completed a disposal site survey in which it was determined that any new disposal site must have an impermeable

lining along its bottom and sides. The cost of landfilling in such a site was estimated to be $6.50/ton of wastes. The greatest costs were for the liner material and for excavation and lining of the entire site before landfilling. During the survey it was found that sufficient soil for cover material would be available onsite at the existing landfill if the covering sequence were modified to require no daily cover and only a final 2-ft layer over the completed landfill. The local regulatory agency would approve sanitary landfill operations without daily cover if the wastes were deposited in a highly compacted and stable bale.

The following data were developed during the investigation:

Average waste quantity collected, tons/day	500
Maximum allowable depth of landfill, ft	28
Average density of solid wastes in the landfill, lb/yd^3	700
Compaction densities, lb/yd^3	
Home compactor	600
Collection truck	500
High-compression baler	1,800
Wheeled compactor at landfill	900
Cost, $/ton	
Imported earth cover	2.20
Landfilling	3.10
Baling	3.80
Wheeled compactor	0.50

Resolution

The management issues were resolved as follows:

1. The functional elements in which volume reduction was considered feasible were (1) transfer and transport and (2) disposal (see Table 16-1). The element of onsite handling, storage, and processing was eliminated because density achieved with home compactors (600 lb/yd^3) does not exceed the density achieved by the existing landfill procedures (700 lb/yd^3). Moreover, wastes from outside the home would not be compacted. The element of collection was eliminated for the same reason (the density obtainable in a packer truck is only 500 lb/yd^3).

2. The feasible alternative in connection with transfer and transport involved the use of a high-compression baler, and that in connection with disposal involved the use of a special wheeled compactor at the landfill.

   a. High-compression baler. The density factor was calculated as follows:

   $$\text{Density factor} = \frac{\text{bale density}}{\text{landfill density}} = \frac{1,800}{700} = 2.57$$

Only compressible wastes can be processed in a baler. Bales can be landfilled without pushing and compacting. Cover material requirements are reduced significantly. Capital costs are high compared to those for the wheeled-compactor process.

b. Wheeled compactor. The density factor was calculated as follows:

$$\text{Density factor} = \frac{\text{compactor landfill density}}{\text{present landfill density}} = \frac{900}{700} = 1.29$$

All types of wastes can be processed with the wheeled compactor. Earth cover is required daily. Both capital and operating costs are significantly lower than for the process using the high-compression baler.

The high-compression-baler process was selected in spite of the higher costs, which were justified because the life of the landfill would be more than doubled. Ultimately it was decided that the baler should be installed as a part of a transfer facility. All noncompressible wastes were to be hauled directly to the landfill.

3. Assuming that the essential cover material was available onsite, the cost per ton for landfilling baled wastes was estimated as follows:

$$\text{Cost of landfilling} = \$3.80/\text{ton} + (\$3.10/\text{ton} - \$2.20/\text{ton})$$
$$= \$4.70/\text{ton}$$

Comment   The final cost of landfilling ($4.70/ton) was less than the cost of acquiring and operating a new disposal site, which would require lining ($6.50/ton). In most cases the use of a transfer station will reduce collection system costs. If not, then a portion of the transfer station cost should be charged against the disposal operation.

CASE STUDY 16-2   *Selection of a Processing System and a Market for Newspaper Recovery*

A community is starting a newspaper recovery system. The necessary equipment for collecting homeowner-separated newspaper has been selected (newspaper racks have been mounted on collection trucks), and a contract for sale of the recovered newspaper has been advertised. The separation by residents is voluntary; residents are asked to participate by setting out bundled newspaper weekly along with their waste containers. The paper must then be sorted by the collector to be prepared for sale. Sorting involves removal of strings, boxes, and objectionable paper. One buyer responded with the following bids:

Alternative 1: $10/ton for unsorted newspaper

Alternative 2: $15/ton for sorted newspaper

Management Issues

The principal management issues in this situation were as follows:

1. What is the break-even participation rate for alternative 1? For alternative 2?

2. Which alternative is preferred if the community is willing to subsidize a newspaper recovery system to a maximum amount of $75,000/yr? Under what conditions would such a subsidy be used?

3. Should the community's collectors continue to sort the newspaper or should the successful bidder be allowed to sort?

4. How much paper might be lost to theft during the time it is on the curb awaiting collection?

Information and Data

A 6-month pilot program was conducted in an attempt to measure resident participation. The program included one collection per month by separate newspaper trucks. The following data were available from the pilot program:

Cost to community to collect newspaper, $/yr	116,000
Cost for hand-sorting newspaper, $/yr	24,000
Homeowner participation rate, percent/wk	
Range	10–20
Average	14
Total quantity of newspaper available, tons/yr	61,000
Paper lost to theft, percent of that set out over	
the full 6-month pilot program	50

Resolution

Resolution of this case involved the following steps:

1. Net profit curves were developed for both processing options, as shown in Fig. 16-5. The curves were derived in two steps. First, the percent of participation required to develop a zero-profit point was calculated for each alternative. Zero profit occurs when cost equals revenue, as given by the following expression.

$$\frac{\text{Cost}}{\text{Yr}} = \left(\frac{\text{revenue}}{\text{ton}}\right)\left(\frac{\text{available tons}}{\text{yr}}\right)\left(\frac{\text{percent participation}}{100}\right)$$

The expression for percent participation that results in zero profit is

$$\text{Percent participation} = \frac{(\text{cost/yr})(100)}{(\text{revenue/ton})(\text{available tons/yr})}$$

For alternative 1,

$$\text{Percent participation} = \frac{(\$116,000/\text{yr})(100)}{(\$10/\text{ton})(61,000 \text{ tons/yr})} = 19 \text{ percent}$$

For alternative 2,

$$\text{Percent participation} = \frac{(\$140,000/\text{yr})(100)}{(\$15/\text{ton})(61,000 \text{ tons/yr})} = 15.3 \text{ percent}$$

Second, the percent participation values at zero profit were plotted, and a line was extended to connect these points with the net profit values at zero percent participation for each alternative. The break-even participation rates are 19 and 15.3 percent for alternatives 1 and 2, respectively.

2. The best alternative under a subsidized system would depend on the most probable level of homeowner participation. As indicated in Fig. 16-5, alternative 1 is preferred if the participation rate is

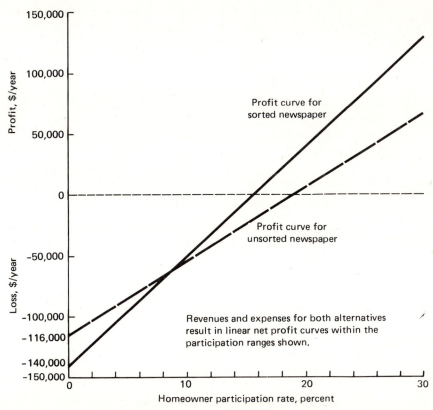

FIG. 16-5   Profit curves for Case Study 16-2.

below 8 percent; alternative 2 is preferred if the participation rate is above 8 percent. The conditions under which a subsidy might be used include the following: (1) a high probability that homeowner participation would increase once the system is operating, (2) a definite plan that the subsidy would be stopped if it is needed beyond a preset future time, and (3) a good chance that market prices for newspaper would increase within a preset future time. The question of subsidies is a difficult one for management; the decision whether to subsidize should be made by elected officials. In this situation alternative 2 was selected because it involved the least subsidy and offered the greatest potential for profitability if the participation rate were to improve in the future.

3. The community was concerned about the theft of 50 percent of the bundled newspaper set out during the pilot program. However, theft was not expected to be a problem for the full-scale system because (1) the pickup of newspaper by the weekly collection vehicles would eliminate the need for accumulation and placement of large quantities of newspaper on community streets, and (2) the weekly quantities per home would be small, and an illegal collector would have to stop at so many homes that it would become uneconomical. The economies would accrue to the waste collection agency because both wastes and newspaper would be collected on the same route.

Comment The decision to sell clean newspaper is based on economics. Those economics cannot be determined until some pilot studies are completed and reliable data on the quantities of newspaper are available.

CASE STUDY 16-3 *Evaluation of Materials Recovery*

In the early 1970s the city of New Orleans was faced with the problem of expanding and improving its waste incinerators [8]. The new and more stringent air pollution standards set at that time exceeded the capability of the existing incinerators and air pollution control equipment. Before incurring the major expenses of upgrading the incinerators, the city evaluated the potential for processing wastes and recovering materials for resale.

Management Issues

The management issues in this case included the following:

1. Would the recovery operations significantly reduce the quantity of wastes requiring disposal?

2. What should be the capacity of the processing and recovery facility?

3. Who should construct and operate the facility?

4. Should the recovery facility be designed on the basis of existing technology, or should research and development be undertaken?

5. Would the operating costs of a new facility be competitive with those of replacement incinerators?

Information and Data

The city stated that its intent was twofold: (1) to dispose of its residential solid wastes in an economic and sanitary manner, and (2) to invest in, and profit from, a recovery module using city wastes. Because markets were essential to the operation of a recovery module, the city contracted with the National Center for Resource Recovery (NCRR), a nonprofit manufacturing industry–based organization, to assist in the development of viable market contracts and the selection of processing equipment. A combined project team comprised of city and NCRR employees was formed to review potential recovery systems for use in New Orleans. The pertinent data were as follows:

Residential wastes, tons/day	1,500
Recoverable wastes, percent by weight of total delivered	
Ferrous metal	7.5
Glass	10.87
Aluminum	0.92
Newspaper	8.0
Nonferrous metal	unknown
Sanitary landfill area, acres	1,495

Resolution

The resolution of this case involved five steps:

1. The project team found that markets existed for ferrous metal, glass, aluminum, and newspaper. The glass market involved the highest risk. The reduction of waste quantities requiring disposal was calculated by applying the given composition by weight values (see Chap. 10). Summing all recoverable wastes resulted in a 27.92 percent reduction in the weight of wastes requiring disposal.

   The project team recognized that all incoming recoverable wastes could not be recovered (see Chap. 9 for typical recovery rates). Therefore, the projected quantity reduction was decreased to 17.1 percent of incoming residential solid wastes. The 17.1 percent reduction might or might not be significant; it certainly was not as significant as the 70 to 80 percent volume reduction obtainable through incineration. However, if adequate landfill space were available and the recovery operations generated a profit, then the reduction would be significant, even when compared to incineration.

2. The question of the capacity of the processing and recovery facility was evaluated in light of available technology and market conditions. A reliable technology data base was not available for facilities that process large quantities of mixed residential wastes. Additionally, there were uncertainties in the stability of materials markets. The future action by other cities throughout the nation in these same recovered materials markets were of special concern.

   After considering these many variables, the project team recommended a facility with a capacity of 750 tons/day (50 percent of the 1,500 tons/day of residential wastes). There were two primary considerations in sizing a new-technology facility such as this: (1) functional reliability, so that the waste management system would continue to operate during periods of breakdowns or routine maintenance, and (2) sufficient capacity to provide the markets with a large and stable source of materials.

3. The city wished to minimize risks and yet take full advantage of the availability of material markets. City agencies had very limited experience in dealing with markets. Thus, it was believed that a publicly owned and operated facility would probably have weak market contacts. Moreover, there would be unknown economic risks because of changing market prices for recovered wastes. It was important to have a firm estimate of system cost for comparison with the cost of incineration. Consequently, it was recommended that the construction and operating entity be a private agency. Performance specifications were written, and a construction-operation contract was advertised. Upon receiving bids, the city was able to compare the bid price with the estimated cost of incineration. (The bid document was among the first of its kind in the United States to encourage materials recovery from municipal wastes.) It was a performance specification which gave the bidder great latitude in selecting equipment type and construction. The uncertainties of market terms made equipment flexibility important to interested bidders. The sales market, an essential part of any recovery operation, was identified and guaranteed by the National Center for Resource Recovery. This also was an innovative attempt to encourage bidders. Many other public agencies may choose to construct publicly owned recovery facilities and sell recovered materials directly to manufacturing industries.

4. Both the city and the National Center for Resource Recovery wished to include some research and development facilities. How-

ever, the city had great difficulty in commiting public funds to that purpose. Research and development facilities were included when NCRR, acting as a coventurer, guaranteed portions of the capital in the form of a loan-grant not to exceed $750,000. A key financial feature was the pledge by the private agency to convert up to one-third of the loan each year to an outright grant, if necessary, to cover operating losses of the recovery module during a 3-yr test and evaluation period.

The recovery system was designed to meet the written market agreements for ferrous metal, newspaper, aluminum, glass, and other nonferrous metals. The major pieces of equipment specified to support that recovery include the following: shredders, air classifier, paper baler, rising-current separator, ferrous magnetic separator, metal baler, heavy media separator, dryer, electrostatic separator, glass transparency and color sorter, and all related conveyors and storage spaces. (Typical equipment layouts and additional descriptions are presented in Chap. 9.)

5. Bids were received on the facilities in late 1974. The lowest bid was for $10.95/ton—the price the city would pay to the contractor for all wastes (up to a maximum of 750 tons/day) delivered by the city to the contractor's processing and recovery facility. The city would also receive 15 percent of any profits derived by the contractor from the sale of recovered materials (the formula for calculating profit was set in the bid proposal). The cost to the city of $10.95/ton compared very favorably to the cost of $8.00/ton to $12.00/ton typically estimated for incineration.

CASE STUDY 16-4 *Evaluation of Energy Recovery*

The city of Nashville, Tennessee, disposes of its solid wastes in landfills. By 1972, however, not one of the city's landfills met state regulations [6]. The agencies responsible for waste management therefore investigated various means of improving the disposal situation. The concept of using the energy available in wastes prior to landfill gained community support; so the investigation concentrated on energy recovery.

Management Issues

The management issues were related to the conversion of wastes to energy and the marketing of energy. The main questions were:

1. In what form should the energy be recovered from wastes, and how should it be delivered to an energy market?

2. Who should construct and operate the energy conversion facility?

3. Could the energy conversion facility be cost-effective when measured against the cost of improved landfills?

Information and Data

The city was reconstructing and expanding its central business district. A nonprofit corporation, the Nashville Thermal Transfer Corporation, was created to construct, own, and operate a central energy facility to heat and cool municipal buildings within the district. The corporation was independent of the solid waste management system, and the fuels to be used were to be selected on an economically competitive basis. It was estimated that the use of the central facility would save the 27 individual users over 20 million kW of electrical power as compared to the continued use of individual heating and cooling systems. Approximately 1,400 tons/day of municipal wastes were being generated.

Resolution

The following steps were taken to answer the questions.

1. The search for an energy market was centered in the urban core of the city. In fact, the Nashville Thermal Transfer Corporation was a prime force in evaluating the potential of using wastes as an energy source. The existence of an energy wholesaler with existing contracts for energy sale was a strong motivation to use wastes as a fuel.

   It was determined that the wastes should be converted to steam, which would be distributed directly to customers for heating or processed through condensing turbines and chillers for air conditioning. A materials flowsheet is shown in Fig. 16-6. The energy system operator and the solid waste system operator were in an excellent position to plan a joint solution to the energy problem. (The solid waste system manager cannot work alone; a cooperative effort is required.)

2. The energy distribution system was already under construction, and it was to be operated by the Nashville Thermal Transfer Corporation. However, the storage, furnace, ash removal, and air pollution control facilities (see Fig. 16-6) could be managed by either the city or the nonprofit corporation. The corporation chose to construct and operate all these facilities as well. Three transfer stations received wastes from local collection trucks. The wastes were transferred to transport trailers that delivered them to the energy facility. Initially, the energy facility was to receive and process 720 tons/day of municipal solid wastes. The wastes were to be delivered free for a period of 30 yr. The logical choice for an operator of the energy facility was the energy wholesaler. That organization's experience with energy conversion and distribution

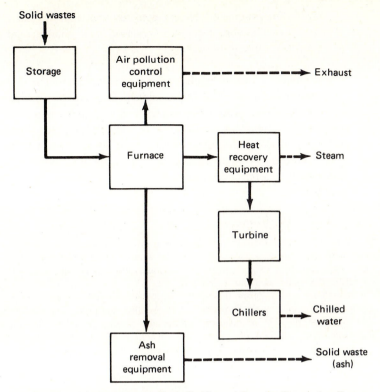

FIG. 16-6   Flowsheet for the Nashville Thermal Transfer Corporation Energy Conversion Facility for Case Study 16-4.

was far more beneficial than the detriment of its inexperience with solid waste fuels.

3.   Costs to the solid waste management system would originate in two places: at the transfer system and at the energy conversion facility. It was the total of these two costs that had to be weighed against the cost of landfills.

The energy conversion system was found to be cost-effective when it was measured against the cost of improved landfills. The projected savings were estimated to be $1,250,000 annually. The savings would result from changed transportation and future landfill costs. The actual maximum savings would be achieved when the energy facility handled the full capacity of 1,400 tons/day.

It is pointed out, however, that utility wholesalers are sensitive to the price of fuel. In this case the fuel was free, but there is still an indirect fuel cost because auxiliary burning equipment is required (especially air pollution control equipment). Poor equipment performance or stricter air discharge standards could require significant equipment changes with resultant cost increases. Thus, the cost-effectiveness could be reversed, and landfill could be the more favorable solution.

Comment   The techniques used in establishing this program were carefully thought out. The key ingredient to ensure success in energy conversion was established by contractual agreement before the waste processing system was constructed. In addition, an energy distribution piping network was constructed economically and was readily available. Another favorable consideration was the apparent low cost ($3/ton) of the transfer and transport system.

16-4   DISCUSSION TOPICS

16-1.   Contact solid waste management agencies in your community and identify where processing equipment is used. List the total number of unit operations used within each applicable functional element. List separately any process that is not operated as a part of the solid waste management system.

16-2.   A community is considering the benefits of installing a processing facility consisting primarily of a solid waste shredder. When installed, the processing facility will add $3.25/ton to the cost of managing community wastes. Revenues from materials separated at the facility and sold to local buyers will increase by $2.65/ton of wastes processed. Should the community construct and operate the shredding facility? Under what conditions would the excess cost be justified?

16-3.   What are the management issues involved in the evaluation and selection of an energy conversion facility that uses a fuel mixture of 10 percent solid wastes and 90 percent pulverized coal?

16-4.   Draw a diagram showing the sequence of the functional elements of solid waste management. Under each element, list some common processing equipment and techniques used to improve management functions.

16-5.   Resource recovery is becoming a high-technology functional element involving significant economic and political variables (see process flow diagrams in Chap. 9). Select a materials recovery process flowsheet from Chap. 9 and evaluate the interactions of solid waste management and manufacturing. Should solid waste management include only front-end systems? What issues discourage manufacturing industries from building front-end systems?

16-6.   A city of 150,000 people is evaluating the economic benefits to be derived from materials recovered from their solid wastes. Assume that the wastes are of the typical physical composition shown in Table 4-2 and that the waste generation rate is 6.4 lb/capita/day (see Table 4-14). The waste processing and recovery equipment to be used is that shown in Fig. 9-2. Assume that all the wastes must be processed and that the annual unit cost for the processing and recovery operations is $15.00/ton. The market value of the recovered materials is as follows: ferrous metal, $25/ton; aluminum, $300/ton; and mixed colored glass, $20/ton. If the recovery efficiency for all the components is 90 percent, determine the revenue that this city can expect from the sale of recovered materials. Is processing and recovery economical for this city? At what unit processing cost will the operation break even? Based on a review of the literature, is the computed break-even unit cost realistic? What are the alternatives available to this city to reduce the unit cost of processing and recovery?

16-7.   What is the single most important factor in the successful implementation of a resource recovery system? What additional factors should a solid waste management system planner consider when evaluating buyers and markets for recovered wastes?

16-8. System reliability is important both to the solid waste manager and to the manufacturer. What administrative and operational techniques does one use to improve reliability? Assume that a community generates 6,000 tons/day of municipal solid wastes. What is your evaluation of the importance of system reliability to the solid waste manager?

16-5 REFERENCES

1. *American Metal Market,* Fairchild Publications, New York.
2. Brunner, D. R. and D. J. Keller: *Sanitary Landfill Design and Operation,* U.S. Environmental Protection Agency, Publication SW-65ts, Washington, D. C., 1972.
3. *Current Status of Resource Recovery Systems and Processes,* State of California, Solid Waste Management Board Bulletin 5, Sacramento, 1975.
4. Drobny, N. L., H. E. Hull, and R. F. Testin: *Recovery and Utilization of Municipal Solid Waste,* Report prepared for Solid Waste Management Office by Battelle Memorial Institute, Columbus Laboratories, U.S. Environmental Protection Agency, Publication SW-10c, Washington, D.C., 1971.
5. Greco, J. R.: Analyzing Source Separation Methods at Residential Level, *Solid Wastes Management,* vol. 17, no. 10, 1974.
6. Nashville's Dual-Purpose Plant Turns Waste into Energy, Energy into Heat and Air Conditioned Comfort, *Resource Recovery,* Wakeman-Walworth, Darien, Conn., 1974.
7. *Official Board Markets,* Magazines for Industry, Inc., Chicago.
8. *Resource Recovery and Disposal Program,* City of New Orleans, New Orleans, 1974.
9. *Resource Recovery and Source Reduction,* First Report to Congress, U.S. Environmental Protection Agency, Publication SW-118, Washington, D.C., 1973.
10. *Resource Recovery and Source Reduction,* Second Report to Congress, U.S. Environmental Protection Agency, Publication SW-122, Washington, D.C., 1974.
11. *Resource Recovery and Source Reduction,* Third Report to Congress, U.S. Environmental Protection Agency, Publication SW-161, Washington, D.C., 1975.

17

Disposal-The "No Alternative" Option

Disposal is the "no alternative" option because it is the last functional element in the solid waste management system and the ultimate fate of all wastes that are of no further value (see Fig. 1-1). The possibility that materials which are of no value today may be of value in the future is not precluded. Because most issues related to disposal evolve from public concerns for the environment and the traditional image the public has of a disposal site as a dump, the planning necessary to build public trust and understanding when a decision to implement a disposal option is made is considered in detail in this chapter.

17-1 MANAGEMENT ISSUES AND CONCERNS

In the implementation of new disposal sites, the single most important issue for management is to find a location that is acceptable to the public and to local regulatory agencies. In the management of existing disposal sites, the major concern is to ensure that proper operational procedures are carefully and routinely followed.

Selection and operation of disposal sites are complicated because people often react in a negative manner when they are asked to pay for the disposal of their solid wastes; the higher the costs, the more negative the reaction. Some people are reluctant to recognize that sanitary landfills are more costly than dumps, even though equipment and labor costs are increasing and land values are rising. In view of the higher costs of mining and the depletion of resources, people also question the merit of burying solid wastes that contain valuable resources.

The basic issues for the planner and manager are (1) justification of the need for a disposal site to the public, (2) evaluation and community acceptance of the disposal site location, (3) cost-effectiveness and control of operations, (4) use of special-purpose disposal sites, and (5) management policies and regulations to prevent environmental damage.

Justification of Need

Landfill disposal is often selected as the most economical and viable solution to the solid waste problem. However, merely stating this will not ensure community support. The suspicion of community residents that a better solution exists can be satisfied only by a well-documented justification, which should consist of more than tables of data showing comparisons of capital costs and equipment. It should also include an evaluation of processing and recovery options and of the potential use of the wastes for land reclamation if reclamation is a consideration. Several examples of documentation are presented in this chapter, and additional examples may be found in Refs. 3 and 12.

In the following discussion, land reclamation, conversion of wastes to energy, and recovery of materials are ranked in terms of the wastes remaining for disposal after the greatest degree of recovery has been accomplished.

Land Reclamation   Land reclamation and landfill disposal operations are normally identical with respect to equipment and facilities. The difference is usually one of semantics, although land reclamation does require special management actions not required at a disposal site. Typically, approximately 95 percent of a community's wastes can be used in land reclamation; the remaining 5 percent must be placed in a disposal site (the actual percentage varies with the amount of hazardous and toxic wastes). Land reclamation thus makes use of the greatest portion of a community's wastes and results in the lowest requirement for disposal of the nonusable remainder.

Land reclamation should not be started until a final land use has been designated for the affected area. That designation normally is not required for a disposal site. The following general rules apply to process selection for the treatment of wastes used in land reclamation: (1) toxic or hazardous wastes should never be used, (2) organic wastes require the greatest processing and control, and (3) inorganic wastes may be used without processing. Disposal sites normally receive completely unprocessed wastes; the only restrictions originate from the class designation approved for the site (discussed later in this chapter). Sound management guidelines for sanitary landfills have been published by the EPA [9] and the state of California [7].

Conversion into Energy   The organic fraction of wastes can be converted into an energy source (fuel, chemical conversion, and anaerobic biological conversion). Approximately 50 percent of the urban wastes in a typical community can be used for this purpose. The remaining wastes must be processed for land reclamation purposes or for recovery of material, or they must be placed in a disposal site. The burning of unprocessed municipal wastes as a fuel will produce an ash residue from which metals can be extracted. When this is done, an additional 7 to 15 percent of the inorganic wastes can be recycled. The remaining wastes (35 to 43 percent of all community wastes), plus all process residues, must be placed in a disposal site.

Conversion of wastes into energy is the second most efficient recovery method (after land reclamation) used to reduce the quantity of wastes needing disposal. The residues that can remain after energy conversion are discussed in Case Study 17-1; additional discussions are found in Ref. 3.

Recovery of Materials   Even with recovery and reuse of materials approximately 50 to 70 percent of wastes will require a disposal site. However, the secondary materials market is unstable over time, and so most agencies must also develop backup disposal sites to receive the significant additional quantities if a materials contract should fail.

Adaptability for Processing and Recovery   Because disposal represents a long-term commitment of wastes to the land, the management plan for disposal should be structured for review and adaptability if certain conditions, such as economics, political organizations, and social standards, should change in the future. The greatly increased cost of petroleum in the mid-1970s caused an economic change that favored the recovery of organic wastes as a fuel substitute. Newly elected officials can cause similar changes by passing legislation that would increase tax incentives for the construction of processing plants to recover materials. Social standards could also change so that consumer products that are not packaged functionally (excess decorative packaging adds more wastes) would be boycotted in retail stores or prohibited by legislation.

The level of documentation needed to show the alternative options will vary with each community. One example is illustrated in Fig. 17-1. As shown, the recommended disposal system involves the use of a sanitary landfill, but four separate opportunities are identified for possible changes in future years, and each opportunity is reevaluated before the recommended facilities are expanded. A graphic presentation is often useful to summarize these options and their relationships.

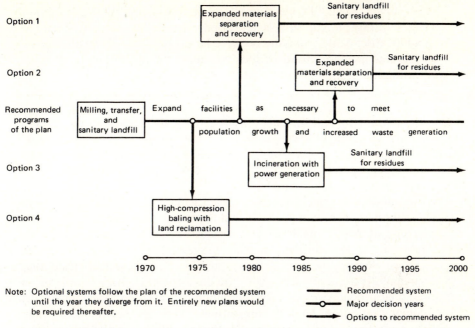

FIG. 17-1   Pictorial documentation of future alternative options to recommended disposal plan.

Evaluation of Site Location

Along with the need for a complete and clear documentation of the evaluation that resulted in the rejection of recovery options, documentation is needed concerning the engineering and management evaluation of the site location. Engineering considerations are presented in detail in Chap. 10. As a minimum, the following four management concerns must be documented in detail:

1. *Environmental impact.* The strongest public objections in most cases are related primarily to environmental concerns. For this reason, whenever possible, a disposal site should be located so that its impact on the surrounding environment is minimized.

2. *Adaptability for future changes.* The site should be located so that it might either be converted to a recovery operation or have processing and recovery equipment installed at a future time. An example would be the recovery of methane gas (see Chap. 10).

3. *Use as temporary storage area.* The disposal site should be evaluated as a temporary storage area for certain wastes in special situations (such as tires and scrap metal storage). A temporary storage area might be acceptable in certain locations where a permanent site would not be acceptable.

4. *Emergency plans*. Disposal sites are often essential during an emergency or after a natural disaster (floods, tornadoes, etc.). Capacity at any location can be an issue because large quantities of wastes can be generated in short time periods. A common resolution is to relax landfill site controls during emergencies in the interests of preserving public health.

The attempt to resolve concerns regarding the location of a landfill disposal site is often time-consuming and may be unsuccessful (as illustrated in Case Study 17-2). The greatest difficulties arise in communities that have been operating dumps. In these cases, the EPA guidelines on closing open dumps should be helpful [2].

A prudent management technique used widely in locating a single site is to identify numerous feasible sites and then to present the pros and cons of each one at public meetings and regulatory agency hearings until an acceptable site is agreed upon. In such sessions, it is important to present all facts as clearly and concisely as possible and to demonstrate that sanitary landfills are compatible with the environment. (An example of an informational brochure written for the public is presented in Appendix B.) The demonstration of landfill compatibility with its surroundings may require converting a local dump to a sanitary landfill. A helpful beginning may be made by installing attractive entrance signs, as shown in Fig. 17-2.

As was mentioned earlier, the use of the oceans to receive and hold wastes was practiced widely at one time, and disposal of sludges from ocean barges is still practiced [8]. However, a moratorium has been placed on any new disposal operations by federal regulatory agencies; so ocean disposal does not now represent a viable managment option for most communities.

Cost-Effectiveness and Control of Operations

The cost of waste disposal operations is objectionable to most community residents because it represents payment for a service whose benefits to the individual are difficult to measure. In addition, when residents deliver wastes to a disposal site, they often see other wastes there that would be of benefit to them. They then conclude that disposal should cost nothing and that the disposal site operator should act as a broker, receiving all wastes without charge and paying for operations by selling wastes to those who would benefit. They fail to recognize that the waste delivery cycle and the buyer cycle do not match. (In effect, plans for processing and recovery represent efforts to make these cycles match.) The issue for the site operator is the sanitary management of wastes; the issue for the community is to minimize the cost of disposal without degrading the environment.

Cost-Effectiveness   Broadly defined, *cost-effectiveness* means getting the job done for the lowest cost. The task for the waste system manager with

(a)

(b)

FIG. 17-2 Disposal site entrance signs. (a) Attractive, low-key sign. (b) Public appeal through identification of future land use and of joint effort of two city departments. (Courtesy Jim Wong, Sanitation Department, City of Phoenix.)

respect to disposal is to find the option that has the lowest cost. The following options should be evaluated before selecting a waste disposal site:

1. The community may choose to own and operate its own disposal site through a public agency. In this way the community maintains control over the operation and the rates charged for disposal. The principal concern is the scale of the operation which affects cost-effectiveness.

2. The community may select a private agency to operate its disposal site. The concern with this option is the potential loss of control over a site containing wastes which might cause environmental damage long after the operator has completed filling and sold the site for other uses.

3. The community may choose to work jointly with surrounding communities to develop a regional disposal site. This option allows the community to benefit from the economy of scale, but the community loses absolute control over rates for disposal. Also, administration under a regional plan is more complex than administration under a local plan.

4. The community may contract for disposal in a completely separate community. In this option, all rights and obligations concerning the disposal site are removed from the community. Again, a major concern is the lack of control over future rates for disposal.

Each of these options carries a different level of responsibility and control for the community in which the wastes are generated. The current trend among regulatory agencies is to require a community to maintain long-term responsibility for disposal of its wastes. Actions by the state of Oregon and the state of Washington in requiring state permits for all disposal sites and the recent disposal standards of the state of California are representative of this trend [7]. It is difficult to generalize about the preferred option since the selection is a matter of local political and social choice.

Control of Operations   The primary concern of the management of a sanitary landfill operation is to maintain cost-effectiveness. The important operating variables that must be controlled are (1) hours of operation, (2) source of cover material, (3) size of the daily unloading and working area, (4) control of the filling sequence so that the capacity of the site is maintained at original estimates, and (5) support facilities, such as water supply, scales, wash racks, and personnel offices. Intrinsically involved in these operating variables are labor and equipment costs.

The resolution of concerns regarding site operations evolves from a detailed disposal plan for the site. Commonly called an "operating plan," its contents will vary depending on the quantity of wastes handled and the topography of the site. In every case, the plan should have a layout of disposal facilities that reflects the arrangement of greatest cost-effectiveness. The development of an operating plan is the responsibility of the disposal site owner. However, some regulatory agencies require such plans as a condition for issuing operating permits for disposal sites [7].

A typical disposal site layout is shown in Fig. 17-3. The important management feature to note is the placement of all permanent facilities (scales, personnel building, etc.) at the highest elevation on the site. This control step allows the full use of all remaining land areas for cut and fill, with final fill elevations to be near the 220-ft contour. The location of the wet-weather operating area and the litter control fence should also be noted. The wet-weather area must be located near all-weather access roads, and a good supply of cover material must be stockpiled nearby. The litter control fence, which is essential to prevent litter and complaints from nearby landowners, should be positioned on the site so that prevailing winds blow across the dumping face first and then through the fence.

Rural and recreational waste disposal systems are a special concern to management and are discussed in special studies reported in Ref. 4.

Use of Special-Purpose Disposal Sites

Special-purpose disposal sites are often used for hazardous wastes and nonprocessible wastes. Only a brief discussion of management concerns regarding hazardous wastes is presented here since this subject is discussed separately in Chap. 11.

Hazardous Wastes   There is nothing simple about the disposal of hazardous wastes, either on land or in water. The sources of hazardous wastes are usually industry and agriculture, although some commercial activities, such as hospitals, also generate small quantities. Hazardous wastes from the production of atomic power must be handled separately. The importance of monitoring and managing hazardous wastes is becoming widely accepted, as exemplified by the recently published California guidelines for the handling and disposal of hazardous wastes [10, 11].

A degree of planning uncertainty surrounds hazardous waste disposal because of a lack of accurate data. Currently, state and federal regulatory agencies have imposed strict environmental monitoring requirements on organizations that handle and dispose of hazardous wastes. The reporting format is becoming standardized for all industries and waste haulers. In the past, industries could arrange for hazardous waste removal without regard for close regulation or complete documentation. As a result, very little accurate historical data are available for planning purposes in most communities.

Nonprocessible wastes   The use of special landfills for nonprocessible (mostly demolition) wastes is beneficial for several reasons: (1) They

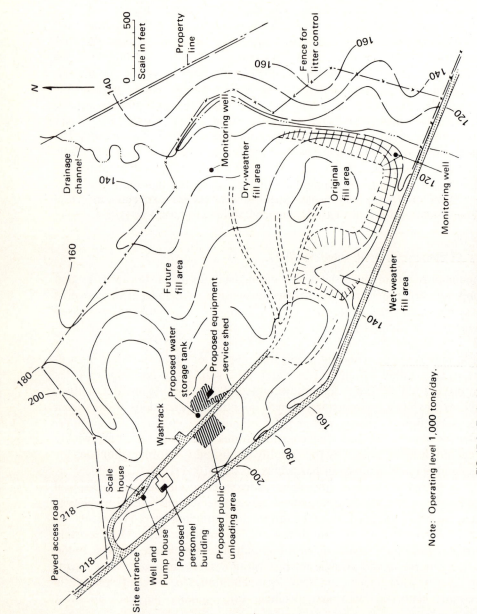

N

Scale in feet
0   500

Property line

Paved access road

218

218

Site entrance

Scale house

Well and Pump house

Proposed personnel building

Proposed public unloading area

Washrack

Proposed water storage tank

Proposed equipment service shed

Drainage channel

Monitoring well

Future fill area

Dry-weather fill area

Fence for litter control

Original fill area

Wet-weather fill area

Monitoring well

140

160

180

200

218

200

180

160

140

160

160

140

120

120

Note: Operating level 1,000 tons/day.

FIG. 17-3   Typical layout of disposal site facilities. (Adapted from Ref. 6.)

normally cannot be processed through solid waste facilities, such as transfer stations, incinerators, or shredding mills; (2) they are not attractive to flies, rodents, or birds and therefore may be acceptable as fill material at locations where ordinary solid wastes would be unacceptable; and (3) volumes and sources cannot be predicted for long periods because they vary with time, depending on local construction and demolition activities.

The resolution of what to do with these wastes involves administrative and operational actions. In most cases the preferred solution is to dispose of them in a privately operated landfill. Because these wastes are not a nuisance, it is often possible to develop a list of private landowners who want nonprocessible wastes for land reclamation purposes. If there are no private landowners willing to accept the wastes, the public agency may choose to acquire and develop a special disposal site for them.

Management Policies and Regulations

Management policies and regulations of interest with respect to landfilling operations usually deal with the classification of disposal sites and wastes discharged to land and the use of the sites. Numerous communities throughout the United States have developed such regulations and ordinances to control solid waste disposal. Comprehensive lists of ordinances are presented in Ref. 5, and a typical ordinance is presented in Ref. 1. The following excerpts from the Administrative Code of the state of California are presented as an example [7].

Classification of Disposal Sites   In subchapter 15, title 23, of the Administrative Code of the state of California, waste disposal sites are defined as any place used for the disposal of solid or liquid wastes and are classified as follows:

Class I disposal sites. Those at which complete protection is provided for all time for the quality of ground and subsurface waters from all wastes deposited therein and against hazard to public health and wildlife resources.

Class II disposal sites. Those at which protection is provided to water quality from Group 2 and Group 3 wastes. The types of physical features and the extent of protection of groundwater quality divides Class II sites into the two following categories:

> Class II-1 sites are those overlying usable groundwater and geologic conditions are either naturally capable of preventing lateral and vertical hydraulic continuity between liquids and gases emanating from the waste in the site and usable surface or groundwaters, or the disposal area has been modified to achieve such capability.

> Class II-2 sites are those having vertical and lateral hydraulic continuity with usable groundwater but for which geological and hydraulic features such as soil type, artificial barriers, depth to groundwater, and other factors will assure protection of the quality of usable groundwater underneath or adjacent to the site.

Class III disposal sites. Those at which protection is provided to water quality from Group 3 wastes by location, construction, and operation which prevents erosion of deposited material.

The Administrative Code also contains additional criteria that must be met for a site to qualify in a particular class. These criteria are related primarily to geological conditions, drainage provisions, and flood control.

Classification of Wastes Discharged to Land   In article 3, subchapter 15, title 23, of the Administrative Code of the state of California, wastes discharged to land are classified as follows:

> *Group 1 Wastes.* Consist of or contain toxic substances (lethal, injurious, or damaging to man or other living organisms including plants, domestic animals, fish and wildlife) which could significantly impair the quality of usable waters. Examples include community incinerator ashes, acids, alkalies, and chemical warfare agents.

> *Group 2 Wastes.* Consist of or contain chemically or biologically decomposable material which does not include toxic substances or those capable of significantly impairing the quality of usable waters. Examples include garbage, paper, metal, glass, yard waste, dead animals, sewage treatment residue and water treatment residue.

> *Group 3 Wastes.* Consist entirely of nonwater-soluble, non-decomposable inert solids. Examples include concrete, inert plastics, vehicle tires, glass and inert slags.

Use of Sites   Disposal of solid or liquid wastes may be accomplished in California only at sites approved by the local regional water quality control board unless a waiver has been granted by the board. Such waivers are conditional and may be terminated by the board at any time.

Disposal of various classes of wastes may be unlimited or limited at disposal sites. Any wastes may be disposed of at an unlimited class I site, but certain wastes are subject to quantity and concentration restrictions in a limited class I site. Any group 2 or group 3 wastes may be disposed of in a class II site unless restrictions are imposed by the regional water quality control board or other appropriate authority. In some cases, certain group I wastes may be allowed in a class II-1 site, when approved by the board. Class III sites have no such special disposal arrangements; only group 3 wastes may be disposed of in a class III site.

Disposal site records are required at class I and class II-1 sites handling group 1 wastes. Records of the volume, type, disposal manner, and disposal location are required for all group 1 wastes received at the site. Forms are provided by the state water quality control board for such records.

System Continuity

Disposal is the "last stop" for wastes, and the relationship between disposal and the other functional elements in the system is best described in terms of flexibility and adaptability. The operational plan for disposal must be flexible enough to receive and to allow for the handling of all the wastes when the operations associated with any of the other elements break down. The disposal site operation must also be cost-effective with decreased waste quantities, as would happen if a recovery program were to be implemented in the future.

Specific discussions of the relationships between disposal and the other functional elements are presented in Chaps. 14 (collection), 15 (transfer and transport), and 16 (processing and recovery). Some typical interaction situations at a disposal site are shown in Figs. 17-4, 17-5, and 17-6.

17-2 INVENTORY AND DATA ACCUMULATION

The inventory for disposal sites is straightforward and involves the identification of sites that provide the greatest system economies. The standards by which land reclamation sites should be evaluated are essentially the same as those for sanitary landfills. Certain states have created additional disposal site standards. For example, the California disposal site standards, scheduled to become effective on July 1, 1976 [7], specify design parameters and requirements for site improvements, health and safety, fill operations, and site controls.

It is important to establish the amount of processible and nonprocessi-

FIG. 17-4 Contents of collection vehicle being unloaded at landfill disposal site.

FIG. 17-5   Crawler tractor pulling wastes out of transport trailer at landfill using cable pulloff system.

FIG. 17-6   Ferrous metal separator located at landfill disposal site. Separated ferrous metal is discharged from belt conveyor into open-top collection truck.

ble wastes in the community because separate disposal sites may be beneficial. The best inventory method is to monitor existing disposal sites or illegal dumps (see Chap. 4). In addition, interviews should be held with representatives of all existing collection and disposal agencies.

Soils and Hydrogeologic Information

The most important factor for effective disposal operations is the availability of cover material. In the inventory step it is important to identify where soils are available, whether the characteristics of the soils are suited to covering wastes, and what quantities are available. Another important factor that must be considered for effective environmental control is the potential for surface and groundwater contamination.

The inventory should begin with a visit to the local office of the Soil Conservation Service, U.S. Department of Agriculture, to obtain areawide soils maps. In addition, the U.S. Geological Survey (USGS) should be contacted. The USGS has developed hydrogeologic maps on many areas of the country that may be helpful. Groundwater studies prepared by state water agencies are also excellent sources of data. This information should be used to identify acceptable areas, which should then be inventoried in more detail. The detailed inventory should include soils borings and test wells in sufficient number to predict accurately the environmental conditions of each site. This information should be documented fully and included in the site operating plan. Detailed engineering data are presented in Chap. 10.

Community Involvement

Typically, the community will become involved in disposal site selection at some time during the selection procedure. In the development of a disposal plan, it will be beneficial in many cases to involve interested groups during site inventory. This action broadens the base of local knowledge that can be brought to focus on the site search. Additionally, the community thereby gains a better understanding of the problems associated with finding acceptable sites.

Regional Options

The inventory of acceptable sites should be extended beyond the political boundaries of a single community. It is often helpful to request the assistance of a state regulatory agency in identifying all acceptable regional sites. Once this preliminary identification is completed, the appropriate disposal agency should be contacted directly to obtain the following: (1) cost of disposal, (2) total available capacity in the disposal site, (3) portion of capacity available to the community, (4) any buy-in cost for that capacity, (5) any restrictions on types of wastes received, and (6) cost escalation at future dates. It is also important to identify the added costs to the local waste collection agency for any extended delivery routes associated with a regional landfill.

Identification of Old Landfills

An attempt should be made during inventory to document the location of all closed landfills, for it may become necessary to monitor leachates at a future time. The most practical inventory technique is to review old records and to discuss the matter with long-time community residents.

17-3   CASE STUDIES IN DISPOSAL

Two case studies involving disposal are presented to illustrate some of the management issues involved. The first case is an evaluation of the need to develop long-term disposal sites. The second case illustrates the management problems associated with opening a new disposal site.

CASE STUDY 17-1   *Assessment of Disposal Site Needs Following Processing and Energy Recovery*

The government of a large city is evaluating its future disposal site needs when processing and energy recovery facilities become operational. Because the waste system manager has received strong objections by citizens to the continuation of existing large sanitary landfills, a plan is being developed to assess long-term needs for disposal sites.

Management Issues
The major issues in this situation are:

1. What quantity of wastes will remain to be disposed of following processing and energy recovery?
2. Should long-term disposal sites be acquired?

Information and Data
Large canyons that are acceptable as disposal sites exist in the surrounding region. Strict local air discharge regulations and the existing large-capacity, inexpensive disposal sites have resulted in the rejection to date of all energy conversion processes. However, energy conversion processes are once again being considered because energy costs in the community have increased to such a level that solid wastes may be an economical fuel. Air pollution control equipment has also

become more efficient. The pertinent data were as follows:

Urban waste quantity, tons/yr	11,500,000
Waste composition, percent of total annual	
Household wastes	31
Commercial wastes	19
Other wastes	50
Residue from energy conversion, percent of household and commercial quantities	15

Resolution

The planning director asked the processing and recovery planning team to work closely with the disposal planning team in evaluating the issues. The following step-by-step procedure was followed:

1. Convert the percentages shown to tons per year:

    Household wastes = 3,565,000 tons/yr

    Commercial wastes = 2,185,000 tons/yr

    Other wastes = 5,750,000 tons/yr

2. The source of organics for energy recovery are the household and commercial wastes.

    Total organic wastes = 5,750,000 tons/yr

3. The wastes requiring disposal are the residues from energy recovery processes plus the other wastes not processed through energy recovery.

Residues	862,500 tons/yr
Other wastes	5,750,000 tons/yr
Total	6,612,500 tons/yr

4. Clearly, energy recovery alone is not an answer to disposal needs. Planning must include the identification of disposal sites for 6,612,-500 tons of wastes, or about 60 percent of the original quantity. Ignoring a need of that magnitude would result in system failure. It was essential for the community to continue to acquire and develop disposal sites.

Comment   Solid waste management officials in many communities are subjected to public pressures to implement energy recovery systems. Many of these pressures originate with environmental groups who object to sanitary landfills. As demonstrated in this case, even if a comprehensive energy conversion system were implemented, a large disposal system must be available to receive residues from processed wastes as well as nonprocessible wastes.

CASE STUDY 17-2   *Opening a New Disposal Site*

The city of Sunny Hills has a population of approximately 30,000. It is located in a county with a total population of approximately 70,000. The city manager has been told by elected members of the city council that the existing city disposal site is a disgrace and must be closed. The city manager found a nearby site, and the owner is willing to sell the land to the city. Before making the purchase, the city retained an engineering consultant to determine the suitability of the land for use as a disposal site. The conclusion was that it would be usable as a class II landfill disposal site. However, on the basis of the environmental impact report for the site and the strong objections of neighboring landowners, the planning commission recommended to the city council that the site be rejected because adverse environmental impacts outweighed the benefits to the community.

Management Issues

The city manager had used federal revenue-sharing funds to pay for the site investigation. No additional funds were available to undertake investigations at other sites. All acceptable sites, including the one that was investigated, were located outside city limits. Consequently, the following management issues had to be resolved:

1. Should the city manager take steps to change public opinion about the site?

2. Is there a sufficient and adequate operating plan presented in the consultant's report?

3. Should the site be acquired over the objections of the planning commission and local landowners?

Information and Data

The existing city disposal site is operated by a private contractor. Wastes have been filled past preset maximum elevations. However, the site is located outside city limits, and city residents are not especially concerned about the deteriorating environment of a remote location. It has been difficult to implement a replacement disposal site because of major differences over environmental issues between the city planning commission and the city council.

The county planning agency has been working for 5 yr to develop a regional landfill. All attempts have been unsuccessful because the elected officials of each city cannot agree on ownership and administrative controls for a regional site. An added complication is the strong political influence of the local private waste collectors who have been developing separate plans for a privately owned long-term canyon disposal site. The regional regulatory agency has refused to grant an operating permit for their site, and the collectors are opposing public agency efforts to establish new sites.

Data concerning waste quantities and the other results of the consultant's investigations are those presented in Example 10-7. All the technical details of that example are available for resolving this case study.

Resolution

The city manager, acting as the planning director since the staff for such activities was limited, took the following steps to resolve this politically sensitive situation:

1. The city manager had attended the planning commission meeting at which objections to the new site were voiced by owners of adjacent lands. Although the time and some limited funds to conduct public hearings and to distribute documents describing the reasons for selecting the site were available, the city manager chose not to use them because none of the objections were voiced by city residents. Since the city's intent was to acquire only the property for the disposal site, the approval of the acquisition by owners of adjacent lands was not needed. Therefore, no steps were taken to change public opinion.

2. The adverse impacts of the site were listed in the environmental impact report, but sufficient benefits were documented so that the city manager assumed site acquisition would proceed. The findings of the planning commission were adverse to the project, however, and the consultant's investigation was reviewed thoroughly in an attempt to find inadequacies that might be corrected. The soils and groundwater data were found to be so limited that questions were raised as to the validity of the conclusions drawn. Nevertheless, the city had no funds for additional studies, and the city manager ordered no new investigations.

3. Having exhausted the means by which public opinion concerning the site could be changed, the city manager was now in a position either to push for the project over the objections of the planning commission or to abandon that site. Pursuing the project was not unusual in this case because the city council and the commission had a history of disagreement on similar issues. In all cases the elected officials make the final decision on a community project. The city manager presented the results of the site investigation to the council and recommended that the site be purchased for use as

a disposal facility. The council approved the manager's recommendation and directed its legal counsel to begin site acquisition proceedings.

At this point, the project was stopped. Immediately after city council action approving site acquisition, the owners of lands adjacent to the site caught the attention of the local newspaper with their objections. Several emotional articles were printed concerning the hazards of wastes and the depressing economic effect of waste disposal sites on adjacent land values. A petition was circulated in the city to place the issue of acquiring the disposal site on a special election ballot. Surprisingly, the petition was successful, and a special election was held. Even more surprising was the result. The site acquisition was turned down by city residents who had not even appeared at earlier hearings!

The problem was not resolved to the satisfaction of the city manager or the city council. The old site remained open for a short time, and then the private operator arranged for additional disposal space in another landfill within the county.

Comment This case is an example of the importance of good public communications. It is not an unusual case; similar situations are found in many communities. What was unusual was the timing of opposition by city residents. The greatest shortcoming in the approach used was the lack of backup sites to offer when strong opposition appeared for the preferred site.

17-4 DISCUSSION TOPICS

17-1. Review Case Study 17-2. What actions would you have taken to determine public attitudes about the disposal site? Did the owners of adjacent lands constitute the only organized opposition to the site? Was the failure to acquire the disposal site a bad result? If not, briefly explain your reasoning.

17-2. Contact a local agency and request data on the disposal site(s) serving your community. How many disposal sites presently exist? Are there separate landfills for municipal wastes, hazardous wastes, and nonprocessible wastes?

17-3. Review Example 10-7. From the data given, select, list, and discuss those factors which should be included in the documentation of disposal site need.

17-4. Assume that you are the city engineer in a city with a population of 42,000. The disposal site has been operated for many years by a private company. A combination of environmental factors and greatly increasing waste quantities has caused the site to be filled more rapidly than expected. It will close in 6 months. You have not been involved in solid waste management in any way up to this time. Last night the city council directed you to solve the problem. What immediate steps would you take to meet the council directive? What technique would you use to find a new disposal site?

17-5. Refer to Discussion Topic 17-4. Assume that the mayor has recently visited the private industry disposal site and saw a discarded lawn mower which was the same model as one he owned and from which he wanted to salvage a replacement part. Although he was unable to obtain the part because salvage at this site is prohibited, the mayor did want salvaging at the replacement landfill. Discuss salvage and its impact on the management of this site.

17-6. Sanitary landfills are responsive to economy of scale. List and discuss the options which a waste system manager might consider to arrive at a least-cost disposal site.

17-7. Why might it be beneficial to maintain separate landfills for nonprocessible wastes? What would be the most desirable operating agency for a nonprocessible waste disposal site? Why?

17-8. What is the greatest public health concern associated with sanitary landfills? What techniques are used to mitigate potential health problems? What inventory methods are used?

17-5 REFERENCES

1. American Public Works Association: "Municipal Refuse Disposal," 3d ed., Public Administration Service, Chicago, 1970.

2. Brunner, D. R., et al.: *Closing Open Dumps,* U.S. Environmental Protection Agency, Publication SW-61ts, Washington, D.C., 1971.

3. Dair, F. R.: Pragmatic Planning for Solid Waste Management, *Proceedings of 27th Annual Conference of the California Institute for Transportation and Traffic Engineering,* University of California, Berkeley, 1975.

4. *Design Criteria for Solid Waste Management in Recreational Areas,* U.S. Environmental Protection Agency, Publication SW-91ts, Washington, D.C., 1972.

5. *Digest of Selected Local Solid Waste Management Ordinances,* U.S. Environmental Protection Agency, Publication SW-38c, Washington, D.C., 1972.

6. Division of Solid Waste Management: *Operating Plan Solid Waste Disposal Site Kiefer Boulevard and Grant Line Road,* Department of Public Works, Sacramento County, Sacramento, Calif., 1975.

7. *Minimum Standards for Solid Waste Handling and Disposal,* State of California, California Administrative Code, Title 14, Division 7, chapter 3, Sacramento, 1974.

8. Smith, D. D. and R. P. Brown: *Ocean Disposal of Barge-Delivered Liquid and Solid Wastes from U.S. Coastal Cities,* U.S. Environmental Protection Agency, Publication SW-19c, Washington, D.C., 1971.

9. Sorg, T. J. and H. L. Hickman, Jr.: *Sanitary Landfill Facts,* U.S. Department of Health, Education, and Welfare, Publication SW-4ts, Washington, D.C., 1970.

10. *Tentative Guidelines for the Safe Handling and Disposal of Used Pesticide Containers in California,* California State Department of Public Health, Berkeley, 1970.

11. *Tentative Guidelines for Hazardous Waste Land Disposal Facilities,* California State Department of Public Health, Sacramento, 1972.

12. Theisen, H. and M. Brown: Hawaii's Environmental Planning Aims at Flexibility in Solid Waste Management, *Public Works,* vol. 103, no. 9, 1972.

18

Plan Development, Selection, and Implementation

Plan development and selection action are the final steps in the management planning cycle, and implementation is the final result of decision making. Planning steps 1 and 2 (problem identification and inventory) provide the planner with the data and information needed to finish the tasks in steps 3 (evaluation and development of alternatives), 4 (program and plan selection), and 5 (development of implementation schedules). Steps 1 and 2 have been stressed in Chaps. 13 through 17. Because the details involved in steps 3, 4, and 5 are common to all the functional elements, they are presented together in this closing chapter.

Each step is described only briefly because commonly accepted procedures for these activities have not yet been developed in the field of solid waste management. The primary emphasis in this chapter is to illustrate the various activities through detailed case studies.

18-1 EVALUATION AND DEVELOPMENT OF ALTERNATIVES

Step 3 in the planning cycle usually is reached after the planning process has been under way for several months (see Fig. 12-2). The results of step 2 (inventory) may affect the original objectives and problem specification (step 1). It is therefore important at this point to review those objectives and any new problems identified in the inventory step and to make whatever adjustments may be necessary prior to the final commitment of funds and staff to the development of alternatives, selection, and implementation (steps 3, 4, and 5).

Waste management programs are presented to decision-makers in the form of alternatives so that the decision-makers can make their own judgments on the probable success of each one. The use of alternatives in the planning cycle is illustrated in Case Study 18-3. Additional discussions of alternatives may be found in Refs. 5, 6, 9, and 10.

Perhaps the most important requirement for an alternative is that it be quantifiable with respect to equipment, disposal sites, economics, etc. An alternative can be as simple as specifying the details of one-person versus two-person collection crews, or it may be as complex as specifying landfill disposal of all wastes versus processing wastes at multiple stations and selling recovered materials to numerous dispersed markets. Every alternative must satisfy the requirement of measurability. Documentation for each alternative, regardless of complexity, must encompass the following: (1) performance, (2) economic analysis, (3) impact assessment, and (4) administration and management and an implementation schedule.

Performance

Performance means getting the job done. The workforce and equipment required to provide the level of service desired by the community must be specified. The details of performance will vary with individual communities, but significant details that must be identified include (1) level of service, (2) equipment reliability and flexibility, (3) equipment and workforce expandability, and (4) program compatibility with other environmental programs (air and water) and with future changes in solid waste technology.

With these details established it is possible to contrast performance functions of a recommended program with performance functions of alternatives without additional planning studies. This is an important part in achieving plan implementation.

Economic Analysis

Once the details of performance have been identified, it is important to analyze the economic impacts of each alternative. The analysis must include estimates of capital cost as well as of operating costs. The cost of an alternative normally will be expressed as an annual cost. When divided by the annual quantity of wastes handled, the cost can also be expressed as a unit cost. Unit costs, such as dollars per ton, are often used to compare the cost-effectiveness of alternatives.

When cost estimates are completed, financing methods can be identified. Some of the available financing methods are reported in Table 18-1. A financial analysis must be made for each program alternative, but the

TABLE 18-1   FINANCING METHODS FOR WASTE MANAGEMENT SYSTEMS

Financing method	Characteristics
Debt	
General obligation bonds	Voter approval required; low interest cost; excellent marketability; primary source of revenue is the local general fund.
Revenue bonds	Voter approval required; moderate interest cost depending on project; do not affect local agency debt capacity; revenues available from user charges only.
Joint-power agency bonds	Voter approval required; moderate interest cost depending on project; a high potential for legal complication and issuance difficulty; revenues available from user charges or contract payments.
Nonprofit-corporation leaseback bonds	No voter approval required; high interest cost; a high potential for legal complications and issuance difficulty; revenues available from rental payments.
Nonpublic	No voter approval required; high interest cost.
Revenue	
Pay-as-you-go	No voter approval required; no interest costs; requires careful long-term planning so advance budgets are identified.
Leasing	No voter approval required; no debt restrictions; revenues obtained from current operating budgets.

details must be limited to those consistent with the planning level and available planning funds. More comprehensive discussions of financing are available in Refs. 4 and 8.

The final objective of many financial analyses is the establishment of service rates. Rates should be equitable and should reflect as closely as possible the actual cost of providing the service.

Impact Assessment

The programs of a waste management plan will have an impact on a community in three ways: (1) through changes to the natural environment, (2) through involvement of the human environment, and (3) through a reordering of the community's socioeconomic structure. An attempt should be made to make quantitative estimates of each impact. Unfortunately, most planning and decision-making must be completed without full benefit of these estimates because the interactions of the natural environment, human environment, and socioeconomic structure are very complex, and the monitoring of our massive resource system is very difficult.

Determining the impact of alternative programs requires information from community agencies and groups not normally involved in solid waste management, including business and environmental groups, regulatory agencies for air and water quality control, legislative bodies, and resources agencies. This information from such diverse sources will help to fill voids caused by unattainable quantitative estimates.

Administration and Management

The administrative functions and organizations for implementation must also be identified for each alternative. It is most practical for the planner to develop details of administration only for the short-term planning period (7 yr into the future). Detailed administrative planning for the long term is meaningless because changes can occur so rapidly in the solid waste management field. Managers who are responsible for operations during the short term will usually establish organizational policies and functions for the long term.

18-2 PROGRAM AND PLAN SELECTION

The development of facility programs and a plan is a major task in achieving effective solid waste management. Before the plan is presented to the community for acceptance, it is first refined through agency and public review. The best way to gain acceptance of the plan is to obtain support of key community groups. Another way is to demonstrate that the plan is compatible with other community goals. Additional means of achieving public support are presented in Ref. 2.

Obtain Community Support

The most positive bases for support are the residents and businesses of the community. Their involvement can take place either during the development of the plan (note the citizens' advisory committee in Case Study 18-3) or during its implementation. A strong public relations effort will be needed to develop community understanding of the plan. A sample of a public relations approach, which represents one part of a comprehensive public information program, is given in Appendix B.

Political support should be tested and developed during the planning study through the presentation of progress reports at regularly scheduled political meetings (city council, commissioners, supervisors, etc.). This procedure removes the element of political surprise from plan recommendations—a wise approach since politicians are the final decision-makers. The impact of political support is illustrated in Case Study 18-2.

Support must also be obtained from state and federal regulatory agencies. The surest means of obtaining their support is to include regulatory standards and controls in the plan. If a variance cannot be avoided, it should be discussed fully with officials of the appropriate agency before the plan is adopted.

Demonstrate Compatibility with Community Goals

The solid waste management programs must be compatible with other community goals. Generally, the higher the level of planning, the greater the need for compatibility. These other goals include land-use zoning goals, environmental goals, and state and federal goals.

Because solid waste management activities are highly visible, all programs must be compatible with community goals as expressed in general plans and land-use zoning. All programs must also be compatible with environmental goals, which are generally community-oriented but might extend beyond community boundaries (for example, leachate movement in surface streams). In most cases, an environmental impact report is required at the time of implementation. Many agencies provide such reports as a part of plan development.

State and federal agencies are taking a greater interest in resource and raw material depletions. The waste management system planner should monitor such interest (which is usually highly visible in the form of legislation) and should make the community plan compatible where it is economically feasible to do so.

18-3 DEVELOPMENT OF IMPLEMENTATION SCHEDULES

The primary objective of an implementation schedule is to establish the organizational structure and a time sequence of actions. The time sequence is normally divided into short-term and long-term actions. Other elements important to implementation are fiscal management and administrative considerations (regulations and standards).

Organization

The organizational structure refers to the agencies legally responsible for performing the tasks set forth in the recommended plan. A logical split of organizations is by functional activity. Thus, both administrative and operational agencies must be defined in the implementation schedule. These responsibilities are often assumed by a single agency. Additional information on organization may be found in Refs. 7 and 11.

Fiscal Management

The implementation schedule must also contain the following details of fiscal management: (1) capital formation, (2) cash-flow requirements, and (3) revenue programs, such as rates or taxes. An important part of fiscal management is the establishment and maintenance of equity among those paying for the recommended program. The matter of equity becomes more difficult to set as solid waste systems become more complex—especially as resources are recovered and sold.

Regulations and Standards

Regulations and standards are the means by which system control is maintained. A time sequence within which designated agencies will establish ordinances, standards, and other means of control must be included in the implementation schedule.

Plan Review and Updating

The primary objective of the implementation schedule is to set actions for short-term programs. However, any solid waste management plan will require periodic updating. Significant technology changes, especially for processing and recovery, are expected within the next 5 to 10 yr. Therefore, to make long-term plans, the manager must (1) monitor developing technology, (2) maintain contact with the community and its resources, (3) monitor existing standards and assess their continued need, and (4) update the community waste management plan. It is often best to assign these responsibilities to a single agency.

18-4 CASE STUDIES

Three case studies have been selected to demonstrate the procedures of program development in solid waste management. The first case deals with the administrative program details that normally are found in a comprehensive plan. The second case deals with the specific program details associated with a local plan, and the third case deals with the broader program details found in a regional plan.

CASE STUDY 18-1   *Develop an Approach to Involve Decision-makers in the Planning Process*

A regional solid waste management plan is being developed for a large city and a heavily populated unincorporated county area by the county staff. The study has progressed through step 2 (inventory). From the inventory it is apparent that significant administrative problems exist in the countywide system. The plan director has requested that planning procedures used in the remaining portions of the study (steps 3, 4, and 5) ensure that decision-makers review all administrative programs that are developed.

Management Issues

The management issues were based on the following problems related to administration:

1. There was no enforcement of mandatory collection service.
2. Records on waste quantities and operating costs were incomplete.

3. Nonuniform procedures were being used for exemptions from mandatory service.

4. Differential rates were charged for identical service along the boundaries of the different political entities.

5. The issuance of permits was inadequate as a number of collectors were operating illegally.

6. There were no resource recovery standards.

Information and Data

The city and the unincorporated area of the county had separate and uncoordinated solid waste management regulations and ordinances. In the city, residential collection service was mandatory and was provided by city crews. Bills were sent to every city resident; nonpayment of a bill resulted in continued service and a tax lien against the property. In the unincorporated area, residential collection service was also mandatory and was provided by public works crews. Bills were sent only to those residents who signed up for service; nonpayment of a bill resulted in termination of collection service and action by the health agency to force compliance with health regulations. Each collection agency maintained separate records, conducted completely separate operations, and charged independent rates.

Collection service to commercial and industrial activities was provided by private collectors operating under permits. The city had an unlimited permit system. The public agency for the unincorporated area had a limited system to allow competition at a level that would encourage capital investments in new equipment. Commercial collectors had to honor political boundaries and obtain a permit to operate in each area. This restriction, based on political boundaries alone, caused operational problems for commercial collectors who had contracted to provide collection service to retail store chains that had stores in both the city and the county.

The public agency in the unincorporated area initiated a resource recovery program involving the home separation of newspapers. Similar programs had been rejected by the city agency as infeasible. Thus, there was a basic difference in resource recovery objectives between incorporated and unincorporated areas of the community.

Resolution

The plan director and the planning staff jointly developed a planning and decision matrix, such as the one shown in Fig. 18-1, to involve the decision-makers in planning. In this situation the planner is charged with the development of a series of sequential action steps, each of which is reviewed jointly with the decision-maker before they move to

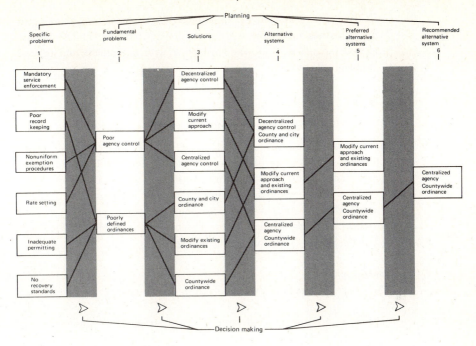

FIG. 18-1 Planning and decision-making matrix for Case Study 18-1.

the next step. The specific details were as follows:

1. The specific problems were listed in column 1.

2. The specific problems were related to the fundamental problems identified in column 2. In this case, agency control and legal regulations were the key factors.

3. The possible solutions were listed in column 3.

4. The solutions were grouped into the three alternatives in column 4. Each alternative must contain at least one of the solutions.

5. The three alternatives were then reduced to the two preferred alternatives in column 5. The decision-makers, who controlled the economic resources needed for implementation, worked closely with the planners during this phase. If column 4 had contained only one alternative, then planning would have moved directly to the development of an implementation schedule for the recommended alternative (column 6).

6. The alternative in column 6 was recommended, and a detailed implementation schedule was prepared.

Comment This example illustrates the relationship of planning to decision making in dealing with an administrative problem. The same procedure should be followed in dealing with equipment and related facilities problems.

CASE STUDY 18-2 *Local Plan Development (Adapted from Ref. 1)*

In 1968, a regional agency was asked to take over the solid waste collection system in an urban area of the Western United States. The agency chose to operate the collection system with public employees. In 1972, a study of the collection system was undertaken to evaluate whether the existing system was being managed properly and to recommend practical alternatives. The existing collection service was to be considered as one of the alternatives.

Management Issues

The key management issues involved in this situation were as follows:

1. What are the most desirable work performance standards and route design factors, and what changes would be necessary to provide the lowest-cost system?

2. If changes are recommended, can an implementation schedule be developed to give a gradual transition that would minimize the impact of changes on personnel and finances?

Information and Data

The agency took over solid waste collection from a number of private contractors. All contractor employees were offered jobs with the agency, and some of the contractors' equipment was purchased for continued use on collection routes. Most of the collection service was being provided to single-family homes, but a number of mobile-home parks, duplexes, apartments, and commercial establishments were also served. The basic service for two containers in a backyard location was expanded to include unlimited pickup at a curbside location of garden trimmings when a ban against burning went into effect in 1971. Collection routes were set up on a task system basis to increase labor productivity. The workday was based on a maximum of 7 h or completion of the assigned route. The following additional data were available:

Population served	270,000
Number of accounts	75,000
Land area served, mi^2	157
Number of daily collection routes	51

Collection frequency, per week	once
Truck size, yd³	25
Crew size, persons	2
Crew working hours per day (task system)	7
Average one-way haul to disposal, mi	15
Quantity of wastes collected per route, tons/day	7.25
Annual waste quantity, tons	123,500
Annual cost of collection, $	3,500,000

Resolution

A consultant was retained to assist in the development of the management report. A study outline was developed, and the key work programs were identified by the following outline headings:

1. Field observations
2. Crew work-load determinations
3. Existing system cost development
4. Alternative systems cost development
5. Implementation factors for alternative systems

The important inventory data and key factors for cost measurement involved labor and equipment. The results are summarized as follows:

1. *Field observations*. A field observation procedure was established to obtain performance data for the three alternatives. Special routes were established so that three types of collection operations could be observed. They were as follows: (1) the existing system (two-person crew), (2) curb collection with a two-person crew, and (3) continued use of the existing container location with a three-person crew. The routes were each observed for 2 wk. The results of the inventory are summarized in Table 18-2. As shown in the table, the most efficient type of service was curb collection using a two-person crew (43 to 45 person-min/ton). The labor requirements were 23 percent lower than for the existing service. This evaluation identified a type of service that could provide greater efficiency than the existing service. Although the evidence in itself is not final justification for management to change collection service, the potential 23 percent labor cost reduction within a high labor cost budget of $3.5 million is significant enough to extend the investigation.

TABLE 18-2   SUMMARY DATA ON PERFORMANCE OF VARIOUS COLLECTION CREWS IN CASE STUDY 18-2

Type of service	Number of units served	Quantity collected, tons	Route collection time, min	Person-min/ton
Existing system, two-person crew	260–292	7.9–8.1	221–235	56–58
Curb collection, two-person crew	281–308	8.6–8.7	185–195	43–45
Existing container location, three-person crew	240–268	8.5–8.7	170–203	60–70

2. *Determination of crew workload.* A constraint that limits the efficiency of collection is the amount of work that reasonably can be performed by each crew. The existing service required each crew to handle 7.25 tons/day. A questionnaire survey was made of four other communities with similar collection systems to determine average crew work loads. The results of the survey, summarized in Table 18-3, indicated that higher crew work loads were assigned in three of the communities. Thus, it became apparent that practical alternatives did exist and were being used in similar communities.

3. *Development of costs for existing system.* The decision was made that several alternatives were practical and should be evaluated for cost development. Prior to initiating that work, it was necessary to develop the cost base for the existing system, as shown in Table 18-4. The cost per ton and cost per account are the economic levels against which alternatives must be measured. In the existing system these costs are financed through capital reserves and a monthly solid waste collection fee billed to each account.

4. *Development of costs for alternative systems.* The primary alternatives presented for cost development are (1) curb collection with one- and two-person crews and (2) backyard and curb collection with three-person crews. The personnel requirements and costs for each alternative are presented in Table 18-5. The types of equipment, capital costs, and amortized annual costs for each alternative are presented in Table 18-6. These very specific operating cost data are then combined to yield the costs shown in Table 18-7 for comparison with existing system costs. The cost per ton of wastes for curb collection is 17 percent lower (than existing costs) with a one-person crew and 16 percent lower with a two-person crew. Thus, the lowest-cost system is curb collection

TABLE 18-3 SURVEY OF COLLECTION CREW WORK LOADS IN OTHER COMMUNITIES IN CASE STUDY 18-2

Location	Population served	Collection frequency	Pickup Location	Crew size	Truck size, yd³	Average one-way haul, mi	Crew work load, tons/day
Partial backyard collection							
City A	90,000	1	Backyard and alley	2	20	13	6.0– 6.5
City B	175,000	2	Backyard and curb	3	25	8	8.5– 9.0
Curb collection							
City C	400,000	1	Curb and alley	2	20	14	15.5–16.0
City D	700,000	1	Curb	2	20	5	14.0–14.5

TABLE 18-4   TOTAL ANNUAL AND UNIT COSTS FOR EXISTING
COLLECTION SYSTEM IN CASE STUDY 18-2*

Cost category	Total cost, $	Unit cost $/ton	Unit cost $/account
Personnel	2,248,300	18.20	29.98
Equipment	730,800	5.92	9.74
Miscellaneous (billing data processing, etc.)	224,000	1.81	2.99
Indirect overheads	137,600	1.11	1.83
Disposal charges	159,300	1.29	2.12
Total	3,500,000	28.33	46.66

* All costs in mid-1975 dollars (ENRCC index = 2,248).

with a one-person crew, but it cannot be implemented immediately because 80 new trucks are required. No means of financing this $2,560,000 capital expenditure was identified.

5. *Implementation factors for alternative systems.* Because of the severe changes necessary to implement one-person-crew curb collection, the decision was made to use two-person curb service as an initial transitional step for improved efficiency. The decision was a policy matter and was made by elected officials.

A strong social resistance to changing the service level had to be overcome. Curb collection required that all residents place wastes at the curb for collection and return the empty containers to the house after pickup. An effective educational program was developed highlighting the improvement in the physical welfare of the workers (resulting from elimination of the need to carry containers) and postponement of the need for future rate increases.

The following implementation schedule was established:

January 1972	Start curb collection program
	Provide public information
	Develop new route criteria
March 1972	Complete work-load negotiations with labor unions as appropriate
	Update route books and accounting records
June 1972	Present complete program to elected officials for formal adoption
August 1972	Complete public hearings
	Switch entire community to curb service

TABLE 18-5 POSITIONS AND PERSONNEL COSTS FOR ALTERNATIVE COLLECTION SYSTEMS IN CASE STUDY 18-2*

| | Curb collection | | | | Backyard/curb collection | |
| | One-person crew | | Two-person crew | | Three-person crew | |
Position category	Positions	Cost, $	Positions	Cost, $	Positions	Cost, $
Foreperson	8	114,700	6	86,000	6	86,000
Collector II	85	1,033,400	55	668,700	55	668,700
Collector I	—	—	67	704,600	134	1,409,300
Extra help	—	51,700	—	68,700	—	103,900
Billing and administrative	20	209,200	20	209,200	20	209,200
Total	113	1,409,000	148	1,737,200	215	2,477,100

* All costs in mid-1975 dollars (ENRCC index = 2,248).

TABLE 18-6   EQUIPMENT COSTS OF ALTERNATIVE SYSTEMS IN CASE STUDY 18-2*

Number of units	Chassis	Body	Model year	Purchase price, $	Annual investment costs,† $
		Curb Collection, One-Person Crew			
1	International	Fast Pack, 13 yd³	1971	11,647	2,600
1	Stake Bed		1972	11,000	2,400
80	International	Shu Pak, 37 yd³	1972	2,560,000	480,000
—					
82				2,582,647	485,000
		Curb Collection, Two-Person Crew and Backyard/Curb Collection, Three-Person Crew			
1	International	Fast Pack, 13 yd³	1971	11,647	2,600
2	White	Heil, 20 yd³	1969	32,550	7,200
2	International	Heil, 20 yd³	1970	37,464	8,200
15	International	Gar Wood, 25 yd³	1969	305,535	67,100
25	International	Gar Wood, 25 yd³	1971	535,449	117,600
8	Kenworth	25 yd³	1972	197,600	43,400
1	New (to be pur-chased)	Stake Bed	1972	11,000	2,400
—					
54				1,131,245	248,500

* All costs in mid-1975 dollars (ENRCC index = 2,248).
† Amortization over 6 yr for 80 new trucks of one-person alternative and 5 yr for other alternative at 6 percent after crediting present worth of 10 percent trade-in allowance.

TABLE 18-7   COMPARISON OF COSTS OF EXISTING AND ALTERNATIVE COLLECTION SYSTEMS IN CASE STUDY 18-2

System	Cost,* $/yr Total	Per ton	Per account
Existing system	3,500,000	28.33	46.66
Alternative systems			
Curb collection			
One-person crew	2,912,000†	23.57	38.84
Two-person crew	2,945,000†	23.85	39.27
Backyard and curb collection			
Three-person crew	3,697,100†	29.94	49.29

* In mid-1975 dollars (ENRCC index = 2,248).
† This total is derived by adding operations and maintenance and other indirect costs to personnel and equipment costs. The sum of operation and maintenance costs, miscellaneous administrative costs, overhead, and disposal charges are as follows:

One-person crew $1,018,900/yr
Two-person crew $959,500/yr
Three-person crew $971,500/yr

| October 1972 | Review collection truck replacement budget to determine timing for purchase of one-person equipment |
| June 1973 | Review curb service and establish a firm schedule for transition to a one-person curb service |

Comment The detailed data presented in this case are essential in the development of a sound operating plan. The agency responsible for waste collection can use the criteria and standards presented in the plan to evaluate most of its day-to-day operations. However, it should be recognized that the use of detailed data in a plan often reduces its long-term credibility as the data are subject to considerable change over time. In this case, a significant increase in the capital cost of one-person collection trucks could delay the implementation schedule for transition to one-person curb collection service.

CASE STUDY 18-3 *Regional Plan Development (Adapted from Ref. 3)*

A regional government has assumed the responsibility for development of a solid waste management plan. The geographic area to be covered by the plan includes four cities and a rural unincorporated community. The plan is being developed by a consultant working under the direction of a project director employed by the regional authority. Waste management presently is fragmented among the four cities, and there is no history of regional cooperation. The impetus for regional planning comes from recently enacted state mandates for regional approaches to management as well as from the imminent closing of three landfills that serve the individual cities.

Management Issues

The major management issues to be addressed during the study are:

1. What is the best arrangement of landfills to receive wastes from all cities within the region?

2. What processing and recovery equipment should be installed so that resource recovery is developed?

Information and Data

Each city presently has its own solid waste management system including collection and direct haul to nearby landfills. Two cities have public agency collection crews. The other two have contracts with private collectors. Each city operates its own landfill, and two cities operate transfer stations. A third transfer station in the rural area serves the unincorporated community. The onsite storage and collection systems are relatively free of problems, and the rates charged for service are considered reasonable. There is no processing equipment now operating, and resource recovery is limited to hand separation of cardboard and bulky metal and wood wastes. The region is generally flat with a high water table and very limited choices for long-term landfill disposal sites. The cities have the normal light industrial and commercial businesses found in a region of 820,000 population. Freeways, railroads, and deep-sea waterways serve the region. The following additional data were developed:

City A

Population	220,000
Collection routes	54
Collection truck size, yd^3	20
Collection crew size, persons	3
Disposal site area, acres	82
Disposal site remaining life, yr	5

City B

Population	570,000
Collection routes	110
Collection trucks, number	
31-yd^3 capacity	30
25-yd^3 capacity	80
Collection crew size	
One person	30
Two persons	80
Transfer station, tons/day	500
Disposal site area, acres	800
Disposal site remaining life, yr	25

City C

Population	20,000
Collection (franchise with private collector, no data available)	
Transfer station, tons/day	450
Disposal site area, acres	15
Disposal site remaining life, yr	2

City D

Population	10,000
Collection (franchise with private collector, no data available)	
Disposal site area, acres	6
Disposal site remaining life, yr	1

Rural

Transfer station, tons/day	13

Resolution

Because of the importance of this planning study, the regional political body appointed a 15-member citizens' advisory committee to work with the planners. The members were selected from lists provided by interested citizens and political groups. This group worked closely with the project director to resolve the community waste management problems.

1. The first step in the solution involved the establishment of procedures to control and complete the planning study within the allotted time and available resources. The consultant and project director worked together to develop the activity diagram shown in Fig. 18-2. The complexity of regional planning is seen by comparing the tasks identified in Fig. 18-2 to the relatively simple outline headings developed as the first solution action for Case Study 18-1. The activity diagram is useful in several ways: (1) to organize the planning study, (2) to report progress on specific work tasks, and (3) to moniter progress of tasks in relation to the time schedule on the diagram.

2. The inventory was set up to accumulate information on every functional element of the system, and special emphasis was placed on landfills and resource recovery. In Fig. 18-2, the inventory highlighted the following details:

    a. *Waste generation and composition.* The cities and unincorporated area did not have good records on what types of wastes existed and where each type was generated. A detailed monitoring, sampling, separation, and weighing inventory was scheduled and completed at every disposal site. Portable scales were rented and set up at each site to weigh incoming trucks. A sampling crew of five people randomly selected loads from which samples, ranging in size from 375 to 1,425 lb, were drawn. The crew then hand-separated waste components into representative categories (see Chap. 4). The entire field inventory took place over a 2-wk period. This depth of inventory is often not necessary. However, if resource recovery alternatives are to be developed in the regional plan, a detailed knowledge of waste components is essential.

    b. *Transportation routes and disposal sites.* A critical problem to the region was the closing of three existing disposal sites. The planners recognized that alternatives would be developed in which (1) new disposal sites would be identified, (2) all wastes would be routed to the existing large-capacity disposal site, and (3) additional transfer stations would be evaluated to make the transportation system more efficient. The site inventory was concentrated on land reclamation, both at the existing sites and at new sites.

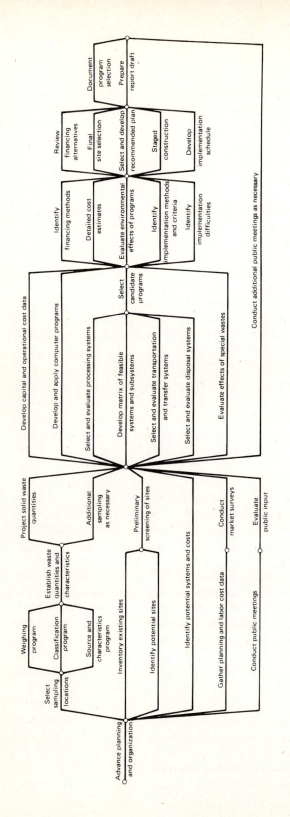

FIG. 18-2 Activity chart for Case Study 18-3.

      c. Materials markets. An effort was undertaken to find markets for recovered waste materials. The citizens' advisory committee was asked to hold public meetings in the community in an attempt to identify resource shortages and to set recovery priorities. Contact was also established with state and national industries.

      d. Public meetings. Four public meetings—one in each city—were scheduled early in the study to help identify social and community problems.

3.   The planners and the advisory committee identified a number of problems related to the existing waste management system (see Table 18-8). A priority was assigned to each problem, and a set of possible solutions was documented for each one. These solutions were grouped to form the programs of the recommended plan. Each solution was then subjected to the following criteria:

Does it have to be done (is it a state or federal mandate)?

Is it politically acceptable?

When does it need to be done (immediate, future)?

Does it increase or decrease costs (capital and operating)?

TABLE 18-8   EXISTING OPERATIONAL AND ADMINISTRATIVE PROBLEMS IN CASE STUDY 18-3

Functional element	Problem
Generation	Records on type, location, and quantity of wastes are inadequate or nonexistent.
	Waste quantities are increasing.
Storage	Local standards are inadequate.
	Containers are inaccessible, overflowing, or underground.
Collection	Collection service area records are incomplete.
	Private industry collection permits are not assigned uniformly.
Transfer	Existing transfer stations do not have capacity for increased quantities or for resource recovery operations. Regional need for transfer stations is not defined.
Processing and recovery	Existing programs do not meet mandated goals.
	Available recovery methods do not meet reliability standards.
	Funding sources must be found to finance a recovery system.
Disposal	No data are available on leachate movement.
	Capacity in three of the four sites is exhausted.
	No final land-use plans are available for sites.
Administration and control	Nine jurisdictions are setting policy.
	Waste generation is unrestricted at the source.
	Local ordinances are uncoordinated and incomplete.

What is the effect on service level?

What are environmental issues (resources and energy)?

Is it practical (on the basis of past experience)?

4. The programs were grouped into four alternatives for detailed analysis, summarized as follows:

Alternative 1. Basically an extension of the existing system, this alternative calls for the construction of transfer stations to replace exhausted landfills with transport of wastes to the remaining city B landfill. Resource recovery would be maintained at existing levels. The movement of wastes is shown in Fig. 18-3.

Alternative 2. A significantly changed system involving a central processing facility is proposed in this alternative. Most wastes

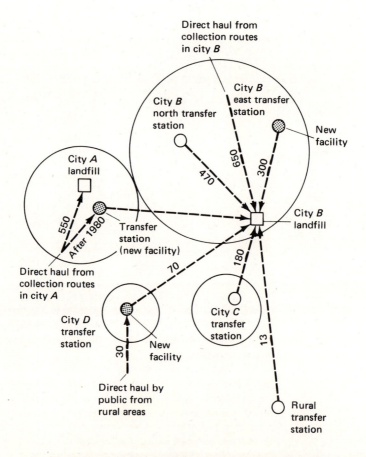

Note: Numbers correspond to solid waste loadings in tons per day. Large circles show approximate location of city boundaries.

FIG. 18-3   Waste movement for Alternative 1 in Case Study 18-3.

would be moved to the processing facility and then to the city B disposal site. Resource recovery at this facility would be expanded to include magnetic separation. The movement of wastes is shown in Fig. 18-4.

Alternative 3. The major feature of this alternative is the addition of air classification, with glass and aluminum recovery, to the facilities of alternative 2. The city B landfill would then be closed completely to unprocessed wastes. The movement of wastes is shown in Fig. 18-5.

Alternative 4. Energy recovery is the primary feature in this alternative. The energy recovery station would accept all organic wastes from the processing station. The energy recovery conversion is to steam, and the steam would be sold in city A and used for heating and air conditioning of downtown buildings. The movement of wastes is shown in Fig. 18-6.

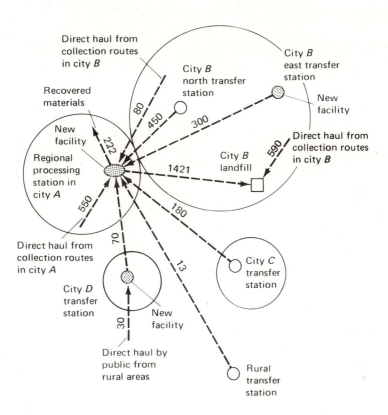

Note: Numbers correspond to solid waste loadings or recovered raw materials in tons per day. Large circles show approximate location of city boundaries.

FIG. 18-4 Waste movement for Alternative 2 in Case Study 18-3.

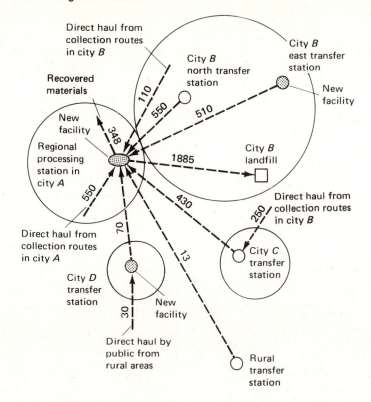

Note: Numbers correspond to solid waste loadings or recovered raw
materials in tons per day. Large circles show approximate
location of city boundaries.

FIG. 18-5   Waste movement for Alternative 3 in Case Study 18-3.

TABLE 18-9   CAPITAL COSTS OF FACILITIES FOR ALL ALTERNATIVES IN CASE STUDY 18-3[a]
($000)

	Alternative			
Facilities	1	2	3	4
Transfer	$2,110	$ 2,110	$ 2,110	$ 2,110
Transport	490	830[b]	1,240[b]	1,240[c]
Processing and materials recovery	—	7,700[d]	15,290[e]	15,290[e]
Energy recovery	—	—	—	33,000[f]
Total cost	$2,600	$10,640	$18,640	$51,640

[a] All costs in mid-1975 dollars (ENRCC index = 2,248).
[b] Includes transport from processing to landfill.
[c] Includes transport from processing to energy recovery and landfill of residue.
[d] Processing and magnetic separation only.
[e] Processing, air classification, full-scale materials recovery.
[f] Includes costs of water-wall incinerator and approximately 5,000 ft of 14-in steam distribution piping from the
incinerator to customers.

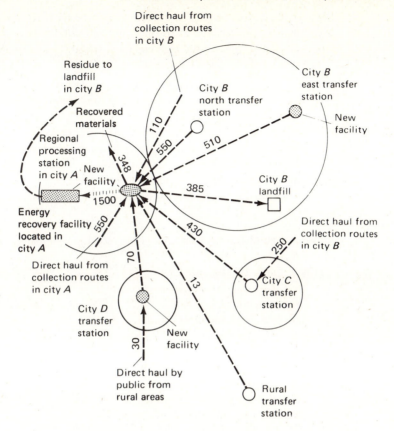

Note: Numbers correspond to solid waste loadings or recovered raw
materials in tons per day. Large circles show approximate
location of city boundaries.

FIG. 18-6 Waste movement for Alternative 4 in Case Study 18-3.

The capital costs for all alternatives are summarized in Table
18-9.

5. Alternative 3 was recommended for implementation. The programs of this alternative began the process of meeting resource
recovery goals mandated by federal legislation and improving the
landfill programs of the region. The required facilities are expand-
able and can be used to produce a processed waste for most
known resource recovery systems. The benefits from the economics of large-scale operations and from the resultant environmental
improvement are significant.

The other alternatives were not recommended for the following reasons. Alternative 1 was rejected because the added transfer
station program increased the disposal costs for all agencies and
offered little or no opportunity for future cost reductions. Alterna-

tive 2 was rejected because the level of resource recovery was not sufficient for materials sales to offset annual costs. Alternative 4 was rejected because of the high capital cost and the uncertainties associated with the energy production.

6. The last step after the selection of an alternative was the development of a more detailed management schedule for implementation, financing, and administration.

Implementation schedule. A task-by-task, year-by-year schedule was developed for the period 1976 to 2000. (Because of the lengthy text needed to present the details of all tasks, the following presentation has been arranged so that activities are listed in a topical format for the functional elements of waste management.)

1976

Task 1   Hold hearings and adopt the solid waste management plan

Task 2   Transfer stations
   City B east transfer station
      Begin preliminary design
      Develop predesign cost estimates
      Begin environmental impact report
      Select a site
   City D transfer station
      Begin preliminary design
      Develop predesign cost estimates
      Begin environmental impact report
      Select a site

Task 3   Resource recovery
   Regional processing facility
      Begin preliminary design
      Develop predesign cost estimates and investigate sites
   Resource recovery studies
      Initiate materials-use policy
      Implement home-separation trial program

Task 4   Disposal
   Develop and adopt disposal site policy
   Begin gas and leachate monitoring system study
   Evaluate use of phased-out or exhausted landfills

Task 5   Administration
   Establish a joint administrative agency (all cities)
   Select solid waste planning committee
   Establish a transfer-disposal enterprise

Task 6   Special projects
   Begin hazardous waste inventory program
   Begin development of storage standards
   Implement inventory program

1977

Task 1   Transfer stations

  City B east transfer station

    Final environmental impact report

    Begin final design

  City D transfer station

    Final environmental impact report

    Begin final design

Task 2   Resource recovery

  Regional processing facility

    Select a site

    Advertise and receive bids from private industry for construction and operation of facility

  Resource recovery studies

    Expand or eliminate home separation

    Finalize a materials-use policy

Task 3   Disposal

  Recommend and install appropriate monitoring systems for gas and leachate

Task 4   Administration

  Review and update plan

  Develop financing plan for the regional processing facility

  Hold planning committee meetings

Task 5   Special projects

  Recommend hazardous waste program

  Complete and adopt storage standards

1978

Task 1   Transfer stations

  City B east transfer station

    Complete final design and award construction contract

  City D transfer station

    Complete final design and award construction contract

Task 2   Resource recovery

  Regional processing facility

    Specify equipment and begin facility construction

  Resource recovery studies

    Select recovery systems that are economically justified

    Update materials-use policy

    Prepare report on existing resource recovery programs and recommend action

Task 3   Disposal

  Update disposal site policy

  Maintain and evaluate gas and leachate system

Task 4   Administration

  Prepare final contract for all cities for regional financing of processing

Revise local ordinances to ensure waste delivery to the regional facility

Establish a processing enterprise

Task 5   Special projects

Implement hazardous waste program

1979

Task 1   Rewrite plan as necessary

Task 2   Transfer stations

Operate all transfer stations and upgrade transport equipment

Task 3   Resource recovery

Regional processing facility

Complete facility construction

Resource recovery studies

Prepare marketing program

Implement recommended resource recovery programs

Task 4   Disposal

Update disposal site policy

Task 5   Administration

Set user rates for processing facility

Establish materials sale contracts

Strengthen enforcement programs

1980–1990 (Medium Term)

Review plan annually

Transfer and disposal

Prepare annual report on the efficiency of the transfer-disposal enterprise and plans for upgrading

Resource recovery

Finish construction of processing system

Prepare annual report on processing enterprise

Administration

Coordinate regional activities

1990–2000 (Long Term)

Review plan annually

Replace and upgrade equipment in all facilities

The complexity of this full-system solid waste management plan is reflected in the comprehensive implementation schedule. The detailed discussion of action through 1979 is essential to successful implementation. Specific details are reduced significantly for the 1980–1990 period, and they are reduced even more for the 1990–2000 period.

Financing. The major facilities of the program are identified in step 4. Preliminary capital costs for construction of the major facilities are presented in Tables 18-10, 18-11, and 18-12. Capital financing for the transfer stations is to be derived from a sinking

TABLE 18-10 PRELIMINARY CAPITAL COST ESTIMATES FOR CITY D TRANSFER STATION IN CASE STUDY 18-3 (Design Capacity = 100 tons/day)

Description	Construction cost,* $
Excavation	30,000
Concrete	27,000
Paving	28,000
Fencing	16,000
Landscaping	30,000
Site, electric service	22,000
Sewer—water (onsite)	25,000
Site work	12,000
Office building	25,000
Operating equipment (internal)	50,000
Subtotal	265,000
Land costs, 5 acres @ $10,000/acre	50,000
Subtotal	315,000
Engineering (10 percent)	32,000
Subtotal	347,000
Contingencies (15 percent)	52,000
Total (excluding transport vehicles)	399,000

* In mid-1975 dollars (ENRCC index = 2,248).

fund and existing capital reserves (financing of $1,854,000). The processing and recovery station will be financed by revenue bonds (financing of $13,893,750 capital will require a bond reserve and issuance cost of $1,396,250 for a total cost of $15,290,000). Operating costs and revenue for repayment of the bonds will be derived from service charges. The effect of increased rates on each city is presented in Table 18-13.

Administration. The major elements of the plan are summarized in Table 18-14. Each element requires administrative action by a local agency. The responsible agencies were existing agencies for all elements except planning and coordination. Planning and coordination are to be accomplished by a joint-powers authority established by all participating cities. Two new enterprises are to be established to handle the joint transfer and disposal facilities and the processing and recovery facilities. The setting up of separate administrative enterprises was a good, workable means of maintaining accountability and equity in the regional system. Accountability becomes more important as revenues are derived from the sale of recovered resources. It should be noted that either a public or a private agency can operate the facilities. Finally, it is extremely important to understand that the entire plan requires administration, not just the facility programs within the plan.

Comment The regional approach to solid waste management is a complex undertaking. Yet, a completed regional plan contains the basic details from which sound decisions can be made and understood. Those decisions will rest primarily with the autonomous local cities even while the economies of a regional plan are achieved.

TABLE 18-11 PRELIMINARY CAPITAL COST ESTIMATES FOR CITY B EAST TRANSFER STATION IN CASE STUDY 18-3 **(Design Capacity = 500 tons/day)**

Description	Construction cost,* $
Excavation	37,000
Concrete	165,000
Paving	103,000
Scale—scalehouse	47,000
Fencing	37,000
Landscaping	58,000
Site, electric service	35,000
Sewer—water (onsite)	29,000
Site work	12,000
Office—locker building	87,000
Pit area building	260,000
Operating equipment (internal)	130,000
Subtotal	1,000,000
Land costs, 15 acres @ $10,000/acre	150,000
Subtotal	1,150,000
Engineering (10 percent)	115,000
Subtotal	1,265,000
Contingency (15 percent)	190,000
Total (excluding transport vehicles)	1,455,000

* In mid-1975 dollars (ENRCC index = 2,248).

TABLE 18-12 PRELIMINARY CAPITAL COST ESTIMATES FOR REGIONAL
PROCESSING FACILITY IN CASE STUDY 18-3
(Design Capacity = 1,500 tons/day)

	Construction cost,* $	
Description	First phase 500	Final phase† 1,500
Mills @ $170,000 each (30 tons/h each)	340,000	850,000
25 percent for equipment installation	85,000	212,500
Air classifiers @ $300,000 each	—	1,500,000
Magnetic separation @ $50,000 each	100,000	250,000
Glass and aluminum separation	—	4,200,000
Conveyors @ $1,500/ft	900,000	2,250,000
10 percent for equipment installation	90,000	225,000
Building for processing facilities	1,000,000	1,000,000
Land	200,000	200,000
Subtotal	2,715,000	10,687,500
Contingencies @ 30 percent	814,500	3,206,250‡
Total	3,529,500	13,893,750

* In mid-1975 dollars (ENRCC index = 2,248).
† Includes all expenditures for the first phase.
‡ The contingency amount is large because the implementation schedule sets construction
for 1978–1979. Costs will change, and more definitive estimates must be made later.

TABLE 18-13 PROJECTED SERVICE CHARGES FOR 1975 TO 1985
IN CASE STUDY 18-3*·†

	Charges per month per household		
Service area	1975 (Current)	1976–1979‡	1980–1985§
City A	$4.00	$4.05	$4.90
City B	4.00	4.40	4.40
City C	3.25	3.55	4.00
City D	4.75	3.80	4.30

* Based on mid-1975 dollars (ENRCC index = 2,248).
† The service charge was projected only to 1985 because inflation and other
operating system modifications will significantly change costs by that time.
Customer service charges pay for collection, transfer, processing, and disposal.
‡ Average monthly charge was set by adding operating cost for the period to
annualized capital costs and dividing that total by the total months in the period.
§ Significant changes in cost for the period caused by opening of the front-end
processing system.

TABLE 18-14   ADMINISTRATIVE ELEMENTS OF THE PLAN AND REQUIRED ACTION IN CASE STUDY 18-3

Element	Major local action
Disposal enforcement/surveillance	
Enforce health standards	Enforce state and local standards
Enforce operating standards	Enforce state and local standards
Control pollution at landfills	Coordinate with state regulatory agencies
Control litter	Develop program
Ongoing planning and coordination	
Publicize plan	Inform public
Review and revise plan	Revise plan in 1978
Storage and collection	
Enforce service standards	Enforce state and local standards
Enforce health standards	Enforce state and local standards
Special waste management	
Monitor hazardous waste	Monitor wastes
New and existing landfills	
Develop landfill-use program	Establish capacities
Develop final use program	Develop final use plans
Plan new landfills	Review use conditions
Facilities plan	
Build transfer stations	Design, construct, operate
Build processing station	Design, construct, operate

18-5   DISCUSSION TOPICS

18-1.   Contact the agencies responsible for solid waste management in your community. Identify all plans and program changes undertaken by each agency within the past year. Evaluate the plans and discuss the success or failure of plan implementation. List the number of policy programs requiring political decisions and the number of programs acted upon without political decisions.

18-2.   Alternatives are normally used in planning documents. Discuss the degree of detail in which various alternatives can and should be presented. List the essential information that should appear in any alternative.

18-3.   A popular concept in solid waste management planning is the development of programs and alternatives in which regional authorities or districts are identified as the preferred administrative agency. Small cities and communities often resist regional plans because they fear the loss of local control, especially the ability to set service rates and conditions. Assume that you are a city council member in a small city of 10,000 people and that your city has been included in a recently completed solid waste management plan for a region containing 600,000 people. The plan has been forwarded to your city council for review and approval. The approval of your city is essential to the acceptance of this plan for the entire region. The major program recommended in the plan is a change from local to regional administration for transfer, processing and recovery, and disposal. What are the key issues for your city? What is the impact on the city-owned solid waste collection system? What future changes in regional waste management objectives might cause your city to lose control of collection operations to the regional authority?

18-4. What public information means are used by management agencies in your community? Randomly select the names of five community residents from your phone book and ask them if they are aware of solid waste management issues on the local level and on state, national, and international levels.

18-5. Obtain a solid waste management plan from a local, state, or federal agency. Review the plan and comment on its contents. Is the presentation logical and easy to follow? How effectively does the plan deal with program implementation? Are all administrative and operational functions and agencies identified?

18-6. Review Case Studies 14-1 and 15-2. Consider each case study separately and answer the following questions for each one: (1) Do the programs as described meet the requirements of a complete plan? Identify any missing components. (2) What difficulties would you anticipate in future plan updating?

18-7. Compare Case Study 18-3 to another solid waste management plan prepared by a state or local (nonregional) agency. How do the problems specifications, goals, and planning approach differ? Why might this difference exist?

18-6 REFERENCES

1. Black & Veatch, Consulting Engineers: *Report on Refuse Collection for Sacramento County, California,* Report prepared for Department of Public Works, Sacramento County, Sacramento, 1972.

2. Buhler, F. (ed.): "Municipal Solid Waste Management Public Relations," National League of Cities, Washington, D. C., 1974.

3. Division of Solid Waste Management: *Sacramento County Solid Waste Management Plan,* Final Report, Department of Public Works, Sacramento County, Sacramento, Calif., 1976.

4. *Financing for Solid Waste Management,* State of California Solid Waste Management Board, Bulletin 6, Sacramento, 1975.

5. Golueke, C. G. and P. H. McGauhey: *Comprehensive Studies of solid Wastes Management,* First Annual Report, Sanitary Engineering Research Laboratory, SERL Research Laboratory, SERL Report 67-7, University of California, Berkeley, 1967.

6. Golueke, C. G. and P. H. McGauhey: *Comprehensive Studies of Solid Waste Management,* Second Annual Report, Sanitary Engineering Research Laboratory, SERL Report 69-1, University of California, Berkeley, 1969.

7. *Guidelines for Local Governments on Solid Waste Management,* U.S. Environmental Protection Agency, Publication SW-17c, U.S. Government Printing Office, Washington, D.C., 1971.

8. Johnson, R. W.: "Financial Management," 3d ed., Allyn and Bacon, Boston, 1966.

9. McFarland, J. M., et al.: *Comprehensive Studies of Solid Wastes Management,* Final Report, Sanitary Engineering Research Laboratory, SERL Report 72-3, University of California, Berkeley, 1972.

10. Theisen, H. and M. Brown: Hawaii's Environmental Planning Aims at Flexibility in Solid Waste Management, *Public Works,* vol. 103, no. 9, 1972.

11. Toftner, R. O. and R. M. Clark: *Intergovernmetal Approaches to Solid Waste Management,* U.S. Environmental Protection Agency, Publication SW-47ts, Washington, D.C., 1971.

A
GLOSSARY

Agricultural Solid Wastes   Wastes produced from the raising of plants and animals for food, including manure, plant stalks, hulls, and leaves.

Ash   The incombustible material that remains after a fuel or solid waste has been burned.

At-Site Time   The time spent unloading and waiting to unload the contents of a collection vehicle or loaded container at a transfer station, processing facility, or disposal site.

Bacteria   Single-cell, microscopic organisms with rigid cell walls. They may be aerobic, anaerobic, or facultative; they can cause disease; and some are important in the stabilization and conversion of solid wastes.

Biodegradable   A compound that can be degraded or converted to simpler compounds by microorganisms.

Bulky Waste   Large wastes such as appliances, furniture, some automobile parts, trees and branches, palm fronds, and stumps.

Carbonaceous Matter   Pure carbon or carbon compounds present in solid wastes.

Carbon Dioxide (CO_2)   A colorless, odorless, nonpoisonous gas that forms carbonic acid when dissolved in water. It is produced during the thermal degradation and microbial decomposition of solid wastes.

Carbon Monoxide (CO)   A colorless, poisonous gas that has an exceedingly faint metallic odor and taste. It is produced during the thermal degradation and microbial decomposition of solid wastes when the oxygen supply is limited.

Collection   The act of picking up wastes at homes, businesses, commercial and industrial plants, and other locations, loading them into a collection vehicle (usually enclosed), and hauling them to a facility for transfer or further processing or to a disposal site.

Collection Routes   The established routes followed in the collection of wastes from homes, businesses, commercial and industrial plants, and other locations.

Collection Systems   Collectors and equipment used for the collection of solid wastes. Solid waste collection systems may be classified from several points of view, such as the mode of operation, the equipment used, and the types of wastes collected. In this text, collection systems have been classified according to their mode of operation into two categories: hauled container systems and stationary container systems.

Combustibles   Various materials in the waste stream that are burnable. In general, they are organic in nature—paper, plastics, wood, and food wastes.

Combustion   The chemical combining of oxygen with a substance that results in the production of heat and usually light.

Commercial Solid Wastes   Wastes that originate in wholesale, retail, or service establishments, such as office buildings, stores, markets, theaters, hotels, and warehouses.

Compactor   Any power-driven mechanical equipment designed to compress and thereby reduce the volume of wastes.

Compactor Collection Vehicle   A large vehicle with an enclosed body having special power-driven equipment for loading, compressing, and distributing wastes within the body.

Component Separation   The arranging or sorting of wastes into components or classes.

Compost   A mixture of organic wastes partially decomposed by aerobic bacteria to an intermediate state. It can be used as a soil conditioner.

Construction Wastes   Wastes produced in the course of construction of homes, office buildings, dams, industrial plants, schools, etc. The materials usually include used lumber, miscellaneous metal parts, packaging materials, cans, boxes, wire, excess sheet metal, etc.

Container   A receptable used for the storage of solid wastes until they are collected.

Conversion   The transformation of wastes into other forms. Transformation by burning or pyrolysis into steam, gas, or oil are examples.

Conversion Products   Products derived from the first-step conversion of solid wastes, such as heat from combustion and gas from biological conversion.

Cover Material   Soil used to cover compacted solid wastes in a sanitary landfill.

Decomposition   The breakdown of organic wastes by bacterial, chemical, or thermal means. Complete chemical oxidation leaves only carbon dioxide, water, and inorganic solids.

Demolition Wastes   Wastes produced from the destruction of buildings, roads, sidewalks, etc. These wastes usually include large, broken pieces of concrete, pipe, radiators, duct work, electrical wire, broken-up plaster walls, lighting fixtures, bricks, and glass.

Dewatering   The removal of water from solid wastes and sludges by various thermal and mechanical means.

Digestion   The biological conversion of processed organic wastes to methane and carbon dioxide under anaerobic conditions.

Disposal   The activities associated with the long-term handling of (1) solid wastes that are collected and of no further use and (2) the residual matter after solid wastes have been processed and the recovery of conversion products or energy has been accomplished. Normally disposal is accomplished by means of sanitary landfilling.

Effluent   Any solid, liquid, or gas that enters the environment as a by-product of human activities.

Energy Recovery   The process of recovering energy from the conversion products derived from solid wastes, such as the heat produced from the burning of solid wastes.

Ferrous Metals   Metals composed predominantly of iron. In the waste materials stream, these metals usually include cans, automobiles, refrigerators, stoves, etc.

Fly Ash   Small solid particles of ash and soot generated when coal, oil, or wastes are burned. With proper equipment, fly ash is collected before it enters the atmosphere. Fly ash residue can be used for building materials (bricks) or in a sanitary landfill.

Food Wastes   Animal and vegetable wastes resulting from the handling, storage, sale, preparation, cooking, and serving of foods; commonly called garbage.

Front-End System   Those processes used for the recovery of materials from solid wastes and the preparation of individual components for subsequent conversion using rear-end systems.

Functional Element   The term functional element is used in this text to describe the various activities associated with the management of solid wastes from the point of generation to final disposal. In general, a functional element represents a physical activity. The six functional elements used throughout this book are waste generation, onsite storage, collection, transfer and transport, processing and recovery, and disposal.

Garbage (see Food Wastes)

Generation (see Waste Generation)

Groundwater   Water beneath the earth's surface and located between saturated soil and rock. It is the water that supplies wells and springs.

Haul Distance   The distance a collection vehicle travels (1) after picking up a loaded container (hauled container system) or from its last pickup stop on a collection route (stationary container system) to the solid waste transfer station, processing facility, or sanitary landfill, and (2) the distance the collection vehicle travels after unloading to the location where the empty container is to be deposited or to the beginning of a new collection route.

Haul Time   The elapsed or cumulative time spent transporting solid wastes between two specific locations.

Hauled Container System   Collection systems in which the containers used for the storage of wastes are hauled to the disposal site, emptied, and returned to either their original location or some other location.

Hazardous Wastes   Wastes that by their nature are inherently dangerous to handle or dispose of. These wastes include radioactive substances, toxic chemicals, biological wastes, flammable wastes, and explosives. They usually are produced in industrial operations or in institutions.

Hog Feeding   Disposing of food wastes by feeding them to hogs. State regulations usually require that the wastes be cooked to kill bacteria before feeding. Some states have regulations making hog feeding of food wastes illegal.

Hydrogen Sulfide (H_2S)   A poisonous gas with the odor of rotten eggs that is produced from the reduction of sulfates in, and the putrefaction of, a sulfur-containing organic material.

Incineration   The controlled process by which solid, liquid, or gaseous combustible wastes are burned and changed into gases, and the residue produced contains little or no combustible material.

Industrial Wastes   Wastes generally discarded from industrial operations or derived from manufacturing processes. A distinction should be made between scrap (those materials which can be recycled at a profit) and solid wastes (those that are beyond the reach of economic reclamation).

Leachate   Liquid containing decomposed wastes, bacteria, and other materials that drains out of landfills.

Litter   That highly visible portion of solid wastes that is generated by the consumer and carelessly discarded outside the regular disposal system. Litter accounts for only about 2 percent of the total solid waste volume.

Manual Separation   The separation of wastes by hand. Sometimes called "hand-picking" or "hand sorting," manual separation is done in the home or office by keeping food wastes separate from newspaper, or in a recovery plant by picking out large cardboard or metal objects.

Materials Balance   An accounting of the weights of materials entering and leaving a processing unit, such as an incinerator, usually on an hourly basis.

Materials Recovery (see Resource Recovery)

Mechanical Separation   The separation of wastes into various components by mechanical means.

Methane (CH₄)   An odorless, colorless, and asphyxiating gas that can explode under certain circumstances and that can be produced by solid wastes undergoing anaerobic decomposition.

Microorganisms   Generally, any living thing microscopic in size and including bacteria, yeasts, simple fungi, some algae, slime molds, and protozoans. They are involved in stabilization of wastes (composting) and in sewage treatment processes.

Moisture Content   The weight loss (expressed in percent) when a sample of solid wastes is dried to a constant weight at a temperature of 100° to 105°C.

Municipal Wastes   The combined residential and commercial wastes generated in a given municipal area. The collection and disposal of these wastes are usually the responsibility of local government.

Nonferrous Metals   Metals that contain no iron. In wastes these are usually aluminum, copper wire, brass, bronze, etc.

Off-Route Time   All time spent by the collectors on activities that are nonproductive from the point of view of the overall collection operation.

Onsite Handling, Storage, and Processing   The activities associated with the handling, storage, and processing of solid wastes at the source of generation before they are collected.

Organic Materials   Chemical compounds of carbon combined with other chemical elements and generally manufactured in the life processes of plants and animals. Most organic compounds are a source of food for bacteria and are usually combustible.

Pickup Time   *For a hauled container system,* it represents the time spent driving to a loaded container after an empty container has been deposited, plus the time spent picking up the loaded container and the time required to redeposit the container after its contents have been emptied. *For a stationary container system,* it refers to the time spent loading the collection vehicle, beginning with the stopping of the vehicle prior to loading the contents of the first container and ending when the contents of the last container to be emptied have been loaded.

Pollution   The contamination of soil, water, or the atmosphere by the discharge of wastes or other offensive materials.

Primary Materials   Virgin or new materials used for manufacturing basic products. Examples include wood pulp, iron ore, and silica sand.

Processing   Any method, system, or other means designated to change the physical form or chemical content of solid wastes.

Putrescible   Subject to decomposition or decay. Usually used in reference to food wastes and other organic wastes.

Pyrolysis   A way of breaking down burnable waste by combustion in the absence of air. High heat is usually applied to the wastes in a closed chamber, and all moisture evaporates and materials break down into various hydrocarbon gases and carbonlike residue.

Rear-End System   Those chemical, thermal, and biological systems and related ancillary facilities used for the conversion of processed solid wastes into various products.

Reclamation   The restoration to a better or more useful state, such as land reclamation by sanitary landfilling, or the extraction of useful materials from solid wastes.

Recoverable Resources   Materials that still have useful physical or chemical properties after serving a specific purpose and can therefore be reused or recycled for the same or other purposes.

Recovery (see Resource Recovery)

Recycling   Separating a given waste material (e.g., glass) from the waste stream and processing it so that it may be used again as a useful material for products which may or may not be similar to the original.

Refuse   A term often used interchangeably with the term *solid wastes*. To avoid confusion, the term *refuse* is not used in this text.

Residential Wastes   Wastes generated in houses and apartments, including paper, cardboard, beverage and food cans, plastics, food wastes, glass containers, and garden wastes.

Residue   The solid materials remaining after completion of a chemical or physical process, such as burning, evaporation, distillation, or filtration.

Resource Recovery   Resource recovery is a general term used to describe the extraction of economically usable materials or energy from wastes. The concept may involve recycling or conversion into different and sometimes unrelated uses.

Reuse   The use of a waste material or product more than once.

Rubbish   A general term for solid wastes—excluding food wastes and ashes—taken from residences, commercial establishments, and institutions.

Sanitary Landfill   A land area where solid wastes are disposed of using sanitary landfilling techniques.

Sanitary Landfilling   An engineered method of disposing of solid wastes on land in a manner that protects the environment, by spreading the wastes in thin layers, compacting it to the smallest practical volume, and covering it with soil by the end of each working day.

Secondary Material   A material that is used in place of a primary or raw material in manufacturing a product.

Separation   To divide wastes into groups of similar materials, such as paper products, glass, food wastes, and metals. Also used to describe the further sorting of materials into more specific categories, such as clear glass and dark glass. Separation may be done manually or mechanically with specialized equipment.

Service Site or Location   A residential unit, business, commercial, or industrial establishment, or other pickup point from which solid wastes are collected periodically.

Shredding   Mechanical operations used to reduce the size of solid wastes. See also Size Reduction (Mechanical).

Size Reduction (Mechanical)   The mechanical conversion of solid wastes into small pieces. In practice, the terms shredding, grinding, and milling are used interchangeably to describe mechanical size-reduction operations.

Solid Waste Management   The purposeful, systematic control of the functional elements of generation, onsite storage, collection, transfer and transport, processing and recovery, and disposal associated with the management of solid wastes from the point of generation to final disposal. *Plan*. Solid waste management plans are developed to define and establish objectives and policies, and they deal with problems at any level—city or county, subregional or regional, state or federal. Typically, a local plan encompasses one or more functional elements and one or more program areas. *Program*. In the field of solid waste management planning, the term program encompasses all the activities associated with the development of a solution to a problem or problems within a functional element of a solid waste management system. Programs dealing with specific problems related to a functional element may or may not involve policy issues and objectives. If they do, they must be presented to the appropriate decision-makers, such as the members of a city council or a county board. *System*. The assemblage of one or more of the functional elements to achieve a given objective or goal.

Solid Wastes   Any of a wide variety of solid materials, as well as some liquids in containers, which are discarded or rejected as being spent, useless, worthless, or in excess. Does not usually include waste solids from treatment facilities. See also agricultural, commercial, construction, demolition, hazardous, industrial, municipal, and residential wastes.

Stationary Container Systems   Collection systems in which the containers used for the storage of wastes remain at the point of waste generation, except for occasional short trips to the collection vehicle.

Transfer   The act of transferring wastes from the collection vehicle to larger transport vehicles.

Transfer Station   A place or facility where wastes are transferred from smaller collection vehicles (e.g., compactor trucks) into larger transport vehicles (e.g., over-the-road and off-road tractor trailers, railroad gondola cars, or barges) for movement to disposal areas, usually landfills. In some transfer operations, compaction or separation may be done at the station.

Transport   The transport of solid wastes transferred from collection vehicles to a facility or disposal site for further processing or action.

Trash   Wastes that usually do not include food wastes but may include other organic materials, such as plant trimmings.

Treatment Process Sludges   Liquid and semisolid wastes resulting from the treatment of domestic waste water and industrial wastes.

Virgin Material   Any basic material for industrial processes which has not previously been used, for example, wood-pulp trees, iron ore, silica sand, crude oil, bauxite. See also Secondary Material, Primary Materials.

Volume Reduction   The processing of wastes so as to decrease the amount of space they occupy. Complete conventional incineration can reduce volume by 90 percent; high-temperature incineration can reduce volume by as much as 98 percent. Compaction systems can also reduce volume by 50 to 80 percent.

Waste Generation   The act or process of generating solid wastes.

Waste Sources   Agricultural, residential, commercial, and industrial activities, open areas, and treatment plants where solid wastes are generated.

Waste Stream   The waste output of an area, location, or facility.

B
PUBLIC INFORMATION PROGRAMS

The purpose of public participation in management is twofold: first, the opportunity is created for managers to inform and instruct the public; second, a channel of communication is opened that allows the public to communicate its needs and desires to managers. The opportunity for two-way communication is of great importance, for the activities, goals, and means of solid waste management presently are not well understood by the public.

There are numerous ways in which to convey the ideas and concepts concerning a technical system to a nontechnical community. The most successful communication channels are developed by using simple terms and objective-oriented graphics in written materials (plans, information booklets, etc.) and an open-door policy for the input of new ideas to management. An example of such a communication technique is the pamphlet used to provide information to the people of Lane County, Oregon, about their solid waste management system. It is reproduced in this appendix. It should be noted that an information program of the type presented here can be developed only when adequate money and time are applied to it. A specific portion of a plan development budget should be set aside for public information programs.

a golden trashery

of terms about solid waste in Lane County

solid waste

(refuse, trash, garbage) - any
useless, unwanted, or discarded
material that is not wet
enough to be free-flowing.

Solid Waste Division -

the agency responsible for
management of the garbage disposal
program in Lane County.

Part of the County's
Department of Environmental
Management.

Solid Waste Management Plan - a comprehensive summary of information on solid waste in Lane County, including plans for systematic control of storage, collection, transport, separation, processing,
recycling, financing, and disposal. Emphasizes flexibility, proven
technology, and **resource recovery**. Adopted by the Board of
County Commissioners in September, 1973.

accelerated action plan - the proposal to speed up the schedule for a county-wide system of **resource recovery** through mechanical as well as voluntary methods.

resource recovery - taking useful materials or energy out of solid waste at any stage before ultimate disposal.

recycling - transforming waste materials into useful items by reprocessing them.

reuse - returning an item to productive use for the same purpose as it was orginally intended, without changing its identity.

energy recovery - the use of solid wastes as fuel,
supplementing wood wastes,
coal, or other fuels, to produce energy in
the form of steam or electricity.

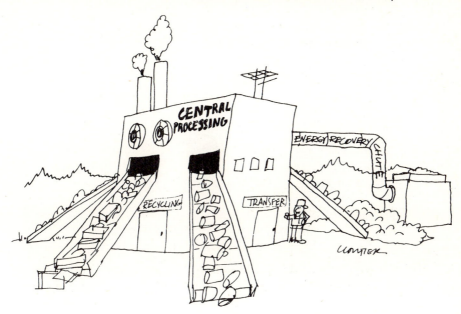

central processing, recycling, and transfer facility - a building where
almost all the solid wastes in the county would be brought and
shredded. Components that would have recovery value would be
separated and shipped to markets, portions suitable for **energy
recovery** would go to furnaces in a nearby plant, and the residue
would go to a landfill. Proposed in the **accelerated action plan.**

EWEB - Eugene Water and Electric Board, supplier of utilities to
central Eugene, Santa Clara, River Road, etc. Has designed tests to
determine whether one of their existing boilers could be modified
to use solid wastes as fuel.

sanitary landfill - a place where wastes are buried in a manner that minimizes the detrimental impact on the environment. Wastes dumped onto the prepared site are spread and compaced in layers 10 to 15 feet deep, and covered at the end of each day with 6 to 12 inches of sand, dirt, or wood chips. DEQ will not approve any landfill in Western Oregon that has not been upgraded to sanitary landfill status by the end of 1974.

major (metropolitan) sanitary landfill - the disposal site for wastes from metropolitan Eugene-Springfield, which constitute at least 80% of the county's solid wastes, and the ultimate disposal site for wastes transferred from several outlying areas. Day Island, current major sanitary landfill, must close this summer; a site to replace it has not been chosen.

interim sanitary landfill - a site to be used temporarily as the **major sanitary landfill** until a long-range, permanent site has been chosen. Bethel-Danebo, near West 11th Ave. in Eugene, will be used for this purpose for 18 to 36 months.

modified landfill - a place where solid wastes are spread and compacted periodically and are covered with sand, dirt, or woodchips. Currently located at Cottage Grove, Creswell, Florence, Franklin, and Oakridge. Must be upgraded to sanitary landfill status or phased out by the end of 1974.

rural transfer station - a place in an outlying section of the county where wastes are dumped into **drop-boxes** for periodic transportation by trailer to a nearby landfill.

Former open dumps are are being converted to transfer stations in 11 locations.

drop-box - a steel box, holding up to 45 cubic yards of wastes, with a screened top to prevent scattering of lightweight materials. Two drop-boxes are at each **rural transfer station.**

demolition wastes - the debris
resulting from tearing down
old buildings.
Includes concrete,
brick,
steel,
wiring,
etc.

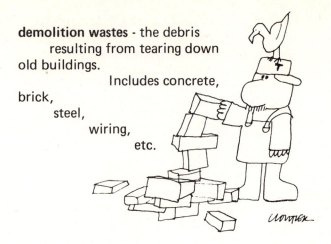

special wastes - any solid wastes
that require special handling, for example
poison containers,
hospital wastes,
automobile tires,
appliances,
dead animals,
etc.

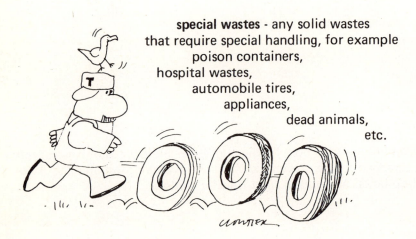

kraft paper - a heavy grade of paper, used in most grocery bags,
postal wrapping bags, manila folders, and carboard.

corrugated cardboard - two layers of flat cardboard
with a ribbed layer in between.

DEQ - the State Department of Environmental Quality, with headquarters in Portland and a regional office in Eugene. The state agency that oversees control of air and water pollution and management of solid waste, including operation of disposal sites.

EPA - the federal Environmental
Protection Agency, with headquarters in
Washington, D.C. and Seattle.
The federal agency responsible for
all aspects of preserving
the environment.

LRAPA - Lane Regional Air Pollution Authority, the agency responsible for maintaining air quality in Lane County.

bond - a formal loan to a public or private agency by investors, normally for long-term capital needs. Repayment is guaranteed within a specified time, when the bond is said to "mature."

general obligation bond - the source of repayment of the loan may be whatever the borrower deems appropriate; it is not a specifically designated revenue source.

capital funds or costs - funds for or costs of construction of a building or purchase of heavy equipment.

operational funds or costs - funds for or costs of ongoing, daily operations of an agency or program, including salaries, supplies, fuel, phone, etc.

lane county

Solid Waste Division
135 East 6th Ave., Eugene
687-4119

C
STATISTICAL ANALYSIS OF SOLID WASTE GENERATION RATES

The purpose of this appendix is to introduce (1) some statistical measures that are commonly used to characterize solid waste generation rates and (2) graphical procedures for obtaining the necessary information from such data.

C-1 STATISTICAL MEASURES

Commonly used statistical measures include frequency and the mean, mode, median, standard deviation, and coefficient of variation.

Frequency The *frequency* of occurrence represents the number of times a given value occurs in a set of observations.

Mean The *mean* is defined as the arithmetic average of a number of individual measurements and is given by

$$\bar{x} = \frac{1}{n} \sum_{i=1}^{n} x_i \tag{C-1}$$

where $\bar{x}$ = mean value
n = number of observations
x_i = the ith observation

Median If a series of observations are arranged in order of increasing value, the middle-most observation, or the arithmetic mean of the two middle-most observations, in a series is known as the *median*. For example, in a set of 15 measurements, the median will be the 8th value, whereas in a set of 16 measurements, the median will be the average of the 8th and 9th values. In a symmetrical set of values, the median will equal the mean.

Mode   The value occurring with the greatest frequency in a set of observations is known as the *mode*. If a continuous graph of the frequency distribution is drawn, the mode is the value of the high point, or hump, of the curve. In a symmetrical set of observations, the mean, median, and mode will be the same value.

Standard Deviation   Because of the laws of chance, there is uncertainty in any set of measurements. The precision of a set of measurements can be assessed in a number of different ways. Most commonly, the error of an individual measurement in a set is defined as the difference between the arithmetic mean and the value of the measurement. If v is used to denote the error $x - \bar{x}$, then the *standard deviation* is defined as

$$\sigma_s = \sqrt{\frac{\Sigma v^2}{n - 1}} \tag{C-2}$$

where σ_s = standard deviation
$\quad\quad\quad v$ = error of individual measurement = $x - \bar{x}$
$\quad\quad\quad n$ = number of observations

From the form of the equation, it can be concluded that the larger the scatter in a set of measurements, the larger the value of σ_s will be. Conversely, as the precision of a set of measurement improves, the value of the standard deviation will decrease. From the theoretical considerations, it can be shown that if the measurements are distributed normally, then 68.27 percent of the observations will fall within plus or minus one standard deviation from the mean $(\bar{x} \pm \sigma_s)$ [C-4, C-5].

Coefficient of Variation   Although the standard deviation can be used as an indication of the absolute dispersion of a set of measured values, in itself it provides little or no information as to whether the value is large or small. To overcome this difficulty, the *coefficient of variation*, as defined in Eq. C-3, is used as a relative measure of dispersion [C-4].

$$CV = \frac{100\sigma_s}{\bar{x}} \tag{C-3}$$

where CV = coefficient of variation, percent
$\quad\quad\quad \sigma_s$ = standard deviation (see Eq. C-2)
$\quad\quad\quad \bar{x}$ = mean value (see Eq. C-1)

Typically, the coefficient of variation for solid waste generation rates will vary from 10 to 60 percent. To judge whether this percentage represents a large or small scatter, it can be compared to values obtained from measurements in other fields. For measurements in the biological field, the coefficient of variation will vary from about 10 to 30 percent. The coefficient of variation for chemical analyses varies from 2 to 10 percent. Clearly, the scatter in solid waste generation data is significant.

C-2   GRAPHICAL REPRESENTATION AND ANALYSIS

The graphical presentation and analysis of observed data can be used to depict and evaluate trends and to determine the reliability of conclusions made from a limited set of observations. Time series, histogram or frequency plots, and frequency-probability plots are used extensively for the presentation and analysis of data.

Time Series   Observations arranged in the order of occurrence in time are often called *time series*. By plotting the observed values versus time, it is often possible to establish trends, cycles or periodicities, and fluctuations that may be of value in understanding the basic nature of the phenomenon under evaluation.

Frequency Distributions   Observations arranged in order of magnitude form an array. If whole numbers are assigned to a magnitude range, then the frequency of occurrence of whole numbers can be plotted against the magnitude ranges. The resulting plots (see Fig. C-1) are called *histograms*. As shown, histograms can be symmetrical (Fig. C-1a) and asymmetrical (Fig. C-1b). The data forming symmetrical histograms are said to be *arithmetically normal,* whereas the data in asymmetrical histograms are said to be *skewed*. Typically, skewed data are geometrically normal (see subsequent discussion). Most solid waste generation data are distributed asymmetrically.

Arithmetic Probability Paper   Although it is possible to express the summation or probability curve in equation form, it has been found more useful to develop graph paper, called "arithmetic probability paper," with special coordinates on which data that are normally distributed (see Fig. C-1a) will plot as a straight line. When using arithmetic probability paper, the origin of the coordinate axis is placed in the center of horizontal axis, and a probability scale is developed which extends on either side. To use this paper, the measurements in a data set are first arranged in order of increasing magnitude, and a corresponding plotting position is determined by using Eq. C-4. The data are then plotted.

$$\text{Plotting position (\%)} = \left(\frac{m}{n+1}\right)100 \qquad \text{(C-4)}$$

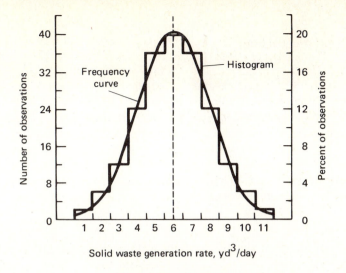

(a) **Symmetrical distribution**
(arithmetically normal)

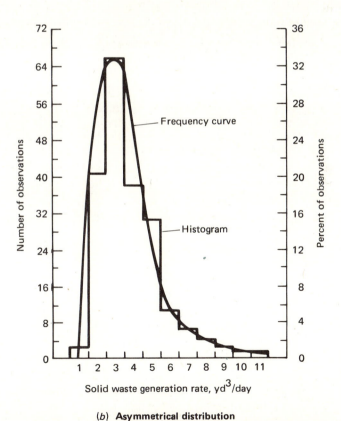

(b) **Asymmetrical distribution**
(geometrically normal)

FIG. C-1 Frequency distribution for the generation of solid wastes.

where m = order number
        n = number of observations

In effect, the plotting position represents the percent or frequency of observations that are equal to, or less than, the indicated value. The term $(n + 1)$ is used to account for the fact that there may be an observation that is either larger or smaller than the largest or smallest in the data set.

By plotting data on arithmetic probability paper, it is possible to determine:

1.   Whether the data are normally distributed by noting if the data can be fit with a straight line. Departure from a straight line can be taken as an indication of skewness.

2.   The approximate magnitude of the arithmetic mean. Usually it will be best to compute the mean and to pass the straight line plotted by eye through the computed value.

3.   The approximate value of the standard deviation by finding the values on the curve at the 84.1 $(50 + 68.27/2)$ and 15.9 $(50 - 68.27/2)$ percent points and noting that these values correspond to $\bar{x} \pm \sigma_s$.

4.   The expected frequency of any observation of a given magnitude.

The use of arithmetic probability paper is illustrated using the solid waste generation data given in Table C-1. As shown, the data in Table C-1 have been arranged in order of increasing magnitude, and plotting positions have been computed by using Eq. C-4. The corresponding arithmetic

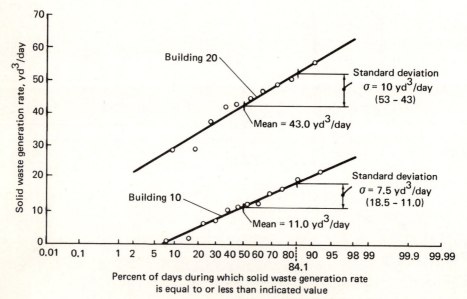

Fig. C-2   Probability plot of solid waste generation data (Ref. C-1).

TABLE C-1 FREQUENCY DISTRIBUTION OF SOLID WASTE
GENERATION RATES FOR TWO TYPICAL INDUSTRIAL FACILITIES

	Building 10		Building 20	
Rank serial number, m	Solid wastes, yd³/day	Plotting position,† %	Solid wastes, yd³/day	Plotting position,† %
1	1.0	7.7	28.5	9.1
2	1.9	15.4	29.1	18.2
3	6.6	23.1	37.4	27.3
4	7.4	30.8	42.0	36.4
5	10.4	38.5	42.7	45.5
6	11.3	46.2	44.6	54.65
7	12.0	53.8	46.8	63.6
8	12.6	61.5	48.6	72.7
9	15.8	69.2	50.5	81.8
10	17.0	76.9	56.0	90.9
11	20.0	84.6		
12	22.2	92.3		
Median	11.7		43.7	
Mean‡	11.0		43.0	
σ_s‡	7.5		10.0	

* Adapted from Ref. C-1.
† Plotting position computed using Eq. C-4.
‡ From Fig. C-2.

probability plot of these data is given in Fig. C-2. As illustrated, the solid waste generation data are distributed normally. The corresponding means and standard deviations for the two distributions are also shown.

Log Normal Probability Paper   When the data are skewed (see Fig. C-1b), logarithmic probability paper can be used. The implication here is that the logarithm of the observed values is normally distributed. On logarithmic probability paper, the straight line of best fit passes through the geometric mean and through the intersection of $\bar{X}_g \times \sigma_g$ at a value of 84.1 percent and $\bar{X}_g/\sigma_g$ at a value of 15.9 percent. The geometric standard deviation can be determined from the following equation:

$$\sigma_g = \frac{P_{84.1}}{\bar{X}_g} = \frac{\bar{X}_g}{P_{15.9}} \tag{C-5}$$

where σ_g = geometric standard deviation
    $P_{84.1}$ = value from curve at 84.1 percent
    $\bar{X}_g$ = geometric mean
    $P_{15.9}$ = value from curve at 15.9 percent

C-3   DISCUSSION

The above presentation dealing with the statistical analysis of solid waste generation rates is intended to serve as a brief introduction to the subject. For a more detailed presentation of the fundamentals of statistical analysis, the reader is referred to Refs. C-2 to C-5.

C-4   REFERENCES

C-1.   Tchobanoglous, G. and G. Klein: *An Engineering Evaluation of Refuse Collection Systems Applicable to the Shore Establishment of the U.S. Navy,* Sanitary Engineering Research Laboratory, University of California, Berkeley, 1962.

C-2.   Velz, C. J.: Graphical Approach to Statistics, *Water and Sewage Works,* Reference and Data Issue, 1952.

C-3.   Velz, C. J.: Appendix A, Statistical Tools, in "Applied Stream Sanitation," Wiley-Interscience, New York, 1970.

C-4.   Waugh, A. E.: "Elements of Statistical Analysis," 2d ed., McGraw-Hill, New York, 1943.

C-5.   Young, H. D.: "Statistical Treatment of Experimental Data," McGraw-Hill, New York, 1962.

D
TYPICAL COST DATA AND COST-ESTIMATING PROCEDURES FOR EQUIPMENT USED IN SOLID WASTE MANAGEMENT SYSTEMS

The cost of acquiring and operating solid waste management equipment and facilities is an important determinant in the design and selection of a waste management system. For a complex system, the costs of structures, roads, equipment, landscaping, utility connection fees, and other items might be included. Because of the many factors that must be considered, development of cost estimates for such systems should be undertaken only by qualified persons. The purpose of this appendix is to provide typical cost data on the purchase price of selected solid waste management equipment. Operating costs are also important, but they are not presented here because their variability makes the use of generalized values highly questionable. Cost estimating for processing and resource recovery systems is also presented.

D-1 COST DATA FOR CONVENTIONAL EQUIPMENT

It is often helpful to have some reference cost data available for use in broadly defining the magnitude of the expenditures associated with equipment acquisition decisions. Representative cost data for selected waste management equipment are presented in Table D-1. The information is not intended to be all-inclusive within the type of equipment shown, nor is the type listed the only equipment available. Because equipment costs are highly variable and dependent on factors such as performance specifica-

TABLE D-1   COST DATA FOR SELECTED SOLID WASTE MANAGEMENT EQUIPMENT

Type of equipment	Size, rated capacity, or gross weight		Approximate cost range,* $
	Unit	Value	
Storage containers			
Small-capacity household			
Metal	gal	32	5–8
Plastic	gal	32	6–10
Medium capacity, used with hoist truck, bottom	yd³	6	1,100– 1,400
unloading	yd³	10	1,400– 1,700
Medium capacity, unloaded mechanically, used in	yd³	2	300– 350
conjunction with front-, rear-, and side-loaded	yd³	6	600– 800
compactors and hoist trucks	yd³	8	800– 900
Large capacity, used with tilt-frame collection	yd³	20	1,800– 2,000
vehicles	yd³	30	2,100– 2,500
	yd³	40	2,600– 3,000
Collection vehicles			
Hoist truck	tons	6	21,000– 26,000
Tilt-frame	tons	15	23,000– 27,000
	tons	25	35,000– 40,000
Front-loaded compactor	yd³	24	46,000– 58,000
	yd³	40	55,000– 65,000
Side-loaded compactor	yd³	20	35,000– 40,000
	yd³	35	45,000– 55,000
Rear-loaded compactor	yd³	20	30,000– 40,000
	yd³	25	40,000– 50,000
Transfer and transport equipment			
Rubber-tired clamshell	tons	10	35,000– 45,000
Trailers			
Hydraulic ejector blade	yd³	65	30,000– 35,000
Movable floor or cable pull	yd³	96	25,000– 30,000
Tractors			
Conventional	hp	320	30,000– 35,000
Cab-over	hp	320	30,000– 35,000
Landfill equipment			
Crawler loader	tons	12–16	50,000– 70,000
	tons	20–24	78,000– 95,000
Rubber-tired loader	tons	7–15	30,000– 65,000
	tons	20–35	85,000–150,000
Crawler tractor	tons	10–16	58,000– 70,000
	tons	24–30	90,000–120,000
	tons	40	200,000–230,000
Rubber-tired tractor	tons	16–20	70,000– 80,000
	tons	30–40	130,000–160,000

TABLE D-1   (Continued)

| Type of equipment | Size, rated capacity, or gross weight | | Approximate cost range,* $ |
	Unit	Value	
Compactor, steel-wheeled	tons	16–20	65,000– 85,000
	tons	24–30	115,000–135,000
Road grader	tons	15	65,000– 80,000
Self-propelled elevating scraper	yd³	14–20	90,000–150,000
Water truck	gal	3,000	25,000– 35,000

* All cost data have been adjusted to an Engineering News-Record Construction Cost index of 2,327 that corresponds to the value of the index on March 25, 1976.

tion, quantity ordered, geographic location, and shipping costs, the data reported are meant to serve only as a guide.

Because costs are changing so rapidly both nationally and locally, it is extremely important that any cost data and/or evaluation be referenced to some cost index. Some typical indices are reported in Table D-2. Although none of the indices listed in Table D-2 are wholly satisfactory for use with solid waste management equipment, the Engineering News-Record Construction Cost (ENRCC) index is the one most commonly used for the preparation of general estimates. Data reported in the literature can be adjusted to a common basis for purposes of comparison with the indices reported in Table D-2 by using the following relationship:

$$\text{Current cost} = \frac{\text{current value of index}}{\text{value of index at time of report}} \times (\text{cost cited in report}) \quad \text{(D-1)}$$

When possible, index values should also be adjusted to reflect current local costs. If the month of the year that equipment was purchased is not given, it is common practice to use the June end-of-month index value. All the costs given in Table D-1 have been adjusted to an ENRCC index value of 2,327, which corresponds to the value of the index on March 25, 1976. To project the costs given in Table D-1 into the future, the following relationship can be used:

$$\text{Future cost} = \frac{\text{projected future value of ENRCC index}}{2,327} \times (\text{cost cited in Table D-1}) \quad \text{(D-2)}$$

TABLE D-2   INDICES FOR ADJUSTING COST DATA

Index	Base year (index = 100)
Chemical engineering	
Plant	1957–1959
Equipment, machinery, and supports	1957–1959
Engineering News-Record	
Building	1913
Construction	1913
Marshall and Swift equipment	1926
U.S. Department of Commerce Industrial Production Indexes	
Machinery	1967
Transportation	1967
U.S. Environmental Protection Agency	
Sewage treatment plant construction	1957–1959
Sewer construction	1957–1959

D-2   COST ESTIMATING FOR PROCESSING EQUIPMENT AND FACILITIES

Cost data are not presented for processing and energy conversion equipment because these systems are in such an early stage of development that any generalized cost data concerning the purchase of system components are rendered meaningless. Equipment costs must be estimated from direct quotations of equipment manufacturers.

The complexity of many proposed solid waste processing and recovery systems has resulted in highly complicated cost-estimating techniques. Many communities are now undertaking these complex programs, and public agency officials may not always have a thorough understanding of what costs should be included in an economic analysis of the proposed systems. As a result, many communities are receiving competitive cost proposals from private industry, and the officials involved may not know whether the basis of cost in one proposal is comparable to that in another. Because of these uncertainties, some states and the U.S. Environmental Protection Agency have published guidelines for the cost analysis of resource recovery systems. The guidelines published by the state of California are presented in Table D-3.

TABLE D-3 COST ANALYSIS OF RESOURCE RECOVERY SYSTEMS[a]

Item	Costing method	Annual depreciation[b]	Cost
LAND			
Acquisition[c]			$
Geological/soil survey	Service fee		
Mapping and plot drawings	Service fee		
Land purchase	Number of acres purchased		
Subtotal			$
Land improvement			
Clearing (including demolition)	Number of acres cleared and demolition costs		$
General grading	Number of acres graded		
Storm drains	Number of culverts, drain inlets, connections or discharge structures, length of drain lines		
Water supply lines	Service connection and length of distribution piping		
Waste-water and sewer lines	Length of piping, number of manholes, service connections		
Landscaping	Number of acres landscaped, irrigation system, gravel cover area, number of acres to which weed killer was applied, special fixtures, planting, etc.		
Gas supply lines	Service connections, length of piping, gas meters		
Electrical lines	Length of conduit, cables, wiring, electric meters		
Subtotal			$
Site preparation			
Access road(s)	Amount of special grading and area paved, curbing		
Circulation road(s)	Amount of special grading and area paved, curbing		
Parking areas	Amount of area paved		
Fencing	Length of fence, number of gates		
Site drainage grading	Number of acres graded		
Traffic signals (if required)			
Subtotal			$
Total land costs			$

TABLE D-3   (Continued)

Item	Costing method	Annual depreciation[b]	Cost
BUILDING[d]			
Structural work	Excavation and foundation on cubic-yard basis; above-ground on square foot		$
Masonry, wood, steel, glazing, and painting	Usually on square-foot basis		
Plumbing, climate control, and fire protection	Length of piping, number of fixtures, system cost for heating and cooling		
Total building costs			$
SUPPORT FACILITIES			
Weighing station	Unit cost		$
Scale house	Unit cost		
Electrical substations	Unit cost		
Pads for exterior equipment[e]	Cubic yards of concrete		
Total support facilities costs			$
FIXED EQUIPMEMT[f]			
Charging conveyor	Each		$
Belt conveyors	Each		
Coarse shredder with motor	Each		
Air classifier	Each		
Cyclone separator	Each		
Magnetic separator	Each		
Induction fan	Each		
Pneumatic conveyor feeder	Each		
Pneumatic conveyor blower	Each		
Air ducts	Each		
Bucket conveyors	Each		
Cooling tower	Each		
Water treatment equipment	System		
Air cleanup equipment	System		
Dust collecting equipment	System		
Scale	Each		
Total fixed equipment costs			$
MOBILE EQUIPMENT			
Facility equipment			
Forklift trucks	Each	*	$
Front-end loaders	Each	*	
Clearing vacuum trucks	Each	*	
Bins and containers	Each	*	
Subtotal			$

TABLE D-3   (Continued)

Item	Costing method	Annual depreciation[b]	Cost
Transport Equipment			
Highway trailers	Each	*	$
Highway tractors	Each	*	
Other vehicles	Each	*	
Subtotal			$
Total mobile equipment costs			$
MISCELLANEOUS			
Spare parts and stores	Mixed	*	$
Maintenance equipment	Mixed	*	
Office equipment	Each	*	
Total miscellaneous costs			$
SUMMARY			
Land			$
Buildings			
Support facilities			
Fixed equipment			
Mobile equipment			
Miscellaneous			
TOTAL COSTS			$

[a] Source: Solid Waste Management Board: *Bulletin No. 7.*, Technical Information Services, State of California, Sacramento, 1975.

[b] Depreciation should be computed on the basis of the estimated life of buildings, support facilities, and each category of fixed and mobile equipment. Improved land value is not depreciable. Fees normally are depreciated over the life of the total facility. A working capital is a renewable fund and is not depreciable. Initial costs for rolling stock can be depreciated over the estimated life of the equipment, or if the equipment is heavily used, the costs can be prorated on a mileage basis to operating costs, whichever provides for more rapid expensing. Therefore, the items marked with * have a shorter depreciation life than that of the facility.

[c] Includes all costs attendant to evaluation and selection of final site.

[d] Includes all structures, such as plant, office, shop, mobile equipment, housing, and service garage.

[e] Such as cooling towers and baghouse filters.

[f] Includes all dynamic and static purchased machinery and equipment for handling and processing wastes, e.g., electrical, control and instrumentation items, fire main pumps, cooling tower, air classifiers, magnetic separator.

E
METRIC CONVERSION FACTORS

APPENDIX E METRIC CONVERSION FACTORS
Factors for the conversion of U.S. Customary Units to the International System (SI) of Units.

Multiply the U.S. customary unit		By	To obtain the corresponding SI unit	
Name	Abbreviation		Name	Symbol
acre	acre	4047	square meter	m²
acre	acre	0.4047	hectare	ha*
British thermal unit	Btu	1.055	kilojoule	kJ
British thermal units per cubic foot	Btu/ft³	37.259	kilojoules per cubic meter	kJ/m³
British thermal units per hour per square foot	Btu/hft²	3.158	joules per second per square meter	J/s·m²
British thermal units per kilowatt-hour	Btu/kWh	1.055	kilojoules per kilowatt-hour	kJ/kW·h
British thermal units per pound	Btu/lb	2.326	kilojoules per kilogram	kJ/kg
British thermal units per ton	Btu/ton	0.00116	kilojoules per kilogram	kJ/kg
degree Celsius	°C	plus 273	kelvin	K
cubic foot	ft³	0.0283	cubic meter	m³
cubic foot	ft³	28.32	liter	l*
cubic feet per minute	ft³/min	0.0004719	cubic meters per second	m³/s
cubic feet per minute	ft³/min	0.4719	liters per second	l*/s
cubic feet per second	ft³/s	0.0283	cubic meters per second	m³/s
cubic yard	yd³	0.7646	cubic meter	m³
day	d	86.4	kilosecond	ks
degree Fahrenheit	°F	0.555(°F − 32)	degree Celsius	°C
foot	ft	0.3048	meter	m
feet per minute	ft/min	0.00508	meters per second	m/s
feet per second	ft/s	0.3048	meters per second	m/s
gallon	gal	0.003785	cubic meter	m³
gallon	gal	3.785	liter	l*

gallons per minute	gal/min	0.0631	liters per second	l*/s
grain	gr	0.0648	gram	g
horsepower	hp	0.746	kilowatt	kW
horsepower-hour	hp·h	2.684	megajoule	MJ
inch	in	2.54	centimeter	cm
inch	in	0.0254	meter	m
kilowatt-hour	kWh	3.600	megajoule	MJ
pound (force)	lb_f	4.448	newton	N
pound (mass)	lb_m	0.4536	kilogram	kg
pounds per acre	lb/acre	0.1122	grams per square meter	g/m^2
pounds per acre	lb/acre	1.122	kilograms per hectare	kg/ha*
pounds per capita per day	lb/capita/day	0.4536	kilograms per capita per day	kg/capita-day
pounds per cubic foot	lb/ft^3	16.019	kilograms per cubic meter	kg/m^3
pounds per cubic yard	lb/yd^3	0.5933	kilograms per cubic meter	kg/m^3
million gallons per day	mgd	0.04381	cubic meters per second	m^3/s
miles	mi	1.609	kilometer	km
miles per hour	mi/h	1.609	kilometers per hour	km/h
miles per hour	mi/h	0.447	meters per second	m/s
miles per gallon	mpg	0.425	kilometers per liter	km/l*
parts per million	ppm	approximately equal to	milligrams per liter	mg/l*
ounce	oz	28.35	gram	g
pounds per square foot	lb/ft^2	47.88	newtons per square meter	N/m^2
pounds per square inch	lb/in^2	6.895	kilonewtons per square meter	kN/m^2
square foot	ft^2	0.0929	square meter	m^2
square mile	mi^2	2.590	square kilometer	km^2
square yard	yd^2	0.8361	square meter	m^2
ton (2000 pounds mass)	ton (2000 lb_m)	907.2	kilogram	kg
watt-hour	Wh	3.60	kilojoule	kJ
yard	yd	0.9114	meter	m

* Not an SI unit, but a commonly used term.

Indexes

Name Index

Ackoff, R. L., 159
Aiba, S., 313
Allis-Chalmers, Inc., 308
American Public Works Association, 31, 33, 37, 76, 102, 158, 159, 199, 253, 432, 457, 482, 527
AMSCO-WASCON Systems, Inc., 101
Applied Management Sciences, Inc., 457
Arnoff, L. D., 159
AVAC Systems, Inc., 84
Aware, Inc., 395

Balintfy, J. L., 159
Bartlett-Snow, 250
Bender, D. F., 7
Black-Clawson Fibreclaim, Inc., 226, 305
Black & Veatch, Consulting Engineers, 559
Blakebrough, N., 313

Bluckman, L. A., 414
Boettcher, R. A., 253
Brandt, H. T., 37, 414
Britton, P. W., 14, 76
Brown, M., 527, 559
Brown, R. P., 374, 396, 527
Brown and Caldwell, Consulting Engineers, 259, 260, 304, 313
Brunner, D. R., 373, 507, 527
Buhler, F., 559
Burdick, D. S., 159
Bureau of Mines, 40

California, state of, Department of Transportation, 385
California Department of Public Health, 19, 20, 41, 395, 527
Carruth, D., 76

Central Contra Costa Sanitary District, 259, 260, 304
Charnes, A., 159
Chilton, C. H., 254, 314, 374
Christensen, H. E., 395
Chu, K., 159
Churchman, C. W., 159
Clark, D. A., 373
Clark, R. M., 559
Coastal Zone Management Act of 1972, 45
COLMIS, 32, 457
Colonna, R. A., 47, 414, 457
Combustion Engineering, Inc., 267, 268
Connelly, J. A., 102
Corey, R. C., 253, 313, 314
COR-MET, 200, 457
Council on Environment Quality, 41–43
Cummins, R. L., 373

Dair, F., 527
Dallavale, J. M., 253
Darnay, A., 14, 76
Davis, S. N., 373
Dawson, G. W., 395
Dean, J. A., 373
Delaney, J. E., 159
DeMarco, J., 253
Dempster Dumpster Systems, 113, 115
DeWiest, R. J. M., 373
Dickason, O. E., 373
Dings Company, 238
Drobny, H. L., 47, 253, 313, 507

Eliassen, R., 14, 38, 47, 373, 414
Energy Research and Development Administration, 44, 45
Engdahl, R. B., 254
Environmental Protection Agency (see U.S. Environmental Protection Agency)
Eriez Magnetics, 237, 238

Fairolden, R. N., 374
Farb, D., 395
Fieldelman, B. P., 432, 458
Fields, T., Jr., 395
Fisher, T. F., 313
Foree, E. G., 374
Foust, A. S., 254
Franklin, W. E., 14, 76
Franklin, Ohio, 305
Friedman, L., 159

Glasstone, S., 374
Golueke, C. G., 159, 414, 559
Gotaas, H. B., 313
Greco, J. R., 507
Greenleaf/Telesca, Planners, Engineers, and Architects, 102, 432

Hadley, G., 200
Ham, R. K., 254, 374
Hanks, T. G., 14, 102, 432
Hardenbergh, W. A., 20, 38
Hart, J., 5
Hegdahl, T. A., 200, 482
Hempstead, New York, 306
Hickman, H. L., 374, 527
Hill, R. M., 254
Hughes, G. M., 374
Hull, H. E., 47, 253, 313, 374, 507
Humphrey, A. E., 313

Interstate Commerce Commission, 46

Jeris, J. S., 313
Johnson, R. W., 559
Jones, B. B., 38
Jumikis, A. R., 374

Kaiser, E. R., 76
Kasbohm, M. L., 313
Keller, D. J., 373, 507
Kirkpatrick, S. D., 254, 314, 374
Klee, A. J., 14, 76
Klein, G., 76, 159, 200
Kohler, M. A., 374
Krauskopf, K. B., 374
Kruse, C. W., 159

La Guardia, F., 20
Landon, R. A., 374
Liebman, J. C., 159
Lindsey, A. W., 395
Linsley, R. K., Jr., 374
Little, H. R., 200, 457
Los Angeles County, 18, 36, 373
    County Sanitation Districts, 489
Lummus Company, The, 281
Lynch, G. E., 414

McCabe, W. L., 254
McFarland, J. M., 313, 414, 559
McGauhey, P. H., 159, 414, 559
McLaren, C., 47, 414, 457
Management Technology, Inc., 457
Mansfield, E., 482
Mantell, C. L., 313
Marks, D. H., 159
Mawhinney, M. H., 254
Maxfield, P. L., 414
Mead, B. E., 374
Meissner, H. G., 102, 313
Meller, F. H., 313
Merz, R. C., 374
Metcalf & Eddy, Engineers, Inc., 200, 254,
    265, 302, 303, 314, 395, 457
Millis, N. F., 313
Milwaukee, Wisconsin, 241
Moyer, J. E., 374
Muhich, A. J., 14, 76

Nashville, Tennessee, 503
National Center for Resource Recovery
    (NCRR), 239, 501
National Environmental Policy Act, 1969,
    41, 43
Naylor, T. H., 159
Nemerow, H. L., 396
New Orleans, Louisiana, 239, 500
New York City, 9, 18, 20, 27-29, 315

Occupational Safety and Health Act of 1970,
    45
Orange County, California, 164, 179, 200
Orning, A. A., 314
Owen, F., 38

Parsons, H. de B., 16, 38, 200, 374
Paulhus, J. L. H., 374
Perry, R. H., 254, 314, 374
Peterson, M. L., 76
Phoenix, city of, Sanitation Department, 513
Pittman, F. K., 47, 396
Powell, M. D., 432, 458
Public Law 89-272, 40
Public Law 92-500, 383
Public Law 95-512, 41

Quon, J. E., 159

Radar, Pneumatics, Inc., 230
Ralph Stone and Company, Inc., 159
Regan, R., 313
Reinfeld, N. V., 200
Reinhardt, J. J., 254, 374
Research-Cotrell, Inc., 219
Resource Planning Associates, 458
Resources Recovery Act, 1970, 41
Ricci, L. J., 314
Rich, L. G., 314
Riggs, J. L., 200
Rivero, J. R., 313
Roe, M. J., 458
Rose, M. J., 432, 458
Ross, R. D., 254

Sacramento County, California, 30, 174, 420
Sacramento Waste Disposal Company, 242
Salvato, J. A., 374
San Francisco, California, 36, 80, 110, 167,
    170-173, 176, 179
San Mateo County, California, 36
Santa Clara County, California, 36
Sasieni, M., 159
Schneider, K. J., 47, 396
Schur, D. A., 159
Schwieger, R. G., 254, 314
Seattle, Washington, 179
Shuckrow, A. J., 395
Shuster, K. A., 159
Smith, D. D., 374, 396, 527
Smith, J. C., 254
Soil Conservation Service, 521
Solid Waste Disposal Act, 1965, 40
Sorg, T. J., 374, 527
Sortex Company of North America, Inc., 247
Spooner, C. S., 432
Stanley Consultants, Inc., 353
Stear, J. R., 254, 314
Stierli, H., 76
Stone, R., 374
Stragier, M. G., 432
Sunn, Low, Tom, and Hara, Inc., 302, 303,
    314
Swift, W. W., 395
Syrek, D. B., 432

Tchobanoglous, G., 76, 159, 200, 314, 586
Testin, R. F., 47, 253, 313, 507
Theisen, H., 414, 527, 559
Toftner, R. O., 414, 559
Trinks, W., 254

Triple/S Dynamics Systems, Inc., 230, 233, 240, 241
Truitt, M. M., 159

Union Carbide Corporation, 279, 306, 307
U.S. Army Corps of Engineers, 20, 39, 42, 45
U.S. Atomic Energy Commission, 41
U.S. Department of Agriculture, 521
U.S. Department of Health, Education, and Welfare, 46, 70
U.S. Department of Labor, 46
U.S. Department of Transportation, 46
U.S. Environmental Protection Agency, 8, 10, 26, 32, 41–45, 375, 402
U.S. Geological Survey, 521
U.S. Public Health Service, 4, 8, 10, 39–43
Universal Vibrating Screen Company, 240
University of California, 159

Velz, C. J., 586

Vincenz, J., 20
Vogel, W. R., 188, 200, 469

Walker, W. H., 254
Wall, T. E., 374
Walt Disney World, Florida, 184, 185
Ward, S. D., 395
Waste Management, Inc., 308
Waugh, A. E., 586
Wenson, S. J., 159
Wilkie, W. G., 374
Williams-Gardner, A., 254
Williams Patent Crusher and Pulverizer Company, Inc., 223, 224
Wong, J., 513
Wood, D. K., 314

Yaspan, A., 159
Young, H. D., 586
Young, J. C., 374

Subject Index

Abandoned vehicles, disposal of, 437, 439
Absorption coefficients for gases, 335
Actinomycetes, 284
Activity chart:
  definition of, 410
  examples of, 411, 546
  use for project control, 410
Advanced techniques of analysis, 151—155
  application of, 153
    design of proposed systems, 154
    existing systems: evaluation of, 153
      modification of, 154
  operations research, 152
  simulation, 153
  systems analysis, 152
Aerobic decomposition:
  in composting, 287
  reactions for, 288
  in sanitary landfills, 327

Aerobic metabolism, 285
Aerobic stabilization, 288
Aesthetic considerations:
  in disposal, 512, 515
  in onsite storage, 417
  in transfer station operation, 464, 465
Agency, collection: private, 434
    public, 434
Agricultural wastes:
  definition of, 54
  sources of, 52
  typical unit generation rates for, 72
Air:
  absorption coefficient for, 335
  density of, 328
  requirements for combustion, 270
Air classification (*see* Air separation)
Air classifier, 231
Air pollutants from incineration, 214

Air pollution control:
  from incineration, 214−216
  from processing operations (*see* Process-
    ing, selection of equipment for)
Air separation, 228−235
  application of, 228
  definition of, 228
  equipment for, 231
    selection of, 232
  theory of, 235
Alley collection service, 104, 105
Alternatives in waste management planning:
  definition of, 35
  evaluation of, 528−531
    administration and management, 531
    economic analysis, 529
    impact assessment, 530
    performance, 529
Aluminum:
  quantities of, in municipal masks, 55
  recovery of, 241, 246, 248
  sale of, 486
Ammonia (NH₃):
  density of, 328
  in landfills, 327
Anaerobic decomposition in landfills, 327
Anaerobic digestion, 292−294
  conversion reactions, 293
  design considerations, 294
  process description, 292
  process microbiology, 293
Anaerobic fermentation, 292
Anaerobic metabolism, 285
Animal wastes, 72
Apartments:
  high-rise, definition of, 77
  medium-rise, definition of, 77
  waste collection from, 81
Area method of sanitary landfilling, 321
Arithmetic probability paper, 582
Ashes:
  density of, 59
  sources of, 52
Assimilatory biological process, 284
At-site time:
  definition of, 123
  use in computations, 125, 129
Automobiles:
  abandoned, 437
  disposal-reclaiming cycle, 439
Autotrophic microrganisms, 284
Auxiliary fuel:
  for incinerators, 216
  use of, 295, 296

Backyard burning, 96
  effects of ban on, 96
  incinerators for, 96
Backyard carry collection service, 104, 105
Bacteria:
  aerobic, 285
  anaerobic, 285
    acid formers, 293
    methane formers, 293
  autotrophic, 284
  composition of, 282
  description of, 282
  dimensions of, 282
  facultative, 285
  growth requirements: environmental, 285
    nutritional, 285
  heterotrophic, 285
  metabolism, 285
  temperature growth ranges, 286
Bacterial assimilatory processes, 284
Bacterial dissimilatory processes, 284
Bacterial metabolism, 285
Bacterial stabilization, 288
Baling, 208
Ballistic inertial separator, 245
Barges for transport of solid wastes, 184
Barrier vent for landfill:
  to control gas movement, 336
  design of, 336, 337
Batch feeding of incinerators, 212
Biological conversion products:
  characteristics of, 283
  recovery of, 282−295
    by anaerobic digestion, 292
    by composting, 286
    by other biological processes, 294
Biological fermentation, 283
Biological processes:
  assimilatory, 284
  dissimilatory, 284
Boilers:
  efficiency of, 297
  oil fired, 298
  solid waste fired, 268, 298
  waste heat, 267
  water-walled, 267
Bonds, financing, 436
Brick-lined furnace, 218
Bridging of wastes, 302
Budgeting for planning, 409, 569
Budgetory constraints in collection, 446
Bulk density:
  of solid waste components, 60
  of solid wastes, 59

Bulky wastes, 52, 53
Burning of solid wastes:
  in back yards, 96
  at open dumps, 18, 19
Buyer, recovered materials and conversion
    products, 493

Caloric value:
  computation of, 62
  of solid waste components, 62
Carbon content of solid waste components,
    61
Carbon dioxide (CO_2):
  absorption coefficient for, 335
  conversion to methane, 281
  density of, 328
  in digestion gas, 293
  in landfills, 327
  removal by scrubbing, 281
  solubility in water, computation of, 335
Carbon monoxide (CO):
  absorption coefficient for, 335
  conversion to methane, 281
  density of, 328
  in landfills, 327
  shift conversion of, 281
  solubility in water, computation of, 335
Carbon-nitrogen ratio in composting, 291
Cardboard:
  baling, 208
  quantities in municipal wastes, 55
  separation of, 242
Case studies:
  in collection, 447 – 456
  in disposal, 522 – 526
  in onsite handling, storage, and processing,
    426 – 431
  in processing and recovery, 495 – 508
  in transfer and transport, 470 – 481
Cell, landfill, 317
    construction, 321 – 326
      conventional, 321 – 325
      for gas recovery, 326
    dimensions, 317
    venting, 326 – 327
Cell carbon, 284
Centrifugation, 252
Chemical composition of solid wastes,
    59 – 62
  proximate analysis, 59
  typical data on, 60 – 62
Chemical conversion products:
  characteristics of, 266

Chemical conversion products:
  recovery of, 265 – 282
    by incineration with heat recovery, 265
    by incineration-pyrolysis, 279
    by other processes, 281
    by pyrolysis, 277
Chemical volume reduction (incineration),
    211 – 221
  air pollution control, 214
  application of, 211
  design considerations, 216
  process description, 212
Chute, solid waste, 80 – 82
Clamshell:
  mobile, 165
  stationary, 172
Class designation for landfills, 517 – 518
Classification of solid waste materials, 52 – 54
Co-disposal:
  definition of, 307
  of treatment plant sludges, 306
Coefficient of variation, 581
Collection:
  case studies in, 447 – 456
  definition of, 25, 104
  inventory and data accumulation, 441 – 447
  management issues and concerns,
    433 – 441
  operations, 442 – 445
    agencies, 442
    crew size and working conditions, 443
    equipment, 444
    special collectors and informal haulers,
      445
    special wastes, 445
    system records, 442
Collection agencies:
  private, 434
  public, 434
Collection crews:
  efficiency of, 444
  size of, 120, 444
  workloads for, 444
Collection management information system
    (COLMIS), 32, 457
Collection points, 104 – 105
Collection routes, 140 – 151, 458
  layout of, 141 – 145
    for hauled container system, 141
    important factors in, 140
    for stationary container system, 143
  schedules for, 145
Collection service, 104 – 108
  commercial-industrial, 108

Collection service:
  community service levels, 434
  for detached residential dwellings, 104
    alley, 105
    backyard carry, 105
    curb, 105
    setout, 105
    setout-setback, 105
  residential, 104
    low-rise detached dwellings, 104
    medium- and high-rise apartments, 108
Collection systems:
  analysis of, 121−155
    advanced techniques for, 151−155
      application of, 153−155
        design of proposed systems, 154
        evaluation of existing systems, 153
        modification of existing systems, 154
      operations research, 152
      simulation, 153
      systems analysis, 152
    definition of terms, 122−124
      at-site, 123
      haul, 122
      off-route, 124
      pickup, 122
    haul speed and time, 125
    hauled container, 124−129
      container utilization factor, 127
      labor requirements for, 127
      time per trip, 124−125
    satellite, 108
    stationary container, 129−140
      systems: with manually leaded vehi-
        cles, 129−134
      with mechanically loaded vehicles,
        134−140
  equipment: for hauled container, 112−115
    data on, 114
    hoist trucks, 113
    tilt-frame container, 113
    trash trailers, 115
  for stationary container, 115−117
    data on, 114
    manually loaded vehicles, 117
    mechanically loaded vehicles, 117
  labor requirements for, 120−121
    hauled container, 120
    stationary container: manually loaded,
      121
    mechanically loaded, 120
  types of, 108−112
    hauled container, 110
    stationary container, 112

Collection vehicles:
  for hauled container systems, 112−115
    hoist truck, 113, 114
    tilt-frame, 113, 114
    truck-tractor, 114
    typical data on, 114
  for stationary container systems, 115−120
    manually loaded compactors, 114, 117
    satellite, 108, 119
    self-loading compactors, 114, 117
    typical data on, 114
Collector injuries, 417
Combustible rubbish, 53
Combustion:
  air requirements for, 272
  chambers, 214, 267
  computations, 269
  gases, 270, 273
    composition of, 275
    temperature of, 275, 276
  heat released, computation of, 271
  reactions, 270
Commercial areas, 52
Commercial collection service, 108
Commercial wastes:
  quantities of, 70
  sources of, 52
Community involvement in disposal site se-
    lection, 521
Community problem-solving cycle, 400−401
  normal sequence in, 401
  role of feedback in, 401
Compaction:
  application of, 203
  equipment for, 203, 204
  high-pressure, 206
  low-pressure, 206
  selection of equipment for, 205, 207
  theory of, 208
Compaction ratio, 208
Compactors:
  classification, by pressure, 206
  mobile, 203
  selection of, 207
  stationary, 203
Component separation, 228−248
  application of, 228
  methods for: air separation, 228
    electrostatic separation, 246
    flotation, 245
    handsorting, 228
    heavy media separation, 246
    inertial separation, 245
    linear induction separation, 248

Component separation:
  magnetic separation, 235
  optical sorting, 246
  screening, 239
  (*See also specific entries*)
Composition of solid wastes, 54−64
  chemical, 59−62
    definition of, 59
    typical data on, 60−62
  physical, 55−59
    definition of, 55
    typical data on, 55, 57, 59, 60
Composting, 286−292
  application of, 95
  design considerations, 290
    carbon-nitrogen ratio, 291
    control of pathogens, 291
    heat evolution, 291
    land area requirements, 291
    mixing/turning, 290
    moisture content, 291
    oxygen consumption role, 291
    oxygen requirements, 291
    particle size, 290
    pH, 291
    seeding and mixing, 290
    temperatures, 291
  environmental concerns, 292
  process description, 287
  process microbiology, 287
Comprehensive plans, preparation of, 35,
    407−410
Construction wastes:
  definition of, 53
  disposal of, 515
  quantities of, 70
Container liners, 89
Container utilization factor:
  definition of, 127
  use in computations, 127, 130
Containers used for solid wastes, 83−90
  application of, 87
  onsite storage locations, 90
  types of, 83−90
    commercial, 90
    low-rise dwellings, 85
    medium- and high-rise apartments, 89
  typical data on, 86
Contamination of groundwater:
  by landfill gases, 334
  by leachate, 343
Continuous feed incinerator, 265
Contours, landfill, 362
Contract collection agencies, 32, 33, 434

Contractual arrangements:
  franchise, 435
  limited permits, 435
  unlimited permits, 435
Conversion of solid wastes:
  biological, 282−295
  chemical, 265−282
  products: biological, 283, 287, 293
    chemical, 266, 277, 278
  thermal, 266
Conversion factors (U.S. customary units to
    metric units), 594−595
Conveyors, problems with, 264
Copper precipitation, 494
Cost data:
  adjustment of, 589
    current cost, 592
    future cost, 593
  for conventional equipment, 587−589
    collection vehicles, 588
    landfill equipment, 588
    storage containers, 588
    transfer and transport equipment, 588
Cost effectiveness, 512
Cost estimating for processing operations,
    593
  building, 591
  fixed equipment, 591
  land, 590
  miscellaneous, 592
Cost indexes, 589
Cover soil for sanitary landfills:
  requirements for, 317, 344
  suitability of various types, 344
Cranes, incinerator, 218
Crew (*see* Collection crews)
Crusher, jaw, 222
Curb collection service, 104, 105
Cyclone separator, 215

Dead animals, collection of, 445
Decision process in solid waste management,
    410−413
  important events, 412
  requirements for, 412
Decomposition:
  biological, 282
    aerobic, 287
    anaerobic, 293
  chemical, 211, 265
  thermal, 211, 270
Demolition wastes:
  definition of, 53

Demolition wastes:
  sources of, 52
Density:
  of gases, 328
  of solid wastes, 58 – 59
    individual components, 60
    wastes from various sources, 58
Depletion allowances, 492
Depression method of sanitary landfilling,
  323
Destructive distillation, 277
Developing concepts in planning, 403
Dewatering, 248 – 249
Direct discharge transfer station, 163
Disposable container:
  application of, 87
  data on, 86
  types of, 89
    paper, 89
    plastic, 89
Disposal:
  case studies in, 522 – 526
  definition of, 27, 315
  early methods of, 17 – 20
  inventory and data accumulation, 519 – 522
  landfill method, 315 – 369
  management issues and concerns,
    508 – 519
  modern methods of, 315 – 370
  (See also Landfill)
Disposal fees, 412, 514
Dissimilatory biological process, 284
Drainage facilities for landfills, 350
Drop cloth, 107
Drying, 248 – 251
  application of, 248
  definition of, 248
  equipment for, 248
  selection of equipment for, 250
  theory of, 249
Dumping:
  effects of, 18
  on land, 18
  in water, 18
  (See also Unloading operations at disposal
    site)
Dumping area for incinerators, 212
Dust:
  classification of, 215
  control of, 215

Economic constraints in collection, 446
Economy of scale, 494

Education programs, public, 569
Efficiency factors:
  for boilers, 298
  for electric generator, 298
  for incinerator/boiler, 298
  for methanation process, 298
  for pyrolysis reactors, 298
  for steam turbine-generator, 298
  for turbines, 298
EIS (see Environmental Impact Statement)
Electrostatic precipitator, 215
Electrostatic separation, 246
Emission control, 215
Energy:
  content of solid wastes, 62
  recovery from solid wastes, 257
Energy balance calculations for incineration
    process, 271 – 273
Energy content of solid wastes:
  computation of, 62
  for individual components, 62
Energy recovery from conversion products,
    295 – 301
  efficiency factors for, 297
  process heat rate: definition of, 296
    typical values for, 297
  systems for: gas turbine-generator, 295
    steam turbine-generator, 295
  typical flowsheets for, 301
Environmental impact assessment, 42
Environmental Impact Report (EIR), 42, 464
Environmental Impact Statement (EIS), 42
Equipment cost data, 587 – 590
  adjustment of, 589
    current cost, 589
    future cost, 589
  for conventional equipment, 588
Equipment selection:
  for air separation, 232
  for compaction, 207
  for drying, 250
  for magnetic separation, 236
  for onsite processing, 92
  for sanitary landfills, 358
  for screening, 241
  for size reduction, 225
Evaluation of alternatives in waste manage-
    ment planning, 528 – 531
  administration and management, 531
  economic analysis, 529
  impact assessment, 530
  performance, 529
Excess air in incineration, 267
Exhaust combustion gases, 273

Explosive wastes:
  classification of, 381
  handling of, 389
  storage of, 388

Fabric filter, 215
  dust collector, 216
Federal laws affecting solid waste manage-
    ment, 39–43
  National Environmental Policy Act, 1969,
    41
  Resource Recovery Act, 1970, 41
  Solid Waste Disposal Act, 1965, 40
Fencing:
  for sanitary landfills, 345, 515, 516
  for transfer stations, 465
Ferrous metals:
  quantities in municipal wastes, 55
  recovery of, 235
  sale of, 486, 491
Filtration:
  fabric, 216
  pressure and vacuum, 252
Final cover for sanitary landfill:
  characteristics of, 344
  requirements for, 317, 344
Financing methods, 436, 460, 530
Fire control at landfills, 345
Firing methods for incinerators, 268
Flammable wastes:
  classification of, 379
  handling of, 389
  storage of, 388
Flotation separation, 245
Flowsheets:
  for anaerobic conversion of solid wastes to
    gas, 308
  for disposal of sludge and solid wastes, 306
  for materials and energy recovery, 302, 304
  for materials recovery, 258, 259, 305
  for production of compost, 307
Fluidized-bed incineration, 266
Flyash from incinerators, 214
Food wastes:
  definition of, 53
  density of, 60
  moisture content of, 57
  sources of, 52
Forced draft fan, 213
Franchise waste collection, 33, 435, 447
Freight rates, differential, 492
Frequency distribution, 580
Front-end system, 255

Fuels, auxiliary, 216, 295, 296
Functional elements, 21–27
  application of, 24
  definition of, 21, 34
  description of individual elements, 24–27
    collection, 25
    disposal, 26
    generation, 24
    onsite handling, storage, and processing,
      24
    processing and recovery, 26
    transfer and transport, 26
  (See also specific entries)
Fungi, 284
Furnaces, incinerator, 213

Garbage grinders, 91
Gas temperatures in incinerators, 273
Gaseous emissions, control of, 215
Gases:
  absorption coefficients for, 335
  control of movement in landfills, 336–338
    by impermeable methods, 338
    by permeable methods, 336
  density of, 328
  digestion, anaerobic, 293
  formation in landfills, 327
  incineration, 270, 273
  movement in landfills, 334
  pyrolysis, 277
Generation of solid wastes:
  future challenges and opportunities, 12
  impacts of, 3
  projections for the future, 10
  quantities of wastes, 7
  in a technological society, 4
  (See also Solid wastes)
Geographic aspects of solid waste manage-
    ment, 71, 423
Geologic considerations in landfill siting, 320
Glass:
  optical sorting of, 246
  recovery of, 241, 246
  specifications for, 257
Governmental agencies:
  Energy Research and Development Ad-
    ministration, 44
  Interstate Commerce Commission, 46
  U.S. Army Corps of Engineers, 45
  U.S. Department of Health, Education, and
    Welfare, 46
  U.S. Department of Labor, 45
  U.S. Department of Transportation, 46

Governmental agencies:
  U.S. Environmental Protection Agency, 43
  (*See also* Legislation affecting solid wastes)
Graphical methods of statistical analysis, 582
Grates for incinerators, 214
Gravel trenches for control of landfill gases,
    336 – 337
Grinder, household, 91
Grinding:
  application of, 91, 221
  equipment for, 222
  (*See also* Mechanical size reduction)
Groundwater:
  contamination, 27
    by leachate, 339
  increase in hardness due to landfill gases,
    334

Hammermills:
  description of, 224
  horizontal shaft, 224
  vertical shaft, 225
Handling, onsite, 78 – 83
    commercial, 83
    equipment for, 79
    residential, 79
Handsorting, 92, 228
Haul:
  definition of, 122 – 124
    for hauled container system, 122, 123
    for stationary container system, 122, 124
  use in computations, 125, 129
Haul time:
  definition of, 122 – 124
    for hauled container system, 122, 123
    for stationary container system, 122, 124
  use in computations, 125, 129
Hauled container collection systems:
  analysis of, 124 – 129
  applications for, 112
  definition of, 110
  types of: hoist-truck, 113
    tilt-frame container, 113
    trash-trailer, 114
Hazardous wastes, 375 – 394
  classification of, 376 – 381
    biological wastes, 379
    chemicals, 379
    explosives, 381
    flammable wastes, 379
    radioactive substances, 376
    typical hazardous compounds, 380
  collection of, 387

Hazardous wastes:
  definition of, 54, 375
  disposal of, 393 – 394
  generation of, 383 – 385
  identification of, 375 – 377
  onsite storage of, 385 – 387
  planning for, 394
  processing of, 389 – 393
  regulations, 381 – 383
    federal, 381
    international, 381
    local, 383
    state, 383
  transfer and transport of, 389
Health hazards, 4, 375
Heat rate for power plants, 296
Heavy media separation, 246
Heavy metal toxicity, 292
Heterotrophic microorganism, 285
Heuristic routing in solid waste collection,
    159
High-pressure compaction, 206
High-rise apartments:
  collection of solid wastes from, 108
  definition of, 77
  onsite activities: handling, 79
    processing, 97
    storage, 89
High-temperature incinerator, 211
Highway:
  as means for surface transport, 177, 437,
    469
  regulations, 177
Hog feeding, 18
Hoist-truck collection vehicle, 113
Home separation, 92, 497
Hospital wastes, 379
Humus, compost, 286
Hydraulic transport systems, 184
Hydrogen (H_2):
  absorption coefficient for, 335
  density of, 328
  in landfills, 327
  in pyrolysis gases, 277
  solubility in water, computation of, 335
Hydrogen sulfide (H_2S):
  absorption coefficient for, 335
  density of, 328
  in landfills, 327
  solubility in water, computation of, 335
Hydrologic considerations in landfill siting,
    320
Hydrolysis, 266
Hydrolytic conversion, 282

Hydropulper:
  application of, 99
  description of, 225
  typical examples of, 101

Incentives:
  for collectors, 437
  for litter control, 417
  for resource recovery, 492
Incineration:
  air pollution control, 214
  application of, 211
  combustion computations, 269
  with heat recovery, 265, 503
  without heat recovery, 211
  process description, 212
Incineration-pyrolysis, 279–282
  conversion efficiency, 280
  conversion products, 280
  process description, 279
  (See also Pyrolysis)
Incinerator:
  combustion chambers, 265
    refractory lined, 265
    water-walled, 265, 268
  design considerations, 218
  firing methods, 268
    double-vortex, 269
    spread stoker, 269
    suspension, 269
  grates, 213, 214
  performance considerations, 216
Inclined inertial separator, 245
Inclined screens, 240
Inducted draft fan, 214
Industrial wastes:
  quantities of, 70
  sources of, 52
Inertial separators, 245
Infiltration potential for leachate, 342
Informal haulers, 445
Inventory and data accumulation for manage-
    ment:
  for collection, 441–447
    collection system constraints, 445
    collection system operations, 442
  for disposal, 519–522
    community involvement, 521
    identification of old landfills, 522
    regional options, 521
    soils and hydrological information, 521
  for onsite storage, 422–426
    geographic areas, 423

Inventory and data accumulation for manage-
    ment:
    identification of equipment, 425
    inventory requirements, 423
    quantities and types of wastes, 423
    service conditions, 426
  for processing, 493–495
    economics, 494
    local needs, 493
    state, national, and international needs,
      494
  for transfer and transport, 466–470
    collection system operation, 467
    disposal sites, 468
    planning levels, 466
    sources, types, and quantities of wastes,
      467
    transfer and transport operations, 469

Jurisdictional constraints in collection, 446

Kitchen grinders, 91

Labor-management relations, 436
Labor organizations, 436
Labor requirements for collection:
  with hauled container systems, 125–129
  with stationary container systems,
    129–140
Land disposal of solid wastes, 129–140
  advantages of, 316
  disadvantages of, 316
  history of, 17, 20
  early methods of, 17
  sanitary landfill, 317, 347
  (See also Landfill)
Land requirements for landfilling, 317, 347
Land use groups in planning, 423, 424
Landfill:
  definition of, 316
  design, 344–369
    drainage and seepage control, 350
    equipment requirements, 354
    example of, 357
    land requirements, 347
    operation plan, 352
    seepage potential, 350
    solid waste filling plan, 353
    types of waste, 350
  designation, by class, 517

Landfill:
  reactions occurring in, 326–333
    decomposition, 327
    gas formation, 327
    leachate formation, 331
  settlement, 333
  site selection, 317–321, 524–526
    available land area, 317
    climatic conditions, 320
    geologic and hydrogeologic conditions, 320
    haul distance, effect of, 319
    impact of resource recovery, 319
    local environmental conditions, 320
    soil conditions, 319
    surface-water hydrology, 320
    ultimate use of site, 321
Landfill gases:
  control of movement, 336–338
    by impermeable methods, 338
    by permeable methods, 336
  movement of, 334
  types of, 327–328
Landfilling methods, 321–326
  in dry areas, 321–325
    area, 321
    depression, 323
    ramp, 322
    trench, 323
  with milled solid wastes, 324
  in wet areas, 325
Leachate:
  chemical composition of, 332
  control of movement of, 343–344
  estimation of vertical seepage, 342–343
  formation in landfills, 331
  interactions, 343
  movement in landfills, 338–343
  seepage of, 338–342
Legal restrictions for recycling, 491
Legislation affecting solid wastes:
  National Environmental Policy Act, 1969, 41
  Resources Recovery Act, 1970, 41
  Solid Waste Disposal Act, 1965, 40
  (See also Governmental agencies)
Licensing:
  of collection operations, 445
  of landfill operations, 514
Linear induction separation, 248
Liners for landfills, 338–339
  clay materials, 339
  synthetic membrane, 339

Liquid wastes:
  hazardous, 384–385
  from treatment plants, 307
Litter, 53
Load-count analysis, 66
Local laws and regulations:
  content of, 382, 517
  influence on solid wastes, 382, 435
Local markets for recovered materials, 486
Log normal probability paper, 585
Long-term planning, 403
Low-rise dwelling:
  collection of wastes from, 104–108
  definition of, 77
  onsite activities: handling, 79
    processing, 91–97
    storage, 85–89

Magnetic separation, 235–239
  application of, 236
  definition of, 236
  equipment for, 236–238
  selection of equipment for, 236
Management issues and concerns:
  for collection, 433–441
    financing, 436
    labor constraints, 436
    management records, 435
    public and private agencies, 434
    service levels, 434
    special wastes, 437
    system continuity, 439
    transportation technology, 437
  for disposal, 508–519
    control of operation, 514
    cost effectiveness, 512
    justification of need, 509
    management policies and regulations, 517
    site location, 511
    system continuity, 518
  for onsite storage, 416–422
    diverse waste source, 416
    ordinances and standards, 417
    public health and aesthetics, 417
    sizes and conditions of wastes, 417
    system continuity, 419
  for processing, 484–493
    identification of markets, 486
    market stability, 491
    material specification, 488
    priorities, 484

Management issues and concerns:
for processing: system continuity, 492
unproved technology, 485
where and when to use, 484
for transfer and transport, 459 – 466
economic feasibility, 460
political and social aspects, 463
system continuity, 464
technical feasibility, 462
users of transfer station, 460
Manually loaded vehicles, 114, 117, 120, 121
Market outlets for recovered materials, 486
Marketing locations for recovered materials:
international, 494
local, 494
national, 494
Materials balance, use of: in design of processing facilities, 262
in determination of waste generation rates, 67
in incineration combustion computations, 271
Materials and energy recovery systems, 301 – 310
flowsheets: using biological conversion processes, 307 –.309
using chemical conversion processes, 301 – 307
review of, 309
Materials processing and recovery:
applications of, 202, 255
methods for: air separation, 228
compaction, 203
drying, 248
electrostatic separation, 246
flotation, 245
handsorting, 228
heavy media separation, 246
incineration, 211
inertial separation, 245
linear induction separation, 248
magnetic separation, 235
optical sorting, 246
screening, 239
size reduction, 221
(See also specific entries)
systems for, 258 – 264
design and layout, 259
equipment limitations, 264
loading rates, 261
materials balance, 261
typical flowsheets, 258

Materials recovery:
impact of resource shortages, 492
legal aspects of, 491
market demands for, 491
political aspects of, 492
Materials specifications, 256 – 258
for fuel source, 257
for land reclamation, 258
for reuse as raw materials, 257
Mean, 580
Mechanical size reduction, 221 – 228
application of, 221
definition of, 221
equipment for, 222
selection of equipment for, 225
Mechanical volume reduction, 203 – 211
application of, 203
compaction equipment for, 203, 204
selection of equipment for, 205, 207
Mechanically loaded vehicles, 114, 117, 118
Median, 581
Medium-rise apartments:
collection of wastes from, 108
definition of, 77
onsite activities: handling, 79
processing, 97
storage, 89
Membrane, synthetic: for control of gas movement in landfills, 339
for leachate control in landfills, 339, 344
Mesophilic microorganisms, 286
Metabolism, bacterial: aerobic, 285
anaerobic, 285
facultative, 285
Metals:
recovery of ferrous, 235
recycling of, 491
Methanation process:
application, 281
flowsheet, 281
reactions, 281
Methane (CH_4):
absorption coefficient for, 335
density of, 328
in landfills, 327
solubility in water, computation of, 335
Metric conversion factors, 594 – 595
Microbial waste conversion process, 283
Microorganisms:
actinomycetes, 284
bacteria, 283
classification of, 283
fungi, 284

Microorganisms:
  yeast, 284
Milling of solid wastes:
  definition of, 324
  before landfilling, 324
Mobile packers, 204
Mode, 581
Moisture:
  effects on compaction density, 211
  leachate formation, 343
Moisture content:
  in compost, 291
  in solid wastes, 56–58
    definition of, 57
    estimation of, 57
    typical data on, 57
Motor vehicle transport, 177, 182
Moving-floor unloading mechanism, 177
Multifamily dwellings, 77
Municipal solid wastes:
  definition of, 52
  generation rates of, 70, 71
  sources of, 52
  (*See also* Solid wastes)

Newspaper recovery, 497
Nitrogen (N_2):
  absorption coefficient for, 335
  density of, 328
  in incineration, 269, 273
  in landfills, 327
  in pyrolysis gases, 277
  solubility in water, computation of, 335
Noise, 89
Nonburnable materials, 53
Noncombustible rubbish, 53
Nonferrous metals:
  quantities in municipal wastes, 55
  recovery of, 241, 246, 248
  sale of, 486

Ocean disposal, 315, 369–370
  of industrial wastes, 369
  of municipal solid wastes, 370
Odor, 78
Off-route time:
  definition of, 124
  use in computations, 125, 129
Oil:
  as a hazardous waste, 379
  from pyrolysis, 277
One-person collection, 121

Onsite handling, 78–83
  commerical, 83
  residential, 79
    low-rise dwellings, 79
    medium- and high-rise apartments, 79
  storage, and processing: definition of, 77
    inventory and data accumulation,
      422–426
    management issues and concerns,
      416–422
Onsite processing, 90–101
  commercial-industrial facilities, 99
  low-rise dwellings, 91–97
    compaction, 94
    composting, 95
    grinding, 91
    incineration, 96
    sorting, 92
  medium- and high-rise apartments, 97–99
    compacting, 97
    grinding, 97
    incineration, 97
    shredding and pulping, 99
    sorting, 97
  selection of equipment for, 92
Onsite storage, 83–90
  containers, 83
    applications of, 87
    commercial, 90
    data on, 86
    low-rise dwellings, 85
    medium- and high-rise apartments, 89
  locations, 90
Open dumps:
  burning at, 18, 19
  problems with, 18
  use of, 18
Open top containers:
  typical data on, 112
  use: with hoist truck, 112
    with self-loading compactor, 112
    with tilt-frame truck, 112
Operating area, sanitary landfill, 345, 514
Operating plan for sanitary landfill, 352
Operations research, 152
Optical sorting, 246, 247
Ordinances:
  for collection, 420, 435
  for disposal, 514, 517
  local, 426
    for hazardous wastes, 383
  for onsite storage, 417, 420
  for private agencies, 435
  (*See also* Regulations)

Organization chart:
    for department of sanitation, city of New
        York, 29
    for division of solid waste management,
        Sacramento County, 30
Overfire air, 213
Oxygen (O₂):
    absorption coefficient for, 335
    density of, 328
    in landfills, 327
    solubility in water, computation of, 335

Packaging wastes, 7, 12
Paper:
    quantities, in municipal wastes, 55
    recovery of, 228, 229
    specifications for, 257
Particle size classification, 215
Particulate matter:
    control of, 215
    from incineration process, 214
    size classification, 215
Pathogenic organisms, control of, in compost,
    291
Pathological wastes, 379
Per capita waste generation rates:
    factors affecting rates, 71–73
    typical values for, 70, 71
Personnel, 436, 443
pH:
    lowering of groundwater, 334
    optimum: for bacterial growth, 286
        for composting, 291
Physical composition of solid waste:
    density, 58
    individual components, 55
    moisture content, 56–57
    typical data on, 55
Pickup service (see Collection service)
Pickup time:
    definition of, 122
        for hauled container system, 122
        for stationary container system, 122
    use in computation, 125, 129
Pipe, vent for landfill, 337
Planning:
    decision process, 410–413, 533
        important events in, 412
        requirements for, 412
    important considerations in planning
        process, 400–404
        concepts and technologies, 403

Planning:
    important considerations in planning
        process: framework for activities,
        400
    planning levels, 402
    time periods, 401
    methodology for planning studies,
        407–410
    organization of work effort, 409
    planning steps, 407
    programs and plans, 404–407
    types of, 405
    public information programs, 569
    for solid waste management, 399–413
Plans, solid waste management: acceptance
    of, 531
    definition of, 35, 405
        local, 406
        regional, 406
        state, 406
    objectives of, 405
    preparation of, 35–37, 405, 536, 543
    uses of, 405
Plant heat rate (see Process heat rate)
Plastic bags, 86, 87, 89
Plastics:
    quantities, in municipal wastes, 55
    trends in use, 64
Pneumatic systems:
    for shredded wastes, 302
    for waste collection, 83
    for waste transport, 84, 184
Political aspects:
    in disposal site selection, 508, 521
    in materials and energy recovery, 492
    in plan development, 408
Pollutants:
    from incineration process, 214
    from landfills: gases, 327
        leachate, 331
Polymer membranes:
    for control of gas movement in landfills,
        338, 339
    for leachate control in landfills, 339,
        344
Power requirements:
    for front-end systems, 301
    for power production facilities, 298
Precipitator, electrostatic, 219
Price-specification range for recovered mate-
    rials, 487
Primary combustion chamber, 212, 214
Priorities, community, 484
Probability paper, 582, 585

Problem-solving cycle, community, 401
    normal sequence, 401
    role of feedback, 401
Process heat rate, 296
    definition of, 296
    typical values for, 297
Processing:
    case studies in, 495 – 506
    cost estimating for, 590
    definition of, 201, 483
    description of equipment for: air separa-
        tion, 231
        compaction, 203
        drying, 248
        magnetic separation, 236
        optical sorting, 246
        screening, 239
        size reduction, 222
    equipment limitations, 264, 485
    inventory and data accumulations,
        493 – 495
    management issues and concerns,
        484 – 493
    onsite, 90 – 101
        commercial/industrial facilities, 99 – 101
        low-rise dwelling, 91 – 97
        medium- and high-rise apartments,
            97 – 99
    purposes of: for improving efficiency, 202
        for materials recovery, 202
        for recovery of conversion products, 202
    selection of equipment for: air separation,
        232
        compaction, 207
        drying, 250
        magnetic separation, 236
        onsite processing, 92
        screening, 241
        size reduction, 225
    (See also Processing and recovery systems)
Processing and recovery systems, 258 – 264
    cost estimating for, 590
    design and layout, 259
    equipment limitations, 264
    loading rates, 262
    materials balances, 262
    typical flowsheets, 258
Program and plan selection, 408, 531
    community support, 531
    compatibility with community goals, 532
Programs, solid waste management: defini-
        tion of, 35, 404
    development of, 408, 528
    uses of, 404, 529

Protists, 282
Proximate analysis:
    definition of, 59
    of municipal solid wastes, 60
    of pyrolytic char, 278
Public acceptance:
    of landfills, 509, 521
    of management plans, 531, 569
Public health:
    considerations: in elimination of dumps, 18,
        20
        in onsite storage, 78
Public information programs, 569
Public meetings, 531
Public participation in solid waste manage-
        ment, 434, 521, 531
Public relations:
    communication programs, 434
    incentive programs, 417
    information programs, 34, 569
Pulverizing (see Size reduction, mechanical)
Pyrolysis:
    cellulose conversion reaction, 277
    conversion products, 277
        char, 278
        energy content of, 279
        gases, 278
        oils and tars, 278
    process description, 277
    product yields, 278
    proximate analysis of char, 278
    (See also Incineration-pyrolysis)

Radiation drying, 248
Radioactive wastes, 377
Railhaul, 470
Railroad transport, 184
Ramp method for landfilling, 322
Raw materials, recovery of (see Recovery, of
        materials)
Reactors:
    fixed-bed catalytic, 281
    fluidized bed, 266, 305
    incineration-pyrolysis, 279
Rear-end systems, 256
Reciprocating grate, 214
Reclamation, land, 488
Recovery:
    of biological conversion products,
        282 – 295
        by anaerobic digestion, 292
        by composting, 286
        by other biological processes, 294

Recovery:
  of chemical conversion products, 265 – 282
    by incineration with heat recovery, 265
    by incineration-pyrolysis, 279
    by other processes, 281
    by pyrolysis, 277
  of energy from conversion products, 295 – 301
    efficiency factors for, 298
    process heat rate, 296
    systems for: gas turbine-generator, 295
      steam turbine-generator, 295
    typical flowsheets for, 301
  of materials, 258 – 264
    for energy or fuel production, 257, 488
    impact of resource shortages, 492
    for land reclamation, 258, 488
    legal aspects of, 491
    market demands for, 491
    methods of: air separation, 228
      electrostatic separation, 246
      flotation, 245
      handsorting, 228
      heavy media separation, 246
      inertial separation, 245
      linear induction separation, 248
      magnetic separation, 235
      optical sorting, 246
      screening, 239
      (See also specific entries)
    political aspects of, 492
    for reuse as raw materials, 257, 486
    specifications for recovered materials, 257, 488
      for fuel source, 257
      for land reclamation, 257
      for reuse, 257
Recycling (see Recovery, of materials)
Reduction, 18
Refractory linings for incinerators, 218, 267
Refuse (see Solid wastes)
Regional facilities:
  for disposal, 521
  for processing, 493
  for transfer and transport, 460, 463
Regulations:
  for collection, 435
  for disposal, 517
  for hazardous wastes, 381 – 383
Remote sites for landfill, 515
Residential dwellings:
  collection service, types of, 104 – 108
  definition of: high-rise apartments, 77
    low-rise dwellings, 77

Residential dwellings:
  definition of: medium-rise apartments, 77
  onsite activities: handling, 78 – 83
    processing, 91 – 99
    storage, 83 – 90
  (See also specific dwellings)
Residential wastes:
  characteristics of, 55
  sources of, 52, 55
  unit generation rates of, 70, 71
Residues:
  definition of, 53
  from incinerators: characteristics of, 62
    computation of quantities of, 218
  sources of, 52, 53
Resource recovery (see Recovery, of materials)
Respiratory quotient for composting, 291
Rocking grate, 214
Rotary drum screen:
  for separation: of cardboard, 242
    of glass, 240, 241
    of solid waste components, 239
Routes:
  scheduling of, 145
  for transfer vehicles, 185
  for waste collection, 140
Routing, collection system, 140 – 145
Rubbish:
  definition of, 53
  density of, 59
  sources of, 52
Rural solid waste management, 166, 515

Safety hazards to collectors, 418
Salvaging, 174
Sanitary landfill:
  area method, 321
  cover soil for, 317, 344
  definition of, 316
  design of, 344 – 369
  equipment requirements for, 354
  historical development of, 20
  regulations for, 517
  (See also Landfill)
Sanitation requirements for transfer stations, 176
Satellite vehicles, 109
  collection systems, 108
Scales:
  at disposal sites, 345
  at incinerators, 218
  at transfer stations, 176

Scavengers, 445
Scavenging and presorting, 462
Schedules:
  for collection routes, 145
  for implementation of plans, 409
  for plan review and update, 533
Screening:
  application of, 239
  for cardboard separation, 242
  definition of, 239
  equipment for, 239
  for glass separation, 241
  selection of equipment for, 241
  theory of, 243
Seasonal variations:
  in waste composition, 9, 11
  in waste quantities, 72
Secator inertial separator, 245
Secondary combustion chamber, 215
Secondary manufacturing, 6
Secondary materials industries, 495
Seepage from landfills, estimation of, 342
Self-loading compactors, 117–119
Separation of solid wastes:
  application of, 228
  methods of: air separation, 228
    electrostatic separation, 246
    flotation, 245
    handsorting, 228
    heavy media separation, 246
    inertial separation, 245
    linear induction separation, 248
    magnetic separation, 235
    optical sorting, 246
    screening, 239
    (*See also specific entries*)
Setout collection service, 104, 105
Setout-setback collection service, 104, 105
Settlement in landfills, 333
Settling chambers in incinerators, 215
Sewage sludge:
  co-disposal with solid wastes, 306
  disposal in landfills, 326, 372
Shift gas, 281
Shredding (*see* Size reduction, mechanical)
Simulation:
  application of, 153
  description of, 153
Site selection:
  for sanitary landfills, 317–321, 514
  for transfer stations, 185, 469
Size reduction, mechanical: application of, 221
  definition of, 221

Size reduction, mechanical: equipment for, 222
    selection of equipment for, 225
Slagging type incinerator, 211
Social constraints in solid waste management, 446
Soils borings:
  typical example of, 361
  use in landfill design, 350
Solid waste chute, 80–82
Solid waste collection systems (*see* Collection; Collection systems)
Solid waste management:
  definition of, 15
  early developments of, 20–21
  objectives, 15
Solid waste management planning, 34–37, 399–413
  definition of, 34, 399
  definition of terms: alternatives, 35
    functional element, 34
    plans, 35, 405
    program, 35, 404
    system, 35
    (*See also specific entries*)
  preparation of comprehensive plans, 35–37, 407–410
    objectives and goals, 35
    overlapping responsibilities, 36
    regional management problems, 36
Solid waste management systems, 27–34
  contract administration, 32
  equipment management, 30
  financing, 28
  operations, 28
  ordinances and guidelines, 32
  organizational structure, 28
  personnel, 30
  public communications, 34
  reporting, cost accounting, and budgeting, 32
Solid waste materials, classification of, 52–54
Solid wastes:
  burning of (*see* Burning of solid wastes)
  chemical composition of, 59–62
    energy content, 62
    inert residue, 62
    proximate analysis, 59
    typical data on, 60–62
    ultimate analysis, 61
  definition of, 3, 51
  factors affecting generation rates, 71–73
    characteristics of population, 73

Solid wastes:
  factors affecting generation rates: extent
      of salvage and recycling, 73
    frequency of collection, 72
    geographic location, 71
    legislation, 73
    public attitudes, 73
    season of the year, 72
    use of home grinders, 73
  future changes in composition of, 62−64
    food wastes, 63
    paper, 64
    plastics, 64
  generation of: effects of technological ad-
      vances, 7
    future challenges and opportunities, 12
    impacts of, 3
    in a technological society, 4
  generation rates, 64−73
    determination of, 66
    expressions for unit rates, 65
    factors affecting generation rates, 71−73
    measure of quantities, 64
    statistical analysis of, 65
    typical data on, 70, 71
  milling of, 324
  physical composition, 55−59
    density, 58
    individual components, 55
    moisture content, 56
    typical data on, 55
  projections for the future, 10
  quantities of: estimated per capita, 9, 70, 71
    estimated total for U.S., 7
    monthly and seasonal variations, 9
  sources of, 51−52
    agricultural, 52
    commercial, 52
    industrial, 52
    municipal, 52
    open areas, 52
    residential, 52
    treatment plants, 52
  types of, 53−54
    agricultural, 54
    ashes and residue, 53
    demolition and construction, 53
    food, 53
    hazardous, 54
    rubbish, 53
    special, 53
    treatment plant, 54
    (*See also specific entries*)

Sorting (*see* Separation of solid wastes)
Sources of solid wastes, 51−52
  agriculture, 52
  commercial, 52
  industrial, 52
  municipal, 52
  open areas, 52
  residential, 52
  treatment plant, 52
  (*See also specific entries*)
Special wastes, 52, 53, 437
Specifications for materials, 256, 488
  for fuel source, 257
  for land reclamation, 257
  for reuse as raw materials, 257
Stack, 212, 214
Stack effluents, 273
Standard deviation, 581
Station service allowance, 298, 300
Stationary compactors:
  application of, 203
  description of, 203, 206
Stationary container collection systems:
  analysis of, 129−140
  applications for, 115
  definition of, 112
  types of: with manually loaded vehicles,
      117
    with mechanically loaded vehicles, 117
Statistical analysis:
  graphical methods of, 582−585
    frequency distributions, 582
    time series, 582
    use: of arithmetic probability paper, 582
      of log normal probability paper, 585
  measures commonly used, 579−582
    coefficient of variation, 581
    frequency, 580
    mean, 580
    median, 580
    mode, 581
    standard deviation, 581
  of waste generation rates, 65
Steam production from solid wastes, 267
Storage:
  bins, 112
  containers, 83, 112
  onsite, 83−90
    commercial, 90
    medium- and high-rise apartments, 89
    residential, 85
Storage discharge transfer station, 167−170,
    172

Storage pits for incinerators, 212, 218
Supplementary fuel, 216, 295, 296
Surface transportation, 437
Suspension firing in incinerators, 269
Synthetic membrane:
　for control: of gas movement in landfills,
　　　338
　　of leachate control in landfills, 343
Systems:
　analysis, 152
　engineering, 152
　(*See also* Solid waste management systems)

Tariffs, 492
Tax incentives, 510
Temperature:
　for anaerobic digestion, 294
　for composting, 291
　ranges for microorganisms, 286
Textiles, quantities in municipal wastes, 55
Thermophilic organisms, 286
Tilt-frame:
　collection vehicle, 114–115
　container, 112
Time series, 582
Tin cans:
　quantities in municipal wastes, 55
　recovery of, 236
　use of, 494
Tipping, controlled (landfilling), 20
Tote containers, 106–107
Toxic chemical wastes:
　classification of, 379
　handling of, 387
　storage of, 385
　(*See also* Hazardous wastes)
Traffic flow conditions, impact on transport
　　system, 470
Transfer and transport:
　case studies in, 470–481
　definition of, 26, 459
　equipment for: containers: types of, 183
　　　typical data on, 183
　　trailers, 177–182
　　　typical data on, 179
　　　unloading methods, 177, 179
　inventory and data accumulation, 466–470
　management issues and concerns,
　　　459–466
　need for, 160–163
　　excessive haul distances, 161
　　remote disposal sites, 163

Transfer stations, 163–176
　classification by size, 163
　design considerations: capacity, 175
　　equipment, 176
　　sanitation, 176
　location of, 185–197
　　factors to consider in siting, 185
　　method of analysis for, 186
　types of, 163–175
　　combination, 170
　　direct discharge, 163, 166
　　storage discharge, 167
Transitional concepts in solid waste manage-
　　ment, 403
Transport methods and means, 177–182
　motor vehicles, 177
　　compactors, 182
　　trailers and semitrailers, 177
　pneumatic, hydraulic, and others, 184
　water, 184
Transport trailers:
　typical data on, 179
　unloading methods, 177, 179
Trash-trailers, 115
Traveling grate stoker, 213, 214
Treatment plant wastes:
　co-disposal of, 306, 326
　definition of, 53
　sources of, 52
Trench method of sanitary landfilling, 323
Types of solid wastes, 52–54
　agricultural, 54
　ashes and residue, 53
　demolition and construction, 53
　food, 53
　hazardous, 54
　rubbish, 53
　special, 53
　treatment plant, 54
　(*See also* Solid wastes; *and specific entries*)

Unaccounted heat losses, 300
Underfire air, 213
Unloading area:
　for incinerators, 212
　for landfills, 345
　for transfer stations, 175
Unloading operations at disposal site:
　collection vehicle, 519
　trailer with cable pull-off system, 520
　trash-trailer, 520

Vacuum systems:
  for waste collection, 83, 84
  for waste transport, 184
Vapor pressure of water, 335
Vector control, 20
Vehicles:
  abandoned, 437, 439
  collection: for hauled container systems,
      112–115
    for stationary container systems,
      115–117
    typical data on, 112, 114
    (*See also* Collection vehicles)
  transfer and transport: trailers,
      177–182
    typical data on, 179
Vent pipes for landfills, 336, 338
Venting, landfill: objectives, 336
  types of vents, 337
Vertical seepage, estimation of, 342
Vibrating screens, 240
Volume reduction:
  by compaction, 203, 495
  by incineration, 211
  by mechanical size reduction, 221
  (*See also specific entries*)

Waste allocation problem:
  definition of, 186
  solution of, 187–197
Waste bridging, 302
Waste chutes, 80–82
Waste collection systems (*see* Collection;
    Collection systems)
Waste combustion gases, 273
  analysis of, 64–69
  typical data on, 70–71
  variations in, 9, 71
  (*See also* Solid wastes, generation of)
Waste composition, 54–64
  chemical, 59–64

Waste composition;
  chemical; definition of, 59
    typical data on, 60–62
  physical, 55–59
    definition of, 55
    typical data on, 55, 57, 59
Waste density:
  of individual components, 60
  of wastes from various sources, 59
Waste disposal:
  case studies in, 522–526
  definition of, 26–27, 508
  early methods of, 17–20
    dumping: on land, 18
      in water, 18
    feeding to hogs, 18
    incineration, 20
    plowing into the soil, 18
    reduction, 18
  inventory and data accumulation, 519–522
  management issues and concerns,
      508–519
  modern methods of, 315–370
Waste heat recovery, 265
Wastes:
  definition of, 3, 51
  hazardous (*see* Hazardous wastes)
  solid (*see* Solid wastes)
Wastewater treatment plant sludge, 306,
    326
Water transport, 184
Weight-volume analysis, 67
Well vent for landfills, 337
Wet pulping:
  definition of, 225
  equipment for, 101, 225
  for onsite processing, 99, 101
Wet scrubbers, 215

Yard wastes, 57, 59
Yeasts, 284